Graduate Texts in Mathematics   **246**

T0211709

For other titles published in this series, go to
www.springer.com/series/136

Eberhard Kaniuth

A Course in Commutative
Banach Algebras

 Springer

Eberhard Kaniuth
Institute of Mathematics
Paderborn University
Paderborn, Germany
kaniuth@math.uni-paderborn.de

ISBN: 978-1-4419-2479-7   e-ISBN: 978-0-387-72476-8
DOI: 10.1007/978-0-387-72476-8

Dedicated to my wife Ursula

Preface

Banach algebras are Banach spaces equipped with a continuous multiplication. In rough terms, there are three types of them: algebras of bounded linear operators on Banach spaces with composition and the operator norm, algebras consisting of bounded continuous functions on topological spaces with pointwise product and the uniform norm, and algebras of integrable functions on locally compact groups with convolution as multiplication. These all play a key role in modern analysis. Much of operator theory is best approached from a Banach algebra point of view and many questions in complex analysis (such as approximation by polynomials or rational functions in specific domains) are best understood within the framework of Banach algebras. Also, the study of a locally compact Abelian group is closely related to the study of the group algebra $L^1(G)$.

There exist a rich literature and excellent texts on each single class of Banach algebras, notably on uniform algebras and on operator algebras. This work is intended as a textbook which provides a thorough introduction to the theory of commutative Banach algebras and stresses the applications to commutative harmonic analysis while also touching on uniform algebras. In this sense and purpose the book resembles Larsen's classical text [75] which shares many themes and has been a valuable resource. However, for advanced graduate students and researchers I have covered several topics which have not been published in books before, including some journal articles.

The reader is expected to have some basic knowledge of functional analysis, point set topology, complex analysis, measure theory, and locally compact groups. However, many of the prerequisites are collected (without proofs) in the appendix. Here the reader may also find (including proofs) some facts about convolution of functions on locally compact groups, the Pontryagin duality theorem and some of its consequences, and a description of the closed sets in the coset ring of an Abelian topological group.

The book is divided into five chapters, the contents of which can be described as follows. The first chapter introduces the basic concepts and constructions and provides a comprehensive treatment of the spectrum of a Banach algebra element.

Chapter 2 begins with Gelfand's fundamental theorem on representing a commutative Banach algebra A as an algebra of continuous functions on a locally compact Hausdorff space, the structure space $\Delta(A)$ of A, which is defined to be the set of all homomorphisms from A onto $\mathbb{C}$, equipped with the w^*-topology. This Gelfand homomorphism turns out to be an isometric isomorphism onto $C_0(\Delta(A))$ if and only if A is a commutative C^*-algebra. Applications of this basic result include proofs for the existence of the Stone-Čech compactification of a completely regular topological space and of the Bohr compactification of a locally compact Abelian group. The structure space of a finitely generated algebra identifies canonically with the joint spectrum of the set of generators and this leads to a description of the Gelfand representation of several uniform algebras, such as the closure of algebras of polynomial and of rational functions on compact subsets of $\mathbb{C}^n$. Following our intention to emphasize the connection with commutative harmonic analysis, we extensively study the Gelfand representation of algebras associated with locally compact groups. This concerns, in the first place, the convolution algebra $L^1(G)$ of integrable functions on a locally compact Abelian group, but also weighted algebras $L^1(G, \omega)$ and Fourier algebras. Chapter 2 concludes with determining the structure spaces of tensor products of two commutative Banach algebras and a discussion of semisimplicity of the projective tensor product.

In Chapter 3 we focus on some important problems which evolve from the Gelfand representation theory and concern the structure space $\Delta(A)$ and the structure of A itself. The new tools required are holomorphic functional calculi for Banach algebra elements. These are developed in Section 3.1 and subsequently applied to study the topological group of invertible elements of a unital commutative Banach algebra A and the problem of which elements of $\Delta(A)$ extend to elements of $\Delta(B)$ for any commutative Banach algebra B containing A as a closed subalgebra. This latter question is linked with the Shilov boundary which we investigate thoroughly. One of the major highlights in the theory of commutative Banach algebras is Shilov's idempotent theorem. This rests on the multivariable holomorphic functional calculus and is established in Section 3.5, followed by several applications that illustrate the power of the idempotent theorem.

The concept of regularity and its role in ideal theory is the main subject of Chapter 4. The relevance of regularity is due to the fact that it is equivalent to coincidence of the Gelfand topology and the hull-kernel topology on $\Delta(A)$. We show the existence of a greatest regular subalgebra of any commutative Banach algebra and study permanence properties of regularity. One of the most profound results in commutative harmonic analysis is regularity of the group algebra $L^1(G)$. To prove this, we have chosen an approach which is based on the Gelfand theory of commutative C^*-algebras. Recently, certain properties related to, but weaker than, regularity have been investigated. We give a detailed account and comparison of these so-called spectral exten-

sion properties and the unique uniform norm property. Finally, we establish Domar's result which asserts that $L^1(G, \omega)$ is regular whenever the weight ω is nonquasianalytic.

The last chapter is devoted to ideal theory of regular semisimple commutative Banach algebras and to spectral synthesis problems in particular. The basic notions are that of a spectral set and of a Ditkin set in $\Delta(A)$. It is customary to say that spectral synthesis holds for the algebra A if every closed subset of $\Delta(A)$ is a spectral set (equivalently, every closed ideal of A is the intersection of the maximal ideals containing it). In Section 5.2 we present a number of results on generating spectral sets and Ditkin sets, some of which cannot be found elsewhere in this generality. Subsequently, these results are applied to $L^1(G)$. In this context we point out that a famous theorem of Malliavin states that spectral synthesis fails to hold for $L^1(G)$ whenever G is a noncompact locally compact Abelian group. We also present a complete description of all the closed ideals in $L^1(G)$ with bounded approximate identities. Spectral synthesis also fails for the algebra $C^n[0, 1]$ of n-times continuously differentiable functions on the interval $[0, 1]$ and even for a certain algebra with discrete structure space, the Mirkil algebra. Both of these algebras are discussed in detail: $C^n[0, 1]$ because it nevertheless admits a satisfactory ideal structure and the Mirkil algebra because it serves as a counterexample to several conjectures in spectral synthesis.

Each chapter is accompanied by an extensive set of exercises, ranging from simple and straightforward applications of concepts and results developed in the chapter in question to more sophisticated supplements to the theory. These exercises add numerous examples to those already given in the text. In several cases hints are provided, and the reader is strongly encouraged to solve and work out as many of these exercises as possible.

There are various options for using the material as a text for courses, depending on the instructor's intention and inclination. Any one-semester course, however, has to cover Sections 1.1 and 1.2 and Sections 2.1 to 2.4, and might then continue with

- Sections 2.5 and 2.6 and the Shilov boundary if the main emphasis is on uniform algebras,
- Sections 1.5 and 2.11 and the corresponding passages of Chapters 3, 4 and 5 when concentrating on projective tensor products,
- Selected topics from Chapter 3 if the focus is on general Banach algebras rather than group algebras or uniform algebras,
- Sections 2.7 and 4.4 and, if time permits, parts of Chapter 5 whenever applications in commutative harmonic analysis is the preferred objective.

Major portions of the book grew out of graduate courses taught at the University of Heidelberg, the Technical University of Munich and the University of Paderborn.

I owe a great deal to two colleagues and friends. Robert J. Archbold and Ali Ülger have both taken up the onerous burden of reading substantial parts of the text and made many helpful suggestions for improvement. I am also indebted to Bert Schreiber for his help concerning the coset ring of an Abelian group. Finally, I would like to express my appreciation to the editorial staff of Springer-Verlag for their professional support.

Paderborn   *Eberhard Kaniuth*
June, 2008

Contents

1

General Theory of Banach Algebras

This introductory chapter contains several topics which in one way or the other are basic for everything that follows. Although this is a book on commutative Banach algebras, we do not assume commutativity until Chapter 2. In Section 1.1, after giving the definition of a complex Banach algebra, we present a number of examples which play an important role later. We outline the standard construction of adjoining an identity to a Banach algebra and prove some results on approximate identities. The fundamental concept of the spectrum of a Banach algebra element is defined in Section 1.2. The spectrum is shown to be nonempty and the spectral radius formula is given. We also treat the important issue of when the spectrum of an element of a Banach algebra A remains unchanged when embedding A into a larger Banach algebra. Because strong emphasis is placed on algebras appearing in harmonic analysis, we consider in Section 1.3 the convolution algebras $L^1(G)$ and, more generally, Beurling algebras $L^1(G, \omega)$ on locally compact groups G. Both are investigated in subsequent chapters under various aspects. In Section 1.4 we study ideals and quotients of Banach algebras and also introduce the multiplier algebra. A complete description is given of all closed ideals in $C_0(X)$, the algebra of continuous functions on a locally compact Hausdorff space X which vanish at infinity, and the closed ideals in $L^1(G)$ are shown to be precisely the closed translation invariant subspaces of $L^1(G)$. Finally, in Section 1.5 tensor products of two Banach algebras A and B are discussed. As examples we identify $L^1(G, A)$, the L^1-algebra of A-valued integrable functions on a locally compact group G, as the projective tensor product of $L^1(G)$ and A and $C_0(X, A)$ as the injective tensor product of $C_0(X)$ and A.

1.1 Basic definitions and examples

A normed linear space $(A, \|\cdot\|)$ over the complex number field $\mathbb{C}$ is called a *normed algebra* if it is an algebra and the norm is submultiplicative; that is, $\|xy\| \leq \|x\| \cdot \|y\|$ for all $x, y \in A$. A normed algebra A is said to be a *Banach*

E. Kaniuth, *A Course in Commutative Banach Algebras*, Graduate Texts in Mathematics,
DOI 10.1007/978-0-387-72476-8_1, © Springer Science+Business Media, LLC 2009

algebra if the normed space $(A, \| \cdot \|)$ is a Banach space. It is easy to see that the completion $(\widetilde{A}, \| \cdot \|)$ of a normed algebra $(A, \| \cdot \|)$ is a Banach algebra. Indeed, if $x, y \in \widetilde{A}$ and $(x_n)_n$ and $(y_n)_n$ are sequences in A converging in $\widetilde{A}$ to x and y, respectively, then $(x_n y_n)_n$ is a Cauchy sequence in $\widetilde{A}$, and the product of x and y can be defined to be $xy = \lim_{n \to \infty} x_n y_n$.

An algebra A is called *commutative* if $xy = yx$ for all $x, y \in A$. Moreover, A is called *unital* or an algebra with identity if there exists $e \in A$ such that $ex = x = xe$ for all $x \in A$. If A is a unital normed algebra and $A \neq \{0\}$, then e is unique and $\|e\| \geq 1$ as follows from submultiplicativity of the norm. We do not assume that $\|e\| = 1$ although this is the case in all examples. Note, however, that there always exists an equivalent norm $|\cdot|$ on A such that $|e| = 1$. In fact, this is a consequence of the following proposition.

Proposition 1.1.1. *Let A be an algebra with identity e and with a norm $\| \cdot \|$ under which it is a Banach space. Suppose that the multiplication is continuous in each factor separately. Then there exists a norm $\|\cdot\|_0$ on A that is equivalent to $\| \cdot \|$ and for which $\|xy\|_0 \leq \|x\|_0 \|y\|_0$ holds for all $x, y \in A$.*

Proof. For $x \in A$, let $L_x : A \to A$ be the left translation operator $y \to xy$. By the continuity assumption, L_x is continuous. Moreover, since $x = L_x(e)$, the map $x \to L_x$ is an isomorphism of A into $\mathcal{B}(A)$, the algebra of bounded linear operators on A. Let

$$\|x\|_0 = \|L_x\| = \sup\{\|xy\| : \|y\| \leq 1\}, \quad x \in A.$$

Clearly, $\| \cdot \|_0$ is a norm on A satisfying $\|xy\|_0 \leq \|x\|_0 \|y\|_0$. We claim that $\| \cdot \|_0$ is a complete norm. For that, note first that

$$\|x\|_0 \geq \|x(\|e\|^{-1}e)\| = \|e\|^{-1}\|x\|.$$

Thus, if $(x_n)_n$ is a Cauchy sequence with respect to $\| \cdot \|_0$, then it is also a Cauchy sequence in $\|\cdot\|$ and hence $x_n \to x$ for some $x \in A$. On the other hand, because $(L_{x_n})_n$ is a Cauchy sequence in $\mathcal{B}(A)$, $L_{x_n} \to T$ for some $T \in \mathcal{B}(A)$. By continuity of the product in the first variable, $L_{x_n} y \to L_x y$ for each $y \in A$. So $T = L_x$, which proves the claim. Thus both $\| \cdot \|$ and $\| \cdot \|_0$ are complete norms on A. Since $\| \cdot \| \leq \|e\| \cdot \| \cdot \|_0$, the closed graph theorem implies that the two norms are equivalent. □

There are three main classes of Banach algebras which may be described roughly as algebras of functions (with pointwise multiplication), algebras of operators (with composition of operators), and group algebras (with convolution product). We now give examples of Banach algebras in each of the first two classes, whereas convolution algebras are postponed to Section 1.3. Additional, somewhat more elaborate, examples are presented later in the text and also in the exercises.

Example 1.1.2. Let X be a locally compact Hausdorff space. We denote by $C^b(X), C_0(X)$, and $C_c(X)$, respectively, the algebras of all continuous complex-valued functions on X that are bounded, vanish at infinity, or have compact support. The algebra operations are the usual ones of pointwise addition, multiplication, and scalar multiplication. With the supremum norm

$$\|f\|_\infty = \sup_{x \in X} |f(x)| \quad (f \in C_0(X)),$$

the algebras $C^b(X)$ and $C_0(X)$ are commutative Banach algebras, whereas $C_c(X)$ is complete only when X is compact. If X is noncompact, then only $C^b(X)$ is unital.

Example 1.1.3. Let X be a compact subset of $\mathbb{C}$. We introduce three unital closed subalgebras of $(C(X), \|\cdot\|_\infty$ as follows.

The first one, denoted $A(X)$, is the algebra of all functions $f : X \to \mathbb{C}$ which are continuous on X and holomorphic on the interior X° of X. Clearly, $A(X)$ is complete since the uniform limit of a sequence of holomorphic functions is holomorphic. A particularly interesting special case is the *disc algebra* $A(\mathbb{D})$, where $\mathbb{D}$ denotes the closed unit disc $\mathbb{D} = \{z \in \mathbb{C} : |z| \le 1\}$ in the plane.

The second one, $P(X)$, is the subalgebra of $C(X)$ consisting of all functions which are uniform limits of polynomial functions on X. Finally, $R(X)$ is the subalgebra of $C(X)$ of all functions which are uniform limits on X of rational functions p/q, where p and q are polynomials and q has no zero on X.

Note that we always have $P(X) \subseteq R(X) \subseteq A(X)$ and that equality holds at either place can be interpreted as a result in approximation theory.

Example 1.1.4. Let $a, b \in \mathbb{R}$ such that $a < b$ and $n \in \mathbb{N}$, and let $C^n[a, b]$ be the space of all complex-valued functions on [a,b] which are n-times continuously differentiable. With pointwise operations, $C^n[a, b]$ becomes a commutative algebra. We define a norm on $C^n[a, b]$ by

$$\|f\| = \sum_{k=0}^{n} \frac{1}{k!} \|f^{(k)}\|_\infty$$

for $f \in C^n[a, b]$. This norm is submultiplicative. Indeed, for $f, g \in C^n[a, b]$,

$$\|fg\| = \sum_{k=0}^{n} \frac{1}{k!} \|(fg)^{(k)}\|_\infty = \sum_{k=0}^{n} \frac{1}{k!} \left\| \sum_{j=0}^{k} \binom{k}{j} f^{(j)} g^{(k-j)} \right\|_\infty$$

$$= \sum_{k=0}^{n} \left\| \sum_{j=0}^{k} \frac{1}{j!(k-j)!} f^{(j)} g^{(k-j)} \right\|_\infty$$

$$\le \sum_{k=0}^{n} \sum_{j=0}^{k} \frac{1}{j!} \|f^{(j)}\|_\infty \frac{1}{(k-j)!} \|g^{(k-j)}\|_\infty$$

$$\leq \sum_{l=0}^{n} \sum_{j=0}^{n} \frac{1}{j!} \left\| f^{(j)} \right\|_{\infty} \frac{1}{l!} \|g^{(l)}\|_{\infty}$$
$$= \|f\| \cdot \|g\|.$$

We claim that $C^n[a,b]$ is complete. To verify this, let $(f_m)_m$ be a Cauchy sequence in $C^n[a,b]$. Then, by definition of the norm, for each $0 \leq j \leq n$, $(f_m^{(j)})_m$ is a Cauchy sequence in $C[a,b]$ with respect to the $\| \cdot \|_\infty$-norm. Let $g_j = \lim_{m \to \infty} f_m^{(j)}$ for $0 \leq j \leq n$. We show by induction on j that g_0 is j-times differentiable and $g_0^{(j)} = g_j$. As nothing has to be shown for $j = 0$, assume that $1 \leq j \leq n$ and that g_0 is $(j-1)$-times differentiable and satisfies $g_0^{(j-1)} = g_{j-1}$.

For all $m \in \mathbb{N}$ and each $t \in [a,b]$ we have

$$f_m^{(j-1)}(t) = f_m^{(j-1)}(a) + \int_a^t f_m^{(j)}(s)ds.$$

Because $f_m^{(j)}$ and $f_m^{(j-1)}$ converge uniformly with limit g_j and g_{j-1}, respectively, the inductive hypothesis gives

$$g_0^{(j-1)}(t) = g_{j-1}(t) = g_{j-1}(a) + \int_a^t g_j(s)ds = g_0^{(j-1)}(a) + \int_a^t g_j(s)ds$$

for all $t \in [a,b]$. It follows that $g_0^{(j-1)}$ is differentiable and $(g_0^{(j-1)})'(t) = g_j(t)$ for all $t \in [a,b]$. This finishes the inductive proof. Thus, taking $f = g_0$, f is n-times differentiable and $f^{(j)} = g_j$ for $0 \leq j \leq n$. Since $f_m^{(j)} \to g_j$ uniformly for each j, it follows that $f_m \to f$ in $C^n[a,b]$.

Example 1.1.5. For $f \in C(\mathbb{T})$ and $n \in \mathbb{Z}$, the nth Fourier coefficient $c_n(f)$ is defined by

$$c_n(f) = \frac{1}{2\pi} \int_0^{2\pi} f(e^{it})e^{-int}dt.$$

Let $AC(\mathbb{T})$ denote the space of all functions $f \in C(\mathbb{T})$ the Fourier series $\sum_{n \in \mathbb{Z}} c_n(f)e^{int}$ of which is absolutely convergent; that is, $f \in AC(\mathbb{T})$ if and only if $(c_n(f))_n \in l^1(\mathbb{Z})$. Conversely, given $(c_n)_n \in l^1(\mathbb{Z})$, define $f \in C(\mathbb{T})$ by

$$f(e^{it}) = \sum_{k \in \mathbb{Z}} c_k e^{ikt}, \ t \in [0, 2\pi].$$

Then, for each $n \in \mathbb{Z}$,

$$c_n(f) = \frac{1}{2\pi} \int_0^{2\pi} \left(\sum_{k \in \mathbb{Z}} c_k e^{ikt} \right) e^{-int}dt$$
$$= \frac{1}{2\pi} \sum_{k \in \mathbb{Z}} c_k \int_0^{2\pi} e^{i(k-n)t}dt$$
$$= c_n.$$

Thus, equipped with the norm $\|f\| = \sum_{n \in \mathbb{Z}} |c_n(f)|$, $AC(\mathbb{T})$ is a Banach space and isometrically isomorphic to $l^1(\mathbb{Z})$. For $f, g \in AC(\mathbb{T})$ and $n \in \mathbb{Z}$,

$$c_n(fg) = \frac{1}{2\pi} \int_0^{2\pi} f(e^{it})e^{-int} \left(\sum_{k \in \mathbb{Z}} c_k(g)e^{-ikt} \right) dt$$

$$= \frac{1}{2\pi} \sum_{k \in \mathbb{Z}} c_k(g) \int_0^{2\pi} f(e^{it})e^{-i(n-k)t} dt$$

$$= \sum_{k \in \mathbb{Z}} c_k(g)c_{n-k}(f).$$

This implies

$$\sum_{n \in \mathbb{Z}} |c_n(fg)| \leq \sum_{j \in \mathbb{Z}} |c_j(f)| \cdot \sum_{k \in \mathbb{Z}} |c_k(g)|,$$

and hence $AC(\mathbb{T})$ is a commutative Banach algebra under pointwise multiplication. Note that the product on $l^1(\mathbb{Z})$, which makes the map $f \to (c_n(f))_n$ from $AC(\mathbb{T})$ to $l^1(\mathbb{Z})$ an algebra isomorphism, is then given by the following formula. For $(\alpha_n)_n$ and $(\beta_n)_n$ in $l^1(\mathbb{Z})$, let $(\alpha\beta)_n = \sum_{k \in \mathbb{Z}} \alpha_{n-k}\beta_k$, $n \in \mathbb{Z}$. This is the convolution on $l^1(\mathbb{Z})$ (see Section 1.3).

Definition 1.1.6. Let A be a $\mathbb{C}$-algebra. An *involution* on A is a mapping $* : x \to x^*$ from A into A satisfying the following conditions.

(1) $(x + y)^* = x^* + y^*$ and $(\lambda x)^* = \bar{\lambda}x^*$,
(2) $(xy)^* = y^*x^*$ and $(x^*)^* = x$,

for all $x, y \in A$ and $\lambda \in \mathbb{C}$. A is then called a **-algebra* or an *algebra with involution*. A normed algebra (Banach algebra) with involution is called a *normed *-algebra (Banach *-algebra)* if the involution is isometric; that is, $\|x^*\| = \|x\|$ for all $x \in A$.

Example 1.1.7. (1) Let A be an algebra under pointwise operations consisting of complex-valued functions. Suppose that A contains with every function f its complex conjugate $\bar{f}$. Then $f \to \bar{f}$ defines an involution on A.

(2) Complex conjugation does not define an involution on the disc algebra $A(\mathbb{D})$ because the function $z \to \bar{z}$ is not holomorphic. However, one can define an involution on $A(\mathbb{D})$ by setting $f^*(z) = \overline{f(\bar{z})}$ for $f \in A(\mathbb{D}), z \in \mathbb{D}$. With this involution, $A(\mathbb{D})$ becomes a Banach *-algebra (Exercise 1.6.15).

(3) Let H be a Hilbert space, and for $T \in \mathcal{B}(H)$ let $T^* \in \mathcal{B}(H)$ denote the adjoint operator. Then $T \to T^*$ defines an involution on $\mathcal{B}(H)$ making $\mathcal{B}(H)$ a Banach *-algebra.

An algebra A can always be embedded into an algebra with identity as follows. Let A_e denote the set of all pairs $(x, \lambda), x \in A, \lambda \in \mathbb{C}$, that is, $A_e = A \times \mathbb{C}$. Then A_e becomes an algebra if the linear space operations and multiplication are defined by

$$(x, \lambda) + (y, \mu) = (x + y, \lambda + \mu), \ \mu(x, \lambda) = (\mu x, \mu \lambda)$$

and

$$(x, \lambda)(y, \mu) = (xy + \lambda y + \mu x, \lambda \mu)$$

for $x, y \in A$ and $\lambda, \mu \in \mathbb{C}$. A simple calculation shows that the element $e = (0, 1) \in A_e$ is an identity for A_e. Moreover, the mapping $x \to (x, 0)$ is an algebra isomorphism of A onto an ideal of codimension one in A_e. Obviously, A_e is commutative if and only if A is commutative.

Now suppose that A is a normed algebra. We introduce a norm on A_e by

$$\|(x, \lambda)\| = \|x\| + |\lambda|, \quad x \in A, \quad \lambda \in \mathbb{C}.$$

It is straightforward that this turns A_e into a normed algebra and that A_e is a Banach algebra provided A is complete. As $(x, \lambda) = (x, 0) + \lambda(0, 1)$, it is customary to write elements (x, λ) as $x + \lambda e$. The above process is usually referred to as that of *adjoining an identity* to A and A_e is called the *unitisation* of A . The utility of A_e is due to the fact that algebras with identity are often easier to deal with than algebras without identity.

If A is a $*$-algebra, then A_e becomes a $*$-algebra by simply defining the involution by $(x + \lambda)^* = x^* + \overline{\lambda}e$. It is also obvious that if A is a normed $*$-algebra (Banach $*$-algebra), then so is A_e.

If A lacks an identity, then an approximate identity often serves as a good substitute. We proceed by introducing this notion.

Definition 1.1.8. Let A be a normed algebra. A *left (right) approximate identity* for A is a net $(e_\lambda)_\lambda$ in A such that $e_\lambda x \to x$ $(x e_\lambda \to x)$ for each $x \in A$. An *approximate identity* for A is a net $(e_\lambda)_\lambda$ which is both a left and a right approximate identity. A (left or right) approximate identity $(e_\lambda)_\lambda$ is *bounded* by $M > 0$ if $\|e_\lambda\| \leq M$ for all λ.

Definition 1.1.9. A has *left (right) approximate units* if, for each $x \in A$ and $\epsilon > 0$, there exists $u \in A$ such that $\|x - ux\| \leq \epsilon$ $(\|x - xu\| \leq \epsilon)$, and A has an *approximate unit* if, for each $x \in A$ and $\epsilon > 0$, there exists $u \in A$ such that $\|x - ux\| \leq \epsilon$ and $\|x - xu\| \leq \epsilon$. A has a (left, right) approximate unit *bounded* by $M > 0$, if the elements u can be chosen such that $\|u\| \leq M$.

Lemma 1.1.10. *Let $(e_\lambda)_\lambda$ and $(f_\mu)_\mu$ be bounded left and right approximate identities for A, respectively. Then the net*

$$(e_\lambda + f_\mu - f_\mu e_\lambda)_{\lambda, \mu}$$

is a bounded approximate identity for A.

Proof. Let $g_{\lambda, \mu} = e_\lambda + f_\mu - f_\mu e_\lambda$. Then, for any $x \in A$,

$$\|g_{\lambda, \mu} x - x\| = \|(e_\lambda x - x) + f_\mu(x - e_\lambda x)\| \leq (1 + \|f_\mu\|)\|e_\lambda x - x\|$$

and, similarly,
$$\|xg_{\lambda,\mu} - x\| \le (1 + \|e_\lambda\|)\|x - xf_\mu\|.$$
Thus $(g_{\lambda,\mu})_{\lambda,\mu}$ is an approximate identity. Of course,
$$\|g_{\lambda,\mu}\| \le \|e_\lambda\| + \|f_\mu\| + \|e_\lambda\| \cdot \|f_\mu\|,$$
so that $(g_{\lambda,\mu})_{\lambda,\mu}$ is bounded. $\qquad\square$

Clearly, if a normed algebra has a left approximate identity (bounded by M), then it has left approximate units (bounded by M). Actually, as the following proposition shows, these two properties are equivalent.

Proposition 1.1.11. *Let A be a normed algebra and let $M \ge 1$. Then the following three conditions are equivalent.*

(i) *A has left approximate units bounded by M.*
(ii) *Given finitely many elements $x_1, \ldots, x_n$ in A and $\epsilon > 0$, there exists $u \in A$ such that $\|u\| \le M$ and $\|x_j - ux_j\| \le \epsilon$ for $j = 1, \ldots, n$.*
(iii) *A has a left approximate identity bounded by M.*

Proof. To prove (i) $\Rightarrow$ (ii), using the formal notation $(1 - x)y = y - xy$ for $x, y \in A$, we successively choose $u_1, \ldots, u_n \in A$ satisfying $\|u_j\| \le M$ and
$$\|(1 - u_j) \cdot \ldots \cdot (1 - u_1)x_j\| \le \frac{\epsilon}{2(1 + M)^{n-j+1}}, \quad 1 \le j \le n.$$
Define $v \in A$ by $1 - v = (1 - u_n) \cdot \ldots \cdot (1 - u_1)$. Then, for $j = 1, \ldots, n$, we have
$$\begin{aligned}
\|x_j - vx_j\| &\le \|(1 - u_n) \cdot \ldots \cdot (1 - u_{j+1})\| \cdot \|(1 - u_j) \cdot \ldots \cdot (1 - u_1)x_j\| \\
&\le (1 + M)^{n-j}\frac{\epsilon}{2(1 + M)^{n-j+1}} \\
&= \frac{\epsilon}{2(1 + M)}.
\end{aligned}$$
Let now $R = \max\{\|x_j\| : 1 \le j \le n\}$, and choose $u \in A$ with $\|u\| \le M$ and $\|v - uv\| \le \epsilon/(2R + 1)$. Then, for each $1 \le j \le n$, it follows that
$$\begin{aligned}
\|x_j - ux_j\| &\le \|x_j - vx_j\| + \|vx_j - uvx_j\| + \|ux_j - uvx_j\| \\
&\le \frac{\epsilon}{2(1 + M)} + \frac{R\epsilon}{2R + 1} + \frac{M\epsilon}{2(1 + M)} < \epsilon,
\end{aligned}$$
as required.

(ii) $\Rightarrow$ (iii) Let Λ be the family of all nonempty finite subsets of A. For $\lambda \in \Lambda$, let $|\lambda|$ denote the number of elements in λ. Ordered by inclusion, Λ is a directed set. For each $\lambda \in \Lambda$, by (ii) there exists $u_\lambda \in A$ with $\|u_\lambda\| \le M$ and $\|x - u_\lambda x\| \le |\lambda|^{-1}$ for all $x \in \lambda$. Then $(u_\lambda)_\lambda$ is a left approximate identity of bound M for A.

Finally, as already noticed, (iii) $\Rightarrow$ (i). $\qquad\square$

1.2 The spectrum of a Banach algebra element

The theme of this section is to introduce the basic concept of the spectrum of an element of a Banach algebra and to establish several important results about spectra. The reader will observe that this concept extends the notion of spectrum of a bounded linear operator on a Hilbert space.

Definition 1.2.1. Let A be a complex algebra with identity e. An element $x \in A$ is called *invertible* if there exists $y \in A$ such that $xy = yx = e$. Then y is unique and called the *inverse*, denoted x^{-1}, of x. Let $G(A)$ denote the set of invertible elements of A. Then $G(A)$ is a group, and $(xy)^{-1} = y^{-1}x^{-1}$ for $x, y \in G(A)$.

For $x \in A$, the set

$$\sigma_A(x) = \{\lambda \in \mathbb{C} : \lambda e - x \notin G(A)\}$$

is called the *spectrum* of x in A, and its complement, $\rho_A(x) = \mathbb{C} \setminus \sigma_A(x)$, the *resolvent set* of x.

When A does not have an identity, we define $\sigma_A(x)$ and $\rho_A(x)$ by $\sigma_A(x) = \sigma_{A_e}(x)$ and $\rho_A(x) = \rho_{A_e}(x)$.

Remark 1.2.2. For any algebra A and $x \in A$, we have $\sigma_A(x) \cup \{0\} = \sigma_{A_e}(x)$. To see this, suppose first that A does not have an identity. Then $0 \in \sigma_A(x)$ because otherwise $x^{-1}x \in A$. Thus, in this case $\sigma_A(x) \cup \{0\} = \sigma_A(x) = \sigma_{A_e}(x)$.

Now suppose that A has an identity u. For $y \in A$, it is then easily verified that $u - y \in G(A)$ if and only if $e - y \in G(A_e)$. Indeed, if $u - y \in G(A)$ and $(u - y)^{-1} = z + u, z \in A$, then $(e - y)(z + e) = e = (z + e)(e - y)$, and conversely, if $e - y \in G(A_e)$ and $(e - y)^{-1} = z + \mu e, z \in A, \mu \in \mathbb{C}$, then it follows that $\mu = 1$ and $(u - y)(z + u) = u = (z + u)(u - y)$. For $\lambda \neq 0$ and $x \in A$, this implies that $\lambda u - x = \lambda(u - (1/\lambda)x) \notin G(A)$ if and only if $\lambda e - x = \lambda(e - (1/\lambda)x) \notin G(A_e)$. Equivalently, $\lambda \in \sigma_A(x)$ if and only if $\lambda \in \sigma_{A_e}(x)$. Hence $\sigma_A(x) \setminus \{0\} = \sigma_{A_e}(x) \setminus \{0\}$. Since $0 \in \sigma_{A_e}(x)$, this shows that $\sigma_A(x) \cup \{0\} = \sigma_{A_e}(x)$.

Most times, whenever the algebra A under consideration is understood, for $x \in A$ we drop the suffix A and simply write $\sigma(x)$ and $\rho(x)$.

Example 1.2.3. (1) Let X be a compact Hausdorff space and $f \in C(X)$. If $\lambda \notin f(X)$, then $x \to \lambda - f(x)$ has no zero on X and hence $x \to (\lambda - f(x))^{-1}$ is a continuous function. This implies that $\sigma_{C(X)}(f)$ equals the range of f.

(2) Let X be a locally compact, noncompact Hausdorff space and let $f \in C_0(X)$. Since $C_0(X)$ does not have an identity, $\{0\} \cup f(X) \subseteq \sigma(f)$. Conversely, let $\lambda \neq 0$ such that $\lambda \notin f(X)$. We show that the function $\lambda - f(x)$ is invertible in $C_0(X)_e = C_0(X) + \mathbb{C} \cdot 1_X$. The function g defined by

$$g(x) = \frac{f(x)}{1 - \frac{1}{\lambda}f(x)}, \quad x \in X,$$

is continuous on X. Moreover, because f vanishes at infinity, so does g. Thus $g \in C_0(X)$, and it is easily verified that

$$\left(\frac{1}{\lambda} + g(x)\right)(\lambda - f(x)) = 1$$

for all $x \in X$. This proves that $\sigma(f) = f(X) \cup \{0\}$.

Definition 1.2.4. Let A be a normed algebra. For $x \in A$, the number

$$r_A(x) = \inf\{\|x^n\|^{1/n} : n \in \mathbb{N}\}$$

is called the *spectral radius* of x.

Obviously, $r(x) \leq \|x\|$ and $r(\lambda x) = |\lambda| r(x)$ for $\lambda \in \mathbb{C}$. The formula in the following lemma is called the *spectral radius formula*. These two labels, spectral radius and spectral radius formula, are justified soon (compare Theorem 1.2.8).

Lemma 1.2.5. *For every $x \in A$, $r(x) = \lim_{n \to \infty} \|x^n\|^{1/n}$.*

Proof. It suffices to show that, given $\epsilon > 0$, there exists $N(\epsilon) \in \mathbb{N}$ such that $\|x^n\|^{1/n} < r(x) + \epsilon$ for all $n \geq N(\epsilon)$. By definition of $r(x)$, there exists $k \in \mathbb{N}$ such that $\|x^k\|^{1/k} < r(x) + \epsilon$. Now, express any $n \in \mathbb{N}$ in the form $n = p(n)k + q(n)$, where $p(n) \in \mathbb{N}_0$ and $0 \leq q(n) \leq k - 1$. It follows that

$$\frac{p(n)}{n} = \frac{1}{k}\left(1 - \frac{q(n)}{n}\right) \to \frac{1}{k},$$

as $n \to \infty$. Hence

$$\|x^n\|^{1/n} \leq \|x^k\|^{p(n)/n}\|x\|^{q(n)/n} \to \|x^k\|^{1/k} < r(x) + \epsilon,$$

and therefore $\|x^n\|^{1/n} < r(x) + \epsilon$ for all sufficiently large n.

Next we have to find conditions that guarantee the invertibility of an element of a unital Banach algebra.

Lemma 1.2.6. *Let A be a Banach algebra with identity e and let $x \in A$ with $r(x) < 1$. Then $e - x$ is invertible in A and*

$$(e - x)^{-1} = e + \sum_{n=1}^{\infty} x^n.$$

Proof. Fix any η such that $r(x) < \eta < 1$. By Lemma 1.2.5, $\|x^n\|^{1/n} \leq \eta$ for all $n \geq N$ for some $N \in \mathbb{N}$. Then $\|x^n\| \leq \eta^n$ for $n \geq N$, and since $\eta < 1$, the series $\sum_{n=1}^{\infty} \|x^n\|$ converges. Since A is complete, the sequence of partial sums $y_m = e + \sum_{n=1}^{m} x^n, m \in \mathbb{N}$, converges in A with limit $y = e + \sum_{n=1}^{\infty} x^n$. Indeed, $\|y - y_m\| \leq \sum_{n=m+1}^{\infty} \|x^n\|$. Now

$$(e - x)y_m = y_m(e - x) = e - x^{m+1}$$

for all m. Because $y_m \to y$ and $x^m \to 0$ as $n \to \infty$, we conclude that $(e-x)y = y(e - x) = e$. □

Lemma 1.2.7. *Let A be a normed algebra with identity e.*

(i) *If $x, y \in G(A)$ are such that $\|y - x\| \leq \frac{1}{2}\|x^{-1}\|^{-1}$, then*

$$\|y^{-1} - x^{-1}\| \leq 2\|x^{-1}\|^2\|y - x\|.$$

In particular, the mapping $x \to x^{-1}$ is a homeomorphism of $G(A)$.

(ii) *Suppose that A is complete. Then $G(A)$ is open in A, and if $x \in A$ is such that $\|x - e\| < 1$, then $x \in G(A)$.*

Proof. (i) If x and y are as in (i), then

$$\|y^{-1}\| - \|x^{-1}\| \leq \|y^{-1} - x^{-1}\| = \|y^{-1}(x - y)x^{-1}\| \leq \frac{1}{2}\|y^{-1}\|,$$

whence $\|y^{-1}\| \leq 2\|x^{-1}\|$ and therefore

$$\|y^{-1} - x^{-1}\| \leq \|y^{-1}\| \cdot \|x - y\| \cdot \|x^{-1}\| \leq 2\|x^{-1}\|^2\|y - x\|.$$

Thus the bijection $x \to x^{-1}$ of $G(A)$ is continuous and hence, being its own inverse, a homeomorphism.

(ii) Let $x \in A$ be such that $\|e - x\| < 1$. Then $r(e - x) < 1$ and therefore $x = e - (e - x) \in G(A)$ by Lemma 1.2.6. Now let x be an arbitrary element of $G(A)$, and let $y \in A$ with $\|y - x\| < \|x^{-1}\|^{-1}$. Then

$$\|e - x^{-1}y\| \leq \|x^{-1}\| \cdot \|x - y\| < 1.$$

By what we have already seen, $x^{-1}y \in G(A)$ and hence $y \in G(A)$. This shows that $G(A)$ is open in A. □

The following theorem is one of the most fundamental results in the theory of Banach algebras. The first proof we give is the standard one involving an application of Liouville's theorem. This use of Liouville's theorem is the first example of how the theory of holomorphic functions of one complex variable enters the study of Banach algebras. We present several other examples of this phenomenon in subsequent chapters.

Theorem 1.2.8. *Let A be a Banach algebra and $x \in A$. Then the spectrum $\sigma(x)$ of x is a nonempty compact subset of $\mathbb{C}$ and*

$$\max\{|\lambda| : \lambda \in \sigma(x)\} = r(x).$$

Proof. Of course we may assume that A has an identity e. Note first that $\sigma(x)$ is closed. In fact, $G(A)$ is open in A by Lemma 1.2.7 and $\mathbb{C} \setminus \sigma(x)$ is the inverse image of the set $G(A)$ with respect to the continuous mapping $\lambda \to \lambda e - x$. Moreover, if $|\lambda| > r(x)$ then $r((1/\lambda)x) < 1$ and hence $\lambda e - x =$

$\lambda(e - (1/\lambda)x)$ is invertible by Lemma 1.2.6. Thus $\sigma(x)$ is contained in the disc $\{\lambda \in \mathbb{C} : |\lambda| \leq r(x)\}$.

We show next that $\sigma(x) \neq \emptyset$. Take any $l \in A^*$ and consider the function f on $\rho(x)$ defined by

$$f(\lambda) = l((\lambda e - x)^{-1}).$$

For $\lambda, \mu \in \rho(x)$ we have

$$
\begin{aligned}
(\lambda e - x)^{-1} &= (\lambda e - x)^{-1}(\mu e - x)(\mu e - x)^{-1} \\
&= (\lambda e - x)^{-1}((\mu - \lambda)e + \lambda e - x)(\mu e - x)^{-1} \\
&= ((\mu - \lambda)(\lambda e - x)^{-1} + e)(\mu e - x)^{-1} \\
&= (\mu - \lambda)(\lambda e - x)^{-1}(\mu e - x)^{-1} + (\mu e - x)^{-1},
\end{aligned}
$$

and therefore, when $\lambda \neq \mu$,

$$\frac{f(\lambda) - f(\mu)}{\lambda - \mu} = -l\left((\lambda e - x)^{-1}(\mu e - x)^{-1}\right).$$

Since l is continuous and the mapping $y \to y^{-1}$ of $G(A)$ into itself is continuous (Lemma 1.2.7), we conclude that f is a holomorphic function on $\rho(x)$. For $|\lambda| > \|x\|$, we have

$$(\lambda e - x)^{-1} = \left(\lambda\left(e - \frac{1}{\lambda}x\right)\right)^{-1} = \frac{1}{\lambda}\left(e - \frac{1}{\lambda}x\right)^{-1} = \frac{1}{\lambda}\sum_{n=0}^{\infty}\lambda^{-n}x^n$$

and hence

$$\|(\lambda e - x)^{-1}\| \leq \frac{1}{|\lambda|}\sum_{n=0}^{\infty}\left(\frac{\|x\|}{|\lambda|}\right)^n = \frac{1}{|\lambda|} \cdot \frac{1}{1 - |\lambda|^{-1}\|x\|},$$

which tends to zero as $|\lambda| \to \infty$. Thus, since $|f(\lambda)| \leq \|l\| \cdot \|(\lambda e - x)^{-1}\|$, f vanishes at infinity. In particular, f is a bounded function.

Assume now that $\sigma(x) = \emptyset$. Then f is a bounded entire function and hence constant by Liouville's theorem. Since f vanishes at infinity, it follows that $f = 0$. Because $l \in A^*$ was arbitrary, we get that $l((\lambda e - x)^{-1}) = 0$ for each $\lambda \in \rho(x)$ and all $l \in A^*$, contradicting the Hahn–Banach theorem. This shows that indeed $\sigma(x)$ is nonempty.

Let $s(x) = \sup\{|\lambda| : \lambda \in \sigma(x)\}$. Then $s(x) \leq r(x)$. Towards a contradiction, assume that $s(x) < r(x)$ and select any μ such that $s(x) < \mu < r(x)$. By what we have shown above, for $l \in A^*$ the function $f(\lambda) = l((\lambda e - x)^{-1})$ is holomorphic on the set $U = \{\lambda \in \mathbb{C} : |\lambda| > s(x)\}$. Now, for $|\lambda| > \|x\|$,

$$f(\lambda) = \sum_{n=0}^{\infty}\lambda^{-(n+1)}l(x^n).$$

Thus this series is nothing but the Laurent series of f on the domain $|\lambda| > \|x\|$. Because f is holomorphic on U, it follows from uniqueness of the Laurent series expansion that the series

$$\sum_{n=0}^{\infty} l(x^n)\mu^{-(n+1)}$$

converges. This implies that $l(\mu^{-(n+1)}x^n) \to 0$ as $n \to \infty$. So, for each $l \in A^*$, the set of complex numbers

$$\{l(\mu^{-(n+1)}x^n) : n \in \mathbb{N}\}$$

is bounded. By the uniform boundedness principle, there exists $C > 0$ such that $\|\mu^{-(n+1)}x^n\| \leq C$ for all $n \in \mathbb{N}$. It follows that

$$r(x) = \lim_{n\to\infty} \|x^n\|^{1/n} \leq \lim_{n\to\infty} (C\mu^{n+1})^{1/n} = \mu.$$

This contradiction shows that $r(x) = s(x)$ and finishes the proof of the theorem. $\qquad\square$

Theorem 1.2.8 can be proved without recourse to the theory of holomorphic functions and Liouville's theorem in particular. For illustration, we also present such a proof. Let A be a Banach algebra and let $x \in A$. We show that there exists $\lambda \in \sigma(x)$ such that $|\lambda| = r(x)$. Of course, we can again assume that A has an identity e.

To begin with, note that if $r(x) = 0$ then $0 \in \sigma(x)$. Indeed, otherwise x is invertible and

$$0 < \|e\| = \|x^n(x^{-1})^n\| \leq \|x^n\| \cdot \|x^{-1}\|^n$$

for all $n \in \mathbb{N}$, which implies $r(x) \geq \|x^{-1}\|^{-1}$. So let $r(x) > 0$. Since $r(\mu x) = |\mu|r(x)$ and $\sigma(\mu x) = \mu\sigma(x)$ for $\mu \neq 0$, we can assume that $r(x) = 1$.

For $n \in \mathbb{N}$, denote by Ω_n the set of all n-th roots of unity. For $\omega \in \Omega_n, \omega \neq 1$, we have $\sum_{j=0}^{n-1} \omega^j = 0$ since $(1 - \omega)\sum_{j=0}^{n-1} \omega^j = 1 - \omega^n = 0$. Thus, for each $1 \leq k \leq n - 1$,

$$\sum_{\omega\in\Omega_n} \omega^k = \sum_{j=0}^{n-1} \left(\exp 2\pi\frac{j}{n}\right)^k = \sum_{j=0}^{n-1} \left(\exp 2\pi\frac{k}{n}\right)^j = 0.$$

Now suppose that $e - \omega x$ is invertible for every $\omega \in \Omega_n$. Then

$$(e - \omega x)^{-1}(e - (\omega x)^n) = e + \omega x + \cdots + (\omega x)^{n-1}$$

for all $\omega \in \Omega_n$, and hence

$$\frac{1}{n}(e - x^n) \sum_{\omega\in\Omega_n} (e - \omega x)^{-1} = \frac{1}{n} \sum_{\omega\in\Omega_n} \left(\sum_{k=0}^{n-1}(\omega x)^k\right)$$

$$= \frac{1}{n} \sum_{k=0}^{n} \left(x^k \sum_{\omega\in\Omega_n} \omega^k\right)$$

$$= e.$$

Thus $e - x^n$ is invertible in A and

$$(e - x^n)^{-1} = \frac{1}{n} \sum_{\omega \in \Omega_n} (e - \omega x)^{-1}.$$

Towards a contradiction, assume now that $\lambda e - x$ is invertible for all $\lambda \in \mathbb{C}$ with $|\lambda| \geq 1 = r(x)$, that is, $e - \lambda x$ is invertible whenever $|\lambda| \leq 1$. Because the function $\lambda \to \|(e - \lambda x)^{-1}\|$ is continuous on $\rho(x)$, it follows that

$$M = \sup\{\|(e - \lambda x)^{-1}\| : \lambda \in \mathbb{C}, |\lambda| \leq 1\} < \infty.$$

For each $|\lambda| \leq 1$ and $\omega \in \Omega_n$, $n \in \mathbb{N}$, the element $e - \omega \lambda x$ is invertible. Therefore the above formula for inverses and Lemma 1.2.7(i) yield

$$\|(e - (\lambda x)^n)^{-1} - (e - x^n)^{-1}\| \leq \frac{1}{n} \sum_{\omega \in \Omega_n} \|(e - \omega \lambda x)^{-1} - (e - \omega x)^{-1}\|$$

$$\leq \frac{2}{n} \sum_{\omega \in \Omega_n} \|(e - \omega x)^{-1}\|^2 \cdot \|\omega \lambda x - \omega x\|$$

$$\leq 2M^2 |\lambda - 1| \cdot \|x\|,$$

provided λ is such that

$$|\lambda - 1| \cdot \|x\| = \|(e - \omega \lambda x) - (e - \omega x)\| \leq \frac{1}{2}\|(e - \omega x)^{-1}\|^{-1}.$$

We claim next that $(e - x^n)^{-1} \to e$ as $n \to \infty$. For that, let $\epsilon > 0$ be given and choose $0 < \lambda < 1$ such that

$$2M^2 |\lambda - 1| \cdot \|x\| \leq \epsilon \text{ and } |\lambda - 1| \cdot \|x\| \leq \frac{1}{2M}.$$

Because $M \geq \|(e - \omega x)^{-1}\|$, we then get

$$\|(e - (\lambda x)^n)^{-1} - (e - x^n)^{-1}\| \leq \epsilon$$

for all $n \in \mathbb{N}$. Now

$$\lim_{n \to \infty} \|(\lambda x)^n\|^{1/n} = r(\lambda x) = \lambda < 1,$$

and hence $(\lambda x)^n \to 0$ as $n \to \infty$. The map $y \to y^{-1}$ being continuous on $G(A)$, we obtain that $(e - (\lambda x)^n)^{-1} \to e$. It follows that

$$\|e - (e - x^n)^{-1}\| \leq 2\epsilon$$

for n large enough. This proves the above claim. Finally, using continuity of $y \to y^{-1}$ again, we conclude that $x^n \to 0$ as $n \to \infty$. This contradicts the fact that $\|x^n\| \geq r(x)^n = 1$ for all $n \in \mathbb{N}$. Thus there exists $\lambda \in \mathbb{C}$ such that $|\lambda| \geq r(x)$ and $\lambda e - x$ is not invertible.

The following theorem, which turns out to be a simple consequence of Theorem 1.2.8, is usually called the *Gelfand–Mazur theorem*. It is basic to much of Gelfand's theory, which is developed in Chapter 2, and it generalises Frobenius' classical theorem which says that every finite-dimensional complex division algebra is isomorphic to the complex number field.

Theorem 1.2.9. *Let A be a Banach algebra with identity e, and suppose that every nonzero element x of A is invertible. Then A is isomorphic to the field of complex numbers.*

Proof. By Theorem 1.2.8, for every $x \in A$, there exists $\lambda_x \in \mathbb{C}$ such that $\lambda_x e - x \notin G(A)$. Since $G(A) = A \setminus \{0\}$ by hypothesis, it follows that $\lambda_x e = x$. Then, of course, λ_x is unique, and the mapping $x \to \lambda_x$ is an isomorphism from A onto $\mathbb{C}$. $\qquad\square$

The following lemma represents a special case of a more general spectral mapping theorem in the context of the holomorphic functional calculus for commutative Banach algebras (Section 3.1).

Lemma 1.2.10. *Let p be a complex polynomial (without constant term if A does not have an identity). Then, for every element $x \in A$,*

$$\sigma_A(p(x)) = p(\sigma_A(x)) = \{p(\lambda) : \lambda \in \sigma_A(x)\}.$$

Proof. Suppose first that A has an identity e. If p is constant, say $p = \alpha$, then $p(x) = \alpha e$ and hence $\sigma_A(p(x)) = \sigma_A(\alpha e) = \{\alpha\} = p(\sigma_A(x))$. So let p be non-constant. Momentarily, fix any $\lambda \in \mathbb{C}$ and let $\lambda_1, \ldots, \lambda_n$ be the roots of the polynomial $q(z) = \lambda - p(z)$. Then

$$\lambda e - p(x) = (\lambda - p(z))(x) = \alpha(\lambda_1 e - x) \cdot \ldots \cdot (\lambda_n e - x),$$

where $0 \neq \alpha \in \mathbb{C}$, and hence $\lambda e - p(x) \in G(A)$ if and only if $\lambda_i e - x \in G(A)$ for all $i = 1, \ldots, n$. It follows that if $\lambda \in \sigma_A(p(x))$ then $\lambda_i \in \sigma_A(x)$ for at least one i and therefore $\lambda = p(\lambda_i) \in p(\sigma_A(x))$. This shows that $\sigma_A(p(x)) \subseteq p(\sigma_A(x))$.

Conversely, let $\mu \in \sigma_A(x)$ and put $\lambda = p(\mu)$. Then $q(\mu) = 0$ and hence $\mu = \lambda_i$ for some i. This means that $\lambda_i \in \sigma_A(x)$ and consequently $\lambda e - p(x)$ is not invertible, whence $\lambda \in \sigma_A(p(x))$.

Finally, suppose that A does not have an identity. Then

$$\sigma_A(p(x)) = \sigma_{A_e}(p(x)) = p(\sigma_{A_e}(x)) = p(\sigma_A(x))$$

by definition and by what we have shown already. $\qquad\square$

Let A be a Banach algebra with identity e and let B a closed subalgebra of A which contains e. It is an important concern to clarify the relationship between $\sigma_A(x)$ and $\sigma_B(x)$ for elements x of B. The result, Theorem 1.2.12 below, is employed several times in Chapter 2.

In the sequel, for any topological space X and subset Y of X, Y° denotes the *interior of Y* and $\partial(Y)$ denotes the *topological boundary* of Y; that is, $Y^\circ = X \setminus \overline{(X \setminus Y)}$ and $\partial(Y) = \overline{Y} \setminus Y^\circ$.

Lemma 1.2.11. *Let A be a Banach algebra with identity e and B a closed subalgebra of A containing e. If $x \in B$, then*

$$\sigma_A(x) \subseteq \sigma_B(x) \text{ and } \partial(\sigma_B(x)) \subseteq \partial(\sigma_A(x)).$$

Proof. The first inclusion is immediate from the fact that if $\lambda e - x \in G(B)$, then $\lambda e - x \in G(A)$. For the second inclusion it suffices to show that $\partial(\sigma_B(x)) \subseteq \sigma_A(x)$, because then

$$\partial(\sigma_B(x)) \subseteq \sigma_A(x) \cap \overline{\rho_B(x)} \subseteq \sigma_A(x) \cap \overline{\rho_A(x)} = \partial(\sigma_A(x)).$$

Let $\lambda \in \partial(\sigma_B(x))$ and set $y = \lambda e - x$. Then $y \notin G(B)$ and there exists a sequence $(\lambda_n)_n$ in $\rho_B(x)$ with $\lambda_n \to \lambda$. Thus, with $y_n = \lambda_n e - x$, we have $y_n \in G(B)$ and $y_n \to y$. Let $z_n = y_n^{-1}, n \in \mathbb{N}$. Then $\|z_n\| \to \infty$ as $n \to \infty$. Indeed, otherwise there exist $M < \infty$ and a subsequence $(z_{n_k})_k$ such that $\|z_{n_k}\| \le M$ for all k and therefore

$$\|e - z_{n_k}y\| = \|z_{n_k}(y_{n_k} - y)\| \le M\|y_{n_k} - y\| \to 0$$

as $k \to \infty$. It follows that $z_{n_k}y$ is invertible for large k, and hence y is invertible in B, a contradiction. So

$$\frac{\|z_n y_n\|}{\|z_n\|} = \frac{\|e\|}{\|z_n\|} \to 0.$$

Because

$$\left| \frac{\|z_n y\|}{\|z_n\|} - \frac{\|z_n y_n\|}{\|z_n\|} \right| \le \frac{1}{\|z_n\|}\|z_n(y - y_n)\| \le \|y - y_n\| \to 0,$$

the elements $w_n = \|z_n\|^{-1}z_n, n \in \mathbb{N}$, of A satisfy $\|w_n\| = 1$ and $\|w_n y\| \to 0$. This implies that y cannot be invertible in A because otherwise

$$1 = \|w_n\| = \|(w_n y)y^{-1}\| \le \|w_n y\| \cdot \|y^{-1}\| \to 0.$$

So $\lambda \in \sigma_A(x)$, as was to be shown. $\qquad \square$

Theorem 1.2.12. *Let A be a Banach algebra with identity e and let $x \in A$. Then the following conditions are equivalent.*

(i) $\rho_A(x)$ *is connected.*
(ii) $\sigma_A(x) = \sigma_B(x)$ *for every closed subalgebra B of A containing x and e.*

Proof. Suppose that (i) holds and let B be any subalgebra of A as in (ii). We have to show that $\sigma_B(x) \subseteq \sigma_A(x)$, equivalently, that $\sigma_B(x) \cap \rho_A(x) = \emptyset$. Assume that there exists $\lambda \in \sigma_B(x) \cap \rho_A(x)$ and select any $\mu \in \rho_B(x) \subseteq \rho_A(x)$. Since $\rho_A(x)$ is connected and open in $\mathbb{C}$, it is path connected. Hence there exists a continuous function $\gamma : [0,1] \to \rho_A(x)$ with $\gamma(0) = \lambda$ and $\gamma(1) = \mu$. Since $\gamma(0) \in \sigma_B(x)$, the set $\{t \in [0,1] : \gamma(t) \in \sigma_B(x)\}$ is nonempty. Let s

be its supremum. Then $s < 1$, because $\gamma(1) \in \rho_B(x)$ and $\rho_B(x)$ is open. By definition of s, $\gamma((s, 1]) \subseteq \rho_B(x)$ and therefore

$$\gamma(s) \in \overline{\gamma((s, 1])} \subseteq \overline{\rho_B(x)} \quad \text{and} \quad \gamma(s) \in \overline{\gamma([0, 1]) \cap \sigma_B(x)} \subseteq \sigma_B(x),$$

whence $\gamma(s) \in \partial(\sigma_B(x))$. Lemma 1.2.11 shows that $\gamma(s) \in \partial(\sigma_A(x)) \subseteq \sigma_A(x)$. This contradicts $\gamma([0, 1]) \subseteq \rho_A(x)$ and so $\sigma_B(x) \cap \rho_A(x) = \emptyset$.

To show that conversely (ii) implies (i), let B be the closed subalgebra of A generated by x and e. Then B is the closure of the subalgebra consisting of all elements of the form $p(x)$, where p is a polynomial. For $z \in \mathbb{C}$ and $\delta > 0$, set $K(z, \delta) = \{\lambda \in \mathbb{C} : |\lambda - z| \leq \delta\}$. Let Γ denote the connected component of $\mathbb{C} \setminus \sigma_A(x)$ that contains $\mathbb{C} \setminus K(0, r(x))$ (recall that $\sigma_A(x) \subseteq K(0, r(x))$). We have to show that $\Gamma = \rho_A(x)$.

The set Γ is open and closed in $\rho_A(x)$ since $\rho_A(x)$ is locally path connected. Thus $\mathbb{C} \setminus \Gamma$ is closed and $\rho_A(x) \setminus \Gamma$ is open in $\mathbb{C}$. As $\mathbb{C} \setminus \Gamma = \sigma_A(x) \cup (\rho_A(x) \setminus \Gamma)$, we get that

$$\partial(\mathbb{C} \setminus \Gamma) = (\mathbb{C} \setminus \Gamma) \setminus (\mathbb{C} \setminus \Gamma)^\circ \subseteq (\mathbb{C} \setminus \Gamma) \setminus (\rho_A(x) \setminus \Gamma) \subseteq \sigma_A(x).$$

Since $\mathbb{C} \setminus \Gamma$ is compact, for any polynomial p there exists $z_0 \in \mathbb{C} \setminus \Gamma$ such that $|p(z)| \leq |p(z_0)|$ for all $z \in \mathbb{C} \setminus \Gamma$. Then, necessarily, $z_0 \in \partial(\mathbb{C} \setminus \Gamma)$ whenever p is nonconstant. In fact, otherwise $K(z_0, \delta) \subseteq \mathbb{C} \setminus \Gamma$ for some $\delta > 0$ and then the maximum modulus principle would yield that p is constant. Since $\partial(\mathbb{C} \setminus \Gamma) \subseteq \sigma_A(x)$, we thus conclude that

$$\max\{|p(z)| : z \in \mathbb{C} \setminus \Gamma\} \leq \max\{|p(z)| : z \in \sigma_A(x)\}$$

for every polynomial p. Now $\sigma_A(p(x)) = p(\sigma_A(x))$ by Lemma 1.2.10, and hence

$$\max\{|p(z)| : z \in \mathbb{C} \setminus \Gamma\} \leq r(p(x)) \leq \|p(x)\|$$

for every polynomial p.

By hypothesis $\rho_A(x) = \rho_B(x)$. Thus it remains to show that $\rho_B(x) \subseteq \Gamma$. To that end, fix $\lambda \in \rho_B(x)$ and let $q(z) = \lambda - z$ for $z \in \mathbb{C}$. Then $q(x) = \lambda e - x$ is invertible in B. Thus there exists a sequence $(p_n)_n$ of polynomials such that $p_n(x) \to (\lambda e - x)^{-1}$. For each $n \in \mathbb{N}$, define a polynomial q_n by $q_n(z) = 1 - q(z)p_n(z)$. Then

$$q_n(x) = e - p_n(x)(\lambda e - x) \to e - (\lambda e - x)^{-1}(\lambda e - x) = 0.$$

Since $|q_n(z)| \leq \|q_n(x)\|$, it follows that $q_n(z) \to 0$ for each $z \in \mathbb{C} \setminus \Gamma$. On the other hand, $q(\lambda) = 0$ and hence $q_n(\lambda) = 1$ for all n. Hence $\lambda \in \Gamma$, as was to be shown. $\square$

In particular, Theorem 1.2.12 applies when $\sigma_A(x) \subseteq \mathbb{R}$ or when $\sigma_A(x)$ is countable. We close this section with showing that the spectral radius is subadditive and submultiplicative on commuting elements.

Lemma 1.2.13. *Let A be a normed algebra and suppose that $x, y \in A$ are such that $xy = yx$. Then $r(xy) \leq r(x)r(y)$ and $r(x + y) \leq r(x) + r(y)$.*

Proof. Since $(xy)^n = x^n y^n$ for all $n \in \mathbb{N}$, Lemma 1.2.5 yields that

$$r(xy) = \lim_{n \to \infty} \|x^n y^n\|^{1/n} \leq \lim_{n \to \infty} \|x^n\|^{1/n} \cdot \lim_{n \to \infty} \|y^n\|^{1/n} = r(x)r(y).$$

The proof of the second inequality requires much more effort. Take any $\alpha > r(x)$ and $\beta > r(y)$ and set $a = (1/\alpha)x$ and $b = (1/\beta)y$. Then $r(a) < 1$ and $r(b) < 1$.

Because x and y commute, we have

$$\|(x + y)^n\|^{1/n} = \left\| \sum_{j=0}^{n} \binom{n}{j} x^j y^{n-j} \right\|^{1/n}$$

$$\leq \left(\sum_{j=0}^{n} \binom{n}{j} \alpha^j \beta^{n-j} \|a^j\| \cdot \|b^{n-j}\| \right)^{1/n}.$$

For each $n \in \mathbb{N}$, choose $n', n'' \in \mathbb{N}_0$ such that $n' + n'' = n$ and

$$\|a^{n'}\| \cdot \|b^{n''}\| = \max_{0 \leq j \leq n} \|a^j\| \cdot \|b^{n-j}\|.$$

With this choice of n' and n'' we have

$$r(x + y) = \lim_{n \to \infty} \|(x + y)^n\|^{1/n}$$

$$\leq (\alpha + \beta) \liminf_{n \to \infty} \|a^{n'}\|^{1/n} \|b^{n''}\|^{1/n}.$$

Now, the sequence $(n'/n)_n \subseteq [0, 1]$ has a convergent subsequence $(n'_k/n_k)_k$ with limit γ, say. If $\gamma \neq 0$, then $n'_k \to \infty$ and hence

$$\lim_{k \to \infty} \|a^{n'_k}\|^{1/n_k} = r(a)^\gamma < 1,$$

whereas if $\gamma = 0$, then

$$\limsup_{k \to \infty} \|a^{n'_k}\|^{1/n_k} \leq \lim_{k \to \infty} \|a\|^{n'_k/n_k} \leq 1.$$

Thus, in either case,

$$\limsup_{k \to \infty} \|a^{n'_k}\|^{1/n_k} \leq 1.$$

Similarly, since $(n''_k/n_k)_k$ converges,

$$\limsup_{k \to \infty} \|b^{n''_k}\|^{1/n_k} \leq 1.$$

The above upper estimate for $r(x + y)$ now shows that $r(x + y) \leq \alpha + \beta$. Since this holds for all $\alpha > r(x)$ and $\beta > r(y)$, it follows that $r(x + y) \leq r(x) + r(y)$. $\square$

Using the Gelfand homomorphism (Section 2.2), a much simpler proof of the preceding lemma can be given. Thus, if A is commutative, r is an algebra seminorm on A. In general, however, it is not a norm (see the examples in Section 2.1).

1.3 L^1-algebras and Beurling algebras

Let G be a locally compact group with left Haar measure and modular function Δ. Then $L^1(G)$ is not only a Banach space under the norm $\|\cdot\|_1$, but also a normed algebra with *convolution*, which we define now, as multiplication.

Let $f \in L^1(G)$ and $g \in L^p(G)$, $1 \le p < \infty$. Then for almost all $x \in G$, the function $y \to f(xy)g(y^{-1})$ is integrable on G, and the function $f * g$, defined almost everywhere on G by

$$(f * g)(x) = \int_G f(xy)g(y^{-1})dy,$$

belongs to $L^p(G)$. Moreover, $\|f * g\|_p \le \|f\|_1\|g\|_p$. If $p = \infty$, then $(f * g)(x)$ is defined for all $x \in G$ and $f * g$ is bounded and uniformly continuous.

With this convolution and the involution defined by $f^*(x) = \overline{f(x^{-1})}\Delta(x^{-1})$, $L^1(G)$ is a Banach $*$-algebra. It contains $C_c(G)$ as a dense subalgebra and is commutative if and only if the group G is Abelian.

For $x \in G$, the left and right translation operators L_x and R_x on functions on G are defined by $L_x f(y) = f(x^{-1}y)$ and $R_x f(y) = f(yx)$, $y \in G$. Then $\|L_x f\|_1 = \|f\|_1$ and the map $x \to L_x f$ from G into $L^1(G)$ is continuous (compare Lemma 1.3.6 for Beurling algebras).

When G is discrete and the Haar measure is counting measure, $(f*g)(x) = \sum_{y \in G} f(xy)g(y^{-1})$, and δ_e, the Dirac function at the neutral element e of G, is an identity for $l^1(G)$. If G is not discrete, then $L^1(G)$ does not have an identity, but it has a two-sided approximate identity with norm bound 1. In fact, for any open relatively compact symmetric neighbourhood V of e in G, choose a functions $u_V \in L^1(G)$ such that $u_V \ge 0$, $\|u_V\| = 1$ and $\operatorname{supp} u_V \subseteq V$. Then, for every $f \in L^1(G)$, $\|u_V * f - f\|_1 \to 0$ as $V \to \{e\}$. Similarly, $\|f * u_V - f\|_1 \to 0$.

Recall that $M(G)$ is the Banach space of all complex valued regular Borel measures μ on G with the norm $\|\mu\| = |\mu|(G)$, where $|\mu|$ denotes the total variation of μ. For $\mu, \nu \in M(G)$, the convolution of μ and ν is defined by

$$\langle \mu * \nu, g \rangle = \int_G \int_G g(xy)d\mu(x)d\nu(y)$$

for $g \in C_c(G)$. Then $\mu * \nu \in M(G)$ and $\|\mu * \nu\| \le \|\mu\| \cdot \|\nu\|$. Note that $L^1(G)$ embeds into $M(G)$ as a subalgebra (in fact, an ideal) by the mapping $f \to \mu_f$, where $\langle \mu_f, g \rangle = \int_G g(x)f(x)dx$ for all $g \in C_c(G)$.

We refrain from proving the preceding facts about $L^p(G)$ and $M(G)$ here and instead refer the reader to the Appendix and the literature given there.

Definition 1.3.1. A positive function ω on a locally compact group G is called a *weight* or *weight function* if it has the following properties.

(i) $\omega(xy) \leq \omega(x)\omega(y)$ for all $x, y \in G$.
(ii) ω is Borel measurable.

Example 1.3.2. (1) For $\alpha > 0$, define ω_α on $\mathbb{Z}$ by $\omega_\alpha(n) = (1 + |n|)^\alpha$. Then ω_α is a weight on $\mathbb{Z}$.

(2) The functions $t \to \exp(|t|^z), z \in \mathbb{T}$, and $t \to (1 + |t|)^\alpha, \alpha \geq 0$, are weights on $\mathbb{R}$. More generally, let $\delta : \mathbb{R}^n \to [0, \infty)$ be any continuous function satisfying $\delta(x + y) \leq \delta(x) + \delta(y)$ for all $x, y \in \mathbb{R}^n$. Then $\omega(x) = (1 + \delta(x))^\alpha$ defines a weight on $\mathbb{R}^n$.

We first note the interesting fact that a weight is bounded away from both zero and infinity on compact sets.

Lemma 1.3.3. *Let C be a compact subset of G. Then there exist positive real numbers a and b such that $a \leq \omega(x) \leq b$ for all $x \in C$.*

Proof. We first establish the existence of b. To that end, for $n \in \mathbb{N}$, let

$$U_n = \{x \in G : \omega(x) < n\}.$$

Then $\bigcup_{n=1}^\infty U_n = G$ and the sets U_n are measurable. Choose $n \in \mathbb{N}$ such that $|U_n| > 0$. Then U_n^2 has a nonempty interior. Fix $z \in (U_n^2)^\circ$ and let $V = (z^{-1}U_n^2)^\circ$. Then V is an open neighbourhood of the identity, and hence by compactness of C there exist $y_1, \ldots, y_m \in C \cup C^{-1}$ such that

$$C \cup C^{-1} \subseteq y_1 V \cup \ldots \cup y_m V.$$

Now, define $b > 0$ by

$$b = n^2 \omega(z^{-1}) \cdot \max\{\omega(y_j) : 1 \leq j \leq m\}.$$

If $x \in C \cup C^{-1}$, then $x = v y_j$ for some $v \in V$ and $j \in \{1, \ldots, m\}$ and hence

$$\omega(x) \leq \omega(v)\omega(y_j) \leq n^2 \omega(z^{-1})\omega(y_j) \leq b,$$

as wanted. Finally, let

$$a = \inf\{\omega(x) : x \in C\},$$

and suppose that $a = 0$. Then there exists a sequence $(x_n)_n$ in C such that $\omega(x_n) \to 0$. Because

$$1 \leq \omega(e) \leq \omega(x_n)\omega(x_n^{-1}),$$

we must have $\omega(x_n^{-1}) \to \infty$, which contradicts the boundedness of ω on the compact set C^{-1}.

Corollary 1.3.4. *Let ω be a weight function on a compact group G. Then $\omega(x) \geq 1$ for all $x \in G$.*

Proof. Assume that $\omega(x) < 1$ for some $x \in G$. Since $\omega(x^n) \leq \omega(x)^n$ for all n, we obtain $\omega(x^n) \to 0$ as $n \to \infty$. However, G being compact, ω is bounded away from zero by Lemma 1.3.3.

The functions f on G such that $f\omega \in L^1(G)$ form a linear space, which is denoted $L^1(G, \omega)$, and $\|f\|_{1,\omega} = \int_G |f(x)|\omega(x)dx$ defines a norm on $L^1(G, \omega)$. If $(f_n)_n$ is a Cauchy sequence in $L^1(G, \omega)$, then $f_n\omega \to g$ for some $g \in L^1(G)$ and hence $g/\omega \in L^1(G, \omega)$ and $f_n \to g/\omega$ in $L^1(G, \omega)$. Thus $L^1(G, \omega)$ is complete. With convolution, $L^1(G, \omega)$ is a Banach algebra. Indeed, for $f, g \in L^1(G, \omega)$,

$$\int_G |(f * g)(x)|\omega(x)dx \leq \int_G \omega(x)\left(\int_G |f(xy)| \cdot |g(y^{-1})|dy\right)dx$$

$$\leq \int_G \int_G \omega(xy)|f(xy)|\omega(y^{-1})|g(y^{-1})|dydx$$

$$= \int_G |g(y^{-1})|\omega(y^{-1})\Delta(y^{-1}) \cdot \int_G |f(x)|\omega(x)dx$$

$$= \|f\|_{1,\omega}\|g\|_{1,\omega},$$

and hence $f * g \in L^1(G, \omega)$ and $\|f * g\|_{1,\omega} \leq \|f\|_{1,\omega}\|g\|_{1,\omega}$. The involution on $L^1(G, \omega)$ is defined in exactly the same way as for $L^1(G)$. The algebra $L^1(G, \omega)$ is called the *Beurling algebra* on G associated with the weight ω.

It can be shown that, given any weight ω on G, there always exists an upper-semicontinuous weight ω' on G such that $L^1(G, \omega) = L^1(G, \omega')$ (Exercise 1.6.35). However, we are not using this fact.

Lemma 1.3.5. *Let G be a locally compact group and ω a weight on G.*

(i) *Every compactly supported function in $L^1(G)$ belongs to $L^1(G, \omega)$.*
(ii) *$C_c(G)$ is dense in $L^1(G, \omega)$.*

Proof. (i) is immediate since ω is bounded on compact subsets of G by Lemma 1.3.3.

(ii) By (i), $C_c(G) \subseteq L^1(G, \omega)$. To show that $C_c(G)$ is dense in $L^1(G, \omega)$, let $f \in L^1(G, \omega)$ and $\epsilon > 0$ be given. Since $f\omega \in L^1(G)$, there exists $h \in C_c(G)$ such that $\|h - f\omega\|_1 \leq \epsilon$. Let S denote the compact support of h and observe that $\omega(x) \geq \delta$ for some $\delta > 0$ and all $x \in S$ (Lemma 1.3.3). Since ω is bounded on S, $\omega|_S \in L^1(S)$ and hence there exists a continuous function $\eta : S \to \mathbb{R}$ such that $\eta(x) \geq \delta$ for all $x \in S$ and

$$\int_S |\eta(x) - \omega(x)|dx \leq \frac{\epsilon\delta}{\|h\|_\infty}.$$

Now define a function g on G by $g(x) = h(x)/\eta(x)$ for $x \in S$ and $g(x) = 0$ for $x \notin S$. Since $1/\eta(x) \leq 1/\delta$ for all $x \in S$, it is easily verified that g is continuous on G. Thus $g \in C_c(G)$ and

$$\|g - f\|_{1,\omega} = \int_S \omega(x)|g(x) - f(x)|dx + \int_{G\backslash S} \omega(x)|f(x)|dx.$$

We estimate the first integral on the right as follows:

$$\int_S \omega(x)|g(x) - f(x)|dx \leq \int_S \omega(x)\left|\frac{h(x)}{\eta(x)} - \frac{h(x)}{\omega(x)}\right|dx$$

$$+ \int_S \omega(x)\left|\frac{h(x)}{\omega(x)} - f(x)\right|dx$$

$$= \int_S \frac{h(x)}{\eta(x)}|\omega(x) - \eta(x)|\,dx$$

$$+ \int_S |h(x) - \omega(x)f(x)|dx$$

$$\leq \frac{\|h\|_\infty}{\delta}\int_S |\omega(x) - \eta(x)|dx$$

$$+ \int_S |h(x) - \omega(x)f(x)|dx$$

$$\leq \epsilon + \int_S |h(x) - \omega(x)f(x)|dx.$$

It follows that

$$\|g - f\|_{1,\omega} \leq \epsilon + \int_S |h(x) - \omega(x)f(x)|dx + \int_{G\backslash S} \omega(x)|f(x)|dx$$

$$= \epsilon + \int_G |h(x) - \omega(x)f(x)|dx \leq 2\epsilon.$$

This shows that $C_c(G)$ is dense in $L^1(G, \omega)$.

Lemma 1.3.6. *Let ω be a weight on G and $f \in L^1(G, \omega)$.*

(i) *For every $x \in G$, $L_x f \in L^1(G, \omega)$ and $\|L_x f\|_{1,\omega} \leq \omega(x)\|f\|_{1,\omega}$.*
(ii) *The map $x \to L_x f$ from G into $L^1(G, \omega)$ is continuous.*

Proof. (i) follows simply from submultiplicativity of ω:

$$\|L_x f\|_{1,\omega} = \int_G |f(x^{-1}t)|\omega(t)dt$$

$$= \int_G |f(x^{-1}t)|\omega(x^{-1}t)\frac{\omega(t)}{\omega(x^{-1}t)}dt$$

$$\leq \omega(x)\int_G |f(x^{-1}t)|\omega(x^{-1}t)dt$$

$$= \omega(x)\|f\|_{1,\omega}.$$

(ii) Assume first that $f \in C_c(G)$ with support S, say. Let $x \in G$ and choose a compact neighbourhood K of x in G. Let

$$C = \sup\{\omega(s) : s \in KS\} < \infty.$$

Then, for $y \in K$,

$$\|L_y f - L_x f\|_{1,\omega} = \int_{KS} |f(y^{-1}t) - f(x^{-1}t)|\omega(t)dt$$

$$\leq C \int_{KS} |f(y^{-1}t) - f(x^{-1}t)|dt$$

$$= C\|L_y f - L_x f\|_1,$$

which tends to zero as $y \to x$.

Finally, let f be an arbitrary element of $L^1(G,\omega)$ and $\epsilon > 0$. By Lemma 1.3.5 there exists $g \in C_c(G)$ such that $\|f - g\|_{1,\omega} \leq \epsilon$. Then, for all x and y in G,

$$\|L_y f - L_x f\|_{1,\omega} \leq (\|L_y\| + \|L_x\|)\|f - g\|_{1,\omega} + \|L_y g - L_x g\|_{1,\omega}$$

$$\leq \epsilon(\omega(y) + \omega(x)) + \|L_y g - L_x g\|_{1,\omega}.$$

Since ω is locally bounded, the preceding paragraph finishes the proof.

Using Lemma 1.3.6 and the local boundedness of ω, it follows easily that the approximate identity of $L^1(G)$ specified above also forms a bounded approximate identity for $L^1(G,\omega)$. If ω is bounded away from zero, then $L^1(G,\omega)$ is a subalgebra of $L^1(G)$, whereas if ω is bounded, then $L^1(G)$ is a subalgebra of $L^1(G,\omega)$. In particular, if G is compact, then $L^1(G,\omega) = L^1(G)$ as algebras.

It follows readily from the corresponding fact for $L^1(G)$ that the dual space of $L^1(G,\omega)$ equals $L^\infty(G,\omega)$, formed by all complex-valued measurable functions g on G such that $g/\omega \in L^\infty(G)$ and equipped with the norm $\|g\|_{\infty,\omega} = \|g/\omega\|_\infty$. So the continuous linear functionals on $L^1(G,\omega)$ are precisely those of the form $f \to \int_G f(x)g(x)dx$, where $g \in L^\infty(G,\omega)$.

1.4 Ideals and multiplier algebras

In this section we study ideals of normed algebras and introduce the concept of a multiplier algebra. As examples, we describe all the closed ideals of $C_0(X)$ and the multiplier algebras of $C_0(X)$ and $L^1(G)$.

Definition 1.4.1. Let A be a complex algebra and I an ideal of A. Then I is called *modular*, if the quotient algebra A/I is unital, that is, there exists $u \in A$ such that the two sets

$$A(1 - u) := \{x - xu : x \in A\} \text{ and } (1 - u)A := \{x - ux : x \in A\}$$

are both contained in I. Such an element u is called an *identity modulo I*. The ideal I is said to be a *maximal modular* ideal if it is modular and also a maximal proper ideal.

Every ideal containing a modular ideal is itself modular. Therefore an ideal I of A is a maximal modular ideal if and only if it is maximal within the set of all modular proper ideals. We henceforth denote by $\mathrm{Max}(A)$ the set of all maximal modular ideals of A.

Lemma 1.4.2. *Every proper modular ideal is contained in a maximal modular ideal.*

Proof. Let I be a proper modular ideal and let $u \in A$ be an identity modulo I. Let $\mathcal{L}$ be the set of all ideals L of A such that $I \subseteq L$ and $u \notin L$. Then $\mathcal{L}$ is nonempty since $I \in \mathcal{L}$. We order $\mathcal{L}$ by inclusion and show that $\mathcal{L}$ satisfies the hypothesis of Zorn's lemma. Thus let $\mathcal{K}$ be a totally ordered subset of $\mathcal{L}$ and put $L = \bigcup \{K : K \in \mathcal{K}\}$. Then $u \notin L$, and L is an ideal since $\mathcal{K}$ is totally ordered. So $L \in \mathcal{L}$ and L is an upper bound for $\mathcal{K}$. Hence, by Zorn's lemma, $\mathcal{L}$ has a maximal element M. Also, if J is a proper ideal containing M, then $u \notin J$ since otherwise $x = (x - ux) + ux \in M + J = J$ for all $x \in A$. Thus $J \in \mathcal{L}$ and hence $M = L$. So $M \in \mathrm{Max}(A)$. $\qquad\square$

Remark 1.4.3. Suppose that A is commutative and has an identity e. Then an element $x \in A$ is invertible if and only if $x \notin M$ for every $M \in \mathrm{Max}(A)$. Indeed, $x \notin G(A)$ if and only if Ax is a proper ideal, and by the preceding lemma this is equivalent to $Ax \subseteq M$ for some $M \in \mathrm{Max}(A)$.

Lemma 1.4.4. *Let A be a normed algebra and I a closed ideal in A. Then A/I, equipped with the quotient norm, is a normed algebra. If A is complete, then so is A/I.*

Proof. Because A/I, with the quotient norm, is a normed space and complete whenever A is, it only remains to observe that the quotient norm is submultiplicative. Now, for $x, y \in A$,

$$
\begin{aligned}
\|(x + I)(y + I)\| = \|xy + I\| &= \inf_{z \in I} \|xy + z\| \\
&\leq \inf_{a,b \in I} \|(x + a)(y + b)\| \\
&\leq \inf_{a,b \in I} \|x + a\| \cdot \|y + b\| \\
&= \|x + I\| \cdot \|y + I\|,
\end{aligned}
$$

as required. $\qquad\square$

Lemma 1.4.5. *Let A be a Banach algebra and I a proper modular ideal of A. If $u \in A$ is an identity modulo I, then*

$$
I \cap \{x \in A : \|x - u\| < 1\} = \emptyset.
$$

In particular, $\overline{I}$ is also a proper ideal, and every maximal modular ideal of A is closed in A.

Proof. Let A' be defined to be A if A has an identity e and $A' = A_e$, the unitisation of A, otherwise. If $x \in A$ is such that $\|x - u\| < 1$, then $e - (u - x)$ is invertible in A' by Lemma 1.2.6. Write $(e - (u - x))^{-1} = y + \lambda e$, where $y \in A$ and $\lambda \in \mathbb{C}$. Then

$$e = \lambda e + y - \lambda u - yu + \lambda x + yx.$$

Towards a contradiction, assume that $x \in I$. If $e \in A$, then

$$e = \lambda e - (\lambda e)u + y - yu + (\lambda e + y)x \in I,$$

which is impossible. If $e \notin A$, then

$$(1 - \lambda)e = y - \lambda u - yu + \lambda x + yx \in A,$$

which forces $\lambda = 1$ and $u = y - yu + x + yx \in I$, a contradiction. Thus x cannot be contained in I.   □

As an application of Urysohn's lemma, we now determine all the closed ideals of $C_0(X)$.

Theorem 1.4.6. *Let X be a locally compact Hausdorff space, and for each subset E of X let*

$$I(E) = \{f \in C_0(X) : f(x) = 0 \text{ for all } x \in E\}.$$

Then the map $E \to I(E)$ is a bijection between the collection of nonempty closed subsets of X and the proper closed ideals of $C_0(X)$. Moreover, $I(E)$ is a modular ideal if and only if E is compact, and $I(E) \in \mathrm{Max}(C_0(X))$ if and only if E is a singleton.

Proof. It is clear that $I(E)$ is a closed ideal of $C_0(X)$. Since, given any point $x \in X$, there exists $f \in C_0(X)$ such that $f(x) \neq 0$, it follows that $I(E)$ is proper whenever $E \neq \emptyset$. Moreover, if E is a closed subset of X and $x \in X \setminus E$, then by Urysohn's lemma there exists $f \in C_0(X)$ such that $f|_E = 0$ and $f(x) \neq 0$. This in particular implies that the assignment $E \to I(E)$ is injective.

Now let I be a proper closed ideal and set

$$E = \{x \in X : f(x) = 0 \text{ for all } f \in I\}.$$

Then E is a closed subset of X and $I \subseteq I(E)$. To prove that actually $I = I(E)$, we show first that every $g \in C_c(G)$ with $E \cap \mathrm{supp}\, g = \emptyset$ belongs to I. To that end, let C be any compact subset of X with $C \cap E = \emptyset$. For every $x \in C$ there exists $h_x \in I$ such that $h_x(x) \neq 0$. Then $|h_x|^2 \in I, |h_x|^2 \geq 0$ and $|h_x|^2(x) > 0$. Because C is compact, there exists a finite subset F of C such that the function h defined by

$$h(y) = \left(\sum_{x \in F} h_x \cdot \overline{h_x} \right)(y) = \sum_{x \in F} |h_x|^2(y), \; y \in X,$$

is strictly positive on C. Note that $h \in I$.

Now, let J be the set of all $g \in C_c(X)$ such that $E \cap \operatorname{supp} g = \emptyset$. By what we have just seen, for any $g \in J$ there exists $h \in I$ such $h(y) > 0$ for all $y \in \operatorname{supp} g$. Define a function f on X by $f(x) = 0$ for $x \in X \setminus \operatorname{supp} g$ and $f(x) = g(x)/h(x)$ for $x \in \operatorname{supp} g$. It is easily verified that f is continuous. Thus $f \in C_0(X)$ and $g = fh \in I$. This shows that $J \subseteq I$, as announced above.

On the other hand, J is dense in $I(E)$. To see this, let $f \in I(E)$ and $\epsilon > 0$ be given and let $C = \{x \in X : |f(x)| \geq \epsilon\}$. Then C is compact and $C \cap E = \emptyset$. Again, by Urysohn's lemma, there exists $h \in C_c(X)$ such that $h(X) \subseteq [0, 1]$, $h|_C = 1$ and $\operatorname{supp} h \subseteq X \setminus E$. Then $g = fh \in J$ and $\|f - g\|_\infty \leq \epsilon$.

Since I is closed, combining what we have shown yields that

$$I(E) \subseteq \overline{J} \subseteq \overline{I} = I \subseteq I(E),$$

so that $I(E) = I$. Clearly, $E \neq \emptyset$ since otherwise $C_c(X) \subseteq I$, whence $I = C_0(X)$.

Finally, if E is compact then there exists $u \in C_0(X)$ with $u(x) = 1$ for all $x \in E$, and this shows that $C_0(X)(1 - u) \subseteq I(E)$. Conversely, if $I(E)$ is modular, there exists $u \in C_0(X)$ such that $C_0(X)(1 - u) \subseteq I(E)$. This implies that $u = 1$ on E and hence E is compact since $u \in C_0(X)$. The remaining assertion concerning maximal modular ideals is now obvious.   □

Let G be a locally compact group. The closed ideals of $L^1(G)$ turn out to be nothing but the closed translation invariant subspaces of $L^1(G)$.

Proposition 1.4.7. *A closed linear subspace I of $L^1(G)$ is an ideal in $L^1(G)$ if and only if I is two-sided translation invariant.*

Proof. Suppose that I is two-sided translation invariant. We have to show that $g * f \in I$ and $f * g \in I$ for each $f \in I$ and $g \in L^1(G)$. Let $\varphi \in L^\infty(G)$ be such that $\int_G f(x)\varphi(x)dx = 0$ for all $f \in I$. Then, for $f \in I$ and any $g \in L^1(G)$,

$$\int_G (g * f)(x)\varphi(x)dx = \int_G \varphi(x)\left(\int_G g(xy)f(y^{-1})dy\right)dx$$

$$= \int_G \varphi(x)\left(\int_G g(y)f(y^{-1}x)dy\right)dx$$

$$= \int_G g(y)\left(\int_G L_y f(x)\varphi(x)dx\right)$$

$$= 0.$$

Since $L^1(G)^* = L^\infty(G)$, the Hahn–Banach theorem implies that $g * f \in I$ for all $f \in I$ and $g \in L^1(G)$. Thus I is a left ideal, and using the right translation invariance of I, it is shown in the same way that I is a right ideal.

Conversely, let I be a closed ideal of $L^1(G)$ and $x \in G$. Let V be a symmetric compact neighbourhood of e in G and let $|V|$ denote the Haar measure of V. Then

$$\|L_x f - |V|^{-1}(1_{xV} * f)\|_1 = \int_G |V|^{-1} \left| \int_V f(x^{-1}y)ds - \int_V f(s^{-1}x^{-1}y)ds \right| dy$$

$$\leq |V|^{-1} \int_V \left(\int_G |f(x^{-1}y) - f(s^{-1}x^{-1}y)|dy \right) ds$$

$$= |V|^{-1} \int_V \|L_x f - L_s(L_x f)\|_1 ds$$

$$\leq \sup_{s \in V} \|L_s f - L_s(L_x f)\|_1.$$

The map $s \to L_s g$ from G into $L^1(G)$ is continuous for any $g \in L^1(G)$. Thus for every $\epsilon > 0$ there exists V such that

$$\sup_{s \in V} \|L_s f - L_s(L_x f)\|_1 \leq \epsilon.$$

As I is a closed ideal, it follows that $L_x f \in I$. Similarly, it is shown that I is right translation invariant. □

The assertion of Proposition 1.4.7 also holds for Beurling algebras $L^1(G, \omega)$. The proof is as for $L^1(G)$ (Exercise 1.6.38). The following two lemmas concern the existence of bounded approximate identities in ideals of normed algebras.

Lemma 1.4.8. *Let I be a closed ideal of a normed algebra A.*

(i) *Suppose that A has a (bounded) left approximate identity. Then A/I has a (bounded) left approximate identity.*

(ii) *Suppose that I and A/I have left approximate identities with bounds M and N, respectively. Then A has a left approximate identity with bound $M + N + MN$.*

Proof. (i) Clearly, if $(e_\lambda)_\lambda$ is a (bounded) left approximate identity for A, then $(e_\lambda + I)_\lambda$ is a (bounded) left approximate identity for A/I.

(ii) Let $x_1, \ldots, x_n \in A$ and $\epsilon > 0$. There exist $u \in A$ with $\|u\| \leq M$ and $y_1, \ldots, y_n \in I$ such that

$$\|x_j - ux_j + y_j\| \leq \frac{\epsilon}{2(1 + N)}, \quad j = 1, \ldots, n.$$

Moreover, there exists $v \in I$ such that $\|v\| \leq N$ and $\|y_j - vy_j\| \leq \epsilon/2$ for $1 \leq j \leq n$. Let $w = u + v - vu$; then $\|w\| \leq M + N + MN$ and

$$\|x_j - wx_j\| \leq \|(x_j - ux_j + y_j) - v(x_j - ux_j + y_j)\| + \|y_j - vy_j\|$$

$$\leq \frac{(1 + N)\epsilon}{2(1 + N)} + \frac{\epsilon}{2} = \epsilon.$$

The statement now follows from Proposition 1.1.10. □

Lemma 1.4.9. *Let I and J be closed ideals with bounded left approximate identities. Then the ideals $I \cap J$ and $\overline{I + J}$ both have bounded left approximate identities.*

Proof. Let $(u_\lambda)_\lambda$ and $(v_\mu)_\mu$ be bounded left approximate identities for I and J, respectively. Then $(u_\lambda v_\mu)_{\lambda,\mu}$ is a bounded left approximate identity for $I \cap J$ since $u_\lambda v_\mu \in I \cap J$ and

$$\|x - u_\lambda v_\mu x\| \le \|x - u_\lambda x\| + \|u_\lambda\| \cdot \|x - v_\mu x\| \to 0$$

for every $x \in I \cap J$.

Let $w_{\lambda,\mu} = u_\lambda + v_\mu - v_\mu u_\lambda$. Then, for $x \in I$ and $y \in J$,

$$\|(x + y) - w_{\lambda,\mu}(x + y)\| \le \|x - u_\lambda x\| + \|y - v_\mu y\|$$
$$+ \|v_\mu\| \cdot \|x - u_\lambda x\| + \|u_\lambda y - v_\mu u_\lambda y\|.$$

This shows that $(w_{\lambda,\mu})_{\lambda,\mu}$ is a bounded left approximate identity for $I + J$. A simple approximation argument shows that the same is true for $\overline{I + J}$.   □

Of course, there are obvious analogues of Lemmas 1.4.8 and 1.4.9 for right approximate identities. We now introduce the important concept of multiplier algebra.

Definition 1.4.10. A Banach algebra A is said to be *faithful* if for every $a \in A$, the condition $Aa = \{0\}$ implies $a = 0$. A mapping $T : A \to A$ is called a *multiplier* of A if $x(Ty) = (Tx)y$ holds for all $x, y \in A$. Let $M(A)$ denote the collection of all multipliers of A. The next proposition shows that $M(A)$ is a commutative Banach algebra, which is called the *multiplier algebra* of A.

Proposition 1.4.11. *Let A be a Banach algebra and suppose that A is faithful. Then $M(A)$ is a commutative closed subalgebra of $\mathcal{B}(A)$ with identity.*

Proof. For $T \in M(A)$, $x, y, z \in A$, and $\alpha, \beta \in \mathbb{C}$, we have

$$xT(\alpha y + \beta z) = (Tx)(\alpha y + \beta z) = \alpha(Tx)y + \beta(Tx)z$$
$$= x(\alpha Ty + \beta Tz).$$

Since A is faithful, this implies that $T(\alpha y + \beta z) = \alpha Ty + \beta Tz$, so T is linear.

Moreover, if $T \in M(A)$, $y, z \in A$, and $(y_n)_n$ is a sequence in A such that $y_n \to y$ and $Ty_n \to z$, then for each $x \in A$,

$$\|xz - x(Ty)\| \le \|x\| \cdot \|z - Ty_n\| + \|(Tx)y_n - (Tx)y\|$$
$$\le \|x\| \cdot \|z - Ty_n\| + \|Tx\| \cdot \|y_n - y\|.$$

Thus $xz = x(Ty)$ for all $x \in A$ and hence, as A is faithful, $z = Ty$. By the closed graph theorem, T is a bounded linear operator.

If $(T_n)_n$ is a sequence in $M(A)$ and $T \in \mathcal{B}(A)$ is such that $\|T_n - T\| \to 0$, then for all $x, y \in A$,

$$\|x(Ty) - (Tx)y\| \le \|x(Ty) - x(T_n y)\| + \|(T_n x)y - (Tx)y\|$$
$$\le 2\|x\| \cdot \|y\| \cdot \|T_n - T\|,$$

and so $x(Ty) = (Tx)y$. Thus $M(A)$ is closed in $\mathcal{B}(A)$.

Finally, to show that $M(A)$ is commutative, observe first that if $T \in M(A)$ and $x, y, z \in A$, then

$$z(x(Ty)) = z((Tx)y) = ((Tz)x)y = (Tz)(xy) = zT(xy),$$

and hence faithfulness implies $x(Ty) = (Tx)y = T(xy)$ for all $x, y \in A$. Therefore, if $T, S \in M(A)$, then

$$(T(Sx))y = T((Sx)y) = (TS)(xy) = T(x(Sy)) = x((TS)y)$$

and also

$$x((ST)y) = (Sx)(Ty) = (T(Sx))y$$

for all $x, y \in A$. Thus $(ST)y = (TS)y$, and hence $M(A)$ is a commutative subalgebra of $\mathcal{B}(A)$. $\square$

It is easy to see that, with minor modifications of the proof, Proposition 1.4.11 remains valid when the hypothesis that $Aa = \{0\}$ implies $a = 0$ is replaced by $aA = \{0\}$ implies $a = 0$.

Suppose that A is commutative and, for $x \in A$, define the multiplication operator $L_x : A \to A$ by $L_x y = xy$. Clearly, $L_x \in M(A)$. If A has an identity e, then there are no other multipliers because every $T \in M(A)$ satisfies

$$Tx = eTx = (Te)x = L_{Te}x$$

for all $x \in A$, which shows that $T = L_{Te}$. However, when A is nonunital, $M(A)$ can be much larger. For instance, if A is an ideal in a larger algebra B, then for every $b \in B$, $L_b|_A$ is a multiplier of A.

Theorem 1.4.12. *Let A be a faithful commutative Banach algebra. Then the mapping $L : x \to L_x$ is a continuous isomorphism of A onto the ideal $L(A) = \{L_x : x \in A\}$ of $M(A)$. If A has an approximate identity bounded by $C > 0$, then $1/C \leq \|L\| \leq 1$.*

Proof. It is obvious that $x \to L_x$ is a norm decreasing homomorphism of A into $M(A)$. The range of A is an ideal of $M(A)$ since, for $T \in M(A)$ and $x, y \in A$,

$$(L_x T)y = x(Ty) = (Tx)y = L_{Tx}y.$$

Let $(e_\alpha)_\alpha$ be an approximate identity for A with norm bound $C > 0$. Then

$$\|L_x\| = \sup\{\|xy\| : y \in A, \|y\| \leq 1\} \geq \frac{1}{C} \sup_\alpha \|xe_\alpha\| \geq \frac{\|x\|}{C}$$

for each $x \in A$, and so $\|L\| \geq \frac{1}{c}$. $\square$

We now identify the multiplier algebra in two concrete cases.

Example 1.4.13. Let X be a locally compact Hausdorff space. We show that the multiplier algebra of $C_0(X)$ can be canonically identified with $C^b(X)$.

Clearly, any $f \in C^b(X)$ defines a multiplier T_f of $C_0(X)$ by $T_f g = fg, g \in C_0(X)$, and $\|T_f\| \le \|f\|_\infty$. Conversely, let T be an arbitrary multiplier of $C_0(X)$. For every $x \in X$ there exists $g \in C_0(X)$ such that $g(x) \ne 0$, and for any two such functions g_1, g_2, we have

$$\frac{Tg_1(x)}{g_1(x)} = \frac{Tg_2(x)}{g_2(x)}$$

since $g_2(Tg_1) = (Tg_2)g_1$. Thus we can define a function f on X by $f(x) = Tg(x)/g(x)$, where $g \in C_0(X)$ is such that $g(x) \ne 0$. This function f is continuous on X because $f(y) = Tg(y)/g(y)$ for all y in a neighbourhood of x. The function f also satisfies $Tg(x) = f(x)g(x)$ for all $g \in C_0(X)$ and $x \in X$ since $(Tg(x))^2 = g(x)(T^2g)(x) = 0$ whenever $g(x) = 0$. Moreover, $\|f\|_\infty \le \|T\|$. In fact, given $x \in X$, by Urysohn's lemma there exists $g \in C_0(X)$ such that $g(x) = 1 = \|g\|_\infty$, and this implies

$$|f(x)| = |Tg(x)| \le \|Tg\|_\infty \le \|T\| \cdot \|g\|_\infty = \|T\|.$$

It follows that the mapping $f \to T_f$ provides an isometric algebra isomorphism between $C^b(X)$ and the multiplier algebra of $C_0(X)$.

Example 1.4.14. Let G be a locally compact Abelian group. We determine the multiplier algebra of $L^1(G)$. Since $L^1(G)$ is an ideal in $M(G)$, it is immediate that for every $\mu \in M(G)$, the convolution operator $T_\mu : L^1(G) \to L^1(G)$, defined by $T_\mu f = \mu * f, \ f \in L^1(G)$, is a multiplier of $L^1(G)$ with $\|T_\mu\| \le \|\mu\|$.

It is less evident that every multiplier of $L^1(G)$ arises in this way. To see this, let $T \in M(L^1(G))$ be given and view T as a continuous linear mapping from $L^1(G)$ into $M(G)$. Let $(u_\alpha)_\alpha$ be an approximate identity for $L^1(G)$ with $\|u_\alpha\|_1 = 1$ for all α, and consider the bounded net $(Tu_\alpha)_\alpha$. Now $M(G) = C_0(G)^*$ and the ball of radius $\|T\|$ in $M(G)$ is w^*-compact by the Banach-Alaoglu theorem. We can therefore assume that the net $(Tu_\alpha)_\alpha$ converges in the w^*-topology to some $\mu \in M(G)$ for which $\|\mu\| \le \|T\|$. We claim that $T = T_\mu$. Once this is shown, it also follows that $\|T_\mu\| = \|\mu\|$.

Since $C_c(G)$ is dense in $L^1(G)$ it suffices to show that $T(f) = \mu * f$ for all $f \in C_c(G)$. Let $\langle g, \nu \rangle = \int_G g(x)d\nu(x)$ for $g \in C_0(G)$ and $\nu \in M(G)$. For all $f, g \in C_c(G)$ we then have

$$\langle g, T(u_\alpha * f) \rangle = \langle g, f * T(u_\alpha) \rangle = \langle f^* * g, T(u_\alpha) \rangle,$$

which converges to $\langle f^* * g, \mu \rangle = \langle g, \mu * f \rangle$. On the other hand,

$$|\langle g, T(u_\alpha * f) \rangle - \langle g, T(f) \rangle| \le \|g\|_\infty \|T\| \cdot \|u_\alpha * f - f\|_1,$$

which tends to zero. It follows that

$$\langle g, T(f) \rangle = \langle g, \mu * f \rangle$$

for all f and g in $C_c(G)$. This implies that $T(f) = \mu * f$ for all $f \in C_c(G)$ and hence $T = T_\mu$.

1.5 Tensor products of Banach algebras

The formation of tensor products, notably the projective and the injective tensor product, of two Banach spaces is one of the most important processes to construct new Banach spaces. In this section we consider these constructions in the context of Banach algebras.

Proposition 1.5.1. *Let A and B be algebras. On the vector space $A \otimes B$ there exists a unique product with respect to which $A \otimes B$ is an algebra, the algebraic tensor product and which satisfies $(a \otimes b)(c \otimes d) = ac \otimes bd$ for all $a, c \in A$ and $b, d \in B$.*

Proof. Given $a \in A$ and $b \in B$, there exists a unique linear operator $\lambda(a, b)$ on $A \otimes B$ such that

$$\lambda(a, b)(c \otimes d) = ac \otimes bd \quad (c \in A, d \in B).$$

The mapping $(a, b) \rightarrow \lambda(a, b)$ is bilinear. Thus there exists a unique linear mapping μ from $A \otimes B$ into the space of linear mappings from $A \otimes B$ into itself such that

$$\mu(a \otimes b) = \lambda(a, b)$$

for all $a \in A$ and $b \in B$. The required product on $A \otimes B$ can then be defined, for $u, v \in A \otimes B$, by

$$uv = \mu(u)(v).$$

It is straightforward to check that this is a product on $A \otimes B$, and it satisfies

$$(a \otimes b)(c \otimes d) = \mu(a \otimes b)(c \otimes d) = \lambda(a, b)(c \otimes d) = ac \otimes bd$$

for all $a, c \in A$ and $b, d \in B$. Finally, it is clear that this last equation determines the product uniquely. □

Let A and B be Banach spaces and let π denote the projective tensor norm on $A \otimes B$ and $A \widehat{\otimes}_\pi B$ the projective tensor product, the completion of $A \otimes B$ with respect to π (Appendix A.2). Recall for later use that every element x of $A \widehat{\otimes}_\pi B$ can be written as a series $x = \sum_{j=1}^{\infty} a_j \otimes b_j$, where $a_j \in A, b_j \in B$, and $\sum_{j=1}^{\infty} \|a_j\| \cdot \|b_j\| < \infty$ (Proposition A.2.9). Now assume that A and B are Banach algebras.

Lemma 1.5.2. *The projective tensor norm on $A \otimes B$ is an algebra norm.*

Proof. Let $x = \sum_{i=1}^{n} a_i \otimes b_i$ and $y = \sum_{j=1}^{m} c_j \otimes d_j$. Then

$$xy = \sum_{i=1}^{n} \sum_{j=1}^{m} a_i c_j \otimes b_i d_j$$

and

$$\sum_{i=1}^{n}\sum_{j=1}^{m}\|a_ic_j\|\cdot\|b_id_j\| \leq \left(\sum_{i=1}^{n}\|a_i\|\cdot\|b_i\|\right)\left(\sum_{j=1}^{m}\|c_j\|\cdot\|d_j\|\right).$$

Taking the infima over all such representations of x and y, we conclude that $\pi(xy) \leq \pi(x)\pi(y)$.   □

Now let A and B be Banach algebras. By Lemma 1.5.2 we can extend the product on $A \otimes B$ to $A \widehat{\otimes}_\pi B$. Then $A \widehat{\otimes}_\pi B$ becomes a Banach algebra, which is commutative if and only if both A and B are commutative. Moreover, if A and B are ∗-algebras, then $A \widehat{\otimes}_\pi B$ is a Banach ∗-algebra for the involution defined by $(a \otimes b)^* = a^* \otimes b^*$ for $a \in A$ and $b \in B$.

In passing we insert a simple result concerning the existence of bounded approximate identities in $A \widehat{\otimes}_\pi B$. We formulate it in terms of left approximate identities, but of course the analogous assertions are true for right and two-sided approximate identities.

Lemma 1.5.3. *Let A and B be Banach algebras having left approximate identities bounded by M and N, respectively. Then $A \widehat{\otimes}_\pi B$ has a left approximate identity bounded by MN.*

Proof. Let $(u_\lambda)_\lambda$ and $(v_\mu)_\mu$ be left approximate identities bounded M and N of A and B, respectively. Let $x = \sum_{j=1}^{\infty} a_j \otimes b_j \in A \widehat{\otimes}_\pi B$ such that $\sum_{j=1}^{\infty}\|a_j\|\cdot\|b_j\| < \infty$, and let $\epsilon > 0$. Choose $n \in \mathbb{N}$ with the property that $\|x - \sum_{j=1}^{n} a_j \otimes b_j\| \leq \epsilon$ and choose $R \geq 1$ so that $\|a_j\|, \|b_j\| \leq R$ for $1 \leq j \leq n$. There exist λ_0 and μ_0 such that $\|a_j - u_\lambda a_j\| \leq \epsilon/(nR)$ for all $\lambda \geq \lambda_0$ and $\|b_j - v_\mu b_j\| \leq \epsilon/(nR)$ for all $\mu \geq \mu_0$ and all $1 \leq j \leq n$. It follows that

$$\|x - (u_\lambda \otimes v_\mu)x\| \leq \epsilon(1 + MN) + \left\|\sum_{j=1}^{n} a_j \otimes b_j - (u_\lambda \otimes v_\mu)\sum_{j=1}^{n} a_j \otimes b_j\right\|$$

$$\leq \epsilon(1 + MN) + \sum_{j=1}^{n}\|a_j - u_\lambda a_j\|\cdot\|b_j\|$$

$$+ \sum_{j=1}^{n}\|a_j\|\cdot\|b_j - v_\mu b_j\|$$

$$+ \sum_{j=1}^{n}\|a_j - u_\lambda a_j\|\cdot\|b_j - v_\mu b_j\|$$

$$\leq \epsilon(1 + MN)2\epsilon + n\left(\frac{\epsilon}{nR}\right)^2$$

$$\leq \epsilon(4 + MN).$$

This shows that the net $(u_\lambda v_\mu)_{\lambda,\mu}$ is a left approximate identity for $A \widehat{\otimes}_\pi B$ bounded by MN.   □

Let A be a Banach algebra and G a locally compact group with left Haar measure dx. It is easily verified that $L^1(G, A)$ is a Banach algebra with convolution product given by

$$(f * g)(x) = \int_G f(xy)g(y^{-1})dy$$

for $f, g \in L^1(G)$, and almost all $x \in G$. Moreover, if A is a Banach $*$-algebra, it is straightforward to check that $f^*(x) = \Delta(x^{-1})f(x^{-1})^*$ defines an isometric involution on $L^1(G, A)$, so that $L^1(G, A)$ is a Banach $*$-algebra as well.

We now realize $L^1(G, A)$ as the projective tensor product of $L^1(G)$ and A.

Proposition 1.5.4. *Let G be a locally compact group and A a Banach algebra. For $f \in L^1(G)$ and $a \in A$ define $fa \in L^1(G, A)$ by $fa(x) = f(x)a$ for $x \in G$. Then there is an isometric algebra isomorphism ϕ of $L^1(G) \widehat{\otimes}_\pi A$ onto $L^1(G, A)$ such that $\phi(f \otimes a) = fa$ for all $f \in L^1(G)$ and $a \in A$.*

Proof. The mapping $(f, a) \rightarrow fa$ from $L^1(G) \times A$ into $L^1(G, A)$ is bilinear. Hence there exists a unique linear mapping

$$\phi : L^1(G) \otimes A \rightarrow L^1(G, A)$$

such that $\phi(f \otimes a)(x) = f(x)a$ for all $f \in L^1(G)$, $a \in A$ and almost all $x \in G$. The map ϕ is a homomorphism since

$$
\begin{aligned}
\phi((f * g) \otimes (ab))(x) &= \left(\int_G f(xy)g(y^{-1})dy \right) ab \\
&= \int_G (f(xy)a)(g(y^{-1})b)dy \\
&= \int_G \phi(f \otimes a)(xy)\phi(g \otimes b)(y^{-1})dy \\
&= (\phi(f \otimes a) * \phi(g \otimes b))(x),
\end{aligned}
$$

for all $f, g \in L^1(G), a, b \in A$ and almost all $x \in G$. For $u = \sum_{i=1}^n f_i \otimes a_i \in L^1(G) \otimes A$ it follows that

$$
\begin{aligned}
\|\phi(u)\|_1 &= \left\| \sum_{i=1}^n f_i a_i \right\| = \int_G \left\| \sum_{i=1}^n f_i a_i(x) \right\| dx \\
&= \int_G \left\| \sum_{i=1}^n f_i(x)a_i \right\| dx \leq \int_G \sum_{i=1}^n |f_i(x)| \cdot \|a_i\| dx \\
&= \sum_{i=1}^n \|f_i\|_1 \|a_i\|.
\end{aligned}
$$

This implies that $\|\phi(u)\|_1 \leq \pi(u)$. Thus ϕ extends uniquely to a norm decreasing homomorphism, also denoted ϕ, from $L^1(G) \widehat{\otimes}_\pi A$ into $L^1(G, A)$.

Now, let $f \in L^1(G, A)$ be a Bochner integrable simple function,

$$f(x) = \sum_{j=1}^{n} 1_{M_j}(x) a_j$$

say, where $a_j \in A$ and the M_j are measurable and pairwise disjoint subsets of G, $1 \le j \le n$. Let

$$u = \sum_{j=1}^{n} 1_{M_j} \otimes a_j \in L^1(G) \otimes A.$$

Then $\phi(u) = f$ and

$$\pi(u) \le \sum_{j=1}^{n} \|1_{M_j}\|_1 \|a_j\| = \sum_{j=1}^{n} |M_j| \cdot \|a_j\| = \|f\|_1 = \|\phi(u)\|_1.$$

The space of integrable simple functions is dense in $L^1(G, A)$. Thus it follows that ϕ is an isometric isomorphism from $L^1(G) \widehat{\otimes}_\pi A$ onto $L^1(G, A)$. $\qquad \square$

If G and H are locally compact groups, their direct product $G \times H$ will be equipped with the product of left Haar measures of G and H.

Proposition 1.5.5. *Let G and H be locally compact groups. Then there exists an isometric $*$-isomorphism ϕ from $L^1(G) \widehat{\otimes}_\pi L^1(H)$ onto $L^1(G \times H)$ such that*

$$\phi(f \otimes g)(x, y) = f(x)g(y)$$

for all $f \in L^1(G)$, $g \in L^1(H)$, and almost all $x \in G$ and $y \in H$.

Proof. We define a linear mapping ψ from $L^1(G \times H)$ onto $L^1(G, L^1(H))$ by $[\psi(F)(x)](y) = F(x, y)$. It is easy to check that $\psi(F^*) = \psi(F)^*$. Using vector-valued integration, we have

$$[\psi(F_1 * F_2)(x)](y) = \int_{G \times H} F_1((x, y)(s, t)^{-1}) F_2(s, t) d(s, t)$$

$$= \int_G \int_H [\psi(F_1)(xs^{-1})](yt^{-1})[\psi(F_2)(s)](t) dt ds$$

$$= \int_G [\psi(F_1)(xs^{-1}) * \psi(F_2)(s)](y) ds$$

$$= \left(\int_G [\psi(F_1) * \psi(F_2)](x) \right)(y)$$

for $F_1, F_2 \in L^1(G \times H)$. Moreover,

$$\|\psi(F)\|_1 = \int_G \|F(x, \cdot)\|_1 dx = \int_G \int_H |F(x, y)| dy dx = \|F\|_1.$$

In Proposition 1.5.4 we have seen that there exists an isometric isomorphism ρ from $L^1(G)\widehat{\otimes}_\pi L^1(H)$ to $L^1(G, L^1(H))$ satisfying $\rho(f \otimes g)(x) = f(x)g$ for all $f \in L^1(G)$, $g \in L^1(H)$, and almost all $x \in G$. Clearly, ρ preserves the involution. Since ρ is surjective and the range of ψ contains $\rho(L^1(G)\otimes L^1(H))$, ψ is also surjective. So

$$\phi = \psi^{-1} \circ \rho : L^1(G)\widehat{\otimes}_\pi L^1(H) \to L^1(G \times H)$$

is the desired isometric $*$-isomorphism. □

It does not seem to be clear at all under what conditions the injective norm on the algebraic tensor product of two Banach algebras A and B is an algebra norm. The following proposition in particular shows that this is the case when $B = C_0(X)$ for some locally compact Hausdorff space X.

Proposition 1.5.6. *Let X be a locally compact Hausdorff space and A a Banach algebra. Then $C_0(X)\widehat{\otimes}_\epsilon A$ is isometrically isomorphic to $C_0(X, A)$.*

Proof. For $f \in C_0(X)$ and $a \in A$, $fa \in C_0(X, A)$ is defined by $fa(x) = f(x)a, x \in X$. The mapping $(f, a) \to fa$ from $C_0(X) \times A$ into $C_0(X, A)$ is bilinear. Hence there exists a unique linear map $\phi : C_0(X) \otimes A \to C_0(X, A)$ such that $\phi(f \otimes a)(x) = f(x)a$ for all $f \in C_0(X), x \in X, a \in A$. Clearly, ϕ is a homomorphism. For $u = \sum_{i=1}^n f_i \otimes a_i, f_i \in C_0(X), a_i \in A$, we have

$$\|\phi(u)\| = \sup\{\|\phi(u)(x)\| : x \in X\}$$

$$= \sup\left\{\left\|\sum_{i=1}^n f_i(x)a_i\right\| : x \in X\right\}$$

$$= \sup\left\{\left|g\left(\sum_{i=1}^n f_i(x)a_i\right)\right| : g \in A_1^*, x \in X\right\}$$

$$= \sup\left\{\left\|\sum_{i=1}^n g(a_i)f_i\right\|_\infty : g \in A_1^*\right\}$$

$$= \sup\left\{\left|\sum_{i=1}^n g(a_i)\mu(f_i)\right| : g \in A_1^*, \mu \in C_0(X)_1^*\right\}$$

$$= \epsilon\left(\sum_{i=1}^n f_i \otimes a_i\right).$$

Thus ϕ is an isometry. It remains to show that $\phi(C_0(X) \otimes A)$ is dense in $C_0(X, A)$. Since $C_c(X, A)$ is dense in $C_0(X, A)$, we can assume that X is compact. Let $f \in C(X, A)$ and $\epsilon > 0$ be given. Choose $x_1, \ldots, x_n \in X$ such that for each $x \in X$ there exists j so that $\|f(x) - f(x_j)\| < \epsilon$. For $1 \leq j \leq n$, define $V_j \subseteq X$ by $V_j = \{x \in X : \|f(x) - f(x_j)\| < \epsilon\}$. Then the sets $V_1, \ldots, V_n$ form an open cover of X. Because X is a compact Hausdorff space, there is a partition of unity subordinate to this cover. That is, there are continuous functions $h_j : X \to [0,1], 1 \le j \le n$, satisfying $h_j(x) = 0$ for $x \notin V_j$ and $\sum_{j=1}^{n} h_j(x) = 1$ for all $x \in X$. For each $x \in X$, follows that

$$\left\| f(x) - \phi\left(\sum_{j=1}^{n} h_j \otimes f(x_j) \right)(x) \right\| = \left\| \sum_{j=1}^{n} h_j(x)(f(x) - f(x_j)) \right\|$$

$$\le \sum h_j(x) \| f(x) - f(x_j) \| < \epsilon,$$

where the last summation extends over all $1 \le j \le n$ such that $x \in V_j$. This shows that the image of $C_0(X) \otimes A$ is dense in $C_0(X, A)$. □

Let X and Y be locally compact Hausdorff spaces. It is easy to see that $C_0(X, C_0(Y))$ is isometrically isomorphic to $C_0(X \times Y)$ (Exercise 1.6.47). Combining this with Proposition 1.5.6 shows that $C_0(X) \widehat{\otimes}_\epsilon C_0(Y)$ is isometrically isomorphic to $C_0(X \times Y)$.

1.6 Exercises

Exercise 1.6.1. A seminorm on an algebra A is a function $s : A \to [0, \infty)$ satisfying $s(x + y) \le s(x) + s(y)$ and $s(\lambda x) = |\lambda| s(x)$ for all $x, y \in A$ and $\lambda \in \mathbb{C}$. Suppose that s has the *square property* $s(x^2) = s(x)^2$, $x \in A$. Show that

$$s(xy + yx) \le 2[s(x)^2 + s(y)^2 + s(x)s(y)]$$

for all $x, y \in A$.

Exercise 1.6.2. Let A be a commutative algebra and $\| \cdot \|$ a norm on A satisfying the square property $\|x^2\| = \|x\|^2$, $x \in A$. Proceed as follows to show that $\| \cdot \|$ is submultiplicative.
  (i) The equation $4xy = (x + y)^2 - (x - y)^2$ implies $2\|xy\| \le (\|x\| + \|y\|)^2$ for all $x, y \in A$.
  (ii) Deduce that $\|xy\| \le 2\|x\| \cdot \|y\|$ by first considering the case where $\|x\|, \|y\| \le 1$.
  (iii) Replace x and y by x^{2^n} and y^{2^n}, $n \in \mathbb{N}$, respectively, and let $n \to \infty$ to conclude that $\|xy\| \le \|x\| \cdot \|y\|$.

Exercise 1.6.3. Let A be a Banach algebra and let x and y be two commuting idempotents in A, that is, $x^2 = x, y^2 = y$, and $xy = yx$. Prove that either $x = y$ or $\|x - y\| \ge 1$. Find an example showing that this may fail if $xy \ne yx$.

Exercise 1.6.4. Let A_λ, $\lambda \in \Lambda$, be a family of Banach algebras and let

$$A = \{ x = (x_\lambda)_\lambda \in \prod_{\lambda \in \Lambda} A_\lambda : \sup_{\lambda \in \Lambda} \|x_\lambda\| < \infty \}.$$

Prove that, with componentwise algebraic operations and the norm $\|x\| = \sup_{\lambda \in \Lambda} \|x_\lambda\|$, A becomes a Banach algebra.

Exercise 1.6.5. Let A be a Banach algebra and let A^* and A^{**} denote the first and the second dual of A. For $a, b \in A$, $f \in A^*$, and $m, n \in A^{**}$ define elements $f \cdot a$ and $m \cdot f$ of A^* and mn of A^{**} by

$$\langle f \cdot a, b \rangle = \langle f, ab \rangle, \quad \langle m \cdot f, a \rangle = \langle m, f \cdot a \rangle \quad \text{and} \quad \langle mn, f \rangle = \langle m, n \cdot f \rangle,$$

respectively. This product $(m, n) \to mn$ on A^{**} is called the *first Arens product*. Show that A^{**}, equipped with the first Arens product, is a Banach algebra and that the canonical embedding $i : A \to A^{**}$ is an isomorphism between A and the subalgebra $i(A)$ of A^{**}.

Exercise 1.6.6. Retain the notation of Exercise 1.6.5 and suppose that A is commutative. Then A^{**} need not be commutative.

(i) Show that $ma = am$ for $m \in A^{**}$ and $a \in A$.

(ii) Show that, for fixed $n \in A^{**}$, the mapping $m \to mn$ is continuous for the w^*-topology on A^{**}.

Exercise 1.6.7. A *derivation* on an algebra A is a linear mapping of A into A such that

$$D(xy) = x(Dy) + (Dx)y \quad (x, y \in A).$$

Show that a derivation D satisfies (with the convention that $D^0 = I$) the Leibniz rule

$$D^n(xy) = \sum_{j=0}^{n} \binom{n}{j} (D^{n-j}x)(D^j y)$$

$(n \in \mathbb{N}, x, y \in A)$.

Exercise 1.6.8. Let D be a continuous derivation on a Banach algebra A. Show that

$$\exp D : A \to A, \quad (\exp D)x = \sum_{n=0}^{\infty} \frac{1}{n!} D^n x,$$

is a continuous automorphism of A.

Exercise 1.6.9. Show that for $1 \le p < \infty$, $l^p = l^p(\mathbb{N})$ with multiplication defined by $(a_n)_n (b_n)_n = (a_n b_n)_n$ and the $\| \cdot \|_p$-norm is a Banach algebra, which has an unbounded, but no bounded approximate identity.

Exercise 1.6.10. Let X be a noncompact locally compact Hausdorff space and denote by $\tilde{X}$ the one-point compactification of X. Show that $C(\tilde{X})$ is algebraically (though not isometrically) isomorphic to the algebra obtained by adjoining an identity to $C_0(X)$.

Exercise 1.6.11. For $0 < \alpha \le 1$, let $\mathrm{Lip}_\alpha[0, 1]$ be the space of all continuous complex valued functions on $[0, 1]$ which satisfy a *Lipschitz condition of order* α. That is, $f \in C[0, 1]$ belongs to $\mathrm{Lip}_\alpha[0, 1]$ if and only if

$$\|f\|_\alpha = \sup_{0 \le t \le 1} |f(t)| + \sup_{0 \le s < t \le 1} \frac{|f(s) - f(t)|}{|s - t|^\alpha} < \infty.$$

Prove that, with pointwise multiplication and the norm $f \to \|f\|_\alpha$, $\mathrm{Lip}_\alpha[0, 1]$ is a Banach algebra.

Exercise 1.6.12. Define a convolution product on $L^1[0, 1]$ by

$$(f * g)(t) = \int_0^t f(t - s)g(s)ds, \quad f, g \in L^1[0, 1], \quad t \in [0, 1].$$

With this product, $L^1[0, 1]$ becomes a commutative Banach algebra, called the *Volterra algebra* and denoted V. Clearly, V does not have an identity. Prove that V has a bounded approximate identity.

Exercise 1.6.13. Let $\mathbb{D}^\circ$ be the open unit disc. For any $1 < p < \infty$, let $H^p(\mathbb{D}^\circ)$ denote the classical *Hardy space*. Recall that $H^p(\mathbb{D}^\circ)$ consists of all holomorphic functions $f : \mathbb{D}^\circ \to \mathbb{C}$ for which the values $\int_\mathbb{R} |f(rz)|^p dz$ are bounded as r varies through $0 < r < 1$. With the norm

$$\|f\|_p = \left(\frac{1}{2\pi} \sup_{0 < r < 1} \int_0^{2\pi} |f(re^{it})|^p dt \right)^{1/p},$$

$H^p(\mathbb{D}^\circ)$ is a Banach space. Show that $H^p(\mathbb{D}^\circ)$ becomes a commutative Banach algebra when endowed with the so-called *Hadamard product*. This product is given by

$$(f \bullet g)(z) = \frac{1}{2\pi i} \int_{\gamma_r} f(w)g(zw^{-1})w^{-1}dw,$$

where $|z| < r < 1$ and $\gamma_r(t) = re^{2\pi i t}, t \in [0, 1]$. Note that because f and g are holomorphic on $\mathbb{D}^\circ$, the value of this integral does not depend on the choice of r with $|z| < r < 1$.

Exercise 1.6.14. Let A be a unital algebra with involution $*$ and let u be an invertible element of A such that $u^* = u$. For $x \in A$, define $\tilde{x}$ by $\tilde{x} = u^{-1}x^*u$. Show that $x \to \tilde{x}$ is an involution on A.

Exercise 1.6.15. Show that an involution can be defined on the disc algebra $A(\mathbb{D})$ by setting $f^*(z) = \overline{f(\bar{z})}$ for $f \in A(\mathbb{D}), z \in \mathbb{D}$. Verify that this involution is isometric, but does not satisfy the C^*-condition.

Exercise 1.6.16. For $z \in \mathbb{D}$, let $\varphi_z \in A(\mathbb{D})^*$ denote the point evaluation at z, that is, $\varphi_z(f) = f(z)$ for all $f \in A(\mathbb{D})$. Let $z, w \in \mathbb{D}$ such that $z \ne w$. Show that $\|\varphi_z - \varphi_w\|_{A(\mathbb{D})^*} = 2$ if and only if $|z| = 1$ or $|w| = 1$.

Exercise 1.6.17. Let A be a Banach algebra and let $(x_n)_n$ be a sequence of invertible elements of A converging to a non-invertible element. Prove that $\lim_{n \to \infty} \|x_n^{-1}\| = \infty$.

Exercise 1.6.18. Let A be a Banach algebra and $r_A : A \to [0, \infty), x \to r_A(x)$ the spectral radius function. Show that r_A is upper semicontinuous, and if $x \in A$ is such that $r_A(x) = 0$, then r_A is continuous at x.

Exercise 1.6.19. Let A be a Banach algebra, let $x \in A$, and let U be an open subset of $\mathbb{C}$ containing $\sigma_A(x)$. Prove that there exists $\delta > 0$ such that $\sigma_A(y) \subseteq U$ for all $y \in A$ with $\|y - x\| < \delta$. Compare this with Exercise 1.6.18.

Exercise 1.6.20. Let $A = \mathcal{B}(l^2(\mathbb{N}))$ and let $T \in A$ be the unilateral shift defined by $(Tx)_1 = 0$ and $(Tx)_n = x_{n-1}$ for $n \geq 2$ and $x = (x_n)_n \in l^2(\mathbb{N})$. Show that $\sigma_A(T^*T) \neq \sigma_A(TT^*)$.

Exercise 1.6.21. Let A be the convolution algebra $l^1(\mathbb{Z})$ and B the closed algebra consisting of all $x = (x_n)_n \in l^1(\mathbb{Z})$ such that $x_n = 0$ for all $n < 0$. Show that $\sigma_A(\delta_1) \neq \sigma_B(\delta_1)$.

Exercise 1.6.22. Let $X = \{z \in \mathbb{C} : 1 \leq |z| \leq 2\}$ and $f(z) = z, z \in X$. Let A be the smallest closed subalgebra of $C(X)$ that contains 1 and f, and let B be the smallest closed subalgebra of $C(X)$ that contains f and $1/f$. Determine the spectra $\sigma_A(f)$ and $\sigma_B(f)$. Do the same when X is a circle.

Exercise 1.6.23. Let $H^\infty(\mathbb{D}^\circ)$ denote the algebra of all bounded holomorphic functions on the open unit disc $\mathbb{D}^\circ$. Equipped with the supremum norm, $H^\infty(\mathbb{D}^\circ)$ is a Banach algebra. Show that $\sigma(f) = \overline{f(\mathbb{D}^\circ)}$ for every $f \in H^\infty(\mathbb{D}^\circ)$.

Exercise 1.6.24. Let A be a Banach algebra with identity e, and let B be a closed subalgebra of A with $e \in B$. Suppose that A is not commutative and B is a maximal commutative subalgebra of A. Show that $\sigma_A(x) = \sigma_B(x)$ for every $x \in B$.

Exercise 1.6.25. Let A be a Banach algebra with identity e and let $x \in A$ be such that $x^2 = x$. Show that $\sigma_A(x) \subseteq \{0, 1\}$ and compute the resolvent function $R(x, \lambda) = (\lambda e - x)^{-1}$.

Exercise 1.6.26. Let A be a Banach algebra with identity e and B a closed subalgebra of A containing e. Show that if $x \in B$ is such that $\sigma_B(x)^\circ = \emptyset$, then $\sigma_A(x) = \sigma_B(x)$.

Exercise 1.6.27. Let X be a locally compact Hausdorff space.
   (i) Prove that $\sigma_{C_0(X)}(f) = \overline{f(X)}$ for every $f \in C_0(X)$.
   (ii) Show that $\sigma_{C_0(X)}(f) = f(X)$ for every $f \in C_0(X)$ if and only if X is not σ-compact.

Exercise 1.6.28. Let A be a Banach $*$-algebra and B a C^*-algebra, and let $\phi : A \to B$ be a $*$-homomorphism. Show that ϕ is continuous and $\|\phi\| \leq 1$.

Exercise 1.6.29. Let G be a discrete group and $0 < p < 1$. Show that $l^p(G)$ with the convolution product is a commutative Banach algebra.

Exercise 1.6.30. (i) Let G be nontrivial discrete group. Show that the $\|\cdot\|_1$-norm on the Banach $*$-algebra $l^1(G)$ fails to be a C^*-norm by considering a linear combination of, say, three Dirac functions.

(ii) Prove the analogous statement for $L^1(\mathbb{R}^n)$.

Exercise 1.6.31. Let $n \in \mathbb{N}$, $1 \le p < \infty$ and

$$A = \{f \in L^1(\mathbb{R}^n) : \widehat{f} \in L^p(\mathbb{R}^n)|\}.$$

Show that A becomes a Banach algebra under the norm $\|f\| = \|f\|_1 + \|\widehat{f}\|_p$.

Exercise 1.6.32. Let $L^1(\mathbb{R}^+)$ denote the subalgebra of $L^1(\mathbb{R})$ consisting of all functions f such that $f(t) = 0$ for all $t < 0$. The purpose of this exercise is to show that $L^1(\mathbb{R}^+)$ is generated by the function f defined by $f(t) = e^{-t}$ for $t \ge 0$ and $f(t) = 0$ for $t < 0$.

(i) Show that $\widehat{f}(y) = 1/(1 + iy)$ for all $y \in \mathbb{R}$. Hence $(\widehat{f})^{(j)}$ is a constant multiple of $(\widehat{f})^{j+1}$ for all $j \in \mathbb{N}_0$.

(ii) Deduce from (i) that the n-fold convolution product f^n satisfies $f^n(t) = c_n t^{n-1} f(t)$ for all $t \in \mathbb{R}$ and $n \in \mathbb{N}$ and some $c_n \neq 0$.

(iii) Let $g \in L^\infty(\mathbb{R}^+) = L^1(\mathbb{R}^+)^*$ be such that

$$\langle f^n, g \rangle = \int_0^\infty f^n(t)g(t)dt = 0$$

for all $n \in \mathbb{N}$. Define a function F on the right half plane by

$$F(z) = \int_0^\infty e^{-tz} g(t)dt.$$

Then F is holomorphic and satisfies $F^{(n)}(1) = 0$ for all $n \in \mathbb{N}_0$. Therefore $F(1 + iy) = 0$ for all $y \in \mathbb{R}$. Conclude that $g = 0$.

Exercise 1.6.33. Let $k \in \mathbb{N}$ and $C = [0,1]^k \subseteq \mathbb{R}^k$ and view $\mathbb{Z}^k$ as the set of all points in $\mathbb{R}^k$ with integer coordinates. For a continuous function f on $\mathbb{R}^k$ define

$$M(f) = \sum_{n \in \mathbb{Z}^k} \max_{x \in C} |f(x + n)|.$$

Let A be the set of all functions f for which $M(f) < \infty$. Prove that A is a subalgebra of $L^1(\mathbb{R}^k)$ and $f \to M(f)$ is a Banach algebra norm on A.

Exercise 1.6.34. Give an example of a weight ω on $\mathbb{Z}$ with the property that if $f \in l^1(\mathbb{Z}, \omega)$, then the function f^*, defined by $f^*(n) = \overline{f(-n)}$, need not belong to $l^1(\mathbb{Z}, \omega)$.

Exercise 1.6.35. Let G be a locally compact group and ω a weight function on G. Let $\mathcal{V}$ be the family of all compact neighbourhoods of the identity of G and define ω' on G by

$$\omega'(x) = \inf_{V \in \mathcal{V}} \{\sup\{\omega(xy) : y \in V\}\}.$$

Prove that ω' is an upper semicontinuous weight on G and that $L^1(G, \omega) = L^1(G, \omega')$.

Exercise 1.6.36. Let G be a locally compact Abelian group and let ω_1 and ω_2 be two weights on G such that $L^1(G, \omega_1) \subseteq L^1(G, \omega_2)$. Show that there exists a constant $c > 0$ such that $\omega_2(x) \leq c\omega_1(x)$ locally almost everywhere on G.
(Hint: $\|f\| = \|f\|_{1,\omega} + \|f\|_{1,\omega_2}$ defines a complete norm on $L^1(G, \omega_1)$.)

Exercise 1.6.37. A weight function ω on $\mathbb{R}^n$ is said to be of *regular growth* if it satisfies the following two conditions.
   (1) $\omega(x/t) \leq \omega(x)$ for all $x \in \mathbb{R}^n$ and $t \in [1, \infty)$.
   (2) There are constants $C \geq 1$ and $\alpha > 0$ such that $\omega(tx) \leq C t^\alpha \omega(x)$ for all $x \in \mathbb{R}^n$ and $t \in [1, \infty)$.
   Observe that condition (2) implies that $\omega(x) \leq D \cdot \|x\|^\alpha$ whenever $\|x\| \geq 1$, where $D = C \cdot \sup\{\omega(x) : \|x\| = 1\}$. Let γ be any measurable function on $\mathbb{R}^n$ satisfying $\gamma(x) \geq 0$, $\gamma(x + y) \leq \gamma(x) + \gamma(y)$ and $\gamma(x/t) \leq \gamma(x)$ for all $x, y \in \mathbb{R}^n$, $t \geq 1$. Show that then $\omega(x) = (1 + \gamma(x))^\alpha$, $\alpha \geq 0$, defines a weight of regular growth on $\mathbb{R}^n$.

Exercise 1.6.38. Let G be a locally compact group and ω a weight on G. Adopt the proof of Proposition 1.4.7 to show that a closed linear subspace of $L^1(G, \omega)$ is an ideal in $L^1(G, \omega)$ if and only if I is two-sided translation invariant.

Exercise 1.6.39. Let H be a Hilbert space and $\mathcal{K}(H)$ the algebra of compact operators on H. Let P be a projection of finite rank in H. Show that $\mathcal{K}(H)P$ is a closed modular left ideal of $\mathcal{K}(H)$.

Exercise 1.6.40. Let H be a Hilbert space. Determine the minimal closed proper left ideals of $\mathcal{K}(H)$ and the maximal modular left ideals of $\mathcal{K}(H)$.

Exercise 1.6.41. Let X be a compact Hausdorff space, Y a closed subspace of X and I the ideal $\{f \in C(X) : f|_Y = 0\}$. Show that $C(X)/I$ is isometrically isomorphic to $C(Y)$.

Exercise 1.6.42. Let A be the space of all sequences $f : \mathbb{N} \to \mathbb{C}$ such that $kf(k) \to 0$ as $k \to \infty$. With pointwise multiplication and the norm

$$\|f\| = \sup\{k|f(k)| : k \in \mathbb{N}\},$$

A is a Banach algebra. Show that the multiplier algebra of A is isometrically isomorphic to $l^\infty(\mathbb{N})$, where $g \in l^\infty(\mathbb{N})$ acts on A by pointwise multiplication.

Exercise 1.6.43. For $1 \leq p < \infty$, let $l^p = l^p(\mathbb{N})$ be as in Exercise 1.6.9. Show that the multiplier algebra of l^p is isometrically isomorphic to l^∞.

Exercise 1.6.44. Let G be a compact Abelian group and let T be an injective multiplier on $L^1(G)$. Use the two facts that $M(L^1(G)) = M(G)$ and that $\widehat{G} \subseteq L^1(G)$ to show that T has dense range.

Exercise 1.6.45. Let A be a faithful Banach algebra. Let $T \in M(A)$ and suppose that T is bijective. Show that $T^{-1} \in M(A)$.

Exercise 1.6.46. Let A be a faithful Banach algebra. Show that the multiplier algebra $M(A)$ is complete in the strong operator topology on $\mathcal{B}(A)$ in which a net $(T_\alpha)_\alpha$ converges to T if and only if $\|T_\alpha x - Tx\| \to 0$ for all $x \in A$.

Exercise 1.6.47. Let X and Y be locally compact Hausdorff spaces. For $f \in C_0(X, C_0(Y))$, define $\phi(f)$ on $X \times Y$ by $\phi(f)(x, y) = f(x)(y)$. Show that the mapping $\phi : f \to \phi(f)$ is an isometric isomorphism from $C_0(X, C_0(Y))$ onto $C_0(X \times Y)$.

Exercise 1.6.48. Let X and Y be nonempty sets and endow $l^1(X)$ and $l^1(Y)$ with the pointwise product. Show that the projective tensor product $l^1(X) \widehat{\otimes}_\pi l^1(Y)$ is isometrically isomorphic to $l^1(X \times Y)$.

Exercise 1.6.49. Let A be a Banach algebra and $\pi_A : A \widehat{\otimes}_\pi A \to A$ the continuous homomorphism satisfying $\pi_A(a \otimes b) = ab$ for all $a, b \in A$. Suppose that A has an identity e. Show that $\ker \pi_A$ equals the closed linear span of all elements of the form $a \otimes b - e \otimes ab$.

Exercise 1.6.50. Let A be a Banach algebra and X a locally compact Hausdorff space. It follows from Proposition 1.5.6 that the injective tensor norm on $A \otimes C_0(X)$ is an algebra norm. Prove this without appealing to Proposition 1.5.6.

1.7 Notes and references

Most of the material collected in this chapter can be found in several books on Banach algebras. Therefore we confine ourselves to only a few references and historical remarks.

Theorem 1.2.8, which is one of the most striking results in Banach algebra theory, is in full generality the work of Gelfand [38]. His proof is based on complex analysis. The more elementary proof, which we have also included and which avoids the use of function theory, is due to Rickart and presented in his monograph [108]. The Gelfand–Mazur theorem was announced in [86] and proven in [38]. The permanence properties of spectra such as Theorem 1.2.12, which are essential for this treatise, appear to be due to Shilov. Concerning approximate identities, we refer to [2], [27], [104], and [138].

Weighted convolution algebras on the real line were introduced by Beurling [12]. They were subsequently defined on arbitrary locally compact groups, termed Beurling algebras, and investigated by many authors. The literature on them is enormous (compare [25]). That weight functions are bounded away from both zero and infinity on compact sets, was observed in [29].

The basic results on ideals in Banach algebras, given in Section 1.4, are all standard, including the description of the closed ideals of $C_0(X)$. A very good reference to the theory of multipliers is [74]. The fact that the multiplier algebra of $L^1(G)$ is isomorphic to the measure algebra $M(G)$, is Wendel's theorem [135]. It is worth mentioning that Wendel's theorem admits a generalisation to Beurling algebras.

2

Gelfand Theory

Our main objective in this chapter is to develop Gelfand's theory for commutative Banach algebras. Associated with any such algebra A is a locally compact Hausdorff space $\Delta(A)$, the structure space of A, and a norm-decreasing homomorphism Γ_A from A into $C_0(\Delta(A))$ (Section 2.2). If A has an identity, $\Delta(A)$ is compact. The converse is true whenever Γ_A is injective (A is semisimple), a fact that can be shown only later (Chapter 3). This representation of A as an algebra of functions on a locally compact Hausdorff space is fundamental to any thorough study of commutative Banach algebras. Thus basic questions are when Γ_A is injective or surjective. It turns out that Γ_A is an isometric isomorphism onto $C_0(\Delta(A))$ precisely when A is a C^*-algebra (Section 2.4).

If A is unital and (topologically) finitely generated by n elements, say, then $\Delta(A)$ can be canonically identified with a compact subset of $\mathbb{C}^n$ (Section 2.3). There is a complete characterisation of subsets of $\mathbb{C}^n$ arising in this way as structure spaces of finitely generated commutative Banach algebras. This leads to the study of uniform algebras, closed unital subalgebras of $C(X)$, for a compact subset X of $\mathbb{C}^n$, which separate the points of X. In Section 2.5 we investigate the algebras $P(X)$ and $R(X)$ of polynomial and rational functions on X, respectively. Comparison of such algebras is interesting from the approximation theory point of view. Considerably more complicated is the algebra $A(X)$ of continuous functions on X which are holomorphic on the interior of X (Section 2.6).

Following our intention to emphasize the connections with commutative harmonic analysis, we extensively study the convolution algebra $L^1(G)$ of integrable functions on a locally compact Abelian group G. The structure space $\Delta(L^1(G))$ turns out to be homeomorphic with the dual group $\widehat{G}$ of G, and the Gelfand homomorphism is injective, but surjective only when G is finite (Section 2.7). Much more subtle are weighted group algebras $L^1(G,\omega)$. We confine ourselves to showing that $L^1(G,\omega)$ is always semisimple and to determining $\Delta(L^1(G,\omega))$ in some special cases (Section 2.8).

E. Kaniuth, *A Course in Commutative Banach Algebras*, Graduate Texts in Mathematics, DOI 10.1007/978-0-387-72476-8_2, © Springer Science+Business Media, LLC 2009

Proceeding further with algebras of functions associated with locally compact groups, in Section 2.9 we elaborate the Gelfand representation of the Fourier algebra $A(G)$ for an arbitrary locally compact group G. Next, applying the Gelfand theory to the C^*-algebra of almost periodic functions, we establish the existence of the Bohr compactification of a locally compact Abelian group (Section 2.10).

Finally, we investigate tensor products of two commutative Banach algebras A and B, especially the projective tensor product $A \widehat{\otimes}_\pi B$. Although $\Delta(A \widehat{\otimes}_\pi B)$ is in the obvious manner homeomorphic to $\Delta(A) \times \Delta(B)$, semisimplicity of $A \widehat{\otimes}_\pi B$ is a very delicate question (Section 2.11). In fact, by using failure of the approximation property for Banach spaces one can construct semisimple commutative Banach algebras A and B such $A \widehat{\otimes}_\pi B$ is not semisimple.

2.1 Multiplicative linear functionals

A linear functional φ on an algebra A is called *multiplicative* if $\varphi(xy) = \varphi(x)\varphi(y)$ for all $x, y \in A$. We start with identifying the multiplicative ones among all linear functionals on A in terms of spectra (Theorem 2.1.2). We do not assume here that A is commutative.

Lemma 2.1.1. *Let A be a real or complex algebra with identity e, and let φ be a linear functional on A satisfying*

$$\varphi(e) = 1 \ \text{ and } \ \varphi(x^2) = \varphi(x)^2$$

for all $x \in A$. Then φ is multiplicative.

Proof. By assumption we have

$$\begin{aligned}
\varphi(x^2) + \varphi(xy + yx) + \varphi(y^2) &= \varphi(x^2 + xy + yx + y^2) \\
&= \varphi((x + y)^2) = (\varphi(x) + \varphi(y))^2 \\
&= \varphi(x^2) + 2\varphi(x)\varphi(y) + \varphi(y^2),
\end{aligned}$$

and therefore

$$\varphi(xy + yx) = 2\varphi(x)\varphi(y)$$

for all $x, y \in A$. Thus it remains to verify that $\varphi(yx) = \varphi(xy)$. Now, for $a, b \in A$, the identity

$$(ab - ba)^2 + (ab + ba)^2 = 2[a(bab) + (bab)a]$$

implies

$$\begin{aligned}
\varphi(ab - ab)^2 + 4\varphi(a)^2\varphi(b)^2 &= \varphi((ab - ba)^2) + \varphi(ab + ba)^2 \\
&= \varphi((ab - ba)^2 + (ab + ba)^2) \\
&= 2\varphi(a(bab) + (bab)a) \\
&= 4\varphi(a)\varphi(bab).
\end{aligned}$$

Taking $a = x - \varphi(x)e$, so that $\varphi(a) = 0$, and $b = y$ we obtain $\varphi(ay) = \varphi(ya)$ and hence $\varphi(xy) = \varphi(yx)$. $\qquad\square$

The following theorem is often called the *Gleason–Kahane–Zelazko theorem*.

Theorem 2.1.2. *Let A be a unital Banach algebra. For a linear functional φ on A the following conditions are equivalent.*

(i) φ *is nonzero and multiplicative.*
(ii) $\varphi(e) = 1$ *and* $\varphi(x) \neq 0$ *for every invertible element x of A.*
(iii) $\varphi(x) \in \sigma_A(x)$ *for every* $x \in A$.

Proof. If φ is nonzero and multiplicative, then $\varphi(e) = 1$ and $1 = \varphi(x)\varphi(x^{-1})$ whenever x is invertible. Thus (i) $\Rightarrow$ (ii). Also, (ii) $\Rightarrow$ (iii) is obvious since if $\lambda \in \rho_A(x)$, then $0 \neq \varphi(x - \lambda e) = \varphi(x) - \lambda$.

Now assume (iii) and note first that $\varphi(e) = 1$. We are going to show that $\varphi(x^2) = \varphi(x)^2$ for all $x \in A$. To that end, let $n \geq 2$ and consider the polynomial

$$p(\lambda) = \varphi((\lambda e - x)^n)$$

of degree n. Denoting its roots by $\lambda_1, \ldots, \lambda_n$, we have for each i,

$$0 = p(\lambda_i) = \varphi((\lambda_i e - x)^n) \in \sigma_A((\lambda_i e - x)^n).$$

This implies that $\lambda_i \in \sigma_A(x)$ and hence $|\lambda_i| \leq r_A(x)$. Now

$$\prod_{i=1}^{n} (\lambda - \lambda_i) = p(\lambda) = \lambda^n - n\varphi(x)\lambda^{n-1} + \binom{n}{2}\varphi(x^2)\lambda^{n-2} + \ldots + (-1)^n\varphi(x^n).$$

Comparing coefficients we see that

$$\sum_{i=1}^{n} \lambda_i = n\varphi(x) \quad \text{and} \quad \sum_{1 \leq i < j \leq n} \lambda_i\lambda_j = \binom{n}{2}\varphi(x^2).$$

On the other hand, by the second equation,

$$\left(\sum_{i=1}^{n} \lambda_i\right)^2 = \sum_{i=1}^{n} \lambda_i^2 + 2\sum_{1 \leq i < j \leq n} \lambda_i\lambda_j = \sum_{i=1}^{n} \lambda_i^2 + n(n-1)\varphi(x^2).$$

Combining these formulae yields

$$n^2|\varphi(x)^2 - \varphi(x^2)| = \left| -n\varphi(x^2) + \sum_{i=1}^{n} \lambda_i^2 \right| \leq n|\varphi(x^2)| + nr_A(x)^2.$$

This being true for all n, we conclude that $\varphi(x^2) = \varphi(x)^2$ for all $x \in A$. It follows now from Lemma 2.1.1 that φ is multiplicative. $\qquad\square$

Throughout the book, for any Banach algebra A, $\Delta(A)$ denotes the set of all nonzero multiplicative linear functionals on A. It is very important to know how $\Delta(A)$ and $\Delta(A_e)$ are related.

Remark 2.1.3. Because $\psi(e) = 1$ for every $\psi \in \Delta(A_e)$, each $\varphi \in \Delta(A)$ has a unique extension $\widetilde{\varphi} \in \Delta(A_e)$ given by

$$\widetilde{\varphi}(x + \lambda e) = \varphi(x) + \lambda, \quad x \in A, \quad \lambda \in \mathbb{C}.$$

Let $\widetilde{\Delta}(A) = \{\widetilde{\varphi} : \varphi \in \Delta(A)\}$. Moreover, let φ_∞ denote the homomorphism from A_e to $\mathbb{C}$ with kernel A, that is, $\varphi_\infty(x + \lambda e) = \lambda$. Then

$$\Delta(A_e) = \widetilde{\Delta}(A) \cup \{\varphi_\infty\}.$$

In fact, if $\psi \in \Delta(A_e)$ and $\psi \neq \varphi_\infty$, then $\psi|_A \in \Delta(A)$ and hence $\psi = \widetilde{\psi|_A}$. Identifying $\Delta(A)$ with $\widetilde{\Delta}(A) \subseteq \Delta(A_e)$ we always regard $\Delta(A)$ as a subset of $\Delta(A_e)$. In this sense, $\Delta(A_e) = \Delta(A) \cup \{\varphi_\infty\}$.

Remark 2.1.4. A simple unitisation argument shows that for a linear functional φ on a Banach algebra A, condition (iii) in Theorem 2.1.2 is equivalent to multiplicativity of φ without assuming that A has an identity.

Lemma 2.1.5. *Let A be a Banach algebra. Every $\varphi \in \Delta(A)$ is a bounded linear functional on A and $|\varphi(x)| \leq r_A(x)$ holds for all $x \in A$. In particular, $\|\varphi\| \leq 1$ and $\|\varphi\| = 1$ if A is unital.*

Proof. We can assume that A has an identity e. If $x \in A$ and $\lambda \in \mathbb{C}$ are such that $|\lambda| > r_A(x)$, then $r_A((1/\lambda)x) < 1$ and hence $\lambda e - x = \lambda(e - (1/\lambda)x)$ is invertible in A by Lemma 1.2.6. This implies $\varphi(x) \neq \lambda$ for all such λ, so that $|\varphi(x)| \leq r_A(x)$. This implies that $\|\varphi\| \leq 1$ and actually $\|\varphi\| = 1$ since $\varphi(e) = 1$. $\qquad \square$

The obvious problem which we have to encounter is the existence of non-zero multiplicative linear functionals on commutative Banach algebras A. To start with let A be a complex Banach space and define a product on A by setting $xy = 0$ for all $x, y \in A$. If φ is a multiplicative linear functional on A, then

$$\varphi(x)^2 = \varphi(xx) = \varphi(0) = 0.$$

Less trivial examples showing that nonzero multiplicative linear functionals need not exist are the following two. Note that, for any commutative Banach algebra A, $\Delta(A) = \emptyset$ whenever $r_A(x) = 0$ for every $x \in A$.

Example 2.1.6. Let $A = P(\mathbb{D})$ as a Banach space. For $f, g \in A$ define a function $f \circ g$ on $\mathbb{D}$ by

$$f \circ g(z) = z \int_0^1 f(z - tz)g(tz)dt,$$

$z \in \mathbb{D}$. We claim that $f \circ g \in A$. For that notice first that if polynomials

$$p(z) = \sum_{j=0}^{n} a_j z^j \quad \text{and} \quad q(z) = \sum_{k=0}^{m} b_k z^k,$$

$a_j, b_k \in \mathbb{C}$, are given then, for any $z \in \mathbb{D}$,

$$(p|_{\mathbb{D}} \circ q|_{\mathbb{D}})(z) = \sum_{j=0}^{n} \sum_{k=0}^{m} a_j b_k z^{j+k+1} \int_0^1 t^k (1-t)^j \, dt,$$

so that $p|_{\mathbb{D}} \circ q|_{\mathbb{D}}$ equals the restriction to $\mathbb{D}$ of a polynomial. Now, given arbitrary f and g in A and $\varepsilon > 0$, let p and q be polynomials such that $\|f - p|_{\mathbb{D}}\|_\infty \leq \epsilon$ and $\|g - q|_{\mathbb{D}}\|_\infty \leq \epsilon$. Then, for any $z \in \mathbb{D}$,

$$
\begin{aligned}
|f \circ g(z) - p|_{\mathbb{D}} \circ q|_{\mathbb{D}}(z)| &\leq \int_0^1 |f(z-tz)g(tz) - p(z-tz)q(tz)| \, dt \\
&\leq \int_0^1 |g(tz)| \cdot |f(z-tz) - p(z-tz)| \, dt \\
&\quad + \int_0^1 |p(z-tz)| \cdot |g(tz) - q(tz)| \, dt \\
&\leq \|g\|_\infty \|f - p|_{\mathbb{D}}\|_\infty + \|p|_{\mathbb{D}}\|_\infty \|g - q|_{\mathbb{D}}\|_\infty \\
&\leq \epsilon(\|f\|_\infty + \|g\|_\infty + \epsilon).
\end{aligned}
$$

Hence $f \circ g$ is the uniform limit of polynomial functions on $\mathbb{D}$, whence $f \circ g \in A$. Clearly, $\|f \circ g\|_\infty \leq \|f\|_\infty \|g\|_\infty$. Moreover, the multiplication $(f, g) \to f \circ g$ is commutative, associative, and distributive. In fact, this is straightforward from the definition of $f \circ g$. Thus A with product $\circ$ is a commutative Banach algebra.

We proceed to show by induction that

$$|f^n(z)| \leq \frac{1}{(n-1)!} \|f\|_\infty^n |z|^{n-1}$$

for every $f \in A$ and all $z \in \mathbb{D}$ and $n \in \mathbb{N}$. The case $n = 1$ being obvious, assume the estimate to hold for n. Let $z \in \mathbb{D}$ and write $z = re^{i\varphi}, 0 \leq r \leq 1, \varphi \in \mathbb{R}$. Then

$$
\begin{aligned}
f^{n+1}(z) &= re^{i\varphi} \int_0^1 f(z - tre^{i\varphi}) f^n(tre^{i\varphi}) \, dt \\
&= e^{i\varphi} \int_0^r f(z - se^{i\varphi}) f^n(se^{i\varphi}) \, ds,
\end{aligned}
$$

and hence by the inductive hypothesis,

$$|f^{n+1}(z)| \le \|f\|_\infty \int_0^r |f^n(se^{i\varphi})|ds$$

$$\le \frac{1}{(n-1)!}\|f\|_\infty^{n+1} \int_0^r s^{n-1}ds$$

$$= \frac{1}{n!}\|f\|_\infty^{n+1}r^n$$

$$= \frac{1}{n!}\|f\|_\infty^{n+1}|z|^n,$$

as required. Thus, for every $f \in A$ and $n \in \mathbb{N}$,

$$\|f^n\|_\infty \le \frac{\|f\|_\infty^n}{(n-1)!}$$

and hence, by the spectral radius formula,

$$r_A(f) = \lim_{n\to\infty} \|f^n\|_\infty^{1/n} \le \|f\|_\infty \lim_{n\to\infty} \left(\frac{1}{(n-1)!}\right)^{1/n} = 0.$$

This shows that $\sigma_A(f) = \{0\}$ for all $f \in A$, and therefore $\Delta(A) = \emptyset$.

Example 2.1.7. Define a bounded linear operator T on $C[0,1]$ by

$$Tf(t) = \int_0^t f(s)ds, \quad f \in C[0,1], \quad t \in [0,1].$$

Let A be the norm closure in $\mathcal{B}(C[0,1])$ of the set of all polynomials in T of the form

$$\sum_{i=1}^n a_iT^i, a_1, \ldots, a_n \in \mathbb{C}, n \in \mathbb{N}.$$

A is a commutative Banach algebra which does not have an identity. A straightforward induction argument shows

$$|T^nf(t)| \le \|f\|_\infty \frac{t^n}{n!}$$

for all $t \in [0,1]$ and $n \in \mathbb{N}$. Hence

$$\|T^nf\|_\infty \le \frac{1}{n!}\|f\|_\infty,$$

and this inequality gives

$$\|T^n\|^{1/n} \le \left(\frac{1}{n!}\right)^{1/n}$$

for all $n \in \mathbb{N}$. Since $(n!)^{1/n} \to \infty$ as $n \to \infty$, we get $r_A(T) = 0$. The spectral radius r_A is subadditive and submultiplicative (Lemma 1.2.13) and continuous. Therefore it follows that $r_A(S) = 0$ for all $S \in A$.

Of course, the algebra A is reminiscent of the Volterra algebra which was introduced in Exercise 1.6.12. In fact, restricting the convolution of Example 1.6.12 to $C[0,1]$ and denoting by u the constant one function on $[0,1]$, we have

$$Tf(t) = \int_0^t u(t-s)f(s)ds = u * f(t)$$

for all $f \in C[0,1]$ and $t \in \mathbb{R}$. It is now easily verified that u^n, the n-fold convolution product of u, is given by $u^n(t) = t^{n-1}/(n-1)!$, $n \in \mathbb{N}$. Since the polynomials are uniformly dense in $C[0,1]$, it follows that A equals the closure of in $\mathcal{B}(C[0,1])$ of the algebra of all convolution operators

$$T_g : C[0,1] \to C[0,1], \ f \to g * f,$$

$g \in C[0,1]$. Because of this similarity, A is also sometimes termed *Volterra algebra*.

If A is a unital commutative Banach algebra, the anomaly just discussed cannot occur. This is an immediate consequence of the next theorem which forms the basic link between $\Delta(A)$ and ideals in A.

Theorem 2.1.8. *For a commutative Banach algebra A, the mapping*

$$\varphi \to \ker \varphi = \{x \in A : \varphi(x) = 0\}$$

is a bijection between $\Delta(A)$ and $\mathrm{Max}(A)$, the set of all maximal modular ideals in A.

Proof. For $\varphi \in \Delta(A)$, $\ker \varphi$ is an ideal and a closed linear subspace of codimension one in A. To verify that $\ker \varphi$ is modular simply choose $u \in A$ such that $\varphi(u) = 1$. Then, for any $x \in A$,

$$\varphi(ux - x) = \varphi(u)\varphi(x) - \varphi(x) = 0,$$

whence $ux - x \in \ker \varphi$. Thus u is an identity modulo $\ker \varphi$, and hence $\ker \varphi$ is a maximal modular ideal.

Let now that $\varphi_1, \varphi_2 \in \Delta(A)$ be such that $\ker \varphi_1 = \ker \varphi_2$ and denote this ideal by I. Let u be an identity modulo I. Then, since the codimension of I is one, each $x \in A$ can be uniquely expressed as

$$x = \lambda u + y, \quad y \in I, \quad \lambda \in \mathbb{C}.$$

As $\varphi(u) = 1$ for every homomorphism φ with $\ker \varphi = I$, we get

$$\varphi_1(x) = \lambda \varphi_1(u) + \varphi_1(y) = \lambda = \lambda \varphi_2(u) + \varphi_2(y) = \varphi_2(x).$$

Finally, let $M \in \mathrm{Max}(A)$ and let u be an identity modulo M. We already know that M is closed in A, so A/M is a Banach algebra. Suppose there exists

$x \in A \setminus M$ such that $x + M$ is not invertible in A/M. Then $A/M(x + M)$ is a proper nonzero ideal in A/M since

$$x + M = (u + M)(x + M) \in A/M(x + M)$$

is nonzero. This contradicts the maximality of M. Thus A/M is a Banach division algebra and hence, by the Gelfand–Mazur theorem (Theorem 1.2.9), isomorphic to the field of complex numbers. Clearly, this isomorphism defines a homomorphism $\varphi : A \to \mathbb{C}$ with $\ker \varphi = M$. □

Definition 2.1.9. Let A be a commutative Banach algebra. The *radical* of A, $\mathrm{rad}(A)$, is defined by

$$\mathrm{rad}(A) = \bigcap \{M : M \in \mathrm{Max}(A)\} = \bigcap \{\ker \varphi : \varphi \in \Delta(A)\},$$

where $\mathrm{rad}(A)$ is understood to be A if $\Delta(A) = \emptyset$. Clearly, $\mathrm{rad}(A)$ is a closed ideal of A. The algebra A is called *semisimple* if $\mathrm{rad}(A) = \{0\}$ and *radical* if $\mathrm{rad}(A) = A$.

In Examples 2.1.6 and 2.1.7 we have already seen examples of radical Banach algebras with nontrivial multiplication. On the other hand, it will follow from Theorem 2.2.5 in the next section that A is semisimple if and only if for every $x \in A$, $r_A(x) = 0$ implies that $x = 0$. Because the spectral radius is subadditive and submultiplicative, this means that A is semisimple if and only if r_A is an algebra norm on A. Thus $\Delta(A) \neq \emptyset$.

Returning to the existence of nonzero multiplicative linear functionals, assume that A is a commutative Banach algebra with identity. Then the proper ideal $\{0\}$ is contained in some maximal ideal which, by Theorem 2.1.8, is the kernel of a homomorphism from A onto $\mathbb{C}$.

We continue with a number of interesting applications of Lemma 2.1.5.

Corollary 2.1.10. *Let ϕ be a homomorphism from a commutative Banach algebra A into a semisimple commutative Banach algebra B. Then ϕ is continuous.*

Proof. By the closed graph theorem it suffices to show that if $x_n \in A, n \in \mathbb{N}$, are such that $x_n \to 0$ and $\phi(x_n) \to b$ for some $b \in B$, then $b = 0$. Let $\varphi \in \Delta(B)$. Then $\varphi \circ \phi \in \Delta(A) \cup \{0\}$ and hence both, φ and $\varphi \circ \phi$, are continuous by Lemma 2.1.5. It follows that

$$\varphi(b) = \lim_{n \to \infty} \varphi(\phi(x_n)) = \lim_{n \to \infty} (\varphi \circ \phi)(x_n) = 0.$$

Since this holds for all $\varphi \in \Delta(B)$ and B is semisimple we get $b = 0$. □

Corollary 2.1.11. *On a semisimple commutative Banach algebra all Banach algebra norms are equivalent.*

Proof. Suppose A is a semisimple commutative Banach algebra, and let $\| \cdot \|_1$ and $\| \cdot \|_2$ be two Banach algebra norms on A. The statement follows by applying Corollary 2.1.10 with ϕ the identity mappings $(A, \| \cdot \|_1) \to (A, \| \cdot \|_2)$ and $(A, \| \cdot \|_2) \to (A, \| \cdot \|_1)$. $\square$

Corollary 2.1.12. *Every involution on a semisimple commutative Banach algebra A is continuous.*

Proof. Let $\| \cdot \|$ be the given norm an A. We define a new norm $| \cdot |$ on A by $|x| = \|x^*\|$. It is clear that $| \cdot |$ is submultiplicative. If $x_n \in A, n \in \mathbb{N}$, form a Cauchy sequence for $| \cdot |$, then $(x_n^*)_n$ is a Cauchy sequence for $\| \cdot \|$. Consequently, $\|x_n^* - x\| \to 0$ for some $x \in A$, and hence $|x_n - x^*| \to 0$. This shows that $(A, | \cdot |)$ is complete. By Corollary 2.1.11 there exists $c > 0$ such that

$$\|x^*\| = |x| \le c\|x\|$$

for all $x \in A$, as was to be shown. $\square$

Let $C^\infty[0,1]$ denote the algebra of all infinitely many times differentiable functions on $[0,1]$.

Corollary 2.1.13. *The algebra $C^\infty[0,1]$ admits no Banach algebra norm.*

Proof. Suppose there is a Banach algebra norm $\| \cdot \|$ on $C^\infty[0,1]$. Applying Corollary 2.1.10 to the identity mapping from $C^\infty[0,1]$ into $C[0,1]$ we see that there exists $c > 0$ such that

$$\|f\|_\infty \le c\|f\|$$

for all $f \in C^\infty[0,1]$. Using this inequality, we prove that the differentiation mapping $D : f \to f'$ from $C^\infty[0,1]$ into itself is continuous. Thus, let $f_n \in C^\infty[0,1], n \in \mathbb{N}$, be such that

$$\lim_{n \to \infty} \|f_n\| = 0 \quad \text{and} \quad \lim_{n \to \infty} \|f_n' - g\| = 0$$

for some $g \in C^\infty[0,1]$. Then

$$\lim_{n \to \infty} \|f_n\|_\infty = 0 \quad \text{and} \quad \lim_{n \to \infty} \|f_n' - g\|_\infty = 0.$$

Since for each $x, y \in [0,1]$,

$$\left| \int_x^y g(t)dt \right| \le |f_n(y) - f_n(x)| + \left| \int_x^y (f_n'(t) - g(t))dt \right|$$

$$\le 2\|f_n\|_\infty + |y - x| \cdot \|f_n' - g\|_\infty,$$

it follows that $\int_x^y g(t)dt = 0$. Hence $g = 0$ because x and y are arbitrary. By the closed graph theorem, D is continuous. Thus there exists $d > 0$ such that

$$\|f'\| \leq d\|f\|$$

for all $f \in C^\infty[0,1]$. Now, let $f(t) = e^{2dt}$, $t \in [0,1]$. Then

$$2d\|f\| = \|f'\| \leq d\|f\|.$$

This contradiction shows that there cannot exist a Banach algebra norm on $C^\infty[0,1]$. □

2.2 The Gelfand representation

In this section we develop the basic elements of Gelfand's theory which represents a (semisimple) commutative Banach algebra as an algebra of continuous functions on a locally compact Hausdorff space.

Definition 2.2.1. Let A be a commutative Banach algebra and, as before, $\Delta(A)$ the set of all nonzero (hence surjective) algebra homomorphisms from A to $\mathbb{C}$. We endow $\Delta(A)$ with the weakest topology with respect to which all the functions

$$\Delta(A) \to \mathbb{C}, \quad \varphi \to \varphi(x), \quad x \in A,$$

are continuous. A neighbourhood basis at $\varphi_0 \in \Delta(A)$ is then given by the collection of sets

$$U(\varphi_0, x_1, \ldots, x_n, \epsilon) = \{\varphi \in \Delta(A) : |\varphi(x_i) - \varphi_0(x_i)| < \epsilon, 1 \leq i \leq n\},$$

where $\epsilon > 0, n \in \mathbb{N}$, and $x_1, \ldots, x_n$ are arbitrary elements of A. This topology on $\Delta(A)$ is called the *Gelfand topology*. There are several names in use for the space $\Delta(A)$, equipped with the Gelfand topology: The *structure space*, the *spectrum* or *Gelfand space* of A, and the *maximal ideal space*, the latter notion being justified through the bijective correspondence between $\Delta(A)$ and $\mathrm{Max}(A)$ (Theorem 2.1.8).

Remark 2.2.2. We have seen in Lemma 2.1.5 that $\Delta(A)$ is contained in the unit ball of A^*. The Gelfand topology obviously coincides with the relative w^*-topology of A^* on $\Delta(A)$. When adjoining an identity e to A, $\Delta(A_e) = \Delta(A) \cup \{\varphi_\infty\}$ (Remark 2.1.3) and according to the following theorem the topology on $\Delta(A)$ is the one induced from $\Delta(A_e)$.

Theorem 2.2.3. *Let A be a commutative Banach algebra. Then*

(i) $\Delta(A)$ *is a locally compact Hausdorff space.*
(ii) $\Delta(A_e) = \Delta(A) \cup \{\varphi_\infty\}$ *is the one-point compactification of $\Delta(A)$.*
(iii) $\Delta(A)$ *is compact if A has an identity.*

Proof. It is easy to see that $\Delta(A)$ is a Hausdorff space. Indeed, if φ_1 and φ_2 are distinct elements of $\Delta(A)$, then for some $x \in A, \delta = \frac{1}{2}|\varphi_1(x) - \varphi_2(x)| > 0$ and hence

$$U(\varphi_1, x, \delta) \cap U(\varphi_2, x, \delta) = \emptyset.$$

To prove that $\Delta(A)$ is compact if A has an identity e, let

$$C = \prod_{x \in A} \{z \in \mathbb{C} : |z| \leq \|x\|\}.$$

Equipped with the product topology, C is a compact space by Tychonoff's theorem. Since $|\varphi(x)| \leq \|x\|$ for all $\varphi \in \Delta(A)$ and $x \in A$, we can define a mapping ϕ from $\Delta(A)$ into C by

$$\phi(\varphi) = (\varphi(x))_{x \in A}.$$

Then ϕ is injective and, by definition of the Gelfand topology, a homeomorphism from $\Delta(A)$ onto $\phi(\Delta(A))$. Thus, in order to establish that $\Delta(A)$ is compact it remains to show that $\phi(\Delta(A))$ is closed in C. To this end, let $\lambda = (\lambda_x)_{x \in A} \in C$ lie in the closure of $\phi(\Delta(A))$ and let $x, y \in A, \alpha, \beta \in \mathbb{C}$ and $\varepsilon > 0$ be given. If $\varphi \in \Delta(A)$ is such that $|\varphi(a) - \lambda_a| \leq \varepsilon$ for $a \in \{x, y, xy, \alpha x + \beta y\}$, then

$$\begin{aligned}
|\alpha\lambda_x + \beta\lambda_y - \lambda_{\alpha x + \beta y}| &\leq |\alpha| \, |\lambda_x - \varphi(x)| + |\beta| \, |\lambda_y - \varphi(y)| \\
&\quad + |\varphi(\alpha x + \beta y) - \lambda_{\alpha x + \beta_y}| \\
&\leq \epsilon(|\alpha| + |\beta| + 1)
\end{aligned}$$

and

$$\begin{aligned}
|\lambda_{xy} - \lambda_x\lambda_y| &\leq |\lambda_{xy} - \varphi(xy)| + |\varphi(y)| \, |\varphi(x) - \lambda_x| \\
&\quad + |\lambda_x| \, |\varphi(y) - \lambda_y| \\
&\leq \epsilon(1 + \|y\| + \|x\|).
\end{aligned}$$

Since $\epsilon > 0$ was arbitrary, it follows that $\psi : x \to \lambda_x, A \to \mathbb{C}$ is a homomorphism. Moreover, $\psi \in \Delta(A)$ because $\psi(e) = \lambda_e = 1$. This completes the proof of statement (iii).

Now we drop the hypothesis that A be unital and consider $\Delta(A_e)$ and $\Delta(A) \subseteq \Delta(A_e)$. We denote the basic neighbourhoods in $\Delta(A)$ and $\Delta(A_e)$ by U and U_e, respectively. Then, for $\varphi \in \Delta(A), \epsilon > 0$ and a finite subset F of A,

$$U_e(\varphi, F, \epsilon) = \begin{cases} U(\varphi, F, \epsilon) \cup \{\varphi_\infty\} & \text{if } |\varphi(x)| < \epsilon \text{ for all } x \in F, \\ U(\varphi, F, \epsilon) & \text{otherwise.} \end{cases}$$

It follows that the Gelfand topology on $\Delta(A)$ coincides with the relative Gelfand topology of $\Delta(A_e)$. However, the singleton $\{\varphi_\infty\}$ is closed in $\Delta(A_e)$, so that $\Delta(A)$ is open in $\Delta(A_e)$ and hence is locally compact. This proves (i).

Finally, for $x \in A$ and $\epsilon > 0$,

$$U_\varepsilon(\varphi_\infty, x, \epsilon) = \{\varphi_\infty\} \cup \{\varphi \in \Delta(A) : |\varphi(x)| < \epsilon\}$$
$$= \Delta(A_e) \setminus \{\psi \in \Delta(A_e) : |\psi(x)| \geq \epsilon\}.$$

Now, the sets $\{\psi \in \Delta(A_e), |\psi(x)| \geq \epsilon\}$, $x \in A$, are closed in $\Delta(A_e)$ and hence compact. The complement of a basic neighbourhood of φ_∞ is a finite union of such compact sets. Therefore it follows that $\Delta(A_e)$ is the one-point compactification of $\Delta(A)$. □

A natural question arising in view of the preceding theorem is whether a semisimple commutative Banach algebra A has to possess an identity if $\Delta(A)$ is compact. Actually, this is true. This turns out to be a consequence of Shilov's idempotent theorem, the proof of which utilises the several-variable functional calculus. A considerably simpler proof is available when A is regular (Corollary 4.2.11).

Definition 2.2.4. For $x \in A$, we define $\widehat{x} : \Delta(A) \to \mathbb{C}$ by $\widehat{x}(\varphi) = \varphi(x)$. Then $\widehat{x}$ is a continuous function, which is called the *Gelfand transform* of x. It is easily checked that the mapping

$$\Gamma_A : A \to C(\Delta(A)), \quad x \to \widehat{x}$$

is a homomorphism, the *Gelfand homomorphism* or *Gelfand representation* of A. We quite often denote $\Gamma_A(A)$ by $\widehat{A}$.

Fundamental properties of the Gelfand transform and the Gelfand representation are given in the next theorems.

Theorem 2.2.5. *Let A be a commutative Banach algebra. For each $x \in A$,*

$$\sigma_A(x) \setminus \{0\} \subseteq \widehat{x}(\Delta(A)) = \{\varphi(x) : \varphi \in \Delta(A)\} \subseteq \sigma_A(x).$$

If A is unital, then $\widehat{x}(\Delta(A)) = \sigma_A(x)$.

Proof. Suppose first that A has an identity e. Then $\varphi(x) \in \sigma_A(x)$ for every $\varphi \in \Delta(A)$ (see Theorem 2.1.2). Conversely, if $\lambda \in \sigma_A(x)$, then

$$I = (\lambda e - x)A$$

is a proper ideal in A and hence contained in ker φ for some $\varphi \in \Delta(A)$ (Lemma 1.3.2 and Theorem 2.1.8). It follows that $\varphi(\lambda e - x) = 0$, so that $\lambda \in \widehat{x}(\Delta(A))$.

If A fails to be unital, then by the preceding paragraph and the definition of the spectrum,

$$\sigma_A(x) \setminus \{0\} = \sigma_{A_e}(x) \setminus \{0\} = \widehat{x}(\Delta(A_e)) \setminus \{0\}$$
$$\subseteq \widehat{x}(\Delta(A)) = \widehat{x}(\Delta(A_e)) = \sigma_{A_e}(x)$$
$$= \sigma_A(x),$$

as was to be shown. □

The following corollary is an immediate consequence of Theorem 2.2.5 and the spectral radius formula.

Corollary 2.2.6. *For $x \in A, \widehat{x} = 0$ if and only if*

$$r_A(x) = \lim_{n \to \infty} \|x^n\|^{1/n} = 0.$$

Theorem 2.2.7. *Let A be a commutative Banach algebra and Γ the Gelfand representation of A.*

(i) *Γ maps A into $C_0(\Delta(A))$ and is norm decreasing.*
(ii) *$\Gamma(A)$ strongly separates the points of $\Delta(A)$.*
(iii) *Γ is isometric if and only if $\|x\|^2 = \|x^2\|$ for all $x \in A$.*

Proof. (i) Since, by Theorem 2.2.3, $\Delta(A_e)$ is the one-point compactification of $\Delta(A)$ and $\widehat{x}(\varphi_\infty) = 0$ for $x \in A$, we have $\widehat{x} \in C_0(\Delta(A))$. Moreover, by Theorem 2.2.5,

$$\|\widehat{x}\|_\infty = r_A(x) \le \|x\|.$$

(ii) It is clear that $\Gamma(A)$ strongly separates the points of $\Delta(A)$, that is, $\Gamma(A)(\varphi) \ne \{0\}$ for each $\varphi \in \Delta(A)$, and if $\varphi_1 \ne \varphi_2$, then $\widehat{x}(\varphi_1) \ne \widehat{x}(\varphi_1)$ for some $x \in A$.

(iii) If $\|y\|^2 = \|y^2\|$ for all $y \in A$, then $\|x^{2^n}\| = \|x\|^{2^n}$ for every $x \in A$ and $n \in \mathbb{N}$. Hence

$$\|\widehat{x}\|_\infty = r_A(x) = \lim_{n \to \infty} \|x^{2^n}\|^{1/2^n} = \|x\|.$$

Conversely, $\|x^2\| = \|\widehat{x}^2\|_\infty = \|\widehat{x}\|_\infty^2 = \|x\|^2$ when Γ is an isometry. $\square$

We now present three simple examples. More difficult and challenging ones are discussed in subsequent sections.

Example 2.2.8. Let X be a locally compact Hausdorff space. The closed ideals in $C_0(X)$ have been completely determined in Theorem 1.3.6. In particular,

$$x \to M_x = \{f \in C_0(X) : f(x) = 0\}$$

sets up a one-to-one correspondence between the points of X and the maximal modular ideals of $C_0(X)$. On the other hand, by Theorem 2.1.8, we have a bijection

$$\Delta(C_0(X)) \to \mathrm{Max}(C_0(X)), \quad \varphi \to \ker \varphi.$$

This yields a bijection $X \to \Delta(C_0(X)), x \to \varphi_x$ where $\varphi_x(f) = f(x)$ for $f \in C_0(X)$. The map $x \to \varphi_x$ is a homeomorphism. Indeed, given $x \in X$ and an open neighbourhood V of x, by Urysohn's lemma there exists $f \in C_0(X)$ such that $f(x) \ne 0$ and $f|_{X \setminus V} = 0$, and hence V contains the Gelfand neighbourhood $\{y : |\varphi_y(f) - \varphi_x(f)| < |f(x)|\}$ of x. After identifying X with $\Delta(C_0(X))$, the Gelfand homomorphism of $C_0(X)$ is the identity mapping.

Example 2.2.9. Let $A = C^n[a, b]$, and for each $t \in [0, 1]$ define $\varphi_t \in \Delta(A)$ by $\varphi_t(f) = f(t)$. We claim that

$$\phi : [a, b] \to \Delta(A), \quad t \to \varphi_t$$

is a homeomorphism. Obviously, ϕ is injective and continuous. Let M be any maximal ideal in A. Then, by the same reasoning as in the proof of Theorem 1.3.6, we find $s \in [a, b]$ such that $M = \{f \in A : f(s) = 0\}$. It follows that $M = \ker \varphi_s$. Hence ϕ is a homeomorphism since $[a, b]$ is compact and $\Delta(A)$ is Hausdorff. As in the previous example, after identifying $[a, b]$ with $\Delta(A)$, the Gelfand homomorphism of A is the identity mapping.

Example 2.2.10. We determine the structure space of $l^1(\mathbb{Z})$. For $z \in \mathbb{T}$, define $\varphi_z : l^1(\mathbb{Z}) \to \mathbb{C}$ by

$$\varphi_z(f) = \sum_{n \in \mathbb{Z}} f(n) z^{-n}.$$

Then, for $f, g \in l^1(\mathbb{Z})$,

$$\varphi_z(f * g) = \sum_{n \in \mathbb{Z}} \left(\sum_{m \in \mathbb{Z}} f(n - m) g(m) \right) z^{-n}$$

$$= \sum_{n,m \in \mathbb{Z}} f(n) g(m) z^{-(n+m)}$$

$$= \varphi_z(f) \varphi_z(g).$$

Thus $\varphi_z \in \Delta(l^1(\mathbb{Z}))$ and the map $z \to \varphi_z$ is clearly injective. Conversely, every $\varphi \in \Delta(l^1(\mathbb{Z}))$ is of this form. Indeed, let $z = \varphi(\delta_{-1})$. Then

$$\varphi(\delta_{-n}) = \varphi(\delta_{-1} * \ldots * \delta_{-1}) = \varphi(\delta_{-1})^n = z^n$$

and hence also $\varphi(\delta_n) = 1/\varphi(\delta_{-n}) = z^{-n}$ for all $n \in \mathbb{N}$. Since the finite linear combinations of Dirac functions δ_n, $n \in \mathbb{N}$, are dense in $l^1(\mathbb{Z})$, it follows that $\varphi = \varphi_z$. By routine arguments it is shown that the map $z \to \varphi_z$ is a homeomorphism.

We have seen earlier (Example 1.1.5) that the commutative Banach algebra $AC(\mathbb{T})$ is isomorphic to $l^1(\mathbb{Z})$, the isomorphism being given by $f \to (c_n(f))_n$, where

$$c_n(f) = \frac{1}{2\pi} \int_0^{2\pi} f(e^{it}) e^{-int} dt$$

for $n \in \mathbb{Z}$. Thus, by the preceding example, $\Delta(AC(\mathbb{T}))$ can be identified with $\mathbb{T}$ as follows. For $z \in \mathbb{T}$, let

$$\varphi_z(f) = \sum_{n \in \mathbb{Z}} c_n(f) z^n, \quad f \in AC(\mathbb{T}).$$

Then $z \rightarrow \varphi_z$ is a homeomorphism between $\mathbb{T}$ and $\Delta(AC(\mathbb{T}))$. On making this identification,

$$\widehat{f}(z) = \sum_{n \in \mathbb{Z}} c_n(f) z^n = f(z)$$

for all $f \in AC(\mathbb{T})$, so that the Gelfand representation of $AC(\mathbb{T})$ is the identity. As a simple consequence we obtain the following classical result due to Wiener.

Theorem 2.2.11. *If $f \in AC(\mathbb{T})$ is such that $f(z) \neq 0$ for all $z \in \mathbb{T}$, then $1/f \in AC(\mathbb{T})$; that is, $1/f$ has an absolutely convergent Fourier series.*

Proof. With the previous identification of $\Delta(AC(\mathbb{T}))$ with $\mathbb{T}$, the assumption on f means that f belongs to no maximal ideal of $AC(\mathbb{T})$. Thus f is invertible in $AC(\mathbb{T})$ and so $1/f \in AC(\mathbb{T})$. $\qquad\qquad\square$

Lemma 2.2.12. *Let A and B be commutative Banach algebras. If A and B are algebraically isomorphic, then $\Delta(A)$ and $\Delta(B)$ are homeomorphic.*

Proof. Suppose $\phi : A \rightarrow B$ is an algebra isomorphism. Let $\phi^* : \Delta(B) \rightarrow \Delta(A)$ be the dual mapping; that is,

$$\phi^*(\varphi)(a) = \varphi(\phi(a)), \quad a \in A, \quad \varphi \in \Delta(B).$$

It is easily checked that ϕ^* is a bijection. ϕ^* is continuous provided that all functions

$$\Delta(B) \rightarrow \mathbb{C}, \quad \varphi \rightarrow \phi^*(\varphi)(a), \quad a \in A,$$

are continuous. However, that such functions are continuous follows immediately from the definition of ϕ^* and the definition of the topology on $\Delta(B)$. $(\phi^*)^{-1}$ is continuous on the same grounds. $\qquad\qquad\square$

Corollary 2.2.13. *For locally compact Hausdorff spaces X and Y the following conditions are equivalent.*

(i) $C_0(X)$ *and* $C_0(Y)$ *are isometrically isomorphic.*
(ii) $C_0(X)$ *and* $C_0(Y)$ *are algebraically isomorphic.*
(iii) X *and* Y *are homeomorphic.*

Proof. (i) $\Rightarrow$ (ii) is trivial, and (ii) $\Rightarrow$ (iii) is a consequence of the preceding lemma and Example 2.2.8. Finally, if $\phi : X \rightarrow Y$ is a homeomorphism, then $f \rightarrow f \circ \phi$ is an isometric algebra isomorphism from $C_0(Y)$ to $C_0(X)$. $\qquad\square$

We continue with a proposition which often can efficiently be used to identify the Gelfand topology.

Proposition 2.2.14. *Let X be a locally compact Hausdorff space and let A be a family of functions in $C_0(X)$ which strongly separates the points of X. Then the topology of X equals the weak topology with respect to the functions $x \rightarrow f(x), f \in A$.*

Proof. The given topology on X is stronger than the weak topology. Thus it suffices to show that given $x \in X$ and an open neighbourhood U of x in X, there exists a set V such that $x \in V \subseteq U$ and V is open in the weak topology. Let $\widetilde{X}$ be X if X is compact, and let $\widetilde{X} = X \cup \{\infty\}$ be the one-point compactification of X if X is noncompact. Every $f \in C_0(X)$ extends continuously to $\widetilde{X}$ by setting $f(\infty) = 0$. Since A strongly separates the points of X, for every $y \in \widetilde{X} \setminus U$ there exists $f_y \in A$ such that

$$\epsilon_y = |f_y(y) - f_y(x)| > 0.$$

Then, for every $y \in \widetilde{X} \setminus U$,

$$V_y = \{z \in \widetilde{X} : |f_y(z) - f_y(y)| < \epsilon_y/2\}$$

is an open neighbourhood of y in $\widetilde{X}$, and because $\widetilde{X} \setminus U$ is compact there are finitely many $y_1, \ldots, y_n \in \widetilde{X} \setminus U$ such that $\widetilde{X} \setminus U \subseteq \bigcup_{j=1}^n V_{y_j}$. Let

$$V = \{z \in X : |f_{y_j}(z) - f_{y_j}(x)| < \epsilon_{y_j}/2 \text{ for all } 1 \le j \le n\}.$$

Then $x \in V$ and V is contained in U. Indeed, if $z \in V$ and $z \notin U$, then $z \in V_{y_j}$ for some j, and hence

$$|f_{y_j}(x) - f_{y_j}(y_j)| \le |f_{y_j}(x) - f_{y_j}(z)| + |f_{y_j}(z) - f_{y_j}(y_j)| < \epsilon_{y_j}.$$

This contradicts the definition of ϵ_{y_j}. □

For a closed ideal I of a commutative Banach algebra A, we now relate the Gelfand topologies on $\Delta(I)$ and on $\Delta(A/I)$ to the Gelfand topology on $\Delta(A)$. For a subset M of A, the *hull* $h(M)$ of M is defined to be

$$h(M) = \{\varphi \in \Delta(A) : \varphi(M) = \{0\}\}.$$

Lemma 2.2.15. *Let I be a closed ideal of A and $q : A \to A/I$ the quotient homomorphism.*

(i) *The map $\varphi \to \varphi \circ q$ is a homeomorphism from $\Delta(A/I)$ onto $h(I)$.*
(ii) *The map $\varphi \to \varphi|_I$ is a homeomorphism from $\Delta(A) \setminus h(I)$ onto $\Delta(I)$.*

Proof. (i) It is obvious that the map is a bijection. It is a homeomorphism since

$$U(\varphi, x + I, \epsilon) \circ q = \{\psi \circ q : \psi \in \Delta(A/I), |\psi(x + I) - \varphi(x + I)| < \epsilon\}$$
$$= \{\rho \in h(I) : |\rho(x) - \varphi \circ q(x)| < \epsilon\}$$
$$= U(\varphi \circ q, x, \epsilon)$$

for all $\varphi \in \Delta(A/I), x \in A$ and $\epsilon > 0$.

(ii) If $\varphi_1, \varphi_2 \in \Delta(A) \setminus h(I)$ are such that $\varphi_1|_I = \varphi_2|_I$, then choosing $x \in I$ such that $\varphi_1(x) = 1$, it follows that

$$\varphi_1(y) = \varphi_1(yx) = \varphi_2(yx) = \varphi_2(y)$$

for all $y \in A$. So the map $\varphi \to \varphi|_I$ is injective, and it is clearly continuous. Given $\psi \in \Delta(I)$, again choose $x \in I$ with $\psi(x) = 1$ and define φ on A by $\varphi(y) = \psi(yx)$, $y \in A$. Then φ extends ψ, and it is easily verified that $\varphi \in \Delta(A) \setminus h(I)$. Finally, let $\varphi \in \Delta(A) \setminus h(I)$, $y \in A$, $y \neq 0$, and $\epsilon > 0$ be given and let $\delta = \min\{\epsilon/2, \epsilon/2\|y\|\}$. Then, if $\rho \in \Delta(A)$ is such that $\rho|_I \in U(\varphi|_I, x, yx, \delta)$, it follows that

$$|\rho(y) - \varphi(y)| \leq |\rho(y)| \cdot |\varphi(x) - \rho(x)| + |\rho(yx) - \varphi(yx)|$$
$$< \delta\|y\| + \delta \leq \epsilon,$$

whence $\rho \in U(\varphi, y, \epsilon)$. Thus the map $\varphi \to \varphi|_I$ is also open, hence a homeomorphism. □

By Lemma 2.2.15, for each $y \in A$ there is a unique continuous function f_y on $\Delta(I)$ such that $\widehat{yx}(\varphi) = f_y(\varphi)\widehat{x}(\varphi)$ for all $\varphi \in \Delta(I)$ and $x \in A$. This in particular applies when a commutative Banach algebra A has a bounded approximate identity and hence can be considered as a closed ideal of its multiplier algebra $M(A)$ (Theorem 1.4.12). The following proposition, however, shows that this same conclusion holds if A is merely assumed to be faithful (see Proposition 1.4.11).

Proposition 2.2.16. *Let A be a commutative Banach algebra and let $T \in M(A)$. Then there exists a unique continuous function f on $\Delta(A)$ such that $\widehat{Tx}(\varphi) = f(\varphi)\widehat{x}(\varphi)$ for all $\varphi \in \Delta(A)$ and $x \in A$. Furthermore, f is bounded and $\|f\|_\infty \leq \|T\|$.*

Proof. If $\varphi \in \Delta(A)$ and $x, y \in A$ are such that $\widehat{x}(\varphi) \neq 0$ and $\widehat{y}(\varphi) \neq 0$, then it follows from $(Tx)y = x(Ty)$ that

$$\frac{\widehat{Tx}(\varphi)}{\widehat{x}(\varphi)} = \frac{\widehat{Ty}(\varphi)}{\widehat{y}(\varphi)}.$$

For each $\varphi \in \Delta(A)$ choose $x \in A$ with $\widehat{x}(\varphi) \neq 0$, and define

$$f(\varphi) = \frac{\widehat{Tx}(\varphi)}{\widehat{x}(\varphi)}.$$

The above equation shows that this definition is independent of the choice of x, and hence f is a well-defined continuous function on $\Delta(A)$. Moreover, if $\widehat{x}(\varphi) = 0$ then $\widehat{Tx}(\varphi) = 0$. Indeed, this follows from

$$\widehat{Tx}(\varphi)\widehat{y}(\varphi) = \widehat{x}(\varphi)\widehat{Ty}(\varphi)$$

by choosing y such that $\widehat{y}(\varphi) \neq 0$. Thus the equation $\widehat{Tx}(\varphi) = f(\varphi)\widehat{x}(\varphi)$ holds for all $x \in A$ and $\varphi \in \Delta(A)$.

If g is a second continuous function on $\Delta(A)$ satisfying $\widehat{Tx} = g\,\widehat{x}$ for all $x \in A$, then $(f(\varphi) - g(\varphi))\widehat{x}(\varphi) = 0$ for all $x \in A$ and $\varphi \in \Delta(A)$, and this implies $f(\varphi) = g(\varphi)$. So f is unique.

To show that f is bounded, observe that

$$|f(\varphi)\widehat{x}(\varphi)| = |\widehat{Tx}(\varphi)| \le \|\varphi\| \cdot \|Tx\| \le \|\varphi\| \cdot \|T\| \cdot \|x\|$$

for all $x \in A$ and $\varphi \in \Delta(A)$. Taking $x \in A$ with $\|x\| = 1$, we obtain

$$|f(\varphi)| \cdot \sup\{|\widehat{x}(\varphi)| : \|x\| = 1\} \le \|\varphi\| \cdot \|T\|,$$

for all $\varphi \in \Delta(A)$ and hence $\|f\|_\infty \le \|T\|$. $\square$

2.3 Finitely generated commutative Banach algebras

Many naturally occuring Banach algebras are generated (in the sense of the following definition) by finitely many elements. Such algebras admit a particularly satisfying description of their structure spaces and this is the theme of the present section.

Definition 2.3.1. Let A be a commutative Banach algebra with identity e. A subset E of A is said to *generate* A if every closed subalgebra of A containing E and e coincides with A. Equivalently, the set of all finite linear combinations of elements of the form

$$x_1^{n_1} x_2^{n_2} \cdots x_r^{n_r}, \quad x_i \in E, \quad n_i \in \mathbb{N} \cup \{0\}, \quad r \in \mathbb{N},$$

is dense in A. A is called *finitely generated* if there exists a finite subset of A that generates A.

As a very simple example, recall that $l^1(\mathbb{Z})$ is generated by the two Dirac functions δ_1 and δ_{-1}.

Definition 2.3.2. Let A be a commutative Banach algebra with identity and let $x_1, \ldots, x_n \in A$. Then the *joint spectrum* of $x_1, \ldots, x_n$ is the subset $\sigma_A(x_1, \ldots, x_n)$ of $\mathbb{C}^n$ defined by

$$\sigma_A(x_1, \ldots, x_n) = \{(\varphi(x_1), \ldots, \varphi(x_n)) : \varphi \in \Delta(A)\}.$$

Since $\Delta(A)$ is compact and the mapping

$$\Delta(A) \to \mathbb{C}^n, \quad \varphi \to (\varphi(x_1), \ldots, \varphi(x_n))$$

is continuous, $\sigma_A(x_1, \ldots, x_n)$ is a compact subset of $\mathbb{C}^n$. It is also evident from Theorem 2.2.5 that the joint spectrum of a single element x reduces to the spectrum $\sigma_A(x)$ of x.

Lemma 2.3.3. *Let A be a unital commutative Banach algebra, and suppose that $E \subseteq A$ generates A. Then the mapping*

$$\phi : \Delta(A) \to \prod_{x \in E} \sigma_A(x), \quad \varphi \to (\varphi(x))_{x \in E}$$

is a homeomorphism between $\Delta(A)$ and $\phi(\Delta(A)) \subseteq \prod_{x \in E} \sigma_A(x)$. In particular, if E is finite, say $E = \{x_1, \ldots, x_n\}$, then we have a homeomorphism

$$\Delta(A) \to \sigma_A(x_1, \ldots, x_n), \quad \varphi \to (\varphi(x_1), \ldots, \varphi(x_n)).$$

Proof. Assume first that $\varphi_1, \varphi_2 \in \Delta(A)$ are such that $\varphi_1(x) = \varphi_2(x)$ for all $x \in E$. Let B denote the smallest subalgebra of A containing E and the identity. Then B is dense in A, and $\varphi_1(y) = \varphi_2(y)$ for all $y \in B$. Since elements in $\Delta(A)$ are continuous it follows that $\varphi_1 = \varphi_2$. Hence ϕ is injective.

Now $\prod_{x \in E} \sigma_A(x)$ carries the weak topology with respect to the projections

$$p_y : \prod_{x \in E} \sigma_A(x) \to \sigma_A(y), \quad y \in E.$$

Therefore ϕ is continuous provided that all the functions $p_y \circ \phi, y \in E$, are continuous. However, this is clear from $p_y \circ \phi(\varphi) = \varphi(y)$. Thus

$$\phi : \Delta(A) \to \phi(\Delta(A)), \quad \varphi \to (\varphi(x))_{x \in E}$$

is a continuous bijection between a compact space and a Hausdorff space, and hence is a homeomorphism. □

We now aim at characterizing those compact subsets of $\mathbb{C}^n$ which arise in this way as structure spaces of commutative Banach algebras generated by n elements, $n \in \mathbb{N}$ (Theorem 2.3.6). The relevant geometrical notion is that of polynomial convexity.

Definition 2.3.4. A compact subset K of $\mathbb{C}^n, n \in \mathbb{N}$, is said to be *polynomially convex* if for every $z \in \mathbb{C}^n \setminus K$ there exists a polynomial p such that $p(z) = 1$ and $|p(w)| < 1$ for all $w \in K$.

Lemma 2.3.5. *Every compact convex subset K of $\mathbb{C}^n$ is polynomially convex.*

Proof. We view $\mathbb{C}^n$ as a $2n$-dimensional real vector space. Then, given $w \in \mathbb{C}^n \setminus K$, there exist a real linear functional ψ on $\mathbb{C}^n = \mathbb{R}^{2n}$ and $\alpha \in \mathbb{R}$ such that

$$\psi(w) > \alpha \text{ and } \psi(z) < \alpha \text{ for all } z \in K.$$

Let $z = (z_1, \ldots, z_n) \in \mathbb{C}^n$, with $z_j = x_j + iy_j, x_j, y_j \in \mathbb{R}$. Then ψ has the form

$$\psi(z) = \sum_{j=1}^{n} (a_j x_j + b_j y_j),$$

where $a_j, b_j \in \mathbb{R}, 1 \leq j \leq n$. Let $c_j = a_j - ib_j, 1 \leq j \leq n$, and consider the function

$$f(z) = \exp\left(\sum_{j=1}^{n} c_j z_j\right)$$

on $\mathbb{C}^n$. Then

$$|f(z)| = \exp\left(\operatorname{Re}\left(\sum_{j=1}^{n} c_j z_j\right)\right) = \exp\left(\sum_{j=1}^{n}(a_j x_j + b_j y_j)\right) = \exp\psi(z)$$

and hence $|f(w)| > e^\alpha$ and $|f(z)| < e^\alpha$ for all $z \in K$. It follows that, for a suitable $N \in \mathbb{N}$, the polynomial q defined by

$$q(z) = \prod_{j=1}^{n}\left(\sum_{k=0}^{N} \frac{1}{k!} c_j^k z_j^k\right)$$

satisfies $|q(w)| > e^\alpha$ and $|q(z)| < e^\alpha$ for all $z \in K$. Finally, the polynomial $p = |q(w)|^{-1} q$ has the properties required in Definition 2.3.4. $\qquad\square$

Theorem 2.3.6. *For a compact subset K of $\mathbb{C}^n$ the following conditions are equivalent.*

(i) *There exists a unital commutative Banach algebra A which is generated by n elements $x_1, \ldots, x_n$ such that $K = \sigma_A(x_1, \ldots, x_n)$.*
(ii) *K is polynomially convex.*

Proof. To prove (i) $\Rightarrow$ (ii), let e denote the identity of A and let

$$\lambda = (\lambda_1, \ldots, \lambda_n) \in \mathbb{C}^n \setminus \sigma_A(x_1, \ldots, x_n).$$

Then, given any $\varphi \in \Delta(A), \varphi(x_j) \neq \lambda_j$ for some $1 \leq j \leq n$. Equivalently, for each $M \in \operatorname{Max}(A)$ there exists j such that $x_j - \lambda_j e \notin M$. Consider the ideal

$$I = \left\{\sum_{j=1}^{n}(x_j - \lambda_j e)y_j : y_j \in A\right\}$$

of A. If I were a proper ideal, then $I \subseteq M$ for some $M \in \operatorname{Max}(A)$, but $x_j - \lambda_j e \in I$ and $x_j - \lambda_j e \notin M$ for some j. Thus $I = A$, and hence there exist $y_1, \ldots, y_n \in A$ such that

$$\sum_{j=1}^{n}(x_j - \lambda_j e)y_j = e.$$

Choose $\delta > 0$ such that $\delta \sum_{j=1}^{n} \|x_j - \lambda_j e\| < 1$. Since A is generated by $x_1, \ldots, x_n$, there exist polynomials $p_1, \ldots, p_n$ in n variables such that

$$\|p_j(x_1,\ldots,x_n) - y_j\| \leq \delta$$

for $1 \leq j \leq n$. It follows that

$$\left\| e - \sum_{j=1}^{n}(x_j - \lambda_j e)p_j(x_1,\ldots,x_n) \right\| \leq \sum_{j=1}^{n}\|x_j - \lambda_j e\| \cdot \|y_j - p_j(x_1,\ldots,x_n)\| < 1.$$

Now, define a polynomial p on $\mathbb{C}^n$ by

$$p(z_1,\ldots,z_n) = 1 - \sum_{j=1}^{n}(z_j - \lambda_j)p_j(z_1,\ldots,z_n).$$

Then $p(\lambda_1,\ldots,\lambda_n) = 1$, and for every $\varphi \in \Delta(A)$

$$|p(\varphi(x_1),\ldots,\varphi(x_n))| = \left| 1 - \sum_{j=1}^{n}(\varphi(x_j) - \lambda_j)p_j(\varphi(x_1),\ldots,\varphi(x_n)) \right|$$

$$= \left| \varphi(e) - \sum_{j=1}^{n}\varphi(x_j - \lambda_i e)\varphi(p_j(x_1,\ldots,x_n)) \right|$$

$$\leq \left\| e - \sum_{j=1}^{n}(x_j - \lambda_j e)p_j(x_1,\ldots,x_n) \right\|$$

$$< 1.$$

This proves that $\sigma_A(x_1,\ldots,x_n)$ is polynomially convex.

Conversely, suppose that $K \subseteq \mathbb{C}^n$ is polynomially convex. Let $A = P(K)$, the algebra of all functions $f : K \to \mathbb{C}$ that are uniform limits of polynomial functions on K. Then A is generated by the functions

$$f_j(z) = z_j, \quad z = (z_1,\ldots,z_n) \in K, \quad 1 \leq j \leq n.$$

We are going to show that $K = \sigma_A(f_1,\ldots,f_n)$. For $z \in K$, define $\varphi_z \in \Delta(A)$ by $\varphi_z(f) = f(z)$. As distinct points can be separated by the functions f_j, the mapping

$$\phi : K \to \Delta(A), \quad z \to \varphi_z$$

is injective. ϕ is also continuous since $\Delta(A)$ carries the weak topology with respect to the functions $\varphi \to \varphi(f), f \in A$, and $z \to \varphi_z(f) = f(z)$ is continuous on K. Thus ϕ is a homeomorphism from K onto $\phi(K) \subseteq \Delta(A)$. We claim that $\phi(K) = \Delta(A)$. Towards a contradiction, suppose there exists $\varphi \in \Delta(A) \setminus \phi(K)$ and put

$$\lambda_j = \varphi(f_j), \quad 1 \leq j \leq n, \quad \text{and} \quad \lambda = (\lambda_1,\ldots,\lambda_n).$$

Then $\lambda \notin K$ since otherwise $\varphi_\lambda(f_j) = f_j(\lambda) = \lambda_j = \varphi(f_j), 1 \leq j \leq n$, and hence $\varphi = \varphi_\lambda$ as A is generated by $f_1,\ldots,f_n$. Because K is polynomially convex, we can choose a polynomial p in n variables such that $|p(z_1,\ldots,z_n)| < 1$ for all $z = (z_1,\ldots,z_n) \in K$ and $p(\lambda) = 1$. Then, as K is compact,

$$\|p|_K\|_\infty = \sup_{z \in K} |p(z)| < 1,$$

and hence $|\psi(p|_K)| < 1$ for all $\psi \in \Delta(A)$. Now, $p|_K$ is a finite linear combination of functions of the form

$$z \to z_1^{m_1} z_2^{m_2} \cdot \ldots \cdot z_n^{m_n} = f_1(z)^{m_1} f_2(z)^{m_2} \cdot \ldots \cdot f_n(z)^{m_n}.$$

As $\varphi(f_j) = \lambda_j, 1 \leq j \leq n$, we obtain $\varphi(p|_K) = p(\lambda) = 1$, which is a contradiction. It follows that $\phi(K) = \Delta(A)$, and hence

$$\sigma_A(f_1, \ldots, f_n) = \{(\varphi_z(f_1), \ldots, \varphi_z(f_n)) : z \in K\}$$
$$= \{(z_1, \ldots, z_n) : z \in K\} = K.$$

This shows (ii) $\Rightarrow$ (i).   □

It is worth emphasising that the proof of (ii) $\Rightarrow$ (i) in Theorem 2.3.6 shows that $\Delta(P(K)) = K$ when K is polynomially convex.

The following theorem provides a topological description of polynomially convex subsets of $\mathbb{C}$.

Theorem 2.3.7. *A compact subset K of $\mathbb{C}$ is polynomially convex if and only if $\mathbb{C} \setminus K$ connected.*

Proof. We first assume that K is polynomially convex and that nevertheless $\mathbb{C} \setminus K$ is not connected. Then $\mathbb{C} \setminus K$ has a bounded connected component $S \neq \emptyset$. Then S is closed in $\mathbb{C} \setminus K$ and also open $\mathbb{C} \setminus K$, since $\mathbb{C} \setminus K$ is locally connected. Hence S is also open in $\mathbb{C}$, and therefore its boundary $\partial(S)$ is contained in K.

By Theorem 2.3.6 there exists a commutative Banach algebra A with identity that is generated by some element $a \in A$ such that $K = \sigma_A(a)$. For every $x \in A$ there is a sequence $p_n, n \in \mathbb{N}$, of polynomials such that $\|p_n(a) - x\| \to 0$. Because

$$|p_n(\varphi(a)) - \varphi(x)| = |\varphi(p_n(a)) - \varphi(x)| \leq \|p_n(a) - x\|$$

for all $\varphi \in \Delta(A), (p_n)_{n \in \mathbb{N}}$ converges uniformly on $K = \sigma_A(a) = \hat{a}(\Delta(A))$ with limit $\hat{x}$. Since $\partial(S) \subseteq K$, $(p_n)_{n \in \mathbb{N}}$ converges uniformly on all of S by the maximum modulus principle. We now fix some $\lambda \in S$. Note that $\lim_{n \to \infty} p_n(\lambda)$ does not depend on the particular choice of polynomials p_n with $p_n(a) \to x$. Indeed, if $(q_n)_n$ is a second sequence of polynomials such that $q_n(a) \to x$, then for each $\varphi \in \Delta(A)$

$$|p_n(\varphi(a)) - q_n(\varphi(a))| \leq |p_n(\varphi(a)) - \varphi(x)| + |q_n(\varphi(a)) - \varphi(x)| \to 0,$$

so that $p_n - q_n$ converges uniformly to zero on K, and hence on S. It follows that

$$\lim_{n \to \infty} p_n(\lambda) = \lim_{n \to \infty} q_n(\lambda).$$

This allows us to define $\psi : A \to \mathbb{C}$ by setting

$$\psi(x) = \lim_{n \to \infty} p_n(\lambda),$$

where $(p_n)_n$ is any sequence of polynomials with $p_n(a) \to x$. It is now easily verified that ψ is a homomorphism. For example, if $p_n(a) \to x$ and $q_n(a) \to y$, then $(p_n q_n)(a) \to xy$ and therefore

$$\psi(xy) = \lim_{n \to \infty} (p_n q_n)(\lambda) = \lim_{n \to \infty} p_n(\lambda) \cdot \lim_{n \to \infty} q_n(\lambda) = \psi(x)\psi(y).$$

With $p_n \equiv 1, n \in \mathbb{N}$, we get $\psi(e) = 1$, so that $\psi \in \Delta(A)$. Finally, choosing $p_n(z) = z$ for all $z \in \mathbb{C}, n \in \mathbb{N}$, we obtain $p_n(a) = a$ and hence

$$\psi(a) = \lim_{n \to \infty} p_n(\lambda) = \lambda.$$

Thus $\lambda \in \hat{a}(\Delta(A)) = K$, contradicting the fact that $\lambda \in S \subseteq \mathbb{C} \setminus K$.

Conversely, suppose that $\mathbb{C} \setminus K$ is connected, and consider $A = P(K)$ as in the proof of Theorem 2.3.6, (ii) $\Rightarrow$ (i). Then A is generated by the function $f(z) = z$. Moreover, $\sigma_{C(K)}(f)$, the spectrum of f in $C(K)$, equals K since $z \to \lambda - f(z)$ is invertible in $C(K)$ if and only if $\lambda \notin K$. As $\mathbb{C} \setminus K$ is connected, Theorem 1.2.12 implies $K = \sigma_A(f)$, and hence K is polynomially convex by Theorem 2.3.6, (i) $\Rightarrow$ (ii). □

Remark 2.3.8. More generally, it is true for arbitrary $n \in \mathbb{N}$, that if $K \subseteq \mathbb{C}^n$ is polynomially convex, then $\mathbb{C}^n \setminus K$ is connected. This is proved analogously by employing the maximum modulus principle for polynomials of several complex variables. However, the following example shows that for $n \geq 2$ there exist compact subsets of $\mathbb{C}^n$ which fail to be polynomially convex, even though $\mathbb{C}^n \setminus K$ is connected.

Example 2.3.9. Let $n \geq 2$ and

$$K = \{z = (z_1, \ldots, z_n) \in \mathbb{C}^n : |z_j| = 1, 1 \leq j \leq n\}.$$

Assuming that K is polynomially convex we find a polynomial p in n variables such that $|p(z)| < 1$ for all $z \in K$ and $p(0, 1, \ldots, 1) = 1$. Define a polynomial q in one variable by

$$q(w) = p(w, 1, \ldots, 1), \quad w \in \mathbb{C}.$$

Then $|q(w)| < 1$ for all $w \in \mathbb{C}$ with $|w| = 1$ and $q(0) = 1$. This contradicts the maximum modulus principle. Nevertheless, $\mathbb{C}^n \setminus K$ is connected. To see this, let

$$A_j = \{z = (z_1, \ldots, z_n) \in \mathbb{C}^n : |z_j| > 1\}$$

and

$$B_j = \{z = (z_1, \ldots, z_n) \in \mathbb{C}^n : |z_j| < 1\},$$

$1 \leq j \leq n$, we see that $\mathbb{C}^n \setminus K = \bigcup_{j=1}^{n} (A_j \cup B_j)$. The sets A_j and B_j are arcwise connected, $A_j \cap A_k \neq \emptyset$, $B_j \cap B_k \neq \emptyset$, and, for $j \neq k$, $A_j \cap B_k \neq \emptyset$. It follows that $\mathbb{C}^n \setminus K$ is connected.

2.4 Commutative C^*-algebras

In this section we investigate the question of when the Gelfand homomorphism of a commutative Banach algebra A is an isometric isomorphism onto $C_0(\Delta(A))$. We start with the relevant definition.

Definition 2.4.1. Let A be a Banach algebra with involution $x \to x^*$. Then A is called a C^*-*algebra*, if its norm satisfies the equation $\|x^*x\| = \|x\|^2$ for all $x \in A$. The definition of a C^*-subalgebra is evident.

Note that a C^*-algebra is a Banach $*$-algebra since the equation $\|x\|^2 = \|x^*x\|$ implies $\|x\| \le \|x^*\|$ and hence $\|x\| = \|x^*\|$ for all $x \in A$.

Now let A be a commutative Banach algebra for which the Gelfand homomorphism is an isometric isomorphism onto $C_0(\Delta(A))$. Notice first that in this case for every $x \in A$ there is a unique element $x^* \in A$ such that $\widehat{x^*} = \overline{\widehat{x}}$. Obviously, the mapping $x \to x^*$ is an involution. Moreover,

$$\|x^*\| = \|\widehat{x^*}\|_\infty = \|\overline{\widehat{x}}\|_\infty = \|x\|,$$

and hence

$$\|x^*x\| = \|\widehat{x^*x}\|_\infty = \|\overline{\widehat{x}}\,\widehat{x}\|_\infty = \|\widehat{x}\|_\infty^2 = \|x\|^2.$$

Thus A is a C^*-algebra. The main purpose of what follows is to show that conversely for each commutative C^*-algebra A the Gelfand homomorphism is an isometric $*$-isomorphism onto $C_0(\Delta(A))$. This is one of the most striking results in Gelfand's theory.

Example 2.4.2. (1) Let X be an arbitrary topological space. With the involution given by $f^*(x) = \overline{f(x)}$ and the supremum norm $\| \cdot \|_\infty$, $C^b(X)$ is a commutative C^*-algebra. If X is a locally compact Hausdorff space, then $C_0(X)$ is a C^*-subalgebra of $C^b(X)$.

(2) Let H be a complex Hilbert space, and recall that for $T \in \mathcal{B}(H)$, T^* denotes the adjoint operator of T. Then $\mathcal{B}(H)$ is a C^*-algebra since $\|T^*T\| = \|T\|^2$ holds for all $T \in \mathcal{B}(H)$. However, $\mathcal{B}(H)$ is not commutative whenever $\dim H \ge 2$. $\mathcal{K}(H)$, the closed ideal consisting of all compact operators in H, is a C^*-subalgebra of $\mathcal{B}(H)$ because T^* is compact whenever T is.

(3) Suppose $T \in \mathcal{B}(H)$ is normal, that is, $T^*T = TT^*$, and let $A(T)$ denote the smallest closed subalgebra of $\mathcal{B}(H)$ containing T, T^* and the identity operator of H. Then $A(T)$ is a commutative C^*-algebra with identity.

(4) The Gelfand–Naimark theorem [39] states that for every C^*-algebra A there exists a Hilbert space H such that A is isometrically $*$-isomorphic to some C^*-subalgebra of $\mathcal{B}(H)$.

(5) Let G be a locally compact Abelian group. Then $L^1(G)$ is a commutative Banach $*$-algebra. However, whenever $G \ne \{e\}$, the L^1-norm fails to be a C^*-norm. In fact, it is not difficult to construct $f \in L^1(G)$ such that

$$\|f^* * f\|_1 \ne \|f\|_1^2$$

(Exercise 2.12.25).

(6) The assignment $f \to f^*$, where $f^*(z) = \overline{f(\bar{z})}$, defines an involution on the disc algebra $A(\mathbb{D})$ (Example 1.1.7(2)). However, $A(\mathbb{D})$ fails to be a C^*-algebra (Exercise 1.6.15).

If A is a $*$-algebra, then so is A_e once we define

$$(a + \lambda e)^* = a^* + \bar{\lambda} e, \quad a \in A, \quad \lambda \in \mathbb{C}.$$

Then A_e is a normed $*$-algebra with $\|a + \lambda e\| = \|a\| + |\lambda|$, yet in general not a C^*-algebra if A is. The following lemma, where we do not assume A to be commutative, shows that nevertheless a different norm can be introduced on A_e which extends the norm on A and turns A_e into a C^*-algebra.

Lemma 2.4.3. *Let A be a C^*-algebra without identity. There exists a norm $\| \cdot \|_0$ on A_e such that $\|a\|_0 = \|a\|$ for all $a \in A$ and $(A_e, \| \cdot \|_0)$ becomes a C^*-algebra.*

Proof. Let $\| \cdot \|$ denote the above norm on A_e; that is,

$$\|a + \lambda e\| = \|a\| + |\lambda|, \quad a \in A, \quad \lambda \in \mathbb{C}.$$

For $x \in A_e$, let $L_x : A \to A$ be defined by $L_x(a) = xa$, $a \in A$. Then

$$\|L_x a\| \leq \|x\| \cdot \|a\|,$$

so that L_x is bounded and $\|L_x\| \leq \|x\|$.

We claim that $\|x\|_0 = \|L_x\|$ defines a C^*-norm on A_e extending the given norm on A. Note first that, for $a \in A$,

$$\|L_a(a^*)\| = \|aa^*\| = \|a\|^2 = \|a\| \cdot \|a^*\|$$

and hence $\|L_a\| \geq \|a\|$ and therefore $\|L_a\| = \|a\|$. Now, $x \to \|x\|_0$ is a norm on A_e as soon as we have seen that $L_x = 0$ implies $x = 0$. To this end let

$$x = b + \lambda e, \quad b \in A, \quad \lambda \in \mathbb{C},$$

be such that $xa = 0$ for all $a \in A$. If $\lambda \neq 0$, then $a = (-(1/\lambda)b)a$ for all $a \in A$, that is, $u = -(1/\lambda)b$ is a left identity for A. Since

$$u^* = uu^* = (uu^*)^* = (u^*)^* = u,$$

and hence, for all $a \in A$,

$$au = au^* = (ua^*)^* = (a^*)^* = a,$$

u is also a right identity for A. This contradiction yields $x = b \in A$, and therefore $x = 0$ as $\|b\| = \|L_b\|$. Moreover, $\| \cdot \|_0$ is an algebra norm since

$$\|xy\|_0 = \|L_{xy}\| = \|L_x \circ L_y\| \le \|L_x\|\|L_y\| = \|x\|_0\|y\|_0,$$

and A_e is complete because A is complete and A_e/A is one-dimensional. Finally, $\|\cdot\|_0$ is a C^*-norm on A_e. Indeed, from

$$\begin{aligned}\|L_x(a)\|^2 &= \|xa\|^2 = \|(xa)^*(xa)\| \\ &= \|a^*(x^*x)a\| \le \|a^*\| \cdot \|L_{x^*x}a\| \\ &\le \|a\|^2\|L_{x^*x}\|\end{aligned}$$

it follows that

$$\|x\|_0^2 = \|L_x\|^2 \le \|L_{x^*x}\| = \|x^*x\|_0 \le \|L_{x^*}\|\|L_x\| = \|x^*\|_0\|x\|_0,$$

and this in turn gives

$$\|x\|_0 \le \|x^*\|_0 \quad \text{and} \quad \|x^*\|_0 \le \|x^{**}\|_0 = \|x\|_0.$$

Thus $\|x^*\|_0 = \|x\|_0$, and $\|x^*x\|_0 \le \|x^*\|_0\|x\|_0 = \|x\|_0^2$. $\qquad\square$

Lemma 2.4.4. *Let A be a commutative C^*-algebra. Then the Gelfand homomorphism is a $*$-homomorphism; that is, $\widehat{x^*} = \overline{\widehat{x}}$ for all $x \in A$.*

Proof. We have to show that $\varphi(x^*) = \overline{\varphi(x)}$ for $\varphi \in \Delta(A)$ and $x \in A$. Of course, we can assume that A has an identity e. Let

$$\varphi(x) = \alpha + i\beta \quad \text{and} \quad \varphi(x^*) = \gamma + i\delta,$$

$\alpha, \beta, \gamma, \delta \in \mathbb{R}$. Towards a contradiction, assume that $\beta + \delta \ne 0$ and let

$$y = (\beta + \delta)^{-1}(x + x^* - (\alpha + \gamma)e) \in A.$$

Then $y = y^*$ and

$$\varphi(y) = (\beta + \delta)^{-1}(\alpha + i\beta + \gamma + i\delta - (\alpha + \gamma)) = i.$$

This implies that, for all $t \in \mathbb{R}$,

$$\varphi(y + tie) = \varphi(y) + ti = (t + 1)i,$$

and hence $|t + 1| \le \|y + tie\|$. Since $y = y^*$, the C^*-norm property gives

$$\begin{aligned}(t+1)^2 &\le \|y + tie\|^2 = \|(y + tie)(y + tie)^*\| \\ &= \|(y + tie)(y - tie)\| = \|y^2 + t^2 e\| \\ &\le \|y^2\| + t^2.\end{aligned}$$

However, this inequality cannot hold for large t. This shows that $\delta = -\beta$ and therefore

$$\varphi((ix)^*) = \varphi(-ix^*) = -i\varphi(x^*) = -i(\gamma + i\delta) = -\beta - i\gamma.$$

On the other hand $\varphi(ix) = i(\alpha + i\beta) = -\beta + i\alpha$. Applying what we have seen so far with ix in place of x, we obtain $\gamma = \alpha$ and hence $\varphi(x^*) = \overline{\varphi(x)}$. $\qquad\square$

We are now ready to prove the first main result of this section.

Theorem 2.4.5. *For a commutative C^*-algebra A the Gelfand homomorphism is an isometric $*$-isomorphism from A onto $C_0(\Delta(A))$.*

Proof. To prove that $x \to \widehat{x}$ is isometric, note first that if $y = y^* \in A$, then $\|y\|^2 = \|y^*y\| = \|y^2\|$ and hence by induction $\|y\|^{2^n} = \|y^{2^n}\|$ for all $n \in \mathbb{N}$, so that

$$r_A(y) = \lim_{n \to \infty} \|y^{2^n}\|^{1/2^n} = \|y\|.$$

If now $x \in A$ is arbitrary, then by what we have just seen,

$$r_A(x^*x) = \|x^*x\| = \|x\|^2.$$

Recalling that $\widehat{x^*} = \overline{\widehat{x}}$ (Lemma 2.4.4) and $\sigma_A(x) \setminus \{0\} \subseteq \widehat{x}(\Delta(A)) \subseteq \sigma_A(x)$ (Theorem 2.2.5) we conclude that

$$\|\widehat{x}\|_\infty^2 = \|\overline{\widehat{x}}\,\widehat{x}\|_\infty = \|(x^*x)^\wedge\|_\infty = r_A(x^*x) = \|x\|^2.$$

Thus $x \to \widehat{x}$ is isometric and, in particular, the image $\widehat{A}$ of A is complete with respect to the supremum norm and hence closed in $C_0(\Delta(A))$. On the other hand, $\widehat{A}$ is a $*$-subalgebra of $C_0(\Delta(A))$ which strongly separates the points of $\Delta(A)$ (Theorem 2.2.7). Thus $\widehat{A}$ is dense in $C_0(\Delta(A))$ by the Stone–Weierstrass theorem. This proves that $\widehat{A} = C_0(\Delta(A))$.   □

The preceding theorem, together with the following corollary, sets up a bijection between the homeomorphism classes of locally compact Hausdorff spaces and the isomorphism classes of commutative C^*-algebras.

Corollary 2.4.6. *For two commutative C^*- algebras A and B the following are equivalent.*

(i) *$\Delta(A)$ and $\Delta(B)$ are homeomorphic.*
(ii) *There exists an isometric $*$-isomorphism between A and B.*
(iii) *There exists an algebra isomorphism between A and B.*

Proof. The implication (ii) $\Rightarrow$ (iii) is trivial and, as we have seen earlier (Lemma 2.2.10), the implication (iii) $\Rightarrow$ (i) holds even for general commutative Banach algebras A and B. To prove (i) $\Rightarrow$ (ii), note first that if $\phi : \Delta(A) \to \Delta(B)$ is a homeomorphism, then $f \to f \circ \phi$ is an isometric isomorphism from $C_0(\Delta(B))$ onto $C_0(\Delta(A))$ satisfying $\overline{f} \to \overline{f \circ \phi}$. On the other hand, by Theorem 2.4.5, A and B are isometrically $*$-isomorphic to $C_0(\Delta(A))$ and $C_0(\Delta(B))$, respectively. It follows that A and B are isometrically $*$-isomorphic.   □

Corollary 2.4.7. *Let A be a commutative C^*-algebra. For $x \in A$ consider the following conditions.*

(i) $x = x^*$.

(ii) $\sigma_A(x) \subseteq \mathbb{R}$.

(iii) $\hat{x}$ is real-valued.

(iv) $x = y^*y$ for some $y \in A$.

(v) $\sigma_A(x) \subseteq [0, \infty)$.

(vi) $\hat{x} \geq 0$.

Then (i), (ii), and (iii) are equivalent, and so are (iv), (v), and (vi).

Proof. The equivalence of (ii) and (iii) and of (v) and (vi) follows immediately from

$$\hat{x}(\Delta(A)) \cup \{0\} = \sigma_A(x) \cup \{0\}.$$

The Gelfand homomorphism is injective and satisfies $\widehat{x^*} = \overline{\hat{x}}$. Therefore (i) and (iii) are equivalent. If (iv) holds, then $\hat{x} = \widehat{y^*y} = \overline{\hat{y}}\hat{y} \geq 0$. Conversely, if $\hat{x} \geq 0$, let $f \in C_0(\Delta(A))$ be the positive square root of $\hat{x}$. The Gelfand homomorphism being surjective, there exists $y \in A$ such that $\hat{y} = f$. Now y satisfies $\widehat{y^*y} = \hat{x}$ and hence $y^*y = x$.   □

In the sequel we present two applications of Theorem 2.4.5. The first one (Theorem 2.4.9) is the construction of a functional calculus in which continuous functions act on elements of a commutative C^*-algebra, and the second (Theorem 2.4.12) concerns the existence of a Stone–Čech compactification for a completely regular topological space.

We know that in general the spectrum of an element in a Banach algebra may become larger upon passing to a subalgebra. We need that for C^*-algebras this cannot happen as we observe next.

Lemma 2.4.8. *Let A be a commutative C^*-algebra with identity e and B a C^*-subalgebra of A containing e. Then $\sigma_A(x) = \sigma_B(x)$ for each $x \in B$.*

Proof. It suffices to show that if $y \in B$ is invertible in A, then y is already invertible in B. Let $y \in B \cap G(A)$ and note first that $y^* \in G(A)$ since

$$(y^{-1})^*y^* = (yy^{-1})^* = e^* = ee^* = (e^*e)^* = e^{**} = e.$$

Thus $yy^* \in G(A)$ and, by Theorem 2.2.5, $\widehat{yy^*}(\Delta(A)) = \sigma_A(yy^*)$. On the other hand, by Lemma 2.4.4,

$$\widehat{yy^*}(\Delta(A)) = \{\varphi(yy^*) : \varphi \in \Delta(A)\} = \{|\varphi(y)|^2 : \varphi \in \Delta(A)\}.$$

Hence $\sigma_A(yy^*) \subseteq [0, \infty)$, so that $\rho_A(yy^*) = \mathbb{C} \backslash \sigma_A(yy^*)$ is connected. Theorem 1.2.12 now yields

$$\sigma_B(yy^*) = \sigma_A(yy^*).$$

Therefore, yy^* is invertible in B and hence so is y.   □

Theorem 2.4.9. *Let A be a commutative C^*-algebra with identity e and $x \in A$. Let $A(x)$ denote the smallest C^*-subalgebra of A containing x and e. There exists a unique isometric $*$-isomorphism*

$$\phi : C(\sigma_A(x)) \to A(x), \quad f \to f(x)$$

with the property that ϕ maps the constant function 1 onto e and the function $\lambda \to \lambda$ onto x.

Proof. Because $\sigma_A(x) = \sigma_{A(x)}(x)$ by Lemma 2.4.8, we can assume that $A = A(x)$. This means that the set of all polynomials in x, x^*, and e is dense in A. Let $f \in C(\sigma_A(x))$ denote the function $f(\lambda) = \lambda$, and suppose that ϕ_1 and ϕ_2 are isometric $*$-isomorphisms from $C(\sigma_A(x))$ onto $A = A(x)$ with $\phi_j(1_{\sigma_A(x)}) = e$ and $\phi_j(f) = x$. Then

$$\phi_j^{-1}(x^*) = \overline{\phi_j^{-1}(x)} = \bar{f}, \quad j = 1, 2,$$

so that ϕ_1^{-1} and ϕ_2^{-1} coincide on all polynomials in x, x^* and e. Since $A = A(x)$ and ϕ_1^{-1} and ϕ_2^{-1} are continuous, we conclude that $\phi_1^{-1} = \phi_2^{-1}$.

To prove the existence of ϕ, we show first that $\varphi \to \varphi(x)$ defines a homeomorphism between $\Delta(A)$ and $\sigma_A(x)$. Every $\varphi \in \Delta(A)$ is determined by its value at x since φ is continuous, A is generated by x, x^*, and e, and $\varphi(x^*) = \overline{\varphi(x)}$ and $\varphi(e) = 1$. Thus $\varphi \to \varphi(x)$ is injective. On the other hand, $\widehat{x}(\Delta(A)) = \sigma_A(x)$ by Theorem 2.2.5. Clearly, the map $\varphi \to \varphi(x)$ from $\Delta(A)$ onto $\sigma_A(x)$ is continuous, and hence it is a homeomorphism since $\Delta(A)$ is compact and $\sigma_A(x)$ is a Hausdorff space. Let ψ denote the associated isometric $*$-isomorphism between $C(\sigma_A(x))$ and $C(\Delta(A))$; that is,

$$\psi(g)(\varphi) = g(\varphi(x)), \quad g \in C(\sigma_A(x)), \quad \varphi \in \Delta(A).$$

By Theorem 2.4.5 the Gelfand homomorphism $y \to \widehat{y}$ is an isometric $*$-isomorphism from $A = A(x)$ onto $C(\Delta(A))$. Composing its inverse with ψ, we obtain an isometric $*$-isomorphism $\phi : C(\sigma_A(x)) \to A = A(x)$ given by

$$\phi(g) = y \text{ if and only if } \widehat{y}(\varphi) = g(\varphi(x)) \text{ for all } \varphi \in \Delta(A).$$

Then ϕ has the required properties since $1_{\sigma_A(x)}(\varphi(x)) = 1 = \widehat{e}(\varphi)$ and $f(\varphi(x)) = \varphi(x) = \widehat{x}(\varphi)$ for all $\varphi \in \Delta(A)$. $\qquad\square$

Remark 2.4.10. Returning to Example 2.4.2, let T be a normal operator in a Hilbert space H and $A(T)$ the closed subalgebra of $\mathcal{B}(H)$ generated by T, T^* and the identity operator I on H. According to the preceding theorem, there is a unique isometric $*$-isomorphism from $C(\sigma(T))$ onto $A(T)$ which maps the function $f(\lambda) = \lambda$ to T and the constant one function to I. This result can be used to derive the spectral theorem for normal operators in Hilbert spaces. Because of this, Theorem 2.4.9 is often referred to as the *abstract spectral theorem*.

For the second application of Theorem 2.4.5 mentioned above we first recall some notions from topology.

Definition 2.4.11. Let X be a Hausdorff space. A pair (Y, β), consisting of a compact Hausdorff space Y and a mapping $\beta : X \to Y$, is called a *Stone–Čech compactification* of X, if the following conditions are satisfied.

(i) $\beta(X)$ is dense in Y, and $\beta : X \to \beta(X)$ is a homeomorphism.
(ii) Every $f \in C^b(X)$ extends continuously to Y in the sense that there exists $\widetilde{f} \in C(Y)$ such that $\widetilde{f}(\beta(x)) = f(x)$ for all $x \in X$.

Of course, $\widetilde{f}$ is then uniquely determined since $\beta(X)$ is dense in Y.

Suppose now that X possesses a Stone–Čech compactification (Y, β). Then given a closed subset E of X and $x \in X \setminus E$, there exists $f \in C^b(X)$ such that $f|_E = 0$ and $f(x) \neq 0$. In fact, if C is a closed subset of Y with $C \cap \beta(X) = \beta(E)$, then $\beta(x) \notin C$, and hence by Urysohn's lemma we find $g \in C(Y)$ such that $g(\beta(x)) \neq 0$ and $g|_C = 0$. Now, $f = g \circ \beta \in C^b(X)$ has the desired properties. A Hausdorff space X for which $C^b(X)$ shares this separation property is called *complelety regular*.

Stone and Čech proved that every completely regular space admits a Stone–Čech compactification, which is uniquely determined up to homeomorphisms. We conclude this section by showing that the existence of a Stone–Čech compactification can be obtained as an application of Gelfand's theory.

Theorem 2.4.12. *Let X be a completely regular topological space. Let $Y = \Delta(C^b(X))$ and define $\beta : X \to Y$ by $\beta(x) = \varphi_x$, where φ_x denotes the evaluation of functions in $C(Y)$ at x. Then (Y, β) is a Stone–Čech compactification of X.*

Proof. $C^b(X)$ is a commutative C^*-algebra with identity. Therefore, $Y = \Delta(C^b(X))$ is compact, and the Gelfand homomorphism $f \to \widehat{f}$ is an isometric $*$-isomorphism from $C^b(X)$ onto $C(Y)$. The map

$$\beta : X \to Y, \quad x \to \varphi_x$$

is one-to-one because given distinct points x_1 and x_2 in X, the complete regularity of X guarantees the existence of some $f \in C^b(X)$ with

$$\varphi_{x_1}(f) = f(x_1) \neq f(x_2) = \varphi_{x_2}(f).$$

Condition (ii) of Definition 2.4.11 is satisfied with $\widetilde{f} = \widehat{f}$ since, by definition of β,

$$\widehat{f}(\beta(x)) = \widehat{f}(\varphi_x) = \varphi_x(f) = f(x)$$

for all $f \in C^b(X)$ and $x \in X$.

To verify that $\beta : X \to \beta(X)$ is a homeomorphism, for $x_0 \in X, \epsilon > 0$, and $f_1, \ldots, f_n \in C^b(X)$ consider the sets

$$V = \{x \in X : |f_i(x) - f_i(x_0)| < \epsilon, 1 \leq i \leq n\} \subseteq X$$

and

$$U = \{\varphi_x \in \beta(X) : |\varphi_x(f_i) - \varphi_{x_0}(f_i)| < \epsilon, 1 \leq i \leq n\} \subseteq \beta(X).$$

Then $V = \beta^{-1}(U)$ and V is open in X. These sets U form an open basis for the relative topology on $\beta(X) \subseteq Y = \Delta(C^b(X))$. Hence β is continuous, and for β to be open it suffices to show that such sets V form a basis for the topology on X. For that, let W be an open subset of X containing x_0. Then, since X is completely regular, there exists $f \in C^b(X)$ such that $f(x_0) \neq 0$ and $f|_{X\setminus W} = 0$. It follows that

$$x_0 \in \{x \in X : |f(x) - f(x_0)| < |f(x_0)|\} \subseteq W.$$

To complete the proof of the theorem it remains to show that $\beta(X)$ is dense in Y. Assuming the contrary, there exists $g \in C(Y)$ such that $g \neq 0$, but $g|_{\beta(X)} = 0$. The Gelfand homomorphism maps $C^b(X)$ onto $C(Y)$. Thus we find $f \in C^b(X)$ such that $\widehat{f} = g$. Then

$$0 = g(\varphi_x) = \widehat{f}(\varphi_x) = \varphi_x(f) = f(x)$$

for all $x \in X$. However, $f = 0$ implies $g = 0$. This contradiction shows that $\beta(X)$ is dense in Y. $\qquad\qquad\square$

2.5 The uniform algebras $P(X)$ and $R(X)$

The next two sections centre around elaborating the Gelfand representation of certain algebras of continuous functions on compact spaces.

Definition 2.5.1. Let X be a compact Hausdorff space. A closed subalgebra A of $C(X)$, equipped with the $\|\cdot\|_\infty$-norm, is called a *uniform algebra* if A separates the points of X and contains the constant functions.

In Example 1.1.2 we have already introduced, for X a compact subset of $\mathbb{C}$, the uniform algebras $P(X), R(X)$, and $A(X)$. The definitions in the more general case of a compact subset of $\mathbb{C}^n$ are analogous. Instead of polynomials, rational functions, and holomorphic functions in one variable we simply have to take such functions in n complex variables.

Remark 2.5.2. If A is a uniform algebra on X then, because X is compact and $\Delta(A)$ is a Hausdorff space, the mapping $\phi : x \to \varphi_x$, where $\varphi_x(f) = f(x)$ for $f \in A$, is a homeomorphism of X onto its range $\phi(X) \subseteq \Delta(A)$. In general, however, $\phi(X)$ is a proper subset of $\Delta(A)$.

Our goal is to determine the structure spaces of $P(X)$, $R(X)$, and $A(X)$. In this section we treat $P(X)$ and $R(X)$ for $X \subseteq \mathbb{C}^n$ and in the next section $A(X)$ for $X \subseteq \mathbb{C}$. Moreover, we study the problem of when equality holds for any of the inclusions $P(X) \subseteq R(X)$ and $R(X) \subseteq A(X)$.

Example 2.5.3. Let $\mathbb{D} = \{z \in \mathbb{C} : |z| \leq 1\}$ and $\mathbb{T} = \{z \in \mathbb{C} : |z| = 1\}$, the boundary of $\mathbb{D}$.

(1) The algebra $P(\mathbb{D})$ is generated by the function $f(z) = z, z \in \mathbb{D}$. Now, $\sigma_{P(\mathbb{D})}(f) = \mathbb{D}$. In fact, if $|\lambda| > 1$, then the function

$$z \longrightarrow \frac{1}{\lambda - f(z)} = \frac{1}{\lambda} \frac{1}{1 - \frac{z}{\lambda}} = \frac{1}{\lambda} \sum_{n=0}^{\infty} \left(\frac{z}{\lambda}\right)^n$$

is a uniform limit of polynomials on $\mathbb{D}$, and hence the function $\lambda - f$ is invertible in $P(\mathbb{D})$. Thus, by Lemma 2.3.3, the mapping $z \to \varphi_z$, where $\varphi_z(g) = g(z)$ for $g \in P(\mathbb{D})$, is a homeomorphism between $\mathbb{D}$ and the structure space of $P(\mathbb{D})$.

By the maximum modulus principle, the mapping $r : g \to g|_{\mathbb{T}}$ is an isometric isomorphism from $P(\mathbb{D})$ onto $P(\mathbb{T})$. It follows that $\Delta(P(\mathbb{T})) = \mathbb{D}$ via the mapping $z \to \varphi_z, \varphi_z(h) = r^{-1}(h)(z)$ for $h \in P(\mathbb{T})$.

(2) We claim that $P(\mathbb{D}) = A(\mathbb{D}) \neq C(\mathbb{D})$. Since the function $z \to \bar{z}$ fails to be holomorphic, $A(\mathbb{D}) \neq C(\mathbb{D})$. To show that $P(\mathbb{D}) = A(\mathbb{D})$, let $f \in A(\mathbb{D})$ and for $0 < t < 1$, define f_t by $f_t(z) = f(tz)$. Then f_t is a holomorphic function on $\{z \in \mathbb{C} : |z| < 1/t\}$, and $f_t \to f$ uniformly on $\mathbb{D}$ as $t \to 1$ because f is uniformly continuous on $\mathbb{D}$. Finally, f_t admits a power series representation and hence can be uniformly approximated by polynomials on $\mathbb{D}$. Thus f is a uniform limit of polynomials on $\mathbb{D}$, as required.

Definition 2.5.4. Let X be a compact subset of $\mathbb{C}^n$. The *polynomially convex hull*, $\widehat{X}_p$, of X is the set

$$\widehat{X}_p = \{z \in \mathbb{C}^n : |p(z)| \leq \|p|_X\|_\infty \text{ for all polynomials } p\}.$$

Then, by Definition 2.3.4, X is polynomially convex if and only if $X = \widehat{X}_p$. The *rational convex hull* $\widehat{X}_r$ of X is the set of all $z \in \mathbb{C}^n$ such that

$$|p(z)| \leq |q(z)| \cdot \left\|\frac{p}{q}|_X\right\|_\infty$$

for all polynomials p and q with $q \neq 0$ on X. Finally, X is said to be *rationally convex* if $\widehat{X}_r = X$.

We continue with some simple observations concerning $\widehat{X}_p$ and $\widehat{X}_r$.

Remark 2.5.5. (1) Clearly, $X \subseteq \widehat{X}_r \subseteq \widehat{X}_p$. In particular, if X is polynomially convex, then it is rationally convex.

(2) Each compact subset of $\mathbb{C}$ is rationally convex. Indeed, if $z_0 \in \mathbb{C} \setminus X$, then $q(z) = z - z_0$ satisfies $1 > 0 = |q(z_0)| \cdot \|(1/q)|_X\|_\infty$. On the other hand,

recall that X is polynomially convex if and only if $\mathbb{C}\backslash X$ is connected (Theorem 2.3.7).

(3) Both $\widehat{X}_p$ and $\widehat{X}_r$ are compact. To verify this, since these sets are closed and $\widehat{X}_r \subseteq \widehat{X}_p$, it is enough to show that $\widehat{X}_p$ is bounded. Now, with $p_j(z) = z_j, 1 \leq j \leq n$, for every $z \in \widehat{X}_p$,

$$\|z\|^2 = \sum_{j=1}^{n} |p_j(z)|^2 \leq \sum_{j=1}^{n} \|p_j|x\|_\infty^2.$$

Lemma 2.5.6. *For any compact subset X of $\mathbb{C}^n$,*

$$\widehat{X}_r = \{z \in \mathbb{C}^n : p(z) \in p(X) \text{ for every polynomial } p\}.$$

Proof. Let $z \in \mathbb{C}^n$ and suppose that there is a polynomial p such that $p(z) \notin p(X)$. Then $q(w) = p(w) - p(z)$ is non-zero on X and

$$1 > 0 = |q(z)| \cdot \left\| \frac{1}{q} |x \right\|_\infty,$$

so that $z \notin \widehat{X}_r$. Conversely, if $z \notin \widehat{X}_r$, then there are polynomials p and q, with $q \neq 0$ on X, such that

$$|p(z)| > |q(z)| \cdot \left\| \frac{p}{q} |x \right\|_\infty.$$

In particular, $p(z) \neq 0$. If $q(z) = 0$, we are done since $0 \notin q(X)$. Otherwise, replacing p by $g = q(z)p(z)^{-1}p$, we get that $g(z) = q(z)$ and

$$\left\| \frac{g}{q} |x \right\|_\infty = \left| \frac{q(z)}{p(z)} \right| \cdot \left\| \frac{p}{q} |x \right\|_\infty < 1.$$

Then the polynomial $f = q - g$ satisfies $f(z) = 0$ and $0 \notin f(X)$, for if $x \in X$ and $f(x) = 0$, then $(g/q)(x) = 1$ contradicting $\|(g/q)|x\|_\infty < 1$.   □

We can now work out the Gelfand representation of $P(X)$ and $R(X)$.

Theorem 2.5.7. *Let X be a compact subset of $\mathbb{C}^n$.*

(i) *The restriction map $\phi : f \to f|_X$ is an isometric isomorphism from $P(\widehat{X}_p)$ onto $P(X)$. Moreover, for $x \in \widehat{X}_p$, define $\varphi_x : P(X) \to \mathbb{C}$ by*

$$\varphi_x(f) = \phi^{-1}(f)(x), \quad f \in P(X).$$

Then $x \to \varphi_x$ is a homeomorphism from $\widehat{X}_p$ onto $\Delta(P(X))$.

(ii) *The map $\phi : f \to f|_X$ is an isometric isomorphism from $R(\widehat{X}_r)$ onto $R(X)$, and $x \to \varphi_x$, where*

$$\varphi_x(f) = \phi^{-1}(f)(x), \quad f \in R(X), \quad x \in \widehat{X}_r,$$

is a homeomorphism between $\widehat{X}_r$ and $\Delta(R(X))$.

Proof. (i) The map $q|_{\widehat{X}_p} \to q|_X$ takes the dense subalgebra of $P(\widehat{X}_p)$ consisting of all polynomial functions on $\widehat{X}_p$ homomorphically onto the corresponding subalgebra of $P(X)$. This map preserves the norm since

$$|q(z)| \leq \|q|_X\|_\infty$$

for all polynomials q and all $z \in \widehat{X}_p$. It follows that ϕ is an isometric isomorphism from $P(\widehat{X}_p)$ onto $P(X)$. For each $x \in \widehat{X}_p, \varphi_x$ as defined above belongs to $\Delta(P(X))$, and the mapping $x \to \varphi_x, \widehat{X}_p \to \Delta(P(X))$ is injective. It is continuous since

$$x \to \varphi_x(f) = \phi^{-1}(f)(x)$$

is a continuous function on $\widehat{X}_p$ for every $f \in P(X)$. Hence $x \to \varphi_x$ maps $\widehat{X}_p$ homeomorphically onto its image in $\Delta(P(X))$. It remains to show that given $\varphi \in \Delta(P(X))$, there exists $x \in \widehat{X}_p$ such that $\varphi = \varphi_x$. To that end, let $x_j = \varphi(p_j|_X)$, where $p_j(z) = z_j, 1 \leq j \leq n$. We claim that $x = (x_1, \ldots, x_n) \in \widehat{X}_p$ and $\varphi = \varphi_x$. For any polynomial q,

$$q(x) = q(\varphi(p_1|_X), \ldots, \varphi(p_n|_X)) = \varphi(q|_X),$$

and hence $|q(x)| \leq \|q|_X\|_\infty$. This proves $x \in \widehat{X}_p$, and $\varphi = \varphi_x$ follows from

$$\varphi_x(p_j|_X) = p_j(x) = x_j = \varphi(p_j|_X),$$

$1 \leq j \leq n$, since the functions $p_j|_X$ generate $P(X)$.

   (ii) is proved in very much the same way as (i). Note first that if $f = (p/q)|_{\widehat{X}_r}$, where p and q are polynomials with $q \neq 0$ on $\widehat{X}_r$, then $\|f\|_\infty = \|f|_X\|_\infty$ since for each $x \in \widehat{X}_r$,

$$|p(x)| \leq |q(x)| \cdot \left\|\frac{p}{q}|_X\right\|_\infty.$$

Consequently, $f \to f|_X$ maps the dense subalgebra of rational functions in $R(\widehat{X}_r)$ homomorphically and isometrically onto a dense subalgebra of $R(X)$. This yields the first statement in (ii).

   Clearly, for each $x \in \widehat{X}_r$,

$$\varphi_x(f) = \phi^{-1}(f)(x), \quad f \in R(X),$$

defines an element of $\Delta(R(X))$, and the mapping $x \to \varphi_x, \widehat{X}_r \to \Delta(R(X))$ is injective and continuous. What is left to be shown is that every $\varphi \in \Delta(R(X))$ is of the form $\varphi = \varphi_x$ for some $x \in \widehat{X}_r$. Given φ, as in (i) define $x = (x_1, \ldots, x_n) \in \mathbb{C}^n$ by

$$x_j = \varphi(p_j|_X), \quad 1 \leq j \leq n.$$

Now, for every polynomial q,

$$q(x) = q(\varphi(p_1|x), \ldots, \varphi(p_n|x)) = \varphi(q|x) = \widehat{q|x}(\varphi) \in \sigma_{R(X)}(q|x) = q(X).$$

According to Lemma 2.5.6 this shows that $x \in \widehat{X}_r$. Finally, $\varphi_x(p_j|x) = x_j = \varphi(p_j|x)$ implies $\varphi_x(p|x) = \varphi(p|x)$ and hence

$$\varphi_x\left(\frac{p}{q}\Big|X\right) = \varphi_x(p|x)\varphi_x(q|x)^{-1} = \varphi(p|x)\varphi(q|x)^{-1} = \varphi\left(\frac{p}{q}\Big|X\right)$$

for all polynomials p and q with $q \neq 0$ on X. It follows that $\varphi = \varphi_x$.   □

In the proof of part (i) of Theorem 2.5.7, for surjectivity of the map $x \to \varphi_x$ from $\widehat{X}_p$ to $\Delta(P(X))$ we could alternatively have appealed to the proof of Theorem 2.3.6. We now obtain the following approximation result.

Theorem 2.5.8. *If X is a compact subset of $\mathbb{C}^n$, then $P(X) = R(X)$ if and only if $\widehat{X}_p = \widehat{X}_r$. In particular, for a compact subset X of $\mathbb{C}$, $P(X) = R(X)$ if and only if $\mathbb{C} \setminus X$ is connected.*

Proof. Suppose first that $P(X) = R(X)$, and let $x \in \widehat{X}_p$. Then the function $\varphi_x : f \to (\phi^{-1}f)(x)$, where ϕ is as in part (i) of Theorem 2.5.7, defines an element of $\Delta(P(X)) = \Delta(R(X))$. By Theorem 2.5.7(ii), $\varphi_x = \varphi_y$ for some $y \in \widehat{X}_r$. It follows that

$$q(y) = \varphi_y(q|x) = \varphi_x(q|x) = q(x)$$

for all polynomials q, so that $x = y$. This shows $\widehat{X}_p \subseteq \widehat{X}_r$ and hence $\widehat{X}_p = \widehat{X}_r$ (Remark 2.5.2).

Conversely, let $\widehat{X}_p = \widehat{X}_r$. To prove $R(X) \subseteq P(X)$ it suffices to show that if q is a polynomial such that $q(z) \neq 0$ for all $z \in X$, then $q|x$ is invertible in $P(X)$. Now, by Lemma 2.5.6, q has no zero on $\widehat{X}_r$. Since $\widehat{X}_r = \widehat{X}_p = \Delta(P(X))$, this implies $\varphi(q|x) \neq 0$ for every $\varphi \in \Delta(P(X))$. Therefore $q|x$ is contained in no maximal ideal of $P(X)$, and therefore is invertible in $P(X)$.

Finally, suppose that $n = 1$. If $\mathbb{C} \setminus X$ is connected, then X is polynomially convex (Theorem 2.3.7) and hence $\widehat{X}_r = \widehat{X}_p$. Conversely, if $\widehat{X}_p = \widehat{X}_r$ then, because every compact subset of $\mathbb{C}$ is rationally convex (Remark 2.5.5), X is polynomially convex and hence $\mathbb{C} \setminus X$ is connected.   □

Next we show an interesting result about generation of $R(X)$.

Theorem 2.5.9. *If X is a compact subset of $\mathbb{C}^n$, then $R(X)$ is generated by $n + 1$ elements.*

Proof. The set of $n + 1$ generators we produce consists of the coordinate functions $p_j(z) = z_j, z \in X, 1 \leq j \leq n$, and an additional function f which has to be constructed. Notice first that since $P(X)$ contains a countable dense subset, there exists a sequence of polynomials $q_m, m \in \mathbb{N}$, such that $q_m \neq 0$ on X and the set

$$\left\{ \frac{p}{q_m}\Big|_X : m \in \mathbb{N}, p \text{ a polynomial} \right\}$$

is dense in $R(X)$. Let $g_m = q_m|_X$, and by induction define positive real numbers $c_m, m \in \mathbb{N}$, so that

$$c_m \left\| g_m^{-1} \right\|_\infty < 2^{-m} \quad \text{and} \quad c_m \left\| g_m^{-1} g_k \right\|_\infty < 2^{-m} c_k$$

for $1 \le k \le m - 1$. Then the series $\sum_{k=1}^{\infty} c_k g_k^{-1}(z)$ converges uniformly on X and hence defines an element f of $R(X)$. We claim that A, the unital closed subalgebra of $R(X)$ generated by f and all the $p_j, 1 \le j \le n$, coincides with $R(X)$.

The set of functions of the form $(pg_m^{-1})|_X$, $m \in \mathbb{N}$, p a polynomial, is dense in $R(X)$. Therefore it is enough to show that $g_m^{-1} \in A$ for every $m \in \mathbb{N}$. Let

$$f_m = \sum_{k=m}^{\infty} c_k g_k^{-1} \in R(X).$$

Next, observe that, for each $m \in \mathbb{N}$, $g_m^{-1} \in A$ provided that $f_m \in A$. Indeed, this can be seen as follows. If $f_m \in A$, then $f_m g_m \in A$ and, by the choice of c_k,

$$\left\| f_m g_m - c_m \right\|_\infty = \left\| \sum_{k=m+1}^{\infty} c_k g_m g_k^{-1} \right\|_\infty \le \sum_{k=m+1}^{\infty} c_k \left\| g_m g_k^{-1} \right\|_\infty$$

$$\le c_m \sum_{k=m+1}^{\infty} 2^{-k} < c_m.$$

Thus $f_m g_m$ is invertible in A, and hence so is g_m. It now follows by induction that $f_m \in A$ for all $m \in \mathbb{N}$. Indeed, $f_1 = f \in A$, and supposing that $f_1, \ldots, f_m \in A$, by the preceding paragraph, $g_1^{-1}, \ldots, g_m^{-1} \in A$. It follows that

$$f_{m+1} = f - \sum_{k=1}^{m} c_k g_k^{-1} \in A.$$

This finishes the proof of the theorem.   □

It is worth pointing out that we have not proved that $R(X)$ admits a system of $n + 1$ generators, each of which is a rational function. In fact, this strengthened version is false, as can already be seen in the plane: if X is a compact subset of $\mathbb{C}$ and $\mathbb{C} \backslash X$ has infinitely many connected components, then $R(X)$ cannot be generated by a finite family of rational functions (Exercise 2.12.41) even though it is doubly generated as a Banach algebra.

A nice geometric consequence of Theorem 2.5.9 and the previous results is the following

Corollary 2.5.10. *Every rationally convex compact subset of $\mathbb{C}^n$ is homeo-morphic to some polynomially convex subset of $\mathbb{C}^{n+1}$.*

Proof. If X is a compact subset of $\mathbb{C}^n$ and rationally convex, then $X = \Delta(R(X))$ by Theorem 2.5.7. On the other hand, $R(X)$ is generated by $n+1$ elements $f_1, \ldots, f_{n+1}$, and hence, by Lemma 2.3.3, $\Delta(R(X))$ is homeomorphic to the joint spectrum

$$\sigma_{R(X)}(f_1, \ldots, f_{n+1}) \subseteq \mathbb{C}^{n+1},$$

which is polynomially convex by Theorem 2.3.6. $\qquad\square$

We proceed by constructing a compact subset X of $\mathbb{C}$ with empty interior such that $R(X) \neq C(X)$. This example is usually called *Swiss cheese*, a label which becomes apparent from the construction.

Example 2.5.11. As before, let $\mathbb{D}$ denote the closed unit disc. We are going to show the existence of a sequence of closed discs $\Delta_j, j \in \mathbb{N}$, of radii $r_j > 0$ with the following properties.

(1) $\Delta_j \subseteq \mathbb{D}° = \{z \in \mathbb{C} : |z| < 1\}$ and $\Delta_j \cap \Delta_k = \emptyset$ for $j \neq k$.
(2) $\sum_{j=1}^{\infty} r_j < 1$.
(3) $\mathbb{D} \setminus \bigcup_{j=1}^{\infty} \Delta_j°$ has an empty interior.

Let $y_1, y_2, \ldots$ be a numbering of the countable set of complex numbers $\alpha + i\beta \in \mathbb{D}°$ with α, β rational. We construct by induction on n a sequence $(\Delta_n)_n$ of closed discs such that (1) holds for $1 \leq j \leq k \leq n$, $0 < r_j < 2^{-j}$ for $1 \leq j \leq n$ and

$$y_1, \ldots, y_n \in \bigcup_{j=1}^{n} \Delta_j = \overline{\bigcup_{j=1}^{n} \Delta_j°}.$$

For $y \in \mathbb{C}$ and $r > 0$, let $B(y, r)$ denote the closed disc of radius r around y. Choose $0 < r_1 < \frac{1}{2}$ such that $\Delta_1 = B(y_1, r_1) \subseteq \mathbb{D}°$. Suppose that $\Delta_1, \ldots, \Delta_n$ with the required properties have been found. Then $y_m \notin \bigcup_{j=1}^{n} \Delta_j$ for some $m \geq n + 1$. Indeed, otherwise $y_k \in \bigcup_{j=1}^{n} \Delta_j$ for all k and hence, because the set $\{y_k : k \in \mathbb{N}\}$ is dense in $\mathbb{D}°$, $\mathbb{D}° = \bigcup_{j=1}^{n} \Delta_j$, which is impossible. Let m be minimal such that $y_m \notin \bigcup_{j=1}^{n} \Delta_j$, and choose $0 < r_{n+1} < 2^{-(n+1)}$ such that $\Delta_{n+1} = B(y_m, r_{n+1})$ satisfies

$$\Delta_{n+1} \subseteq \mathbb{D}° \quad \text{and} \quad \Delta_{n+1} \cap \left(\bigcup_{j=1}^{n} \Delta_j \right) = \emptyset.$$

This finishes the inductive step. It is obvious that the sequence $(\Delta_j)_j$ has properties (1) and (2).

Now, let $X = \mathbb{D} \setminus \bigcup_{j=1}^{\infty} \Delta_j°$. Then X has empty interior because $y_n \in \overline{\bigcup_{j=1}^{n} \Delta_j°}$ for each n. So (3) holds also.

To prove that $R(X) \neq C(X)$, we construct a bounded linear functional l on $C(X)$ such that $l \neq 0$ and $l|_{R(X)} = 0$. Let z_j denote the centre of $\Delta_j, j \in \mathbb{N}$, and let $\Gamma_j, j \in \mathbb{N}$, be the curve defined by

$$\Gamma_j(t) = z_j + r_j e^{it}, \quad t \in [0, 2\pi].$$

Moreover, define Γ_0 by

$$\Gamma_0(t) = e^{-it}, \quad t \in [0, 2\pi].$$

For $f \in C(X)$, let

$$l(f) = \sum_{j=0}^{\infty} \int_{\Gamma_j} f(z) dz.$$

Note that since $\sum_{j=1}^{\infty} r_j < \infty$ and

$$\left| \int_{\Gamma_j} f(z) dz \right| = \left| r_j \int_0^{2\pi} f(z_j + r_j e^{it}) i e^{it} dt \right| \leq 2\pi r_j \|f\|_\infty,$$

the above series converges absolutely, and therefore l defines a bounded linear functional on $C(X)$.

Now, $\int_{\Gamma_0} \bar{z} dz = -2\pi i$ and

$$\int_{\Gamma_j} \bar{z} dz = i r_j \int_0^{2\pi} (\bar{z}_j + r_j e^{-it}) e^{it} dt = 2\pi i r_j^2$$

for $j \geq 1$. Thus, by property (2),

$$l(z \to \bar{z}) = 2\pi i \left(\sum_{j=1}^{\infty} r_j^2 - 1 \right) \neq 0.$$

It remains to show that $l|_{R(X)} = 0$.

To that end, let p and q be complex polynomials such that $q(z) \neq 0$ for all $z \in X$. Then $q \neq 0$ on some open neighbourhood V of X. Let $X_n = \mathbb{D} \setminus \bigcup_{j=1}^n \Delta_j^\circ$, so that $X_{n+1} \subseteq X_n$ for all n and $X = \bigcap_{n=1}^{\infty} X_n$. It follows that $X_n \subseteq V$ for all $n \geq n_0$ for some $n_0 \in \mathbb{N}$. We want to apply Cauchy's integral formula to the holomorphic function $f = (p/q)|_V$ and the closed curves $\Gamma_0, \Gamma_1, \ldots, \Gamma_n$ in V, $n \geq n_0$. For every point $z \notin \Gamma_j([0, 2\pi])$, let $w(\Gamma_j, z)$ denote the winding number of Γ_j with respect to z. If $z \in \mathbb{C} \setminus V$, then either $z \notin \mathbb{D}$ and hence $w(\Gamma_j, z) = 0$ for all $j \in \mathbb{N}_0$, or $z \in \mathbb{D}$. In the latter case, $z \in \bigcup_{j=1}^n \Delta_j^\circ$ because $\mathbb{D} \setminus \bigcup_{j=1}^n \Delta_j^\circ = X_n \subseteq V$ for $n \geq n_0$, and therefore $z \in \Delta_j^\circ$ for exactly one $j \in \mathbb{N}$. This implies that

$$\sum_{k=0}^n w(\Gamma_k, z) = w(\Gamma_j, z) + w(\Gamma_0, z) = 1 - 1 = 0.$$

Thus we have seen that $\Sigma_{k=0}^{n} w(\Gamma_k, z) = 0$ for all $z \in \mathbb{C} \setminus V$ and $n \geq n_0$. A version of Cauchy's integral formula (see [23, p. 206]) now yields that

$$\sum_{j=0}^{n} \int_{\Gamma_j} f(z)dz = 0$$

for all $n \geq n_0$ and hence $l(f|_X) = 0$. Since l is continuous, it follows that $l|_{R(X)} = 0$.

The next theorem holds more generally for compact subsets of $\mathbb{C}$ of Lebesgue measure zero and in this generality is referred to as the *Hartogs–Rosenthal theorem*.

Theorem 2.5.12. *Let X be a countable compact subset of $\mathbb{C}$. Then $P(X) = C(X)$.*

Proof. We first observe that $\mathbb{C} \setminus X$ is connected. To see this, let $z_1, z_2 \in \mathbb{C} \setminus X$. Since X is countable, there is a ray L emanating from z_1 which does not intersect X. For any point $z \in L$, let $\overline{z, z_2}$ denote the line segment connecting z and z_2. Again, because X is countable, one of them, say $\overline{z, z_2}$, misses X. So z_1 and z_2 are connected in $\mathbb{C} \setminus X$ by $\overline{z_1, z} \cup \overline{z, z_2}$. Theorem 2.5.5 now shows that $P(X) = R(X)$.

It remains to show that $R(X) = C(X)$. Let $\mu \in C(X)^*$, that is, a bounded regular Borel measure on X, and suppose that μ is nonzero and nevertheless annihilates $R(X)$. Note that, for every $z \in \mathbb{C} \setminus X$, the function $w \to 1/(w-z)$ belongs to $R(X)$ and hence $\int_X 1/(w-z)d\mu(w) = 0$. Since $\operatorname{supp}\mu$ is countable and compact, at least one of the points of $\operatorname{supp}\mu$ is open in $\operatorname{supp}\mu$. So there exist z_0 and an open disc U centered at z_0 of radius $R > 0$ such that $\mu(\{z_0\}) \neq 0$ and $U \cap \operatorname{supp}\mu = \{z_0\}$. Since X is countable, we find $0 < r < R$ such that the path $\gamma(t) = z_0 + re^{it}, t \in [0, 2\pi]$, does not meet X. An easy application of Fubini's theorem shows that

$$\int_\gamma \left(\int_X \frac{1}{w-z} d\mu(w) \right) dz = \int_X \left(\int_\gamma 1/(w-z)dz \right) d\mu(w).$$

Now the left-hand side of this equation is zero since $X \cap \gamma[0, 2\pi] = \emptyset$ and $\int_X 1/(w-z)d\mu(w) = 0$ for every $z \notin X$. On the other hand, the right-hand side is nonzero. To see this, note first that if $w \in \operatorname{supp}\mu$, then either $w \notin U$ or $w = z_0$. In the first case, $\int_\gamma 1/(w-z)dz = 0$, whereas $\int_\gamma 1/(w-z)dz = -2\pi i$ in the second case. It follows that

$$\int_X \left(\int_\gamma \frac{1}{w-z} dz \right) d\mu(w) = -2\pi i \mu(\{z_0\}) \neq 0.$$

This contradiction shows that there is no nonzero $\mu \in C(X)^*$ annihilating $R(X)$. Thus $R(X) = C(X)$ by the Hahn–Banach theorem. $\qquad\square$

Since a countable compact subset of $\mathbb{C}$ has empty interior, Theorem 2.5.12 is a very special case of Mergelyan's theorem which states that if X is a compact subset of $\mathbb{C}$ such that $\mathbb{C} \setminus X$ is connected, then $P(X) = A(X)$. It is also worth pointing out that $R(X) = C(X)$ holds more generally whenever X is totally disconnected. In fact, this follows from Corollary 3.5.6 because every compact subset X of $\mathbb{C}$ is rationally convex (Remark 2.5.5) and therefore homeomorphic to $\Delta(R(X))$ (Theorem 2.5.7). However, the proof of Corollary 3.5.6 relies on Shilov's idempotent theorem. A somewhat surprising consequence of Theorem 2.5.12 is the following corollary.

Corollary 2.5.13. *Let X be a countable compact Hausdorff space and let A be a closed subalgebra of $C(X)$. Then A is self-adjoint.*

Proof. Let $f \in A$. Then $f(X) \cup \{0\}$ is a countable compact subset of $\mathbb{C}$. By the preceding theorem there exists a sequence of polynomials $p_n, n \in \mathbb{N}$, such that $p_n(z) \to \overline{z}$ uniformly on $f(X) \cup \{0\}$. Let $q_n = p_n - p_n(0)$. Then each q_n is a polynomial without constant term and $q_n(z) \to \overline{z}$ uniformly on $f(X)$. Thus $q_n(f(x)) \to \overline{f(x)}$ uniformly on X. Since q_n is without constant term, $q_n \circ f \in A$. This proves that $\overline{f} \in A$. $\square$

In concluding this section we present a theorem (Theorem 2.5.15 below), which is usually referred to as *Wermer's maximality theorem*. The proof requires the following lemma.

Lemma 2.5.14. *Let X be a compact Hausdorff space and A a uniform algebra on X. If f and g are functions in A such that $\|1 + f + \overline{g}\|_\infty < 1$, then $f + g$ is invertible in A.*

Proof. Let $h = f + g$ and $c = \|1 + \operatorname{Re} h\|_\infty$. Since $\|1 + f + \overline{g}\|_\infty < 1$ and hence $\|1 + \overline{f} + g\|_\infty < 1$, we have

$$\|1 + \operatorname{Re} h\|_\infty = \frac{1}{2}\|1 + f + \overline{g} + 1 + \overline{f} + g\|_\infty < 1.$$

Thus, for all $x \in X, |1 + \operatorname{Re} h(x)| \leq c < 1$. This means that $h(x)$ lies in the left half-plane for all x, which suggests that, for small $\epsilon > 0, 1 + \epsilon h(x)$ lies in the unit disc for all x. In fact,

$$|1 + \epsilon h(x)|^2 = 1 + 2\epsilon \operatorname{Re} h(x) + \epsilon^2 |h(x)|^2$$
$$\leq 1 + 2\epsilon(c - 1) + \epsilon^2 \|h\|_\infty^2,$$

for all $x \in X$. Since $c < 1$, it follows that $\|1 + \epsilon h\|_\infty < 1$ for sufficiently small $\epsilon > 0$. This ϵh is invertible (Lemma 1.2.6) and hence so is h. $\square$

Theorem 2.5.15. *Let A be a uniform algebra on the unit circle $\mathbb{T}$ such that $P(\mathbb{T}) \subseteq A$. Then either $A = P(\mathbb{T})$ or $A = C(\mathbb{T})$.*

Proof. For $h \in C(\mathbb{T})$ and $k \in \mathbb{Z}$, let

$$c_k(h) = \frac{1}{2\pi} \int_0^{2\pi} h(e^{it}) e^{-ikt} dt,$$

the k-th Fourier coefficient of h. Then $h \in P(\mathbb{T})$ if and only if $c_k(h) = 0$ for all $k < 0$.

Now suppose that $A \neq P(\mathbb{T})$. Then there exists $h \in A$ with $c_k(h) \neq$ for some $k < 0$. Without loss of generality we can assume that $c_{-1}(h) = 1$. Indeed, the function g defined by $g(z) = h(z)z^{-(k+1)}$ belongs to A since $A \supseteq P(\mathbb{T})$ and $-(k+1) \geq 0$ and $c_{-1}(g) = c_k(h) \neq 0$. Choose a trigonometric polynomial r with $\|h - r\|_\infty < \frac{1}{2}$ and define $s \in C(\mathbb{T})$ by

$$s(z) = r(z) + (1 - c_{-1}(r))z^{-1}.$$

Then $c_{-1}(s) = c_{-1}(r) + (1 - c_{-1}(r)) = 1$ and

$$\|s - h\|_\infty \leq \|r - h\|_\infty + |1 - c_{-1}(r)| = \|r - h\|_\infty + |c_{-1}(h - r)|$$
$$\leq 2\|h - r\|_\infty < 1.$$

Thus s is of the form

$$s(z) = \sum_{k=-N}^{-2} c_k(s)z^k + z^{-1} + \sum_{k=0}^{N} c_k(s)z^k$$

for some $N \in \mathbb{N}$. It follows that

$$zs(z) = \sum_{k=-N}^{-2} c_k(s)z^{k+1} + 1 + z\sum_{k=0}^{N} c_k(s)z^k$$
$$= \overline{z}\,\overline{p(z)} + 1 + zq(z),$$

where p and q are polynomials in z. Since $\|s - h\|_\infty < 1$, we obtain that

$$\|1 + z(q - h) + \overline{z}\,\overline{p}\|_\infty = \|zs - zh\|_\infty = \|s - h\|_\infty < 1.$$

Since $q - h \in A$ and $p \in A$, Lemma 2.5.14 shows that the function

$$z \rightarrow z(q(z) - h(z)) + zp(z) = z(q - h + p)(z)$$

is invertible in A. So the function $z \rightarrow z$ is invertible in A, and hence A contains all the functions $z \rightarrow z^m$, $m \in \mathbb{Z}$. Because the linear combinations of these functions are dense in $C(\mathbb{T})$, we conclude that $A = C(\mathbb{T})$. $\qquad\square$

2.6 The structure space of $A(X)$

Let X be a compact subset of $\mathbb{C}$. Recall that $A(X)$ is the closed subalgebra of $C(X)$ consisting of all functions in $C(X)$ which are holomorphic on the interior X° of X. Our aim is to work out the structure space of $A(X)$. Since $A(X)$ is a uniform algebra, the mapping $x \to \varphi_x$, where φ_x is the point evaluation $\varphi_x(f) = f(x)$, $f \in A(X)$, at x, is an embedding of the compact set X into $\Delta(A(X))$. As might be expected, this map is actually surjective, but this is much harder to prove than the corresponding fact for $R(X)$. To establish this result, we need a sequence of preparatory lemmas.

In passing, we mention that in some special cases we already know that $\Delta(A(X)) = X$.

Remark 2.6.1. Clearly, $R(X) \subseteq A(X)$, and $A(X) = C(X)$ whenever $X^\circ = \emptyset$. There exist sets X such that $R(X)$ is strictly contained in $A(X)$. An example is provided by the so-called Swiss cheese (Example 2.5.11) which was obtained by deleting countable many disjoint open discs from the closed unit disc in an appropriate way. On the other hand, $P(\mathbb{D}) = A(\mathbb{D}) \neq C(\mathbb{D})$ (Example 2.5.3).

In the sequel, $\lambda(M)$ denotes the Lebesgue measure of a Borel subset M of $\mathbb{C}$.

Lemma 2.6.2. Let X be a Borel subset of $\mathbb{C}$. Then, for any $z \in \mathbb{C}$,

$$\int_X \frac{1}{|x - z|} dx \le 2(\pi \lambda(X))^{1/2}.$$

In particular, the functions $x \to 1/(x - z)$, $z \in \mathbb{C}$, are integrable on compact subsets of $\mathbb{C}$.

Proof. Nothing has to be shown if $\lambda(X) = 0$ or $\lambda(X) = \infty$. Thus we can assume that $0 < \lambda(X) < \infty$. Let $R = \pi^{-1/2}\lambda(X)^{1/2}$, $S = \{x \in \mathbb{C} : |x-z| \le R\}$ and, for any $\varepsilon > 0$, $S_\varepsilon = \{x \in \mathbb{C} : \varepsilon \le |x - z| \le R\}$. Then, introducing polar coordinates, we get

$$\int_S \frac{1}{|x - z|} dx = \lim_{\varepsilon \to 0} \int_{S_\varepsilon} \frac{1}{|x - z|} dx = \lim_{\varepsilon \to 0} \int_\varepsilon^R \int_0^{2\pi} d\varphi \, dr$$
$$= \lim_{\varepsilon \to 0} 2\pi (R - \varepsilon) = 2\pi R$$
$$= 2(\pi \lambda(X))^{1/2}.$$

It therefore suffices to show that

$$\int_X \frac{1}{|x - z|} dx \le \int_S \frac{1}{|x - z|} dx.$$

To that end, note first that $\lambda(X) = \pi R^2 = \lambda(S)$ and $X = (X \cap S) \cup (X \setminus S)$ and hence

$$\lambda(X \setminus S) = \lambda(X) - \lambda(X \cap S) = \lambda(S) \setminus \lambda(S \cap X) = \lambda(S \setminus X).$$

Now $1/|x - z| \geq 1/R$ on $S \setminus X$ and $1/|x - z| \leq 1/R$ on $X \setminus S$. It follows that

$$
\begin{aligned}
\int_X \frac{1}{|x - z|} dx &= \int_{X \cap S} \frac{1}{|x - z|} dx + \int_{X \setminus S} \frac{1}{|x - z|} dx \\
&\leq \int_{X \cap S} \frac{1}{|x - z|} dx + \frac{1}{R} \lambda(X \setminus S) \\
&= \int_{X \cap S} \frac{1}{|x - z|} dx + \frac{1}{R} \lambda(S \setminus X) \\
&\leq \int_{X \cap S} \frac{1}{|x - z|} dx + \int_{S \setminus X} \frac{1}{|x - z|} dx \\
&= \int_S \frac{1}{|x - z|} dx,
\end{aligned}
$$

as required.   □

Lemma 2.6.3. *Let K be a compact subset of $\mathbb{C}$ and g a bounded Borel measurable function on K. Define a function f on $\mathbb{C}$ by*

$$f(z) = \int_K \frac{g(x)}{(x - z)} dx.$$

Then f vanishes at infinity and f is holomorphic on $\mathbb{C} \setminus K$ and continuous everywhere.

Proof. First of all, the integral exists for all $z \in \mathbb{C}$ because the function $x \to 1/(x - z)$ is integrable on compact sets (Lemma 2.6.2) and g is bounded. If $R > 0$ and the distance from z to K is $\geq R$, then

$$|f(z)| \leq \frac{1}{R} \int_K |g(x)| dx \leq \frac{\lambda(K)}{R} \|g\|_\infty.$$

This shows that f vanishes at infinity.

Next, for $z, z_0 \in \mathbb{C} \setminus K$ with $z \neq z_0$ we have

$$\frac{f(z) - f(z_0)}{z - z_0} = \frac{1}{z - z_0} \int_K \left(\frac{g(x)}{x - z} - \frac{g(x)}{x - z_0} \right) dx = \int_K \frac{g(x)}{(x - z)(x - z_0)} dx.$$

Since $z_0 \notin K$, the function $x \to g(x)/(x - z)(x - z_0)$ converges uniformly on K, as $z \to z_0$, with limit $g(x)/(x - z_0)^2$. Therefore, as $z \to z_0$,

$$\frac{f(z) - f(z_0)}{z - z_0} \to \int_K \frac{g(x)}{(x - z_0)^2} dx.$$

Thus f is holomorphic on $\mathbb{C} \setminus K$.

It remains to show that f is continuous at all points of K. We fix $R > 0$ so that $K \subseteq U = \{x \in \mathbb{C} : |x| < R/2\}$ and prove that f is continuous on U. Since the function $x \to 1/x$ is integrable on any compact subset of $\mathbb{C}$ (Lemma 2.6.2) and $C_c(\mathbb{C})$ is dense in $L^1(\mathbb{C})$, given $\epsilon > 0$, there exists $h \in C_c(\mathbb{C})$ such that

$$\int_{|x| \leq R} \left| h(x) - \frac{1}{x} \right| dx \leq \epsilon.$$

For $y \in U$ we then have

$$\left| f(y) - \int_K g(u)h(u-y)du \right| \leq \int_K |g(u)| \cdot \left| \frac{1}{u-y} - h(u-y) \right| du$$
$$\leq \|g\|_\infty \int_{K-y} \left| \frac{1}{u} - h(u) \right| du$$
$$\leq \|g\|_\infty \int_{|u| \leq R} \left| \frac{1}{u} - h(u) \right| du$$
$$\leq \epsilon \|g\|_\infty.$$

As h is uniformly continuous, there exists $\delta > 0$ such that, for all $x, y \in \mathbb{C}$, $|h(x) - h(y)| \leq \epsilon$ whenever $|x - y| \leq \delta$. For $x, y \in U$ with $|x - y| \leq \delta$ it follows that

$$|f(x) - f(y)| \leq 2\epsilon \|g\|_\infty + \int_K |g(u)| \cdot |h(u-x) - h(u-y)|du$$
$$\leq \epsilon \|g\|_\infty (2 + \lambda(K)).$$

This shows that f is (uniformly) continuous on U. $\square$

Lemma 2.6.4. *Let X and K be compact subsets of $\mathbb{C}$ and let $f \in A(X)$. Extend f to all of $\mathbb{C}$ by setting $f(x) = 0$ for all $x \in \mathbb{C} \setminus X$, and define h on $\mathbb{C}$ by*

$$h(z) = \int_K \frac{f(x) - f(z)}{x - z} dx.$$

Then h is continuous on $\mathbb{C}$ and holomorphic on X°.

Proof. Since $x \to f(x) - f(z)$ is a bounded Borel measurable function on $\mathbb{C}$, $h(z)$ is defined for all $z \in \mathbb{C}$ and h is a continuous function (Lemma 2.6.3). Therefore, to show that h is holomorphic on X°, by Morera's theorem it is enough to verify that $\int_\gamma h(z)dz = 0$ for every triangle path γ which together with its interior is contained in X°. For that, fix γ, let Γ denote the trace of γ, and note first that the function

$$(x, z) \to \frac{f(x) - f(z)}{x - z}$$

is a Borel function on $K \times \Gamma$ satisfying

$$\int_\gamma \left(\int_K \frac{|f(x) - f(z)|}{|x - z|} \, dx \right) dz < \infty$$

(Lemma 2.6.2). Thus we can apply Fubini's theorem to conclude that

$$\int_\gamma h(z)dz = \int_\gamma \left(\int_K \frac{f(x) - f(z)}{x - z} \, dx \right) dz = \int_K \left(\int_\gamma \frac{f(x) - f(z)}{x - z} dz \right) dx.$$

Now the inner integral along γ is zero for all $x \in \mathbb{C} \setminus \Gamma$. In fact, this is so for every $x \in \mathbb{C} \setminus X^\circ$ since then the function $z \to (x - z)^{-1}(f(x) - f(z))$ is holomorphic on X°, whereas for each $x \in X^\circ \setminus \Gamma$,

$$\int_\gamma \frac{f(x) - f(z)}{x - z} \, dz = 2\pi i \left(f(x) - \frac{1}{2\pi i} \int_\gamma \frac{f(z)}{z - x} \, dz \right) = 0$$

by the Cauchy integral formula. Since $K \cap \Gamma$ has Lebesgue measure zero, it follows that $\int_\gamma h(z)dz = 0$, as required.   □

The preceding three lemmas together lead to the following approximation result which is the main tool to prove that $\Delta(A(A)) = X$.

Lemma 2.6.5. *Let X be a compact subset of $\mathbb{C}$ and let $z_0 \in X$ and $f \in A(X)$. Then there exists a sequence $(f_n)_n$ in $A(X)$ such that*

$$f(z) - f(z_0) - (z - z_0)f_n(z) \to 0$$

uniformly on X as $n \to \infty$.

Proof. Replacing X by $X - z_0$ and f by $f - f(z_0)$, we can assume that $z_0 = 0$ and $f(z_0) = 0$. Extend f to all of $\mathbb{C}$ by setting $f(x) = 0$ for $x \in \mathbb{C} \setminus X$. For $n \in \mathbb{N}$, let $K_n = \{x \in \mathbb{C} : |x| \le 1/n\}$ and define f_n on $\mathbb{C}$ by

$$f_n(z) = \frac{n^2}{\pi} \int_{K_n} \frac{f(x) - f(z)}{x - z} \, dx.$$

Then each f_n is continuous on $\mathbb{C}$ and holomorphic on X° (Lemma 2.6.4). Since $\lambda(K_n) = \pi/n^2$, we have for all $z \in \mathbb{C}$,

$$z f_n(z) - f(z) = \frac{n^2}{\pi} \int_{K_n} \left(z \frac{f(x) - f(z)}{x - z} - f(z) \right) dx.$$

We need to estimate the integral on the right. For $r > 0$, let

$$M(r) = \sup\{|f(z)| : z \in X, |z| \le r\}.$$

With this notation, for all $z \in \mathbb{C}$, it follows that

(1) $$|z f_n(z) - f(z)| \le \frac{n^2}{\pi} \left(|z| M \left(\frac{1}{n} \right) + \frac{1}{n} |f(z)| \right) \int_{K_n} \frac{1}{|x - z|} \, dx.$$

Now, if $|z| > 1/n$, then $|x - z| \geq |z| - 1/n$ for all $x \in K_n$ and hence

(2)
$$\frac{n^2}{\pi} \int_{K_n} \frac{1}{|x - z|} dx \leq \frac{1}{|z| - 1/n}.$$

On the other hand, if $q \geq 1$ and $|z| \leq q/n$, then by Lemma 2.6.2

$$\int_{K_n} \frac{1}{|x - z|} dx = \int_{K_n - z} \frac{1}{|x|} dx \leq \int_{|x| \leq 2q/n} \frac{1}{|x|} dx \leq 2\pi \frac{2q}{n},$$

and hence, for all such z,

(3)
$$\frac{n^2}{\pi} \int_{K_n} \frac{1}{|x - z|} dx \leq 4qn.$$

Now let $\epsilon > 0$ be given and choose $q \in \mathbb{N}$, $q > 1$, such that $(q - 1)\epsilon > \|f\|_\infty$. If $|z| \geq q/n > 1/n$, then combining (1) and (2) yields

$$
\begin{aligned}
|z f_n(z) - f(z)| &\leq \frac{|z| M(\frac{1}{n}) + \frac{1}{n}|f(z)|}{|z| - \frac{1}{n}} \\
&\leq \left(1 + \frac{\frac{1}{n}}{|z| - \frac{1}{n}}\right) M\left(\frac{1}{n}\right) + \frac{1}{n}\|f\|_\infty \frac{1}{|z| - \frac{1}{n}} \\
&\leq \frac{q}{q - 1} M\left(\frac{1}{n}\right) + \|f\|_\infty \frac{1}{q - 1} \\
&\leq \frac{q}{q - 1} M\left(\frac{1}{n}\right) + \epsilon.
\end{aligned}
$$

Similarly, if $|z| < q/n$ then combining (1) and (3) gives

$$
\begin{aligned}
|z f_n(z) - f(z)| &\leq 4qn \left(\frac{q}{n} M\left(\frac{1}{n}\right) + \frac{1}{n} M\left(\frac{q}{n}\right)\right) \\
&= 4q \left(q M\left(\frac{1}{n}\right) + M\left(\frac{q}{n}\right)\right).
\end{aligned}
$$

However, $M(r) \to 0$ as $r \to 0$ since f is continuous on X and $f(0) = 0$. It follows that $z f_n(z) - f(z) \to 0$ uniformly on X. $\square$

Theorem 2.6.6. *Let X be a compact subset of $\mathbb{C}$. Then the mapping $x \to \varphi_x$, where $\varphi_x(f) = f(x)$ for all $f \in A(X)$, is a homeomorphism between X and $\Delta(A(X))$. With this identification of $\Delta(A(X))$ and X, the Gelfand homomorphism of $A(X)$ is the identity.*

Proof. We only have to show that given $\varphi \in \Delta(A(X))$, there exists $x \in X$ such that $\varphi(f) = f(x)$ for all $f \in A(X)$.

Let $x = \varphi(\mathrm{id}_X)$. Then $x \in X$ since, for every $\lambda \in \mathbb{C} \setminus X$, the function $z \to 1/(\lambda - z)$ belongs to $A(X)$ and therefore

$$\lambda \notin \sigma_{A(X)}(\mathrm{id}_X) = \widehat{\mathrm{id}_X}(\Delta(A(X))).$$

Now, for any $f \in A(X)$, by Lemma 2.6.5 there exists a sequence $(f_n)_n$ in $A(X)$ such that

$$f(z) - f(x) - (z - x)f_n(z) \to 0$$

uniformly on X. This implies that

$$\varphi(f) - f(x) = \lim_{n \to \infty} (\varphi(\mathrm{id}_X) - x)\varphi(f_n) = 0,$$

as was to be shown.

2.7 The Gelfand representation of $L^1(G)$

In commutative harmonic analysis the central object of study is the L^1-algebra of a locally compact Abelian group. In this section we present its Gelfand representation. Thus, in the sequel, G always denotes a locally compact Abelian group and $L^1(G)$ the convolution algebra of integrable functions on G.

To begin with, we introduce the dual group of G which turns out to be canonically identifiable with $\Delta(L^1(G))$.

Definition 2.7.1. A *character* α of G is a continuous homomorphism from G into the circle group $\mathbb{T}$. Clearly, the pointwise product of two characters is again a character and so is α^{-1} defined by $\alpha^{-1}(x) = \overline{\alpha(x)}$ for all $x \in G$. Thus $\widehat{G}$, the set of all characters of G, forms a group, the *dual group* of G.

We proceed to show that there is a bijection between $\widehat{G}$ and $\Delta(L^1(G))$.

Theorem 2.7.2. *For $\alpha \in \widehat{G}$, let $\varphi_\alpha : L^1(G) \longrightarrow \mathbb{C}$ be defined by*

$$\varphi_\alpha(f) = \int_G f(x)\overline{\alpha(x)}dx, \quad f \in L^1(G).$$

Then $\varphi_\alpha \in \Delta(L^1(G))$ and the mapping $\alpha \to \varphi_\alpha$ is a bijection from $\widehat{G}$ onto $\Delta(L^1(G))$.

Proof. Of course, φ_α is a linear functional. For $f, g \in C_c(G)$, Fubini's theorem and the invariance of Haar measure yield

$$\begin{aligned}
\varphi_\alpha(f * g) &= \int_G \overline{\alpha(x)} \int_G f(y)g(y^{-1}x)dy\,dx \\
&= \int_G \int_G f(y)\overline{\alpha(x)}g(y^{-1}x)dx\,dy \\
&= \int_G \int_G f(y)\overline{\alpha(yx)}g(x)dx\,dy \\
&= \int_G \int_G \overline{\alpha(x)}g(x)\overline{\alpha(y)}f(y)dx\,dy \\
&= \varphi_\alpha(f)\varphi_\alpha(g).
\end{aligned}$$

Since $|\varphi_\alpha(f)| \le \|f\|_1$ for all $f \in L^1(G)$, this formula even holds for all $f, g \in L^1(G)$. Moreover, φ_α is nonzero since for any nonnegative function f in $C_c(G)$, $f \ne 0$, we have

$$\varphi_\alpha(\alpha f) = \int_G f(x)|\alpha(x)|^2 dx > 0.$$

This shows that $\varphi_\alpha \in \Delta(L^1(G))$. Moreover, the map $\alpha \to \varphi_\alpha$ is injective. Indeed, if $\alpha, \beta \in \hat{G}$ are such that

$$0 = \varphi_\alpha(f) - \varphi_\beta(f) = \int_G f(x)\big(\overline{\alpha(x)} - \overline{\beta(x)}\big) dx$$

for all $f \in L^1(G)$, then $\alpha = \beta$ because $L^1(G)^* = L^\infty(G)$ and α and β are continuous functions.

It remains to show that given $\varphi \in \Delta(L^1(G))$, there exists $\alpha \in \hat{G}$ such that $\varphi = \varphi_\alpha$. To that end, choose $g \in L^1(G)$ such that $\varphi(g) = 1$ and observe that since $\varphi \in L^1(G)^*$, there exists $\chi \in L^\infty(G)$ such that $\varphi(f) = \int_G f(x)\chi(x)dx$ for all $f \in L^1(G)$. The function

$$(x, y) \to \chi(x)f(y)g(y^{-1}x)$$

belongs to $L^1(G \times G)$, and hence Fubini's theorem implies that

$$\varphi(f) = \varphi(f * g) = \int_G \chi(x)\left(\int_G f(y)g(y^{-1}x)dy\right) dx$$
$$= \int_G f(y)\left(\int_G g(y^{-1}x)\chi(x)dx\right) dy$$
$$= \int_G f(y)\varphi(L_y g)dy$$

for all $f \in L^1(G)$. Now, define $\alpha : G \to \mathbb{C}$ by $\alpha(y) = \overline{\varphi(L_y g)}$, $y \in G$. The function α is continuous because the map $y \to L_y g$ from G into $L^1(G)$ is continuous and

$$|\alpha(x) - \alpha(y)| = |\varphi(L_x g - L_y g)| \le \|L_x g - L_y g\|_1$$

for all $x, y \in G$. From $g * L_{xy} g = L_x g * L_y g$ it follows that

$$\alpha(xy) = \overline{\varphi(L_{xy} g)} = \overline{\varphi(g)}\,\overline{\varphi(L_{xy} g)} = \overline{\varphi(g * L_{xy} g)}$$
$$= \overline{\varphi(L_x g * L_y g)} = \overline{\varphi(L_x g)}\overline{\varphi(L_y g)}$$
$$= \alpha(x)\alpha(y).$$

We claim that $|\alpha(x)| = 1$ for all $x \in G$. For that, notice that

$$|\alpha(y)| = |\varphi(L_y g)| \le \|L_y g\|_1 = \|g\|_1$$

for all $y \in G$, and hence, by the multiplicativity of α,

$$|\alpha(x)|^n = |\alpha(x^n)| \leq \|g\|_1$$

for all $n \in \mathbb{Z}$. Since $\alpha(e) = \overline{\varphi(g)} = 1$, we conclude that $|\alpha(x)| = 1$ for every $x \in G$. This shows that $\alpha \in \widehat{G}$ and $\varphi_\alpha = \varphi$. $\qquad\square$

After identifying $\Delta(L^1(G))$ as a set with $\widehat{G}$, our next purpose is to describe the Gelfand topology on $\widehat{G}$ in terms of G itself rather than $L^1(G)$.

Lemma 2.7.3. *Let $f \in L^1(G)$ and $\alpha \in \widehat{G}$.*

(i) *For all $x \in G$, $(f * \alpha)(x) = \alpha(x)\widehat{f}(\alpha) = \widehat{L_{x^{-1}}f}(\alpha)$. In particular, $\widehat{L^1(G)}$ is invariant under multiplication with functions of the form $\alpha \to \alpha(x)$, $x \in G$.*

(ii) *If $g \in L^1(G)$ is defined by $g(x) = \alpha(x)f(x)$, then $\widehat{g} = L_\alpha\widehat{f}$. In particular, $\widehat{L^1(G)} \subseteq C_0(\widehat{G})$ is translation invariant.*

(iii) $\widehat{f^*} = \overline{\widehat{f}}$ *and $\widehat{L^1(G)} \subseteq C_0(\widehat{G})$ is norm-dense in $C_0(\widehat{G})$.*

Proof. (i) $f * \alpha$ is a continuous function and

$$(f * \alpha)(x) = \int_G f(y)\alpha(y^{-1}x)dy = \alpha(x)\widehat{f}(\alpha)$$

for all $x \in G$. On the other hand,

$$(f * \alpha)(x) = \int_G f(xy)\overline{\alpha(y)}dy = \widehat{L_{x^{-1}}f}(\alpha).$$

(ii) For all $\beta \in \widehat{G}$, we have

$$\widehat{g}(\beta) = \int_G f(x)\overline{\beta(x)}\alpha(x)dx = \widehat{f}(\alpha^{-1}\beta) = L_\alpha\widehat{f}(\beta),$$

so that $L_\alpha\widehat{f} = \widehat{g} \in \widehat{L^1(G)}$.

(iii) For each $\alpha \in \widehat{G}$, we have

$$\widehat{f^*}(\alpha) = \int_G \overline{f(x^{-1})}\alpha(x)dx = \int_G \overline{f(x)\alpha(x)}dx = \overline{\widehat{f}(\alpha)},$$

so that $\widehat{f^*} = \overline{\widehat{f}}$. Thus $\widehat{L^1(G)}$ is a self-adjoint subalgebra of $C_0(\widehat{G})$ which strongly separates the points of $\widehat{G}$ and therefore is dense in $(C_0(\widehat{G}), \|\cdot\|_\infty)$ by the Stone–Weierstrass theorem. $\qquad\square$

Lemma 2.7.4. *Let $f \in L^1(G)$ and $\epsilon > 0$ and let σ denote the Gelfand topology on $\widehat{G}$. Then there exists a neighbourhood W of e in G with the following property. If $y, x \in G$ and $\beta, \alpha \in \widehat{G}$ are such that $y \in Wx$, $\varphi_\alpha(f) = 1$, and $\beta \in U(\alpha, f, L_x f, \epsilon/3)$, then*

$$|\beta(y) - \alpha(x)| < \epsilon.$$

In particular, the function $(x, \alpha) \to \alpha(x)$ is continuous on $G \times (\widehat{G}, \sigma)$.

Proof. For arbitrary $y, x \in G$ and $\beta, \alpha \in \widehat{G}$ such that $\widehat{f}(\alpha) = 1$ we obtain from Lemma 2.7.3,

$$
\begin{aligned}
|\beta(y) - \alpha(x)| &\leq |\overline{\beta(y)} - \overline{\beta(y)}\widehat{f}(\beta)| + |\overline{\beta(y)}\widehat{f}(\beta) - \overline{\beta(x)}\widehat{f}(\beta)| \\
&\quad + |\overline{\beta(x)}\widehat{f}(\beta) - \overline{\alpha(x)}\widehat{f}(\alpha)| \\
&= |1 - \widehat{f}(\beta)| + |\widehat{L_y f}(\beta) - \widehat{L_x f}(\beta)| + |\widehat{L_x f}(\beta) - \widehat{L_x f}(\alpha)| \\
&\leq |\widehat{f}(\beta) - \widehat{f}(\alpha)| + \|L_y f - L_x f\|_1 + |\widehat{L_x f}(\beta) - \widehat{L_x f}(\alpha)|.
\end{aligned}
$$

Now let W be a neighbourhood of e such that $\|L_s f - L_t f\|_1 < \epsilon/3$ whenever $t^{-1}s \in W$. For all $y \in Wx$ and $\beta \in U(\alpha, f, L_x f, \epsilon/3)$ it then follows that $|\beta(y) - \alpha(x)| < \epsilon$.

For the last statement of the lemma we only have to recall that given $\alpha \in \widehat{G}$, there exists $f \in L^1(G)$ such that $\widehat{f}(\alpha) = 1$. $\square$

We now consider the compact open topology τ on $\widehat{G}$. A τ-neighbourhood basis of $\alpha_0 \in \widehat{G}$ is formed by the collection of sets

$$
V(\alpha_0, K, \epsilon) = \{\alpha \in \widehat{G} : |\alpha(x) - \alpha_0(x)| < \epsilon \text{ for all } x \in K\},
$$

where $\epsilon > 0$ and K is any compact subset of G. Then $(\widehat{G}, \tau)$ is a topological group since $V(\alpha_0, K, \epsilon)^{-1} = V(\alpha_0^{-1}, K, \epsilon)$ and

$$
V(\alpha_0, K, \epsilon)V(\beta_0, K, \epsilon) \subseteq V(\alpha_0\beta_0, K, 2\epsilon).
$$

In fact, the latter inclusion follows from

$$
\begin{aligned}
|\alpha\beta(x) - \alpha_0\beta_0(x)| &\leq |\alpha(x)(\beta(x)) - \beta_0(x))| + |\beta_0(x)(\alpha(x) - \alpha_0(x))| \\
&\leq |\beta(x) - \beta_0(x)| + |\alpha(x) - \alpha_0(x)|.
\end{aligned}
$$

Theorem 2.7.5. *On $\widehat{G}$ the Gelfand topology and the compact open topology coincide.*

Proof. Let 1_G denote the trivial character of G. Note that, for $\alpha \in \widehat{G}$, $\delta > 0$, and $f_1, \ldots, f_n \in L^1(G)$, we have

$$
\alpha\, U(1_G, f_1, \ldots, f_n, \delta) = U(\alpha, f_1\alpha, \ldots, f_n\alpha, \delta).
$$

In fact, for $\beta \in \widehat{G}$ and $f \in L^1(G)$,

$$
\begin{aligned}
\varphi_\beta(f\alpha) - \varphi_\alpha(f\alpha) &= \int_G f(x)\alpha(x)(\overline{\beta(x)} - \overline{\alpha(x)})dx \\
&= \int_G f(x)(\overline{\alpha^{-1}\beta(x)} - 1)dx \\
&= \varphi_{\alpha^{-1}\beta}(f) - \varphi_{1_G}(f).
\end{aligned}
$$

Hence we only have to verify that every τ-neighbourhood of 1_G contains a σ-neighbourhood of 1_G and vice versa.

Let $V(1_G, K, \delta)$ be given and choose $f \in L^1(G)$ such that $\int_G f(x)dx = 1$. By Lemma 2.7.4, there exists a neighbourhood W of e in G such that if $x, y \in G$ satisfy $y \in Wx$ and if $\alpha \in U(1_G, f, L_x f, \delta/3)$, then $|\alpha(y) - 1| < \delta$. Because K is compact, we find $x_1, \ldots, x_r \in K$ so that $K \subseteq \bigcup_{j=1}^{r} Wx_j$. It follows that

$$U(1_G, f, L_{x_1} f, \ldots, L_{x_r} f, \delta/3) \subseteq V(1_G, K, \delta).$$

Conversely, let $U(1_G, f_1, \ldots, f_n, \delta)$ be given. We can assume that $f_j \neq 0$ for all $j = 1, \ldots, n$. For every j, choose $g_j \in C_c(G)$ with $\|f_j - g_j\|_1 < \delta/4$. Set

$$K = \bigcup \{\operatorname{supp} g_j : 1 \leq j \leq n\}$$

and

$$\epsilon = \frac{\delta}{2} \min\{\|f_j\|^{-1} : 1 \leq j \leq n\}.$$

We claim that

$$V(1_G, K, \epsilon) \subseteq U(1_G, f_1, \ldots, f_n, \delta).$$

Indeed, if $\alpha \in V(1_G, K, \epsilon)$ then, for each $j = 1, \ldots, n$,

$$
\begin{aligned}
|\varphi_\alpha(f_j) - \varphi_{1_G}(f_j)| &\leq \int_K |f_j(x)| \cdot |\alpha(x) - 1| dx \\
&\quad + \int_{G \setminus K} |f_j(x)| \cdot |\alpha(x) - 1| dx \\
&< \epsilon \|f_j\|_1 + 2 \int_{G \setminus K} |f_j(x)| dx \\
&= \varepsilon \|f_j\|_1 + 2 \int_{G \setminus K} |f_j(x) - g_j(x)| dx \\
&\leq \epsilon \|f_j\|_1 + 2 \|f_j - g_j\|_1 \\
&\leq \delta.
\end{aligned}
$$

This completes the proof. $\square$

Since $(\widehat{G}, \sigma)$ is locally compact and $(\widehat{G}, \tau)$ is a topological group, Theorem 2.7.5 in particular shows that $\widehat{G}$ is a locally compact group. Identifying $\Delta(L^1(G))$ as a topological space with $\widehat{G}$, the Gelfand representation of $L^1(G)$ is the mapping $f \to \widehat{f}$, where $\widehat{f} \in C_0(\widehat{G})$ is defined by

$$\widehat{f}(\alpha) = \int_G f(x)\overline{\alpha(x)}dx, \quad \alpha \in \widehat{G}.$$

We now present a number of simple examples of dual groups, such as $\widehat{\mathbb{R}}$, $\widehat{\mathbb{Z}}$, and $\widehat{\mathbb{T}}$.

Example 2.7.6. (1) The dual group of the real line $\mathbb{R}$ is topologically isomorphic to $\mathbb{R}$. In fact, for each $y \in \mathbb{R}$, define a character α_y of $\mathbb{R}$ by $\alpha_y(x) = \exp(2\pi i x y)$, $x \in \mathbb{R}$. Then the map $y \to \alpha_y$ from $\mathbb{R}$ into $\widehat{\mathbb{R}}$ is injective and every character of $\mathbb{R}$ arises in this way (see Exercise 2.12.29). In addition, $y \to \alpha_y$ is a homeomorphism. Now,

$$\varphi_{\alpha_y}(f) = \int_{\mathbb{R}} f(x) \exp(-2\pi i x y) dx = \widehat{f}(y).$$

Thus, after identifying $\widehat{\mathbb{R}}$ with $\mathbb{R}$, the Gelfand homomorphism of $L^1(\mathbb{R})$ agrees with the Fourier transformation.

(2) In Example 2.2.10 we have already determined the Gelfand representation of $l^1(\mathbb{Z})$. Implicit in the arguments given there is the fact that the map $z \to \alpha_z$, where $\alpha_z(n) = z^n$ for $n \in \mathbb{Z}$, is a homeomorphism between $\mathbb{T}$ and the dual group $\widehat{\mathbb{Z}}$.

(3) It follows from $\widehat{\mathbb{Z}} = \mathbb{T}$ and the duality theorem for locally compact Abelian groups (a proof of which we present in Theorem A.5.2) that $\widehat{\mathbb{T}}$ is isomorphic to $\mathbb{Z}$. However, this can be seen directly as follows. First, for every $n \in \mathbb{Z}$, the function $z \to z^n$ is a character of $\mathbb{T}$. To show that every character α of $\mathbb{T}$ is of this form, consider the functions f_k on $\mathbb{T}$ defined by $f_k(z) = z^k$ $(k \in \mathbb{Z})$. By the Weierstrass approximation theorem the linear span of these functions f_k is dense in $C(\mathbb{T})$ and hence in $L^1(\mathbb{T})$. Thus $\widehat{f_k}(\alpha) \neq 0$ for at least one k. On the other hand, for arbitrary $k, l \in \mathbb{Z}$,

$$f_k * f_l(z) = \int_{\mathbb{T}} t^k (t^{-1}z)^l dt = z^k = f_k(z)$$

if $k = l$ and $= 0$ otherwise. Thus $\widehat{f_k}(\alpha)\widehat{f_l}(\alpha) = \widehat{f_k}(\alpha)$ if $l = k$ and $\widehat{f_k}(\alpha)\widehat{f_l}(\alpha) = 0$ otherwise. This implies that $\widehat{f_k}(\alpha) = 1$ for exactly one k and $\widehat{f_l}(\alpha) = 0$ for all $l \in \mathbb{Z}$, $l \neq k$. Now, because

$$\widehat{f_l}(\alpha_k) = \int_{\mathbb{T}} z^{l-k} dz = \delta_{kl}$$

for all $l \in \mathbb{Z}$, we obtain that $\alpha = \alpha_k$, as was to be shown. Finally, the relation $\widehat{f_l}(\alpha_k) = \delta_{kl}$ $(k, l \in \mathbb{Z})$ also shows that the Gelfand topology on $\widehat{\mathbb{T}}$ is discrete. Thus $\widehat{\mathbb{T}}$ is topologically isomorphic to $\mathbb{Z}$, and identifying $\widehat{\mathbb{T}}$ with $\mathbb{Z}$, we have

$$\widehat{f}(n) = \int_{\mathbb{T}} f(z) z^{-n} dz$$

for $f \in L^1(\mathbb{T})$ and $n \in \mathbb{Z}$.

(4) Let G_1 and G_2 be two locally compact Abelian groups and $G = G_1 \times G_2$ their direct product. It is not difficult to show that the map

$$(\alpha_1, \alpha_2) \to \alpha, \ \alpha(x_1, x_2) = \alpha_1(x_1)\alpha_2(x_2) \quad (\alpha_j \in \widehat{G}_j, x_j \in G_j, j = 1, 2)$$

furnishes a topological isomorphism from $\widehat{G}_1 \times \widehat{G}_2$ to $\widehat{G}$. Therefore, combining the cases (1), (2), and (3), the Gelfand representation of $L^1(G)$ can be explicitly given for groups of the form $\mathbb{R}^m \times \mathbb{Z}^n \times \mathbb{T}^r$, $m, n, r \in \mathbb{N}_0$.

Our next goal is to show that $L^1(G)$ is semisimple. To achieve this opens the opportunity to introduce the regular representation of $L^1(G)$ and the group C^*-algebra of G. Both are needed anyway in Chapter 4 in our approach to establish regularity of $L^1(G)$.

To start with, recall that for $f \in L^1(G)$ and $g \in C_c(G)$, the convolution product $f * g$ is given by

$$(f * g)(x) = \int_G f(y)g(y^{-1}x)dy$$

for every $x \in G$, and $f * g$ is a continuous function. For $g, h \in C_c(G)$ and $f \in L^1(G)$, using Fubini's theorem and Hölder's inequality, we get

$$\left| \int_G (f * g)(x)h(x)dx \right| = \left| \int_G \int_G f(y)g(y^{-1}x)h(x)dydx \right|$$
$$= \left| \int_G \int_G f(y)g(x)h(yx)dxdy \right|$$
$$\leq \int_G |f(y)| \int_G |g(x)L_{y^{-1}}h(x)|dxdy$$
$$\leq \int_G |f(y)| \cdot \|L_{y^{-1}}h\|_2 \|g\|_2 dy$$
$$\leq \|f\|_1 \|g\|_2 \|h\|_2.$$

Since $C_c(G)$ is dense in $L^2(G)$ it follows that the map

$$h \to \int_G h(x)(f * g)(x)dx$$

extends to a bounded linear functional on $L^2(G)$ the norm of which is at most $\|f\|_1 \|g\|_2$. Since $L^2(G)^* = L^2(G)$, we conclude that $f * g \in L^2(G)$ and $\|f * g\|_2 \leq \|f\|_1 \|g\|_2$ (see also Proposition A.4.7). Thus the linear mapping $g \to f * g$ from $C_c(G)$ into $L^2(G)$ extends uniquely to a bounded linear transformation $\lambda_f : L^2(G) \to L^2(G)$ and $\|\lambda_f\| \leq \|f\|_1$.

Theorem 2.7.7. *The mapping $\lambda : f \to \lambda_f$ from $L^1(G)$ into $\mathcal{B}(L^2(G))$ is an injective $*$-homomorphism.*

Proof. It is clear that λ is linear. For $f_1, f_2 \in L^1(G)$ and $g \in C_c(G)$,

$$\lambda_{f_1 * f_2}(g) = f_1 * (f_2 * g) = \lambda_{f_1}(\lambda_{f_2}(g)).$$

Thus λ is a homomorphism. Moreover, for $f \in L^1(G)$ and $g, h \in C_c(G)$,

$$\langle \lambda_f^*(g), h \rangle = \langle g, \lambda_f(h) \rangle = \int_G g(x) \int_G \overline{f(y)h(y^{-1}x)}dydx$$

$$= \int_G \int_G \overline{f(y^{-1})}g(x)\overline{h(yx)}dxdy$$

$$= \int_G \int_G f^*(y)g(y^{-1}x)\overline{h(x)}dxdy = \int_G (f^* * g)(x)\overline{h(x)}dx$$

$$= \langle \lambda_{f^*}(g), h \rangle.$$

This proves that λ is a $*$-homomorphism.

Finally, λ is injective. Indeed, if $f \in L^1(G)$ is such that $0 = \lambda_f(g) = f * g$ for all $g \in C_c(G)$, then $f = 0$ since $C_c(G)$ contains an approximate identity for $L^1(G)$.   □

Definition 2.7.8. The $*$-homomorphism $\lambda : f \rightarrow \lambda_f$ from $L^1(G)$ into $\mathcal{B}(L^2(G))$ is called the *regular representation* of $L^1(G)$ on the Hilbert space $L^2(G)$. Let $C^*(G)$ denote the closure of $\lambda(L^1(G))$ in $\mathcal{B}(L^2(G))$. Then, by Theorem 2.7.7, $C^*(G)$ is a commutative C^*-algebra, the so-called *group C^*-algebra* of G.

Every commutative C^*-algebra is semisimple and λ is injective (Theorem 2.7.7). Thus we conclude the following

Corollary 2.7.9. $L^1(G)$ *is semisimple.*

We now turn to the interesting and likewise important question of when the Gelfand homomorphism $L^1(G) \rightarrow C_0(\widehat{G})$ is surjective. Clearly, this is the case if G is finite because then $\widehat{L^1(G)}$ is a finite-dimensional dense linear subspace of $C_0(\widehat{G})$. To establish the converse, we first show that surjectivity forces G to be discrete.

Lemma 2.7.10. *Let G be a locally compact Abelian group, and suppose that the Gelfand homomorphism $\Gamma : f \rightarrow \widehat{f}$ from $L^1(G)$ into $C_0(\widehat{G})$ is surjective. Then G has to be discrete.*

Proof. Let $\Gamma^* : M(\widehat{G}) = (C_0(\widehat{G}))^* \rightarrow L^1(G)^* = L^\infty(G)$ denote the dual mapping of Γ. Since Γ is surjective, it is an isomorphism of Banach spaces and hence Γ^* is also an isomorphism. For $\mu \in M(\widehat{G})$, define its inverse Fourier–Stieltjes transform $\check{\mu}$ on G by $\check{\mu}(x) = \int_G \alpha(x)d\mu(\alpha)$. Then, using Fubini's theorem, for any $\mu \in M(\widehat{G})$ and $f \in L^1(G)$,

$$\langle \Gamma^*(\mu), f \rangle = \langle \mu, \Gamma(f) \rangle = \int_{\widehat{G}} \widehat{f}(\alpha)d\mu(\alpha)$$

$$= \int_{\widehat{G}} \left(\int_G f(x)\overline{\alpha(x)}dx \right) d\mu(\alpha)$$

$$= \int_G f(x) \left(\int_{\widehat{G}} \overline{\alpha(x)}d\mu(\alpha) \right) dx$$

$$= \int_G f(x)\check{\mu}(x)dx.$$

It follows that $\Gamma^*(\mu) = \check{\mu}$ locally almost everywhere for every $\mu \in M(\widehat{G})$. Now, using the facts that the function $(x, \alpha) \to \alpha(x)$ is continuous on $G \times \widehat{G}$ (Lemma 2.7.4) and that

$$|\check{\mu}(x) - \check{\mu}(y)| \leq \int_{\widehat{G}} |\alpha(x) - \alpha(y)| d|\mu|(\alpha),$$

it is easily verified that $\check{\mu}$ is continuous. Since Γ^* is onto, this means that every function in $L^\infty(G)$ is equal locally almost everywhere to a continuous function. However, this implies that G is discrete. Indeed, let U be an open, relatively compact subset of G which is not dense in G and let g be a continuous function on G which equals the characteristic function of U locally almost everywhere. Then $g(x) = 1$ for $x \in \overline{U}$, whereas $g(x) = 0$ for $x \in G \setminus U$. It follows that U is closed in G and since this holds for any such set U, we conclude that G is discrete.   □

Lemma 2.7.11. *Let G be a compact Abelian group and let X be an infinite subgroup of $\widehat{G}$. Then there exists $f \in C(G)$ such that $\mathrm{supp}\widehat{f} \subseteq X$ and*
$$\sum_{\chi \in X} |\widehat{f}(\chi)| = \infty.$$

Proof. The key step in achieving the existence of such a function f is to find a sequence $(g_n)_n$ of continuous functions on G with the following properties.

(1) $\|g_n\|_\infty \leq 2^{(n+1)/2}$.
(2) The range of $\widehat{g}_n$ is contained in $\{-1, 0, 1\}$.
(3) $|\widehat{g}_n|$ is the characteristic function of some subset X_n of X having precisely 2^n elements.

Suppose first that such a sequence $(g_n)_n$ exists. Then, because X is infinite and all X_n are finite, we can inductively define a sequence of characters $\chi_1, \chi_2, \ldots$ in X such that, with $\chi_0 = 1_G$, the sets $\chi_n^{-1} X_n, n \in \mathbb{N}_0$, are pairwise disjoint. Then, let $f_n = 2^{-n/2} \chi_n g_n, n \in \mathbb{N}_0$, so that the range of $\widehat{f}_n$ is contained in $\{-2^{-2/n}, 0, 2^{-n/2}\}$ and $2^{n/2} |\widehat{f}_n|$ is the characteristic function of the set $A_n = \chi_n X_n$ which contains exactly 2^n elements. Since $\|f_n\|_\infty \leq 2^{-n/2} \|g_n\|_\infty \leq 2^{1/2}$, we can define a continuous function f on G by setting

$$f(x) = \sum_{n=0}^{\infty} 2^{-n/2} f_n(x).$$

Now, if $\chi \notin \bigcup_{n=0}^\infty A_n$, then

$$\widehat{f}(\chi) = \sum_{n=0}^{\infty} 2^{-n/2} \widehat{f}_n(\chi) = 0,$$

whereas, if $\chi \in \bigcup_{n=0}^\infty A_n$, then $\chi \in A_n$ for exactly one n and hence

$$|\widehat{f}(\chi)| = 2^{-n/2} |\widehat{f}_n(\chi)| = 2^{-n}.$$

It follows that supp $\widehat{f} \subseteq X$ and

$$\sum_{\chi \in X} |\widehat{f}(\chi)| = \sum_{n=0}^{\infty} \left(\sum_{\chi \in A_n} |\widehat{f}(\chi)| \right) = \sum_{n=0}^{\infty} 2^{-n} |A_n| = \infty,$$

as required.

To start the construction of the sequence $(g_n)_n$, first let $\chi_1, \chi_2, \ldots$ be any sequence of characters of G which are specified later. Define sequences $g_0, g_1, \ldots$ and $h_0, h_1, \ldots$ inductively in $C(G)$ by $g_0 = h_0 = 1_G$ and

$$g_{n+1} = g_n + \chi_{n+1} h_n \quad \text{and} \quad h_{n+1} = g_n - \chi_{n+1} h_n.$$

It is straightforward to verify that

$$|g_{n+1}(x)|^2 + |h_{n+1}(x)|^2 = 2(|g_n(x)|^2 + |h_n(x)|^2)$$

for all $x \in G$ and $n \in \mathbb{N}_0$. Hence the supremum norm of $|g_{n+1}|^2 + |h_{n+1}|^2$ bounded by $\leq 2^{n+2}$ whenever the supremum norm of $|g_n|^2 + |h_n|^2$ is bounded by $\leq 2^{n+1}$. Suppose that the ranges of both $\widehat{g}_n$ and $\widehat{h}_n$ are contained in $\{-1, 0, 1\}$. Then the same is true of $\widehat{g}_{n+1}$ and $\widehat{h}_{n+1}$ provided that χ_{n+1} has the property that

$$\operatorname{supp} \widehat{g}_n \cap \chi_{n+1}(\operatorname{supp} \widehat{h}_n) = \emptyset.$$

Moreover, if χ_{n+1} has this property and $|\widehat{g}_n|$ and $|\widehat{h}_n|$ are the characteristic functions of sets E_n and F_n, respectively, such that E_n and F_n each contain precisely 2^n elements, then $|\widehat{g}_{n+1}|$ is the characteristic function of $E_n \cup \chi_{n+1} F_n$, which has 2^{n+1} elements, and similarly for $|\widehat{h}_{n+1}|$.

It is now obvious fairly how the sequence $\chi_1, \chi_2, \ldots$ has to be chosen. Since $\int_G \chi(x) dx = 0$ for every $\chi \neq 1_G$ (Exercise 2.12.30), we have $\widehat{g}_0 = \widehat{h}_0 = \delta_{1_G}$. Hence χ_1 may be any nontrivial character from X. Suppose that $\chi_1, \ldots, \chi_n \in X$ have been chosen such that $g_1, \ldots, g_n$ and $h_1, \ldots, h_n$ have the above properties. Then we simply have to select $\chi_{n+1} \in X$ so that $E_n \cap \chi_{n+1}^{-1} F_n = \emptyset$, and this is possible since E_n and F_n are finite and X is infinite. This completes the construction of a sequence $(g_n)_n$ with properties (1), (2), and (3) above. □

The functions f_n constructed in the proof of Lemma 2.7.11 are analogues of the Rudin-Shapiro trigonometric polynomials on the circle group (see [72, p. 33, Exercise 6]).

Theorem 2.7.12. *Let G be a locally compact Abelian group. Then the Gelfand homomorphism $\Gamma : L^1(G) \to C_0(\widehat{G})$ is surjective if and only if G is finite.*

Proof. Suppose that Γ is surjective. Then G is discrete by Lemma 2.7.10. Towards a contradiction, assume that G is infinite. For each $x \in G$, define a character χ_x of $\widehat{G}$ by $\chi_x(\alpha) = \alpha(x)$. Then $x \to \chi_x$ is a bijection between

G and the subgroup $X = \{\chi_x : x \in G\}$ of the dual group $\widehat{\widehat{G}}$ of $\widehat{G}$. Since $\widehat{G}$ is compact and X is infinite, by Lemma 2.7.11 there exists $f \in C(\widehat{G})$ such that supp $\widehat{f} \subseteq X$ and $\sum_{\chi \in X} |\widehat{f}(\chi)| = \infty$. Since Γ is surjective, there exists $g \in L^1(G)$ with $\widehat{g} = f$. Now recall that $\int_{\widehat{G}} \chi(\alpha)d\alpha = 0$ for every $\chi \in \widehat{\widehat{G}} \setminus \{1_{\widehat{G}}\}$. It follows that

$$
\begin{aligned}
\widehat{f}(\chi_x) &= \int_{\widehat{G}} \overline{\chi_x(\alpha)} \left(\sum_{y \in G} g(y)\overline{\alpha(y)} \right) d\alpha \\
&= \sum_{y \in G} g(y) \left(\int_{\widehat{G}} \overline{\chi_x(\alpha)} \, \overline{\chi_y(\alpha)} d\alpha \right) \\
&= \sum_{y \in G} g(y) \left(\int_{\widehat{G}} \overline{\chi_{xy}(\alpha)} d\alpha \right) \\
&= g(x^{-1})
\end{aligned}
$$

for every $x \in G$. Thus

$$
\sum_{x \in G} |g(x^{-1})| = \sum_{x \in G} |\widehat{f}(\chi_x)| = \sum_{\chi \in X} |\widehat{f}(\chi)| = \infty.
$$

This contradiction shows that G must be finite. $\qquad\square$

To prove Theorem 2.7.12, it is possible to avoid the use of Lemma 2.7.11 and instead only apply Lemma 2.7.10 and the Pontryagin duality theorem. However, we prefer not to utilise the duality theorem although in Appendix A.5 we have presented a proof of it, based on the Plancherel theorem. In addition, we feel the construction performed in the proof of Lemma 2.7.11 is of independent interest.

2.8 Beurling algebras $L^1(G,\omega)$

Let G be a locally compact Abelian group and ω a weight function on G. In Section 1.3 we have introduced the associated Beurling algebra $L^1(G,\omega)$. Extending some of the results of the preceding section, we now describe the structure space $\Delta(L^1(G,\omega))$ of $L^1(G,\omega)$ in terms of so-called ω-bounded generalised characters of G. These generalized characters can be identified explicitly when G is either the additive group of real numbers or the group of integers. We also show that $L^1(G,\omega)$ is always semisimple.

Definition 2.8.1. An ω-bounded generalised character on G is a continuous homomorphism α from G into the multiplicative group $\mathbb{C}^\times$ of nonzero complex numbers satisfying $|\alpha(x)| \leq \omega(x)$ for all $x \in G$. Let $\widehat{G}(\omega)$ denote the set of all such ω-bounded generalised characters on G equipped with the topology of uniform convergence on compact subsets of G.

It is clear from the very definition of $\widehat{G}(\omega)$ that $\widehat{G}$ is contained in $\widehat{G}(\omega)$ if and only if $\omega(x) \geq 1$ for all $x \in G$. Our first result is the analogue of Theorem 2.7.2.

Theorem 2.8.2. *Let G be a locally compact Abelian group and ω a weight on G. For $\alpha \in \widehat{G}(\omega)$, define $\varphi_\alpha : L^1(G, \omega) \to \mathbb{C}$ by*

$$\varphi_\alpha(f) = \int_G f(x)\overline{\alpha(x)}dx, \quad f \in L^1(G, \omega).$$

Then $\varphi_\alpha \in \Delta(L^1(G, \omega))$, and the map $\alpha \to \varphi_\alpha$ is a bijection between $\widehat{G}(\omega)$ and $\Delta(L^1(G, \omega))$.

Proof. It is straightforward to show that φ_α is a nonzero homomorphism and that, since $C_c(G) \subseteq L^1(G, \omega)$, the map $\alpha \to \varphi_\alpha$ is injective (compare the proof of Theorem 2.7.2).

To show that every $\varphi \in \Delta(L^1(G, \omega))$ equals φ_α for some $\alpha \in \widehat{G}(\omega)$, we proceed in a similar manner as in the proof of Theorem 2.7.2. Choose $g \in C_c(G)$ such that $\varphi(g) = 1$ and define $\alpha : G \to \mathbb{C}$ by $\alpha(y) = \overline{\varphi(L_y g)}, y \in G$. Then α is continuous because the map $x \to L_x g$ from G into $L^1(G, \omega)$ is continuous (Lemma 1.3.6) and

$$|\alpha(x) - \alpha(y)| = |\varphi(L_x g - L_y g)| \leq \|L_x g - L_y g\|_{1,\omega}.$$

For all $y \in G$, using Lemma 1.3.6,

$$|\alpha(y)| = |\varphi(L_y g)| \leq \|L_y g\|_{1,\omega} \leq \omega(y)\|g\|_{1,\omega}.$$

Moreover, since φ is a homomorphism, we have $\alpha(xy) = \alpha(x)\alpha(y)$ for all $x, y \in G$ (compare the proof of Theorem 2.7.2) and therefore

$$|\alpha(y)| = |\alpha(y^n)|^{1/n} \leq \omega(y^n)^{1/n}\|g\|_{1,\omega}^{1/n} \leq \omega(y)\|g\|_{1,\omega}^{1/n}$$

for all $y \in G$ and $n \in \mathbb{N}$. It follows that $|\alpha(y)| \leq \omega(y)$ for all $y \in G$. This shows that $\alpha \in \widehat{G}(\omega)$.

Finally, for any $f \in C_c(G)$,

$$\varphi(f) = \varphi(g * f) = \varphi\left(x \to \int_G f(y)L_y g(x)dy\right)$$
$$= \int_G f(y)\varphi(L_y g)dy = \int_G f(y)\overline{\alpha(y)}dy$$
$$= \varphi_\alpha(f).$$

Since φ and φ_α are continuous, we conclude that $\varphi = \varphi_\alpha$. $\qquad\square$

Remark 2.8.3. Suppose that the weight ω on G satisfies $\lim_{n\to\infty} \omega(x^n)^{1/n} = 1$ for all $x \in G$. Then $\widehat{G} = \widehat{G}(\omega)$. In fact, the condition implies that $\omega(x) \geq 1$

for all $x \in G$ and hence $\widehat{G} \subseteq \widehat{G}(\omega)$. Conversely, let $\alpha \in \widehat{G}(\omega)$. We have seen in the proof of Theorem 2.8.2 that

$$|\alpha(x)| \leq \lim_{n \to \infty} \omega(x^n)^{1/n}$$

and hence $|\alpha(x)| \leq 1$ for all $x \in G$. Since α is multiplicative, this implies that $|\alpha(x)| = 1$ for all $x \in G$. Therefore, $\widehat{G}(\omega) \subseteq \widehat{G}$.

Lemma 2.8.4. *Let $f \in L^1(G, \omega)$, $x \in G$, and $\epsilon > 0$. Then there exist a neighbourhood W of e in G and $\delta > 0$ with the following property. If $y \in G$ and $\beta, \alpha \in \widehat{G}(\omega)$ are such that $y \in Wx$, $\varphi_\alpha(f) = 1$ and $\beta \in U(\alpha, f, L_x f, \delta)$, then*

$$|\beta(y) - \alpha(x)| < \epsilon.$$

In particular, the function $(x, \alpha) \to \alpha(x)$ on $G \times \widehat{G}(\omega)$ is continuous.

Proof. Note first that $\widehat{L_z f}(\gamma) = \overline{\gamma(z)} \widehat{f}(\gamma)$ for all $z \in G$ and $\gamma \in \widehat{G}(\omega)$ since γ is multiplicative. For arbitrary $y, x \in G$ and $\beta, \alpha \in \widehat{G}(\omega)$ such that $\varphi_\alpha(f) = 1$, as in the proof of Lemma 2.7.4 we get

$$|\beta(y) - \alpha(x)| \leq |\beta(y)| \cdot |1 - \widehat{f}(\beta)| + \|L_y f - L_x f\|_{1,\omega} + |\widehat{L_x f}(\beta) - \widehat{L_x f}(\alpha)|$$
$$\leq \omega(y)|\widehat{f}(\beta) - \widehat{f}(\alpha)| + \|L_y f - L_x f\|_{1,\omega} + |\widehat{L_x f}(\beta) - \widehat{L_x f}(\alpha)|.$$

Now, fix a compact neighbourhood K of x and let

$$C = \max\{1, \sup\{\omega(t) : t \in K\}\} < \infty$$

(Lemma 1.3.3). Let $\delta = \epsilon(3C)^{-1}$ and let W be a neighbourhood of e such that $Wx \subseteq K$ and

$$\|L_y f - L_x f\|_{1,\omega} < \delta$$

for all $y \in Wx$ (Lemma 1.3.6). Then, if $y \in Wx$ and $\beta \in U(\alpha, f, L_x f, \delta)$, the above estimate shows that $|\beta(y) - \alpha(x)| < \epsilon$.   □

Because weight functions are only locally bounded, in contrast to the case of $L^1(G)$ we cannot expect that W and δ in the preceding lemma can be chosen independently of x.

Theorem 2.8.5. *On $\widehat{G}(\omega) = \Delta(L^1(G, \omega))$ the Gelfand topology coincides with the topology of uniform convergence on compact subsets of G.*

Proof. Let g be a bounded measurable function on G with compact support, K say. Then $g \in L^1(G, \omega)$ (Lemma 1.3.5) and for any $\alpha, \beta \in \widehat{G}(\omega)$,

$$|\varphi_\alpha(g) - \varphi_\beta(g)| \leq \int_K |g(x)| \cdot |\alpha(x) - \beta(x)| dx$$
$$\leq \|g\|_\infty |K| \sup_{x \in K} |\alpha(x) - \beta(x)|,$$

where $|K|$ denotes Haar measure of K. Now, given $f \in L^1(G, \omega)$ and $\epsilon > 0$, there exists such a function g satisfying $\|g - f\|_{1,\omega} \leq \epsilon$. It follows that

$$|\varphi_\alpha(f) - \varphi_\beta(f)| \leq 2\|f - g\|_{1,\omega} + |\varphi_\alpha(g) - \varphi_\beta(g)|$$
$$\leq 2\epsilon + \|g\|_\infty |K| \sup_{x \in K} |\alpha(x) - \beta(x)|.$$

This shows that the Gelfand topology on $\widehat{G}(\omega)$ is coarser than the topology of uniform convergence on compact subsets of G.

Conversely, let $\alpha \in \widehat{G}(\omega)$, a compact subset K of G, and $\epsilon > 0$ be given. Let

$$V(\alpha, K, \epsilon) = \{\beta \in \widehat{G}(\omega) : |\beta(x) - \alpha(x)| < \epsilon \text{ for all } x \in K\}$$

and choose $f \in L^1(G, \omega)$ such that $\widehat{f}(\alpha) = 1$. By Lemma 2.8.4, for every $x \in K$ there exist a neighbourhood W_x of e in G and $\delta_x > 0$ with the following property: If $y \in W_x x$ and $\beta \in U(\alpha, f, L_x f, \delta_x)$, then $|\beta(y) - \alpha(x)| < \epsilon$. Since K is compact, there exist $x_1, \ldots, x_n \in K$ such that $K \subseteq \bigcup_{j=1}^n W_{x_j} x_j$. Let $\delta = \min\{\delta_{x_1}, \ldots, \delta_{x_n}\}$. Then

$$U(\alpha, f, L_{x_1} f, \ldots, L_{x_n} f, \delta) \subseteq V(\alpha, K, 2\epsilon).$$

Indeed, if β is in the set on the left side and $x \in K$, then $x \in W_{x_j} x_j$ for some $j \in \{1, \ldots, n\}$ and $\beta \in U(\alpha, f, L_{x_j} f, \delta_{x_j})$ and therefore

$$|\beta(x) - \alpha(x)| \leq |\beta(x) - \alpha(x_j)| + |\alpha(x_j) - \alpha(x)| < 2\epsilon.$$

This shows that the Gelfand topology on $\widehat{G}(\omega)$ is finer than the topology of uniform convergence on compact subsets of G. $\qquad\square$

Identifying $\Delta(L^1(G, \omega))$ as a topological space with $\widehat{G}(\omega)$, the Gelfand representation of $L^1(G, \omega)$ is given by the map $f \to \widehat{f}(\alpha)$, where $\widehat{f}(\alpha) = \int_G f(x)\overline{\alpha(x)} dx$ for $\alpha \in \widehat{G}(\omega)$.

We now determine $\Delta(L^1(G, \omega))$ for G equal to $\mathbb{R}$ or to $\mathbb{Z}$.

Lemma 2.8.6. *Let ω be a weight function on $\mathbb{R}$ and define nonnegative real numbers R_+ and R_- by*

$$R_+ = \inf\{\omega(t)^{1/t} : t > 0\} \text{ and } R_- = \sup\{\omega(-t)^{-1/t} : t > 0\}.$$

Then $0 < R_- \leq R_+$, and every $z \in \mathbb{C}$ satisfying $-\ln R_+ \leq \operatorname{Re} z \leq -\ln R_-$ defines an element φ_z of $\Delta(L^1(\mathbb{R}, \omega))$ by

$$\varphi_z(f) = \int_{\mathbb{R}} f(t)e^{-zt} dt, \quad f \in L^1(\mathbb{R}, \omega).$$

Proof. We show first that

$$R_+ = \lim_{t \to \infty} \omega(t)^{1/t}.$$

To see this, let $\epsilon > 0$ and choose $t_0 > 0$ such that $\omega(t_0)^{1/t_0} \leq R_+ + \epsilon$. Write any $t > 0$ as $t = mt_0 + s$, where $m \in \mathbb{N}_0$ and $0 \leq s < t_0$. Then

$$\omega(t)^{1/t} \leq \omega(mt_0)^{1/t}\omega(s)^{1/t} \leq \omega(t_0)^{1/t_0}(\omega(t_0)^{-s/t_0})^{1/t}\omega(s)^{1/t}.$$
$$\leq (R_+ + \epsilon)(\omega(t_0)^{-s/t_0})^{1/t}\omega(s)^{1/t}.$$

This inequality shows that $\omega(t)^{1/t}$ converges, as $t \to \infty$, with limit R_+. Similarly, it is shown that

$$R_- = \lim_{t\to\infty} \omega(-t)^{-1/t}.$$

Since $\omega(0) \leq \omega(-t)\omega(t)$ for all $t \in \mathbb{R}$, we obtain

$$0 < R_- = \lim_{t\to\infty} \omega(0)^{1/t}\omega(-t)^{-1/t} \leq \lim_{t\to\infty} \omega(t)^{1/t} = R_+.$$

Now, let $z \in \mathbb{C}$ be such that $-\ln R_+ \leq \operatorname{Re} z \leq -\ln R_-$. We claim that $|e^{-zt}| \leq \omega(t)$ for all $t \in \mathbb{R}$. For this, notice that by definition of R_+,

$$\exp(-t\operatorname{Re} z) \leq \exp(t \ln R_+) \leq \exp(t \ln(\omega(t)^{1/t})) = \omega(t)$$

for all $t > 0$. Similarly, for all $t < 0$,

$$\exp(-t\operatorname{Re} z) \leq \exp(t \ln R_-) \leq \exp(t \ln(\omega(t)^{1/t})) = \omega(t).$$

Thus $|e^{-zt}| = \exp(-t\operatorname{Re} z) \leq \omega(t)$ for all $t \in \mathbb{R}$ and hence the integral

$$\int_{\mathbb{R}} f(t)e^{-zt}dt$$

converges absolutely for each $f \in L^1(\mathbb{R}, \omega)$. Therefore, we can define a bounded linear functional φ_z on $L^1(\mathbb{R}, \omega)$ by

$$\varphi_z(f) = \int_{\mathbb{R}} f(t)e^{-tz}dt.$$

It is then easily verified that $\varphi_z(f * g) = \varphi_z(f)\varphi_z(g)$ for all $f, g \in L^1(\mathbb{R}, \omega)$. Hence $\varphi_z \in \Delta(L^1(\mathbb{R}, \omega))$. ☐

Proposition 2.8.7. *Let ω be any weight on $\mathbb{R}$ and let R_+ and R_- be as in Lemma 2.8.6. Let S_ω be the vertical strip in the complex plane defined by*

$$S_\omega = \{z \in \mathbb{C} : -\ln R_+ \leq \operatorname{Re} z \leq -\ln R_-\}.$$

Then the map $z \to \varphi_z$, where φ_z is as in Lemma 2.8.6, is a homeomorphism from S_ω onto $\Delta(L^1(\mathbb{R}, \omega))$.

Proof. It is clear that the map $z \to \varphi_z$ from S_ω into $\Delta(L^1(\mathbb{R}, \omega))$ is injective. We show that every $\varphi \in \Delta(L^1(\mathbb{R}, \omega))$ arises in this manner. To see this, recall first from Theorem 2.8.2 that there exists a continuous function $\gamma : \mathbb{R} \to \mathbb{C}$ satisfying $\gamma(t + s) = \gamma(t)\gamma(s)$ and $0 < |\gamma(t)| \leq \omega(t)$ for all $t, s \in \mathbb{R}$ and

$$\varphi(f) = \int_{\mathbb{R}} f(t)\overline{\gamma(t)}dt$$

for all $f \in L^1(\mathbb{R}, \omega)$. The functional equation $\gamma(t + s) = \gamma(t)\gamma(s)$ and the continuity of γ imply that there exists $w \in \mathbb{C}$ such that

$$\gamma(t) = e^{iwt}$$

for all $t \in \mathbb{R}$ (Exercise 2.12.29). If $w = a + ib$ with $a, b \in \mathbb{R}$, then $|\gamma(t)| \leq \omega(t)$ implies that $e^{-bt} \leq \omega(t)$ for all $t \in \mathbb{R}$. Since

$$e^{-b} = (e^{-bn})^{1/n} \leq \omega(n)^{1/n} \to R_+$$

as $n \to \infty$, we get $-b \leq \ln R_+$. Similarly

$$e^{-b} = (e^{(-b)(-n)})^{-1/n} = |\gamma(-n)|^{-1/n} \geq \omega(-n)^{-1/n} \to R_-$$

as $n \to \infty$, whence $-b \geq \ln R_-$. Thus $-\ln R_+ \leq b \leq -\ln R_-$ and hence $b + ia \in S_\omega$ and $\varphi = \varphi_{b+ia}$.

By Theorem 2.8.5, the map $\alpha \to \varphi_\alpha$ is a homeomorphism between $\widehat{\mathbb{R}}(\omega)$ and $\Delta(L^1(\mathbb{R}, \omega))$. On the other hand, the map $z \to \alpha_z$, where $\alpha_z(t) = e^{zt}$ for $t \in \mathbb{R}$, from S_ω to $\widehat{\mathbb{R}}(\omega)$ is bijective and obviously a homeomorphism. Combining these two facts shows that $z \to \varphi_z$ is a homeomorphism from S_ω onto $\Delta(L^1(\mathbb{R}, \omega))$. $\square$

The formula of Lemma 2.8.6 is reminiscent of the Laplace transform. In fact, $\varphi_z(f)$ is nothing but the Laplace transform of f at $z \in S_\omega$. We now turn to the group of integers.

Proposition 2.8.8. *Let ω be a weight function on $\mathbb{Z}$ and define positive real numbers R_+ and R_- by*

$$R_+ = \inf\{w(n)^{1/n} : n \in \mathbb{N}\} \quad and \quad R_- = \sup\{w(m)^{1/m} : m \in -\mathbb{N}\}.$$

Then there is a homeomorphism from the annulus

$$K(R_-, R_+) = \{z \in \mathbb{C} : R_- \leq |z| \leq R_+\}.$$

onto $\Delta(l^1(\mathbb{Z}, \omega))$ given by $z \to \varphi_z$, where

$$\varphi_z(f) = \sum_{n=-\infty}^{\infty} f(n)z^n, \quad f \in l^1(\mathbb{Z}, \omega).$$

Proof. The following formulae can be verified in very much the same manner as the spectral radius formula (Lemma 1.2.5):

$$R_+ = \lim_{n \to \infty} w(n)^{1/n} \quad and \quad R_- = \lim_{n \to \infty} w(-n)^{-1/n}.$$

For the reader's convenience we nevertheless include the proof of the second. Let $\varepsilon > 0$ be given and choose $k \in \mathbb{N}$ such that $\omega(-k)^{-1/k} > R_- - \varepsilon$. Write $n \in \mathbb{N}$ in the form $n = p(n)k + q(n)$, where $p(n) \in \mathbb{N}_0$ and $0 \le q(n) < k$. Then

$$\frac{p(n)}{n} = \frac{1}{k}\left(1 - \frac{q(n)}{n}\right) \to \frac{1}{k}$$

as $n \to \infty$. Since $\omega(r+s) \le \omega(r)\omega(s)$ for all $r, s \in \mathbb{Z}$, we have $\omega(-n) \le \omega(-k)^{p(n)}\omega(-q(n))$ and hence, for all $n \in \mathbb{N}$,

$$\omega(-n)^{-1/n} \ge \omega(-k)^{-p(n)/n}\omega(-q(n))^{-1/n}.$$

The right hand side converges to $\omega(-k)^{-1/k}$ as $n \to \infty$. Thus $\omega(-n)^{-1/n} > R_- - \varepsilon$ eventually and therefore $R_- = \lim_{n \to \infty}\omega(-n)^{-1/n}$. Now, the inequality

$$\omega(n)^{1/n}\omega(-n)^{1/n} \ge \omega(0)^{1/n} \ge 1$$

implies that $\omega(n)^{1/n} \ge \omega(-n)^{-1/n}$ for all $n \in \mathbb{N}$. It follows that

$$R_- = \lim_{n \to \infty}\omega(-n)^{-1/n} \le \lim_{n \to \infty}\omega(n)^{1/n} = R_+.$$

For $z \in K(R_-, R_+)$ and $f \in l^1(\mathbb{Z}, \omega)$, by definition of R_+ and R_-, we have

$$\sum_{n \in \mathbb{Z}}|f(n)| \cdot |z|^n = |f(0)| + \sum_{n=1}^{\infty}|f(n)| \cdot |z|^n + \sum_{n=1}^{\infty}|f(-n)| \cdot |z|^{-n}$$

$$\le |f(0)| + \sum_{n=1}^{\infty}|f(n)|\omega(n) + \sum_{n=1}^{\infty}|f(-n)|(\omega(-n)^{-1/n})^{-n}$$

$$\le \sum_{n \in \mathbb{Z}}|f(n)|\omega(n)$$

$$= \|f\|_{1,\omega}.$$

Thus, for every $z \in K(R_-, R_+)$ we can define a bounded linear functional on $l^1(\mathbb{Z}, \omega)$ by

$$\varphi_z(f) = \sum_{n \in \mathbb{Z}}f(n)z^n.$$

Then, for $f, g \in l^1(\mathbb{Z}, \omega)$,

$$\varphi_z(f * g) = \sum_{n \in \mathbb{Z}}z^n\left(\sum_{m \in \mathbb{Z}}f(n-m)g(m)\right)$$

$$= \sum_{m \in \mathbb{Z}}g(m)z^m\left(\sum_{n \in \mathbb{Z}}f(n-m)z^{n-m}\right)$$

$$= \varphi_z(f)\varphi_z(g).$$

So $\varphi_z \in \Delta(l^1(\mathbb{Z}, \omega))$ and since $\varphi_z(\delta_1) = z$. The map $z \to \varphi_z$ from $K(R_-, R_+)$ into $\Delta(l^1(\mathbb{Z}, \omega))$ is injective, and the map is continuous since $z \to \varphi_z(\delta_m) = z^m$ is continuous for each m. Conversely, let $\varphi \in \Delta(l^1(\mathbb{Z}, \omega))$ and set $z = \varphi(\delta_1)$. Then, for all $n \in \mathbb{N}$,

$$|z|^n = |\varphi(\delta_n)| \leq \|\delta_n\|_{1,\omega} = \omega(n),$$

and hence $|z| \leq \inf\{\omega(n)^{1/n} : n \in \mathbb{N}\} = R_+$. Similarly, it is shown that $|z| \geq R_-$. Since $\varphi(\delta_n) = z^n$ for all $n \in \mathbb{Z}$ and the finite linear combinations of the Dirac functions δ_n, $n \in \mathbb{Z}$, are dense in $l^1(\mathbb{Z}, \omega)$, continuity of φ implies that $\varphi = \varphi_z$. Thus $z \to \varphi_z$ is a continuous bijection between the compact space $K(R_-, R_+)$ and the Hausdorff space $\Delta(l^1(\mathbb{Z}, \omega))$ and hence is a homeomorphism. $\qquad\square$

Propositions 2.8.7 and 2.8.8 in particular show that $L^1(\mathbb{R}, \omega)$ and $l^1(\mathbb{Z}, \omega)$ are semisimple for any weight ω. Our intention is to establish semisimplicity of $L^1(G, \omega)$ for arbitrary locally compact Abelian groups. We start with the following dichotomy.

Lemma 2.8.9. $L^1(G, \omega)$ *is either semisimple or radical.*

Proof. Assume that $L^1(G, \omega)$ is not radical, and fix any $\varphi \in \Delta(L^1(G, \omega))$. By Theorem 2.8.2, there exists a continuous function $\gamma : G \to \mathbb{C}$ satisfying $\gamma(xy) = \gamma(x)\gamma(y)$, $0 < |\gamma(x)| \leq \omega(x)$ for all $x, y \in G$ and

$$\varphi(f) = \int_G f(x)\overline{\gamma(x)}dx$$

for all $f \in L^1(G, \omega)$. For each $\alpha \in \widehat{G}$, define $\psi_\alpha \in L^1(G, \omega)^*$ by

$$\psi_\alpha(f) = \int_G f(x)\overline{\alpha(x)\gamma(x)}dx.$$

Then $\psi_\alpha \in \Delta(L^1(G, \omega))$ since $\alpha\gamma \in \widehat{G}(\omega)$. Now, let f be an element of the radical of $L^1(G, \omega)$. Then $f\overline{\gamma} \in L^1(G)$ and

$$\widehat{f\overline{\gamma}}(\alpha) = \psi_\alpha(f) = 0$$

for all $\alpha \in \widehat{G}$. Since $L^1(G)$ is semisimple (Corollary 2.7.9), it follows that $f\overline{\gamma} = 0$ and hence $f = 0$ almost everywhere since $\gamma(x) \neq 0$ for all $x \in G$. This shows that $L^1(G, \omega)$ is semisimple. $\qquad\square$

Theorem 2.8.10. *Let G be a locally compact Abelian group and ω a weight on G. Then the Beurling algebra $L^1(G, \omega)$ is semisimple.*

Proof. By virtue of Lemma 2.8.9, it suffices to show that $L^1(G, \omega)$ is not radical. We construct a function $f \in L^1(G, \omega)$ such that

$$r_{L^1(G,\omega)}(f) = \lim_{n\to\infty} \|f^n\|_{1,\omega}^{1/n} > 0,$$

where f^n denotes the n-fold convolution product of f.

Choose a relatively compact symmetric neighbourhood U of the identity e of G and let $f = 1_U$, the characteristic function of U. Then

$$M = \sup\{\omega(x) : x \in U\} < \infty$$

since ω is locally bounded, and $\omega(x) \leq M^n$ for all $x \in U^n$. Since $f \in L^1(G,\omega)$ and $\omega(x) \geq \frac{\omega(e)}{\omega(x^{-1})}$ for all $x \in G$, it follows that

$$\begin{aligned}
\|f^n\|_{1,\omega} &= \int_G |f^n(x)|\omega(x)dx \\
&\geq \omega(e) \int_G |1_U^n(x)| \frac{1}{\omega(x^{-1})} dx \\
&\geq \frac{\omega(e)}{M^n} \|1_U^n\|_1.
\end{aligned}$$

This inequality implies that

$$\begin{aligned}
r_{L^1(G,\omega)}(f) &= \lim_{n\to\infty} \|f^n\|_{1,\omega}^{1/n} \geq \frac{1}{M} \lim_{n\to\infty} \|1_U^n\|^{1/n} \\
&= \frac{1}{M} r_{L^1(G)}(1_U),
\end{aligned}$$

and hence $r_{L^1(G,\omega)}(f) > 0$, as required. $\square$

2.9 The Fourier algebra of a locally compact group

In this section we present a class of semisimple commutative Banach algebras which is currently a matter of intensive study, the *Fourier algebras* $A(G)$ of locally compact groups G. When G is Abelian, $A(G)$ can be shown to be isometrically isomorphic to $L^1(\widehat{G})$. We introduce $A(G)$ and determine its structure space.

Let G be an arbitrary locally compact group. For functions f and g in $L^2(G)$, the function $f * \check{g} : G \to \mathbb{C}$ is defined by

$$f * \check{g}(x) = \int_G f(xy)g(y)dy.$$

Then $f * \check{g} \in C_0(G)$ and $\|f * \check{g}\|_\infty \leq \|f\|_2 \|g\|_2$. Since the mappings $f \to f * \check{g}$ and $g \to f * \check{g}$ from $L^2(G)$ into $C_0(G)$, respectively, are linear and continuous, there is a unique continuous linear map ϕ from the projective tensor product $L^2(G)\widehat{\otimes}_\pi L^2(G)$ into $C_0(G)$ satisfying $\phi(f\otimes g) = f * \check{g}$ for all f and g in $L^2(G)$).

Definition 2.9.1. Let $A(G)$ denote the range of the map

$$\phi : L^2(G)\widehat{\otimes}_\pi L^2(G) \to C_0(G),$$

and endow $A(G)$ with the quotient norm from $L^2(G)\widehat{\otimes}_\pi L^2(G)$. Then $A(G)$ becomes a Banach space.

Since $C_0(G)$ is dense in $L^2(G)$, $C_c(G) \otimes C_c(G)$ is dense in $L^2(G)\widehat{\otimes}_\pi L^2(G)$ and hence $\pi(C_c(G) \otimes C_c(G))$ is dense in $A(G)$. So $A(G) \cap C_c(G)$ is dense in $A(G)$.

Theorem 2.9.2. *With pointwise multiplication, $A(G)$ is a Banach algebra.*

Proof. Let $f_1, f_2, g_1, g_2 \in C_c(G)$. We first show that

$$\phi(f_1 \otimes g_1)\phi(f_2 \otimes g_2) \in A(G)$$

and that

$$\|\phi(f_1 \otimes g_1)\phi(f_2 \otimes g_2)\| \le \|\phi(f_1 \otimes g_1)\| \cdot \|\phi(f_2 \otimes g_2)\|.$$

To that end, for $y \in G$, define functions F_y and G_y on G by

$$F_y(x) = f_1(xy)f_2(x) \text{ and } G_y(x) = g_1(xy)g_2(x).$$

Then $F_y, G_y \in C_c(G)$ and the map $y \to F_y \otimes G_y$ vanishes outside the compact subset

$$C = (\operatorname{supp} f_2)^{-1} \operatorname{supp} f_1 \cap (\operatorname{supp} g_2)^{-1} \operatorname{supp} g_1$$

of G. Moreover, the map $y \to F_y \otimes G_y$ from G into $L^2(G)\widehat{\otimes}_\pi L^2(G)$ is continuous. Indeed, for $y, y_0 \in G$,

$$\pi(F_y \otimes G_y - F_{y_0} \otimes G_{y_0}) \le \|F_y - F_{y_0}\|_2(\|G_y - G_{y_0}\|_2 + \|G_{y_0}\|_2)$$
$$+ \|F_{y_0}\|_2 \|G_y - G_{y_0}\|_2$$

and

$$\|F_y - F_{y_0}\|_2 \le \|f_2\|_\infty \|R_y f_1 - R_{y_0} f_1\|_2,$$

and similarly for G_y. Thus the vector-valued integral

$$H = \int_C (F_y \otimes G_y)dy = \int_G (F_y \otimes G_y)dy$$

exists and defines an element of $L^2(G)\widehat{\otimes}_\pi L^2(G)$. Then

$$\phi(H) = \phi(f_1 \otimes g_1)\phi(f_2 \otimes g_2).$$

Indeed, for each $x \in G$, we have

$$\phi(f_1 \otimes g_1)(x)\phi(f_2 \otimes g_2)(x) = \int_G f_1(xy)g_1(y)dy \int_G f_2(xz)g_2(z)dz$$

$$= \int_G \int_G f_1(xzy)g_1(zy)f_2(xz)g_2(z)dzdy$$

$$= \int_G \left(\int_G F_y(xz)G_y(z)dz \right) dy$$

$$= \int_G \phi(F_y \otimes G_y)(x)dy$$

$$= \phi(H)(x).$$

This shows that $[\phi(L^2(G) \otimes L^2(G))]^2 \subseteq A(G)$. We now need to estimate the integral $\int_C \pi(F_y \otimes G_y)dy$. Note first that

$$\int_C \|F_y\|_2^2 dy = \int_C \left(\int_C |f_1(xy)f_2(x)|^2 dx \right) dy$$

$$= \int_C |f_2(x)|^2 \left(\int_C |f_1(xy)|^2 dy \right) dx$$

$$= \|f_1\|_2^2\|f_2\|_2^2,$$

and similarly $\int_G \|G_y\|_2^2 dy = \|g_1\|_2^2\|q_2\|_2^2$. Thus, by the Cauchy-Schwarz inequality,

$$\int_C \pi(F_y \otimes G_y)dy = \int_C \|F_y\|_2\|G_y\|_2 dy$$

$$\leq \left(\int_C \|F_y\|_2^2 dy \right)^{1/2} \left(\int_C \|G_y\|_2^2 dy \right)^{1/2}$$

$$= \|f_1\|_2\|f_2\|_2\|g_1\|_2\|g_2\|_2$$

$$= \pi(f_1 \otimes g_1)\pi(f_2 \otimes g_2).$$

Combining this estimate with the above formula for $\phi(H)$ gives

$$\|\phi(f_1 \otimes g_1)\phi(f_2 \otimes g_2)\|_{A(G)} \leq \|\phi(f_1 \otimes g_1)\|_{A(G)}\|\phi(f_2 \otimes g_2)\|_{A(G)}.$$

Thus multiplication on $\phi(L^2(G) \otimes L^2(G))$ is continuous. This implies that $A(G)$ is closed under multiplication and the norm on $A(G)$ is submultiplicative. $\square$

The following lemma will be used to determine $\Delta(A(G))$ and also later in Chapter 5.

Lemma 2.9.3. *Let* $a \in G$ *and* $f \in A(G)$ *such that* $f(a) = 0$. *Then, given* $\epsilon > 0$, *there exists* $h \in A(G) \cap C_c(G)$ *vanishing in a neighbourhood of* a *such that* $\|h - f\|_{A(G)} \leq \epsilon$.

Proof. Notice first that, since $A(G) \cap C_c(G)$ is dense in $A(G)$, without loss of generality we can assume that $f \neq 0$, f has compact support and $\epsilon \leq \|f\|_\infty$ and $\epsilon < 1$. Let

$$W = \{y \in G : \|f - R_y f\|_{A(G)} \leq \epsilon\}.$$

Then W is a compact neighbourhood of e in G. Choose an open neighbourhood V of e such that $V \subseteq W$ and $\sup\{|f(ay)| : y \in V\} \leq \epsilon$, and choose a compact neighbourhood U of e such that $U \subseteq V$ and $|U| \geq |V|(1 - \epsilon)$. Now, define functions u, g and h by setting $u = |U|^{-1} 1_U, g = 1_{aV} f$ and

$$h = (f - g) * \check{u} \in A(G).$$

Then h has compact support since W is compact and f has compact support. For any $x \in G$,

$$h(x) = |U|^{-1} \int_U f(xy)[1 - 1_{aV}(xy)]dy.$$

It follows that, if $x \in G$ satisfies $a^{-1}xU \subseteq V$, then $h(x) = 0$. Thus h vanishes in a neighbourhood of a. Moreover,

$$\|u\|_2 = |U|^{-1/2} \leq |V|^{-1/2} \left(\frac{1}{1-\epsilon}\right)^{1/2},$$

$$\|g\|_2 = \left(\int_{aV} |f(y)|^p dy\right)^{1/2} \leq \epsilon |V|^{1/2},$$

and

$$\|f - f * \check{u}\|_{A(G)} = \left\|f - |U|^{-1} \int_U (R_y f)dy\right\|_{A(G)}$$
$$\leq \sup_{y \in U} \|f - R_y f\|_{A(G)} \leq \epsilon.$$

Combining all these estimates, we obtain

$$\|f - h\|_{A(G)} \leq \|f - f * \check{u}\|_{A(G)} + \|g\|_2 \|\check{u}\|_2 \leq \epsilon + \epsilon \left(\frac{1}{1-\epsilon}\right)^{1/2}.$$

This finishes the proof.   □

Theorem 2.9.4. *Let G be a locally compact group. For $x \in G$, let $\varphi_x : A(G) \to \mathbb{C}$ denote the evaluation at x. Then the map $x \to \varphi_x$ is a homeomorphism from G onto $\Delta(A(G))$.*

Proof. It is obvious that $\varphi_x \in \Delta(A(G))$ and that the map $x \to \varphi_x$ is injective. Now let $\varphi \in \Delta(A(G))$ be given and suppose that $\varphi \neq \varphi_x$ for all $x \in G$. Then, for each $x \in G$ there exists $f_x \in A(G)$ such that $\varphi(f_x) = 1$, but $\varphi_x(f_x) = 0$.

By Lemma 2.9.3, every $g \in A(G)$ vanishing at x is the limit of a sequence $(g_n)_n$ in $A(G)$ with the property that each g_n vanishes in a neighbourhood of x. Therefore we can assume that f_x vanishes in a neighbourhood V_x of x.

Since $A(G) \cap C_c(G)$ is dense in $A(G)$, there exists $f_0 \in C_c(G) \cap A(G)$ such that $\varphi(f_0) = 1$. Choose $x_1, \ldots, x_n \in \operatorname{supp} f_0$ such that

$$\operatorname{supp} f_0 \subseteq \bigcup_{j=1}^{n} V_{x_j}$$

and let

$$f = f_0 f_{x_1} \cdot \ldots \cdot f_{x_n} \in A(G).$$

Then $f(x) = 0$ for every $x \in G$, whereas

$$\varphi(f) = \varphi(f_0) \prod_{j=1}^{n} \varphi(f_{x_j}) = 1.$$

This contradiction shows that $\varphi = \varphi_x$ for some $x \in G$.

Finally, since the subalgebra $A(G)$ of $C_0(G)$ strongly separates the points of G, by Proposition 2.2.14 the topology on G coincides with the weak topology defined by the set of functions $x \to f(x) = \varphi_x(f), f \in A(G)$. Thus the map $x \to \varphi_x$ from G to $\Delta(A(G))$ is a homeomorphism.   □

Of course, after identifying $\Delta(A(G))$ with G, the Gelfand homomorphism of $A(G)$ is nothing but the identity mapping. In particular, $A(G)$ is semisimple.

We close this section with a straightforward result which, in the terminology of Chapter 4, implies that $A(G)$ is regular.

Lemma 2.9.5. *Let G be a locally compact group, K a compact subset of G and U an open subset of G such that $U \supseteq K$. Then there exists $u \in A(G) \cap C_c(G)$ with the following properties: $0 \leq u \leq 1$, $u(x) = 1$ for all $x \in K$ and $u(x) = 0$ for all $x \in G \setminus U$.*

Proof. Since K is compact, there exists a compact symmetric neighbourhood V of the identity such that $KV^2 \subseteq U$. Let

$$u(x) = |V|^{-1} (1_{KV} * 1_V^{\vee})(x) = |V|^{-1} \cdot |xV \cap KV|.$$

Then $0 \leq u \leq 1$. If $x \in K$, then $|xV \cap KV| = |xV| = |V|$, so that $u(x) = 1$, whereas if $x \notin KV^2$, then $xV \cap KV = \emptyset$ and hence $u(x) = 0$. Thus $\operatorname{supp} u \subseteq KV^2$, which is compact. In particular, $u(x) = 0$ for all $x \in G \setminus U$.   □

2.10 The algebra of almost periodic functions

In Theorem 2.4.12 we have seen that the Stone–Čech compactification $\beta(X)$ of a completely regular topological space X arises as the structure space of the commutative C^*-algebra $C^b(X)$. In this section we study, for G a locally compact group, a certain C^*-subalgebra of $C^b(G)$, the algebra $AP(G)$ of almost periodic functions on G, and show that $\Delta(AP(G))$ is homeomorphic to the Bohr compactification of G.

Let G be a topological group. A complex-valued bounded continuous function f on G is called *left almost periodic* (respectively, *right almost periodic*) if the set $C_f = \{L_x f : x \in G\}$ (respectively, the set $D_f = \{R_y f : y \in G\}$) is relatively compact in $(C^b(G), \|\cdot\|_\infty)$. Let $AP(G)$ denote the set of all left almost periodic functions on G.

Example 2.10.1. Let G be a compact group. Then $AP(G) = C(G)$. In fact, for $f \in C(G)$ the map $x \to L_x f$ from G into $C(G)$ is continuous because f is uniformly continuous and

$$\|L_x f - L_y f\|_\infty = \sup_{t \in G} |f(x^{-1}t) - f(y^{-1}t)|.$$

So C_f is a continuous image of the compact group G, hence compact.

Lemma 2.10.2. $AP(G)$ *is a closed $*$-subalgebra of $C^b(G)$.*

Proof. It is clear that $C_{f+g} \subseteq C_f + C_g, C_{\alpha f} = \alpha C_f$ and $C_{fg} \subseteq C_f C_g$ for $f, g \in C^b(G)$ and $\alpha \in \mathbb{C}$. Thus $AP(G)$ is a subalgebra of $C^b(G)$. Also $f \in AP(G)$ implies that $\overline{f} \in AP(G)$. It remains to show that $AP(G)$ is closed in $C^b(G)$.

Let $f \in \overline{AP(G)}$. Since C_f is bounded in $C^b(G)$, by the Arzela–Ascoli theorem it suffices to verify that C_f is equicontinuous. To that end, let $x \in G$ and $\epsilon > 0$ be given. Choose $g \in AP(G)$ such that $\|f - g\|_\infty \leq \epsilon/3$. Since C_g is equicontinuous, there is a neighbourhood V of x such that $|L_a g(y) - L_a g(x)| \leq \epsilon/3$ for all $a \in G$ and $y \in V$. If follows that

$$|L_a f(y) - L_a f(x)| \leq |L_a f(y) - L_a g(y)| + |L_a g(y) - L_a g(x)|$$
$$+ |L_a g(x) - L_a f(x)|$$
$$\leq 2\|f - g\|_\infty + |L_a g(y) - L_a g(x)| \leq \epsilon$$

for all $y \in V$ and $a \in G$. So C_f is equicontinuous. $\qquad\square$

Since $AP(G)$ is a unital commutative C^*-algebra, Theorem 2.4.5 implies the following

Corollary 2.10.3. *Let $\Delta(AP(G))$ denote the structure space of $AP(G)$. Then the Gelfand homomorphism is an isometric $*$-isomorphism from $AP(G)$ onto $C(\Delta(AP(G)))$.*

Each $x \in G$ defines an element $\varphi_x \in \Delta(AP(G))$ by $\varphi_x(f) = f(x), f \in AP(G)$.

Lemma 2.10.4. *The mapping $\phi : x \to \varphi_x$ from G into $\Delta(AP(G))$ is continuous and has dense range.*

Proof. Because $\Delta(AP(G))$ carries the w^*-topology and the functions $x \to \varphi_x(f) = f(x), f \in AP(G)$, are continuous on G, it follows that ϕ is continuous.

Suppose that there exists a nonempty open subset U of $\Delta(AP(G))$ such that $U \cap \phi(G) = \emptyset$. Then, by Urysohn's lemma there exists $g \in C(\Delta(AP(G)))$ with $g \neq 0$ and $g|_{\Delta(AP(G))\setminus U} = 0$. By Corollary 2.10.3, $g = \hat{f}$ for some $f \in AP(G)$. But then

$$f(x) = \varphi_x(f) = \hat{f}(\varphi_x) = g(\varphi_x) = 0$$

for all $x \in G$, contradicting $f \neq 0$. Thus $\phi(G)$ is dense in $\Delta(AP(G))$. $\qquad\square$

Our aim is to introduce a group structure on $\Delta(AP(G))$ which makes $\Delta(AP(G))$ a compact group and ϕ a group homomorphism. Of course, the mapping ϕ is in general not injective and it is not clear at all that the families of points in G which cannot be separated by $AP(G)$ are cosets of some normal subgroup of G and that therefore ϕ defines a group structure on $\phi(G) \subseteq \Delta(AP(G))$. Moreover, supposing that this problem can be satisfactorily settled, there remains the question of extending the group structure on $\phi(G)$ to the whole of $\Delta(AP(G))$. To handle these problems requires us to consider two-sided translates of $f \in AP(G)$ and to show that actually such an f is also right almost periodic.

Lemma 2.10.5. *Let $f \in AP(G)$ and $\epsilon > 0$. Then there exist finitely many $a_1, \ldots, a_n \in G$ with the following property. For every $a \in G$ there exists some $j \in \{1, \ldots, n\}$ such that*

$$|f(xay) - f(xa_jy)| < \epsilon$$

for all $x, y \in G$.

Proof. There exist $b_1, \ldots, b_m \in G$ such that the set $\{L_{b_j}f : 1 \leq j \leq m\}$ forms an $\epsilon/4$-net for C_f. Let Γ be the finite set of all mappings γ from $\{1, \ldots, m\}$ to itself with the property that there exists $a_\gamma \in G$ such that

$$\|L_{b_{\gamma(i)}}f - L_{b_i a_\gamma}f\|_\infty < \frac{\epsilon}{4}$$

for $i = 1, \ldots, m$. For each $\gamma \in \Gamma$, choose such an a_γ. Now, given any $a \in G$, by the choice of $b_1, \ldots, b_m$ for every $1 \leq i \leq m$ there exists some $j(i) \in \{1, \ldots, m\}$ such that

$$\|L_{b_i a}f - L_{b_{j(i)}}f\|_\infty < \frac{\epsilon}{4}.$$

So $i \to j(i)$ defines an element of Γ. It follows that for every $a \in G$ we find some $\gamma \in \Gamma$ such that

$$\|L_{b_i a}f - L_{b_i a_\gamma}f\|_\infty < \frac{\epsilon}{2}$$

for all $1 \leq i \leq m$. Since for every $x \in G$ there exists b_i so that $\|L_x f - L_{b_i}f\|_\infty < \epsilon/4$, we obtain that

$$\|L_{xa}f - L_{xa_\gamma}f\|_\infty \le \|L_{xa}f - L_{b_ia}f\|_\infty + \|L_{b_ia}f - L_{b_ia_\gamma}f\|_\infty$$
$$+ \|L_{b_ia_\gamma}f - L_{xa_\gamma}f\|_\infty$$
$$< \frac{\epsilon}{4} + \frac{\epsilon}{2} + \frac{\epsilon}{4} = \epsilon.$$

Thus we have seen that for every $a \in G$ there exists some a_γ such that

$$|f(xay) - f(xa_\gamma y)| \le \|L_{xa}f - L_{xa_\gamma}f\|_\infty < \epsilon$$

for all $x, y \in G$. Now, enumerate $\{a_\gamma : \gamma \in \Gamma\}$ as $\{a_1, \dots, a_n\}$.   □

Corollary 2.10.6. Let $f \in AP(G)$ and $\epsilon > 0$. Then there exist $a_1, \dots, a_n \in G$ such that the functions $L_{a_i}R_{a_j}f, 1 \le i, j \le n$, form an ϵ-net for the set of all two-sided translates $L_aR_bf, a, b \in G$.

Proof. Choose $0 < \delta < \epsilon/2$. By Lemma 2.10.5, there exist $a_1, \dots, a_n \in G$ with the property that for any $a \in G$ there is $j \in \{1, \dots, n\}$ such that, for all $x, y \in G$, $|f(xay) - f(xa_jy)| < \delta$. Thus, given $a, b \in G$, there exist i and j such that

$$|f(at) - f(a_it)| < \delta \text{ and } |f(sb) - f(sa_j)| < \delta$$

for all $s, t \in G$. It follows that, for all $x \in G$,

$$|f(axb) - f(a_ixa_j)| \le |f(axb) - f(a_ixb)| + |f(a_ixb) - f(a_ixa_j)|$$
$$< 2\delta,$$

whence $\|L_aR_bf - L_{a_i}R_{a_j}f\|_\infty \le 2\delta < \epsilon$.   □

Corollary 2.10.7. *Retain the notation of Corollary 2.10.6. If x and y are elements of G such that*

$$|f(a_ixa_j) - f(a_iya_j)| < \epsilon$$

for all $1 \le i, j \le n$, then

$$|f(axb) - f(ayb)| < 3\epsilon$$

for all $a, b \in G$.

Proof. Given $a, b \in G$, by Corollary 2.10.6 there exist i and j such that

$$\|L_aR_bf - L_{a_i}R_{a_j}f\|_\infty < \epsilon.$$

Combining with the presumed inequality, we get

$$|f(axb) - f(ayb)| \le \|L_aR_bf - L_{a_i}R_{a_j}f\|_\infty$$
$$+ |f(a_ixa_j) - f(a_iya_j)|$$
$$+ \|L_{a_i}R_{a_j}f - L_aR_bf\|_\infty$$
$$< 3\epsilon,$$

as claimed.   □

It follows from Corollary 2.10.6 that every left almost periodic function is automatically right almost periodic. Therefore, in the sequel we simply call the functions in $AP(G)$ *almost periodic* rather than left almost periodic.

Another consequence of Corollary 2.10.6 is that every almost periodic function is uniformly continuous. Now, on every noncompact locally compact group G one can construct a bounded continuous function which fails to be uniformly continuous (Exercise 2.12.55). Thus $AP(G)$ is a proper subalgebra of $C^b(G)$ whenever G is a noncompact locally compact group.

Let $\varphi, \psi \in \Delta(AP(G))$. For neighbourhoods U of φ and V of ψ in $\Delta(AP(G))$, let

$$\Delta_{U,V} = \{\varphi_{xy} : x, y \in G \text{ such that } \varphi_x \in U \text{ and } \varphi_y \in V\}.$$

Then $\Delta_{U,V} \neq \emptyset$ since $\phi(G)$ is dense in $\Delta(AP(G))$. Let $\mathcal{U}$ and $\mathcal{V}$ be the set of all neighbourhoods of φ and ψ, respectively. Because $\Delta_{U_1,V_1} \subseteq \Delta_{U_2,V_2}$ whenever $U_1 \subseteq U_2$ and $V_1 \subseteq V_2$, the collection of all closed subsets $\overline{\Delta}_{U,V}$ of $\Delta(AP(G))$, where $U \in \mathcal{U}$ and $V \in \mathcal{V}$, has the finite intersection property. $\Delta(AP(G))$ being compact, it follows that the set

$$\Delta_{\varphi,\psi} := \bigcap \{\overline{\Delta}_{U,V} : U \in \mathcal{U}, V \in \mathcal{V}\}$$

is nonempty.

We shall see soon (Corollary 2.10.9) that $\Delta_{\varphi,\psi}$ is a singleton for any two elements φ, ψ of $\Delta(AP(G))$. Since $\varphi_{xy} \in \Delta_{\varphi_x,\varphi_y}$ it follows in particular that $\Delta_{\varphi_x,\varphi_y} = \{\varphi_{xy}\}$ for all $x, y \in G$.

Lemma 2.10.8. *Let $\alpha, \beta \in \Delta(AP(G))$ and $f \in AP(G)$. Let $\epsilon > 0$ and let $\{L_{x_1}f, \ldots, L_{x_n}f\}$ be an ϵ-net for C_f and $\{R_{y_1}f, \ldots, R_{y_m}f\}$ an ϵ-net for D_f. Define neighbourhoods U and V of α and β, respectively, by*

$$U = U(\alpha, R_{y_1}f, \ldots, R_{y_m}f, \epsilon) \text{ and } V = U(\beta, L_{x_1}f, \ldots, L_{x_n}f, \epsilon).$$

If $x, a, y, b \in G$ are such that $\varphi_x, \varphi_a \in U$ and $\varphi_y, \varphi_b \in V$, then

$$|\varphi_{xy}(f) - \varphi_{ab}(f)| < 8\epsilon.$$

Proof. Choose $j \in \{1, \ldots, n\}$ and $k \in \{1, \ldots, m\}$ such that

$$\|L_{x^{-1}}f - L_{x_j}f\|_\infty < \epsilon \text{ and } \|R_b f - R_{y_k}f\|_\infty < \epsilon.$$

Then we have

$$
\begin{aligned}
|\varphi_{xy}(f) - \varphi_{ab}(f)| &\leq |f(xy) - f(xb)| + |f(xb) - f(ab)| \\
&= |L_{x^{-1}}f(y) - L_{x^{-1}}f(b)| + |R_b f(x) - R_b f(a)| \\
&\leq |L_{x^{-1}}f(y) - L_{x_j}f(y)| + |L_{x_j}f(y) - L_{x_j}f(b)| \\
&\quad + |L_{x_j}f(b) - L_{x^{-1}}f(b)| + |R_b f(x) - R_{y_k}f(x)| \\
&\quad + |R_{y_k}f(x) - R_{y_k}f(a)| + |R_{y_k}f(a) - R_b f(a)| \\
&\leq 2\|L_{x^{-1}}f - L_{x_j}f\|_\infty + |L_{x_j}f(x) - L_{x_j}f(a)| \\
&\quad + 2\|R_b f - R_{y_k}f\|_\infty + |R_{y_k}f(y) - R_{y_k}f(b)\| \\
&\leq 4\epsilon + |\varphi_y(L_{x_j}f) - \varphi_b(L_{x_j}f)| + |\varphi_x(R_{y_k}f) - \varphi_a(R_{y_k}f)|.
\end{aligned}
$$

Now, since $\varphi_x, \varphi_a \in U$ and $\varphi_y, \varphi_b \in V$,

$$|\varphi_x(R_{y_k}f) - \varphi_a(R_{y_k}f)| < 2\epsilon \text{ and } |\varphi_y(L_{x_j}f) - \varphi_b(L_{x_j}f)| < 2\epsilon.$$

It follows that $|\varphi_{xy}(f) - \varphi_{ab}(f)| < 8\epsilon.$ $\qquad\square$

Corollary 2.10.9. *For each pair of elements* φ, ψ *of* $\Delta(AP(G)), \Delta_{\varphi,\psi}$ *is a singleton.*

Proof. Let $\alpha, \beta \in \Delta_{\varphi,\psi}$ and $f \in AP(G)$. We show that $|\alpha(f) - \beta(f)| < \delta$ for each $\delta > 0$. Fix δ and let $\epsilon = \delta/24$. Let U and V be defined as in Lemma 2.10.8. By definition of $\Delta_{\varphi,\psi}$ there exist $x, a, y, b \in G$ such that $\varphi_x, \varphi_a \in U, \varphi_y, \varphi_b \in V$, and

$$|\alpha(f) - \varphi_{xy}(f)| < \frac{\delta}{3} \text{ and } |\beta(f) - \varphi_{ab}(f)| < \frac{\delta}{3}.$$

From Lemma 2.10.8 we now infer that

$$\begin{aligned}|\alpha(f) - \beta(f)| &\leq |\alpha(f) - \varphi_{xy}(f)| + |\varphi_{xy}(f) - \varphi_{ab}(f)| \\ &\quad + |\varphi_{ab}(f) - \beta(f)| \\ &\leq \frac{\delta}{3} + \frac{\delta}{3} + 8\epsilon = \delta,\end{aligned}$$

as required. $\qquad\square$

Now we are able to introduce a group structure on $\Delta(AP(G))$.

Theorem 2.10.10. *Let* G *be a topological group. For* $\varphi, \psi \in \Delta(AP(G))$, *let* $\varphi\psi$ *denote the unique element of* $\Delta_{\varphi,\psi}$. *Then the assignment*

$$(\varphi, \psi) \to \varphi\psi, \ \Delta(AP(G)) \times \Delta(AP(G)) \to \Delta(AP(G))$$

turns $\Delta(AP(G))$ *into a compact group. Furthermore,* $\varphi_x\varphi_y = \varphi_{xy}$ *for* $x, y \in G$.

Proof. The last statement is clear since $\Delta_{\varphi_x,\varphi_y} = \{\varphi_{xy}\}$. We show next that multiplication on $\Delta(AP(G))$ is continuous. Let α and β be two elements of $\Delta(AP(G))$. It suffices to show that given $\delta > 0$ and $f_1, \ldots, f_n \in AP(G)$, there exist neighbourhoods U of α and V of β in $\Delta(AP(G))$, respectively, such that

$$|\varphi\psi(f_j) - \alpha\beta(f_j)| < \delta$$

for all $\varphi \in U$ and $\psi \in V$ and $j = 1, \ldots, n$.

Let $\epsilon = \delta/10$ and for any $\rho \in \Delta(AP(G))$ let

$$W_\rho = \{\gamma \in \Delta(AP(G)) : |\gamma(f_j) - \rho(f_j)| < \epsilon \text{ for } 1 \leq j \leq n\}.$$

For each $j = 1, \ldots, n$, Lemma 2.10.8 provides neighbourhoods U_j of α and V_j of β such that

$$|\varphi_{xy}(f_j) - \varphi_{ab}(f_j)| < 8\epsilon$$

whenever $x, a, y, b \in G$ are such that $\varphi_x, \varphi_a \in U_j$ and $\varphi_y, \varphi_b \in V_j$. Let $U = \cap_{j=1}^n U_j$ and $V = \cap_{j=1}^n V_j$. Since $\alpha\beta \in \overline{\Delta}_{U,V}$, we have $\Delta_{U,V} \cap W_{\alpha\beta} \neq \emptyset$, and hence there exist $a, b \in G$ such that $\varphi_a \in U, \varphi_b \in V$ and $\varphi_{ab} \in W_{\alpha\beta}$. Now, let $\varphi \in U$ and $\psi \in V$ be arbitrary. Then $\Delta_{U,V} \cap W_{\varphi\psi} \neq \emptyset$ and hence there exist $x, y \in G$ such that $\varphi_x \in U, \varphi_y \in V$, and $\varphi_{xy} \in W_{\varphi\psi}$. Therefore we have

$$|\varphi_{ab}(f_j) - \alpha\beta(f_j)| < \epsilon \text{ and } |\varphi_{xy}(f_j) - \varphi\psi(f_j)| < \epsilon$$

for $j = 1, \ldots, n$. Because $\varphi_x, \varphi_a \in U$ and $\varphi_y, \varphi_b \in V$, $|\varphi_{xy}(f_j) - \varphi_{ab}(f_j)| < 8\epsilon$ for $j = 1, \ldots, n$. Combining these inequalities gives

$$\begin{aligned} |\varphi\psi(f_j) - \alpha\beta(f_j)| &\leq |\varphi\psi(f_j) - \varphi_{xy}(f_j)| + |\varphi_{xy}(f_j) - \varphi_{ab}(f_j)| \\ &\quad + |\varphi_{ab}(f_j) - \alpha\beta(f_j)| < 10\epsilon \\ &= \delta. \end{aligned}$$

Thus multiplication on $\Delta(AP(G))$ is continuous.

It remains to show the existence and continuity of inverses in $\Delta(AP(G))$. Let $\varphi \in AP(G)$ and let $(x_\alpha)_\alpha$ be a net in G such that $\varphi_{x_\alpha} \to \varphi$ in $\Delta(AP(G))$. We show that the net $(\varphi_{x_\alpha^{-1}})_\alpha$ converges to some element of $\Delta(AP(G))$ and that the limit does not depend on the choice of the net $(x_\alpha)_\alpha$ in G but only on the fact that $\varphi_{x_\alpha} \to \varphi$.

Let $f \in AP(G)$ and $\epsilon > 0$. By Corollary 2.10.6 there exist $a_1, \ldots, a_n \in G$ such that the functions $L_{a_i} R_{a_j} f, 1 \leq i, j \leq n$, form an $\epsilon/3$-net for the set of all two-sided translates $L_a R_b f, a, b \in G$. Define a neighbourhood U of φ in $\Delta(AP(G))$ by

$$U = \{\psi \in \Delta(AP(G)) : |\psi(L_{a_i} R_{a_j} f) - \varphi(L_{a_i} R_{a_j} f)| < \epsilon/3, 1 \leq i, j \leq n\}.$$

If x and y are elements of G such that $\varphi_x, \varphi_y \in U$, then

$$|f(a_i x a_j) - f(a_i y a_j)| < \epsilon/3, \quad 1 \leq i, j \leq n,$$

and hence, by Corollary 2.10.7,

$$|f(axb) - f(ayb)| < \epsilon$$

for all $a, b \in G$. Taking $a = x^{-1}$ and $b = y^{-1}$ this becomes $|f(y^{-1}) - f(x^{-1})| < \epsilon$. This shows that, for each $f \in AP(G)$, the net

$$(\varphi_{x_\alpha^{-1}}(f))_\alpha = (f(x_\alpha^{-1}))_\alpha$$

forms a Cauchy net in $\mathbb{C}$ and that

$$\lim_\alpha \varphi_{x_\alpha^{-1}}(f) = \lim_\beta \varphi_{y_\beta^{-1}}(f),$$

where $(y_\beta)_\beta$ is another net in G such that $\varphi_{y_\beta} \to \varphi$ in $\Delta(AP(G))$. Thus we can define a map $\varphi^{-1} : AP(G) \to \mathbb{C}$ by

$$\varphi^{-1}(f) = \lim_\alpha \varphi_{x_\alpha^{-1}}(f), \quad f \in AP(G),$$

by taking $(x_\alpha)_\alpha$ to be any net in G such that $\varphi_{x_\alpha} \to \varphi$. It is clear that $\varphi^{-1} \in \Delta(AP(G))$ and that $\varphi_x^{-1} = \varphi_{x^{-1}}$ for every $x \in G$. Since multiplication in $\Delta(AP(G))$ is continuous and $\varphi_{ab} = \varphi_a \varphi_b$ for all $a, b \in G$, it follows that

$$\varphi \varphi^{-1} = \lim_\alpha \varphi_{x_\alpha} \cdot \lim_\alpha \varphi_{x_\alpha^{-1}} = \lim_\alpha (\varphi_{x_\alpha x_\alpha^{-1}}) = \varphi_e.$$

Consequently, $\Delta(AP(G))$ is a group and φ^{-1} is the inverse of φ.

Finally, the map $\varphi \to \varphi^{-1}$ from $\Delta(AP(G))$ into $\Delta(AP(G))$ is continuous. To see this, let $\psi \in \Delta(AP(G)), f \in AP(G)$, and $\delta > 0$. Define $g \in AP(G)$ by $g(x) = f(x^{-1})$. If $\varphi \in \Delta(AP(G))$ and $x, y \in G$ are such that

$$|\varphi(g) - \psi(g)| < \delta, \quad |\varphi_x(g) - \varphi(g)| < \delta \text{ and } |\varphi_y(g) - \psi(g)| < \delta,$$

then

$$|\varphi_{x^{-1}}(f) - \varphi_{y^{-1}}(f)| = |\varphi_x(g) - \varphi_y(g)| < 3\delta$$

and hence

$$|\varphi^{-1}(f) - \psi^{-1}(f)| \le |\varphi^{-1}(f) - \varphi_{x^{-1}}(f)| + |\varphi_{y^{-1}}(f) - \psi^{-1}(f)| + 3\delta.$$

As we have shown above, $\varphi_{x^{-1}} \to \varphi^{-1}$ and $\varphi_{y^{-1}} \to \psi^{-1}$ whenever $\varphi_x \to \varphi$ and $\varphi_y \to \psi$. Hence it follows that $\varphi \to \varphi^{-1}$ is continuous.   □

We have thus achieved making $\Delta(AP(G))$ a compact group having the following properties.

(1) The map $\phi : G \to \Delta(AP(G))$ is a homomorphism with dense range.
(2) A bounded continuous function f on G is almost periodic if and only if there exists a function $\widehat{f} \in C(\Delta(AP(G)))$ such that $f(x) = \widehat{f}(\phi(x))$ for all $x \in G$.

We remark next that properties (1) and (2) determine the compact group $\Delta(AP(G))$ up to topological isomorphism.

Remark 2.10.11. Let $\Delta = \Delta(AP(G))$ and suppose that Δ' is a second compact group and $\phi' : G \to \Delta'$ is a homomorphism satisfying the analogous properties (1) and (2). Then $\widehat{f}' \to \widehat{f}$ is an algebraic isomorphism of $C(\Delta')$ onto $C(\Delta)$. Let $\delta : \Delta \to \Delta'$ be the associated homeomorphism; that is, $\delta(\varphi)(\widehat{f}') = \varphi(\widehat{f})$ for $\varphi \in \Delta$ and $f \in AP(G)$. Then

$$\delta(\phi(x))(\widehat{f}') = \phi(x)(\widehat{f}) = f(x) = \phi'(x)(\widehat{f}')$$

for all $x \in G$ and $f \in AP(G)$. Thus δ extends the homomorphism $\phi' \circ \phi^{-1} : \phi(G) \to \phi'(G)$. Because δ is a homeomorphism and $\phi(G)$ is dense in Δ, it follows that δ is a topological isomorphism.

Definition 2.10.12. The topological group G is said to be *almost periodic* if the homomorphism $\phi : G \to \Delta(AP(G))$ is injective. Even though in general ϕ need not be injective, $\Delta(AP(G))$ is called the *Bohr* or *almost periodic compactification of* G and usually denoted $b(G)$.

In the remainder of this section we use the notation $b(G)$ in order to emphasize the fact that $b(G)$ is a compact group rather than just the structure space of the algebra $AP(G)$.

We now turn to locally compact Abelian groups. Of course, in that case a major portion of the analysis in this section is superfluous. However, for such G, considerably more can be said about $AP(G)$, and $b(G)$ can be identified in terms of G only.

Let $T(G)$ denote the linear subspace of $C^b(G)$ consisting of all finite linear combinations of characters of G. Functions in $T(G)$ are called *trigonometric polynomials*. Since $\widehat{G}$ is a group, $T(G)$ is a subalgebra of $C^b(G)$. For $\chi \in \widehat{G}$ and $x \in G$ we have $L_x\chi(y) = \overline{\chi(x)}\chi(y)$. Thus

$$C_\chi = \{\chi(x)\chi : x \in G\} \subseteq \mathbb{T} \cdot \chi,$$

which is a compact subset of $C^b(G)$. This implies that $T(G) \subseteq AP(G)$.

Theorem 2.10.13. *Let G be a locally compact Abelian group. The Gelfand isomorphism $f \to \widehat{f}$ from $AP(G)$ onto $C(b(G))$ maps $\widehat{G}$ onto $\widehat{b(G)}$ and hence $T(G)$ onto $T(b(G))$. Moreover, $T(G)$ is norm dense in $AP(G)$.*

Proof. It suffices to show that if $\gamma \in \widehat{G}$, then $\widehat{\gamma} \in \widehat{b(G)}$, and that every character of $b(G)$ arises in this way. For $x, y \in G$, we have

$$\widehat{\gamma}(\varphi_x\varphi_y) = \widehat{\gamma}(\varphi_{xy}) = \gamma(xy) = \gamma(x)\gamma(y) = \widehat{\gamma}(\varphi_x)\widehat{\gamma}(\varphi_y).$$

Since $\widehat{\gamma}$ is continuous on $b(G)$ and $\phi(G)$ is dense in $b(G)$, we conclude that $\widehat{\gamma} \in \widehat{b(G)}$.

Conversely, if $\chi \in \widehat{b(G)}$ then $\chi \circ \phi \in \widehat{G}$ since ϕ is a continuous homomorphism from G into $b(G)$. By the first part of the proof $\widehat{\chi \circ \phi} \in \widehat{b(G)}$. The two characters χ and $\widehat{\chi \circ \phi}$ of $b(G)$ agree on the dense subset $\phi(G)$, whence $\chi = \widehat{\chi \circ \phi}$.

Because the Gelfand homomorphism of $AP(G)$ onto $C(b(G))$ is isometric and, as we have just seen, maps $T(G)$ onto $T(b(G))$. Thus for the last statement of the theorem it is enough to observe that $T(b(G))$ is norm dense in $C(b(G))$. Now, if H is a compact Abelian group, then $T(H)$ is $*$-subalgebra of $C(H)$ which strongly separates the points of H. Thus $T(H)$ is dense in $C(H)$ by the Stone–Weierstrass theorem. $\qquad\square$

Corollary 2.10.14. *Let G be a locally compact Abelian group, and let $\widehat{G}_d$ denote the algebraic group $\widehat{G}$ endowed with the discrete topology. Then the discrete dual group $\widehat{b(G)}$ of $b(G)$ is isomorphic to $\widehat{G}_d$.*

Proof. Being the dual group of the compact group $b(G)$, $\widehat{b(G)}$ is discrete. By Theorem 2.10.13, the Gelfand homomorphism of $AP(G)$ maps $\widehat{G}$ onto $\widehat{b(G)}$ and this map is obviously a group isomorphism. Thus $\widehat{G}_d$ is isomorphic to $\widehat{b(G)}$. □

Employing the Pontryagin duality theorem for locally compact Abelian groups, Corollary 2.10.14 can be rephrased as follows. The group $b(G)$ is topologically isomorphic to the dual group of $\widehat{G}_d$ since it is topologically isomorphic to the dual group of $\widehat{b(G)}$.

2.11 Structure spaces of tensor products

The purpose of this section is to determine the structure space of the tensor product of two commutative Banach algebras and to investigate its semisimplicity. For the basic theory of tensor products of Banach algebras we refer to Section 1.5. We remind the reader that ϵ denotes the injective tensor norm.

Lemma 2.11.1. *Let A and B be commutative Banach algebras and let γ be an algebra cross-norm on $A \otimes B$ such that $\gamma \geq \epsilon$. Given $\varphi \in \Delta(A)$ and $\psi \in \Delta(B)$, there is a unique element of $\Delta(A \widehat{\otimes}_\gamma B)$, denoted $\varphi \widehat{\otimes}_\gamma \psi$, such that*

$$(\varphi \widehat{\otimes}_\gamma \psi)(x \otimes y) = \varphi(x)\psi(y)$$

for all $x \in A$ and $y \in B$. Furthermore, the mapping

$$\Delta(A) \times \Delta(B) \to \Delta(A \widehat{\otimes}_\gamma B), \ (\varphi, \psi) \to \varphi \widehat{\otimes}_\gamma \psi$$

is a bijection.

Proof. Let $\varphi \in \Delta(A)$ and $\psi \in \Delta(B)$ and recall first that there is a unique homomorphism $\omega : A \otimes B \to \mathbb{C}$ such that $\omega(x \otimes y) = \varphi(x)\psi(y)$ for all $x \in A$ and $y \in B$. By definition of ϵ and since $\gamma \geq \epsilon$, for any $x_1, \ldots, x_n \in A$ and $y_1, \ldots, y_n \in B$, we have

$$\left| \omega\left(\sum_{j=1}^n x_j \otimes y_j \right) \right| = \left| \sum_{j=1}^n \varphi(x_j)\psi(y_j) \right| \leq \epsilon\left(\sum_{j=1}^n x_j \otimes y_j \right) \leq \gamma\left(\sum_{j=1}^n x_j \otimes y_j \right).$$

Thus ω is continuous with respect to γ and therefore extends uniquely to an element of $\Delta(A \otimes_\gamma B)$, denoted $\varphi \widehat{\otimes}_\gamma \psi$.

The mapping $(\varphi, \psi) \to \varphi \widehat{\otimes}_\gamma \psi$ is injective. To verify this, let $\varphi_1, \varphi_2 \in \Delta(A)$ and $\psi_1, \psi_2 \in \Delta(B)$ such that $\varphi_1 \widehat{\otimes}_\gamma \psi_1 = \varphi_2 \widehat{\otimes}_\gamma \psi_2$. Fix $b \in B$ such that $\psi_1(b) = 1$. Then, for all $x \in A$,

$$\varphi_1(x) = \varphi_1(x)\psi_1(b) = (\varphi_1 \otimes \psi_1)(x \otimes b) = (\varphi_2 \otimes \psi_2)(x \otimes b) = \varphi_2(x)\psi_2(b).$$

Now, since φ_1 and φ_2 are non-zero homomorphisms, this equation implies that $\psi_2(b) = 1$. Hence $\varphi_1 = \varphi_2$, and this in turn yields that $\psi_1 = \psi_2$.

It remains to show that given $\rho \in \Delta(A \widehat{\otimes}_\gamma B)$, there exist $\varphi \in \Delta(A)$ and $\psi \in \Delta(B)$ such that $\rho(x \otimes y) = \varphi(x)\psi(y)$ for all $x \in A$ and $y \in B$. Choose $a \in A$ and $b \in B$ such that $\rho(a \otimes b) = 1$, and define $\varphi : A \to \mathbb{C}$ and $\psi : B \to \mathbb{C}$ by

$$\varphi(x) = \rho(xa \otimes b) \text{ and } \psi(y) = \rho(a \otimes yb).$$

Clearly, φ and ψ are linear maps and

$$\varphi(x)\psi(y) = \rho(xa^2 \otimes yb^2) = \rho(x \otimes y)\rho(a^2 \otimes b^2) = \rho(x \otimes y)$$

for all $x \in A$ and $y \in B$. In particular, both φ and ψ are nonzero. Finally, for $x_1, x_2 \in A$,

$$\begin{aligned}
\varphi(x_1 x_2) &= \rho(x_1 x_2 a \otimes b) = \rho(x_1 x_2 a \otimes b)\rho(a \otimes b)\\
&= \rho((x_1 a \otimes b)(x_2 a \otimes b)) = \rho(x_1 a \otimes b)\rho(x_2 a \otimes b)\\
&= \varphi(x_1)\varphi(x_2),
\end{aligned}$$

and similarly, $\psi(y_1 y_2) = \psi(y_1)\psi(y_2)$ for all $y_1, y_2 \in B$. Thus φ and ψ have all the required properties. $\qquad\square$

Theorem 2.11.2. *Let A and B be commutative Banach algebras and let γ be an algebra cross-norm on $A \otimes B$ such that $\gamma \geq \epsilon$. Then the mapping*

$$\Delta(A) \times \Delta(B) \to \Delta(A \widehat{\otimes}_\gamma B), \ (\varphi, \psi) \to \varphi \widehat{\otimes}_\gamma \psi$$

is a homeomorphism.

Proof. As to continuity, it suffices to show that for each $c \in A \widehat{\otimes}_\gamma B$, the function $(\varphi, \psi) \to (\varphi \widehat{\otimes}_\gamma \psi)(c)$ is continuous on $\Delta(A) \times \Delta(B)$. For $c \in A \otimes B$, say $c = \sum_{j=1}^n a_j \otimes b_j, \ a_j \in A, b_j \in B, 1 \leq j \leq n$, this follows at once from the equation

$$(\varphi \widehat{\otimes}_\gamma \psi)(c) = \sum_{j=1}^n \varphi(a_j)\psi(b_j).$$

Now, let $z \in A \widehat{\otimes}_\gamma B$ be arbitrary. Since $\|\varphi \widehat{\otimes}_\gamma \psi\| \leq 1$, the function $(\varphi, \psi) \to (\varphi \widehat{\otimes}_\gamma \psi)(z)$ is a uniform limit on $\Delta(A) \times \Delta(B)$ of functions $(\varphi, \psi) \to (\varphi \widehat{\otimes}_\gamma \psi)(c), c \in A \otimes B$, and therefore is continuous.

For openness, it is enough to prove that the mappings $\varphi \widehat{\otimes}_\gamma \psi \to \varphi$ and $\varphi \widehat{\otimes}_\gamma \psi \to \psi$ from $\Delta(A \widehat{\otimes}_\gamma B)$ into $\Delta(A)$ and $\Delta(B)$, respectively, are continuous. To show that the map $\varphi \widehat{\otimes}_\gamma \psi \to \varphi$ is continuous, we check that, for each $a \in A$, the function

$$F_a : \Delta(A \widehat{\otimes}_\gamma B) \to \mathbb{C}, \ \varphi \widehat{\otimes}_\gamma \psi \to \varphi(a)$$

is continuous. Fix $a \in A$ and for every $\rho = \varphi \widehat{\otimes}_\gamma \psi \in \Delta(A \widehat{\otimes}_\gamma B)$ select $a_\rho \in A$ and $b_\rho \in B$ such that $\rho(a_\rho \otimes b_\rho) = 1$. Then

$$F_a(\rho) = \varphi(a)\varphi(a_\rho)\psi(b_\rho) = \varphi(aa_\rho)\psi(b_\rho) = \rho(aa_\rho \otimes b_\rho).$$

Now let $(\rho_\alpha)_\alpha$ be a net in $\Delta(A\widehat{\otimes}_\gamma B)$ converging to some $\rho \in \Delta(A\widehat{\otimes}_\gamma B)$. Then

$$\begin{aligned}
\rho_\alpha(a_\rho \otimes b_\rho)\rho_\alpha(aa_{\rho_\alpha} \otimes b_{\rho_\alpha}) &= \rho_\alpha(aa_\rho \otimes b_\rho)\rho_\alpha(a_{\rho_\alpha} \otimes b_{\rho_\alpha}) \\
&= \rho_\alpha(aa_\rho \otimes b_\rho) \\
&\to \rho(aa_\rho \otimes b_\rho).
\end{aligned}$$

Since $\rho_\alpha(a_\rho \otimes b_\rho) \to \rho(a_\rho \otimes b_\rho) = 1$, we conclude that

$$F_a(\rho_\alpha) = \rho_\alpha(aa_{\rho_\alpha} \otimes b_{\rho_\alpha}) \to \rho(aa_\rho \otimes b_\rho) = F_a(\rho).$$

Thus F_a is a continuous function. Similarly, the map $\varphi\widehat{\otimes}_\gamma\psi \to \psi$ from $\Delta(A\widehat{\otimes}_\gamma B)$ to $\Delta(B)$ is continuous. $\qquad\square$

As the reader will have observed, the last slightly more technical part of the preceding proof can be omitted when A and B are unital. Indeed, in this case the map $(\varphi, \psi) \to \varphi\widehat{\otimes}_\gamma\psi$ is a continuous bijection from the compact space $\Delta(A) \times \Delta(B)$ to the Hausdorff space $\Delta(A\widehat{\otimes}_\gamma B)$ and hence is a homeomorphism.

Corollary 2.11.3. *Let A, B and γ be as before. If $A\widehat{\otimes}_\gamma B$ is semisimple, then so are A and B.*

Proof. Let $a \in A$ such that $\widehat{a} = 0$. Fix any nonzero $b \in B$. Then

$$\widehat{a \otimes b}(\varphi\widehat{\otimes}_\gamma\psi) = \varphi(a)\psi(b) = 0$$

for all $\varphi \in \Delta(A)$ and $\psi \in \Delta(B)$. Since every $\rho \in \Delta(A\widehat{\otimes}_\gamma B)$ is of the form $\rho = \varphi\widehat{\otimes}_\gamma\psi$ for some $\varphi \in \Delta(A)$ and $\psi \in \Delta(B)$, we get that $\widehat{a \otimes b} = 0$. Because $A\widehat{\otimes}_\gamma B$ is semisimple, it follows that $a \otimes b = 0$ and hence $a = 0$. So A is semisimple, and similarly for B. $\qquad\square$

Remark 2.11.4. The converse to Corollary 2.11.3 is false. In fact, Milne [89] has shown that the following two conditions are equivalent.

(i) The projective tensor product of any two semisimple commutative Banach algebras is semisimple.
(ii) Every Banach space has the approximation property.

However, as first shown by Enflo [31], there are Banach spaces which don't share the approximation property (for all this, compare [114]). In this context compare also Theorem 2.11.6 below and Appendix A.2.

Elements of $\Delta(B)$ give rise to certain continuous homomorphisms from $A\widehat{\otimes}_\gamma B$ onto A. These homomorphisms are extremely useful when dealing with tensor products.

Lemma 2.11.5. *Let A and B be commutative Banach algebras and let γ be an algebra cross-norm on $A \otimes B$ such that $\gamma \geq \epsilon$. Let $\psi \in \Delta(B)$. Then there is a unique continuous homomorphism $\phi_\psi : A \widehat{\otimes}_\gamma B \to A$ such that $\phi_\psi(a \otimes b) = \psi(b)a$ for all $a \in A$ and $b \in B$.*

Proof. The map $A \times B \to A, (a, b) \to \psi(b)a$ is bilinear. Hence there is a unique linear map $\phi_\psi : A \otimes B \to A$ satisfying $\phi_\psi(a \otimes b) = \psi(b)a$ for all $a \in A$ and $b \in B$. Now, let $a_1, \ldots, a_n \in A$ and $b_1, \ldots, b_n \in B$. Then, with $x = \sum_{j=1}^n a_j \otimes b_j$,

$$\|\phi_\psi(x)\| = \left\| \sum_{j=1}^n \psi(b_j)a_j \right\| = \sup\left\{ \left| f\left(\sum_{j=1}^n \psi(b_j)a_j \right) \right| : f \in A_1^* \right\}$$

$$\leq \sup\left\{ \left| \sum_{j=1}^n f(a_j)g(b_j) \right| : f \in A_1^*, g \in B_1^* \right\} = \epsilon(x)$$

$$\leq \gamma(x).$$

Thus ϕ_ψ is norm decreasing for the norm γ on $A \otimes B$ and therefore extends uniquely to a continuous linear map, also denoted ϕ_ψ, from $A \widehat{\otimes}_\gamma B$ to A. Finally, ϕ_ψ is a homomorphism since

$$\phi_\psi((a \otimes b)(a' \otimes b')) = \phi_\psi(aa' \otimes bb') = \psi(bb')aa'$$
$$= (\psi(b)a)(\psi(b')a')$$
$$= \phi_\psi(a \otimes b)\phi_\psi(a' \otimes b')$$

for $a, a' \in A$ and $b, b' \in B$. $\qquad\square$

Of course, starting with $\varphi \in \Delta(A)$, we obtain an analogous homomorphism $\phi_\varphi : A \widehat{\otimes}_\gamma B \to B$. We proceed with two applications of Lemma 2.11.5 which concern the projective tensor product. The first one settles the important question of when $A \widehat{\otimes}_\pi B$ is semisimple.

Theorem 2.11.6. *Let A and B be commutative Banach algebras. Then the projective tensor product $A \widehat{\otimes}_\pi B$ is semisimple if and only if the following two conditions are satisfied.*

(i) *A and B are semisimple.*
(ii) *The natural homomorphism $A \widehat{\otimes}_\pi B \to A \widehat{\otimes}_\epsilon B$ is injective.*

Proof. Suppose first that $A \widehat{\otimes}_\pi B$ is semisimple. Then A and B are semisimple by Corollary 2.11.3. Let ϕ be the natural homomorphism from $A \widehat{\otimes}_\pi B$ into $A \widehat{\otimes}_\epsilon B$ and let $c = \sum_{j=1}^\infty a_j \otimes b_j$, where $\sum_{j=1}^\infty \|a_j\| \cdot \|b_j\| < \infty$, be an element of $A \widehat{\otimes}_\pi B$ such that $\phi(c) = 0$. Then $\sum_{j=1}^\infty f(a_j)g(b_j) = 0$ for all $f \in A^*$ and $g \in B^*$. In particular,

$$(\varphi \widehat{\otimes}_\pi \psi)(c) = \sum_{j=1}^\infty \varphi(a_j)\psi(b_j) = 0$$

for all $\varphi \in \Delta(A)$ and $\psi \in \Delta(B)$. Since $A \widehat{\otimes}_\pi B$ is semisimple and every element of $\Delta(A \widehat{\otimes}_\pi B)$ is of the form $\varphi \widehat{\otimes}_\pi \psi$, it follows that $c = 0$. Thus ϕ is injective.

Conversely, suppose that conditions (i) and (ii) hold. Let c be an element in the radical of $A \widehat{\otimes}_\pi B$, say $c = \sum_{j=1}^\infty a_j \otimes b_j$. Since A is semisimple, from Lemma 2.11.5 we get that $\sum_{j=1}^\infty \psi(b_j)a_j = 0$ for all $\psi \in \Delta(B)$. This implies that, for every $f \in A^*$ and all $\psi \in \Delta(B)$,

$$0 = f\left(\sum_{j=1}^\infty \psi(b_j)a_j\right) = \sum_{j=1}^\infty \psi(b_j)f(a_j) = \psi\left(\sum_{j=1}^\infty f(a_j)b_j\right).$$

Since B is semisimple, it follows that $\sum_{j=1}^\infty f(a_j)b_j = 0$ for every $f \in A^*$. This in turn gives

$$0 = g\left(\sum_{j=1}^\infty f(a_j)b_j\right) = \sum_{j=1}^\infty f(a_j)g(b_j) = (f \widehat{\otimes}_\epsilon g)(c)$$

for every $f \in A^*$ and $g \in B^*$. Now, condition (ii) yields $c = 0$. So $A \widehat{\otimes}_\pi B$ is semisimple. $\qquad\square$

Proposition 2.11.7. *Let A and B be commutative Banach algebras. Then $A \widehat{\otimes}_\pi B$ is unital if and only if both A and B are unital.*

Proof. It is apparent that if e_A and e_B are identities of A and B, respectively, then $e_A \otimes e_B$ is an identity of $A \widehat{\otimes}_\pi B$.

Conversely, let $\sum_{j=1}^\infty a_j \otimes b_j$, where $a_j \in A$ and $b_j \in B$, represent an identity for $A \widehat{\otimes}_\pi B$. Then

$$a \otimes b = \sum_{j=1}^\infty a_j a \otimes b_j b$$

for all $a \in A$ and $b \in B$. Since $A \widehat{\otimes}_\pi B$ is unital, $\Delta(A \widehat{\otimes}_\pi B) \neq \emptyset$ and hence $\Delta(A)$ and $\Delta(B)$ are both nonempty by Theorem 2.11.2. Choose $\psi \in \Delta(B)$ and $b \in B$ with $\psi(b) = 1$, and let $\phi_\psi : A \widehat{\otimes}_\pi B \to A$ be the homomorphism of Lemma 2.11.5. Then

$$a = \phi_\psi(a \otimes b) = \phi_\psi\left(\sum_{j=1}^\infty a_j a \otimes b_j b\right)$$

$$= \sum_{j=1}^\infty \phi_\psi(a_j a \otimes b_j b) = \sum_{j=1}^\infty \psi(b_j b)a_j a$$

$$= \left(\sum_{j=1}^\infty \psi(b_j)a_j\right) a$$

for all $a \in A$. Thus $\sum_{j=1}^\infty \psi(b_j)a_j$ is an identity for A. Similarly, it is shown that B is unital. $\qquad\square$

To conclude this section, let A be a semisimple commutative Banach algebra and G a locally compact Abelian group. Since L^1-spaces do have the approximation property, we could use Theorem 2.11.6 and Corollary 2.7.9 to deduce that $L^1(G, A) = L^1(G) \widehat{\otimes}_\pi A$ is semisimple. However, this can be shown directly without appealing to Theorem 2.11.6 as follows.

Theorem 2.11.8. *Let G be a locally compact Abelian group and A a semisimple commutative Banach algebra. Then $L^1(G, A) = L^1(G) \widehat{\otimes}_\pi A$ is semisimple.*

Proof. Let $\phi : L^1(G) \widehat{\otimes}_\pi A \to L^1(G, A)$ be the isometric isomorphism satisfying $\phi(f \otimes a)(x) = f(x)a$ for all $f \in L^1(G)$ and $a \in A$ and almost all $x \in G$ (Proposition 1.5.4). For $\alpha \in \widehat{G}$, let φ_α be the corresponding element of $\Delta(L^1(G))$ and recall that $\Delta(L^1(G) \widehat{\otimes}_\pi A) = \Delta(L^1(G)) \times \Delta(A)$ (Theorem 2.11.2). Let $f \in L^1(G), \alpha \in \widehat{G}, a \in A$ and $\psi \in \Delta(A)$. Then

$$(\varphi_\alpha \widehat{\otimes}_\pi \psi)(f \otimes a) = \widehat{f}(\alpha)\psi(a) = \psi(a) \int_G f(x)\overline{\alpha(x)}dx$$

$$= \int_G \overline{\alpha(x)}\psi(f(x)a)dx$$

$$= \int_G \overline{\alpha(x)}\psi(\phi(f \otimes a)(x))dx$$

$$= \psi(\phi(f \otimes a))\widehat{}(\alpha).$$

By linearity and continuity, this implies

$$(\varphi_\alpha \widehat{\otimes}_\pi \psi)(u) = \widehat{\psi(\phi(u))}(\alpha)$$

for all $u \in L^1(G) \widehat{\otimes}_\pi A, \alpha \in \widehat{G}$ and $\psi \in \Delta(A)$. Since $L^1(G)$ is semisimple, this equation shows that if $u \in L^1(G) \widehat{\otimes}_\pi A$ is such that $\widehat{u} = 0$, then $\psi(\phi(u)) = 0$ for all $\psi \in \Delta(A)$. Thus $\phi(u) = 0$ since A is semisimple and hence $u = 0$ as ϕ is injective. □

2.12 Exercises

Exercise 2.12.1. The following example shows that the Gleason–Kahane–Zelazko theorem (Theorem 2.1.2) fails to hold for real Banach algebras. Let $A = C^{\mathbb{R}}([0, 1])$ be the algebra of all real valued continuous functions on $[0, 1]$ with the supremum norm. Define $\varphi : A \to \mathbb{R}$ by $\varphi(f) = \int_0^1 f(t)dt$. Show that $\varphi(f) \neq 0$ whenever f is invertible, but φ is not multiplicative.

Exercise 2.12.2. Find an example of a real commutative Banach algebra with identity which does not admit a nonzero real multiplicative linear functional.

Exercise 2.12.3. Let V denote the Volterra integral operator on $L^2[0,1]$ defined by

$$Vf(s) = \int_0^s f(t)dt, \ f \in L^2[0,1], s \in [0,1],$$

and let A be the closed subalgebra of $\mathcal{B}(L^2[0,1])$ generated by V. Show that A has precisely one maximal ideal.

Exercise 2.12.4. For $1 \leq p < \infty$, consider the non-unital commutative Banach algebra $l^p(\mathbb{N})$. Identify the maximal modular ideals of $l^p(\mathbb{N})$. Show that $l^p(\mathbb{N})$ has maximal ideals which are not modular.

Exercise 2.12.5. Let A be a non-unital commutative Banach algebra and M a maximal ideal of A. Show that M is modular if and only if M has codimension one and does not contain A^2.

Exercise 2.12.6. Let A be the algebra of entire functions in the complex plane endowed with the norm $\|f\| = \sup\{|f(z)| : |z| = 1\}$. Then A is a non-complete commutative normed algebra. Prove that A contains maximal ideals of infinite codimension.

Exercise 2.12.7. Let A be the algebra of all continuously differentiable functions $f : [0,1] \to \mathbb{C}$ with pointwise multiplication and the norm $\|f\| = \|f\|_\infty + \|f'\|_\infty$. Let

$$I = \{f \in A : f(0) = f'(0) = 0\}.$$

Show that A/I is a two-dimensional algebra which has a one-dimensional radical. Thus A is an example of a semisimple commutative Banach algebra which admits a non-semisimple quotient.

Exercise 2.12.8. Find examples showing that Corollaries 2.1.10 and 2.1.11 are no longer true without assuming semisimplicity.

Exercise 2.12.9. Let A be a commutative Banach algebra and $\Gamma : A \to \Gamma(A) \subseteq C_0(\Delta(A))$ its Gelfand homomorphism. Show that Γ is a topological isomorphism (if and) only if there exists $c > 0$ such that $\|a^2\| \geq c \|a\|^2$ for all $a \in A$.

Exercise 2.12.10. Let A be a semisimple commutative Banach algebra with norm $\| \cdot \|$, and let B be a subalgebra of A which is a Banach algebra with some norm $| \cdot |$. Show that there exists a constant $c > 0$ such that $\|x\| \leq c|x|$ for all $x \in B$.

Exercise 2.12.11. Let A and B be commutative Banach algebras and $A \oplus B$ their direct sum with the norm $\|(a,b)\| = \max(\|a\|, \|b\|)$. Show that there is a canonical homeomorphism between $\Delta(A \oplus B)$ and the topological disjoint union of $\Delta(A)$ and $\Delta(B)$.

Exercise 2.12.12. Let A be a unital commutative Banach algebra, and let I_1 and I_2 be nontrivial closed ideals of A such that $A = I_1 \oplus I_2$. Show that $\Delta(A)$ is not connected.

Exercise 2.12.13. Let A be a semisimple commutative Banach algebra and $\widehat{A} = \{\widehat{a} : a \in A\}$. Let $\phi : \Delta(A) \to \Delta(A)$ be a homeomorphism. We say that ϕ *is induced from a homomorphism* $h : A \to A$ if $\phi(\varphi)(x) = \varphi(h(x))$ for all $x \in A$ and $\varphi \in \Delta(A)$.

(i) Prove that ϕ is induced from a homomorphism $h : A \to A$ if and only if $f \in \widehat{A}$ implies $f \circ \phi \in \widehat{A}$.

(ii) Find an analogous condition on ϕ which is equivalent to ϕ being induced from an automorphism of A.

Exercise 2.12.14. In Exercise 2.12.13, take $A = l^1(\mathbb{Z})$ and identify $\Delta(A)$ with $\mathbb{T}$. Conclude that a homeomorphism $\phi : \mathbb{T} \to \mathbb{T}$ is induced from a homomorphism of A if and only if $\phi \in \widehat{A}$.

Exercise 2.12.15. Let A and B be commutative Banach algebras and let $h : A \to B$ be a homomorphism with dense range. Show that

$$h^* : \Delta(B) \to \Delta(A), \ h^*(\psi)(a) = \psi(h(a)),$$

$a \in A, \psi \in \Delta(B)$, defines an injective continuous mapping from $\Delta(B)$ into $\Delta(A)$. If B is unital, then h^* maps $\Delta(B)$ homeomorphically onto $h^*(\Delta(B))$.

Exercise 2.12.16. Construct examples of semisimple commutative Banach algebras A and B and a homomorphism $h : A \to B$ with dense range such that the corresponding mapping $h^* : \Delta(B) \to \Delta(A)$ (see Exercise 2.12.15)

(i) is not surjective,

(ii) not a homeomorphism onto its range.

Exercise 2.12.17. Let X and Y be nonempty compact Hausdorff spaces and $\phi : C(X) \to C(Y)$ a unital homomorphism, and let $\phi^* : \Delta(C(Y)) \to \Delta(C(X))$ be the map $\varphi \to \varphi \circ \phi$. Show

(i) ϕ^* is injective if and only if ϕ is surjective.

(ii) ϕ^* is surjective if and only if ϕ is injective.

Exercise 2.12.18. Let X be a compact Hausdorff space and let A be a uniform algebra on X. Let $\varphi : A \to \mathbb{C}$ be a homomorphism. Show that there exists a probability measure μ on X such that $\varphi(f) = \int_X f(x)d\mu(x)$ for all $f \in A$.

Exercise 2.12.19. Let A and B be semisimple and unital commutative Banach algebras. Let ϕ be a linear map of A onto B. Prove that ϕ is an algebra isomorphism between A and B if and only if $\sigma_B(\phi(x)) = \sigma_A(x)$ for all $x \in A$.

Exercise 2.12.20. Let $A \subseteq A_1 \subseteq A_2$ be commutative Banach algebras with norms $\| \cdot \|, \| \cdot \|_1$ and $\| \cdot \|_2$ respectively. Assume that A is dense in A_1 and in A_2 in their respective norms and that $\Delta(A) = \Delta(A_2)$ (that is, every element of $\Delta(A)$ is continuous with respect to $\| \cdot \|_2$). Show that $\Delta(A_1) = \Delta(A_2)$.

Exercise 2.12.21. Let A be a semisimple and faithful commutative Banach algebra. For any $T \in M(A)$, let f_T denote the continuous function on $\Delta(A)$ satisfying $\widehat{Tx}(\varphi) = f_T(\varphi)\widehat{x}(\varphi)$ for all $\varphi \in \Delta(A)$ (Proposition 2.2.16). Show that the mapping $T \to f_T$ is a continuous isomorphism from $M(A)$ onto the subalgebra
$$B = \{f \in C^b(\Delta(A)) : f \cdot \widehat{x} \in \widehat{A} \text{ for all } x \in A\}$$
of $C^b(\Delta(A))$.

Exercise 2.12.22. Let A be a semisimple commutative Banach algebra, $T : A \to A$ a bounded linear operator and T^* the adjoint of T. Prove that $T \in M(A)$ if and only if for each $\varphi \in \Delta(A)$ there exists a constant $c(\varphi)$ such that $T^*(\varphi) = c(\varphi)\varphi$.

Exercise 2.12.23. Let A be a commutative Banach algebra such that $\Delta(A)$ is infinite. Prove that there exists $x \in A$ such that $\sigma_A(x)$ is infinite.
(Hint: Let $\varphi_n \in \Delta(A)$, $n \in \mathbb{N}$, such that $\varphi_n \neq \varphi_m$ for $n \neq m$. For $m, n \in \mathbb{N}$, $n \neq m$, let
$$V_{m,n} = \{x \in A : \varphi_m(x) \neq \varphi_n(x)\}$$
and show that $V_{m,n}$ is dense in A. Conclude that $\cap\{V_{m,n} : m \neq n\} \neq \emptyset$).

Exercise 2.12.24. Consider the disc algebra $A(\mathbb{D})$ and view $\mathbb{D} = \Delta(A(\mathbb{D}))$ as a subset of $A(\mathbb{D})^*$. Show that the topology on $\mathbb{D}$ induced by the norm topology of $A(\mathbb{D})^*$ coincides with the complex plane topology on $\mathbb{D}^\circ$ and with the discrete topology on $\mathbb{T}$.
(Hint: For the first part of the assertion, use Schwarz' lemma which states that if $f : \mathbb{D}^\circ \to \mathbb{D}$ is a holomorphic function vanishing at $z_0 \in \mathbb{D}^\circ$, then $|f(z)| \leq |z - z_0|/|1 - z \overline{z_0}|$ for all $z \in \mathbb{D}^\circ$.)

Exercise 2.12.25. Let A be a closed subalgebra of $C(\mathbb{D})$ satisfying the following two conditions:
   (1) The function $z \to z$ belongs to A.
   (2) For every $f \in A$, $\|f\|_\infty = \|f|_\mathbb{T}\|_\infty$.
Then $A \subseteq A(\mathbb{D})$. To prove this, proceed as follows.
   (i) Apply Wermer's maximality theorem (Theorem 2.5.15) to conclude that $A|_\mathbb{T} = \{f|_\mathbb{T} : f \in A\}$ is equal to either $P(\mathbb{T})$ or $C(\mathbb{T})$.
   (ii) By (2), every $g \in A|_\mathbb{T}$ extends uniquely to some $\widetilde{g} \in A$. Consider the homomorphism $g \to \widetilde{g}(0)$ from $A|_\mathbb{T}$ to $\mathbb{C}$ to exclude the possibility that $A|_\mathbb{T} = C(\mathbb{T})$.
   (iii) Show that if $f \in A$ and $g \in A(\mathbb{D})$ are such that $f|_\mathbb{T} = g|_\mathbb{T}$, then $f = g$.

Exercise 2.12.26. Let A be a commutative Banach algebra with identity e. Prove that the following two conditions are equivalent.

(i) For $x, y \in A, \exp x = \exp y$ implies that $x - y = (2k\pi i)e$ for some $k \in \mathbb{Z}$.

(ii) $\Delta(A)$ is connected.

(Hint: Show that the equation $\exp x = e$ has no nonzero solution in the radical of A and that it has solutions different from $(2k\pi i)e, k \in \mathbb{Z}$, if and only if $\Delta(A)$ is not connected.)

Exercise 2.12.27. Let $l^p(\mathbb{N})$ be as in Exercise 1.6.9. Determine $\Delta(l^p(\mathbb{N}))$.

Exercise 2.12.28. Let $\mathrm{Lip}_\alpha[0, 1]$ be the Banach algebra of Lipschitz functions of order α (see Exercise 1.6.11). For $t \in [0, 1]$, let $\varphi_t(f) = f(t), f \in \mathrm{Lip}_\alpha[0, 1]$. Show that the map $t \to \varphi_t$ is a homeomorphism of $[0, 1]$ onto $\Delta(\mathrm{Lip}_\alpha[0, 1])$. (Hint: If $f \in \mathrm{Lip}_\alpha[0, 1]$ is such that $f(t) \neq 0$ for all $t \in [0, 1]$, then $\frac{1}{f} \in \mathrm{Lip}_\alpha[0, 1]$.)

Exercise 2.12.29. Let γ be a continuous homomorphism of $\mathbb{R}$ into the multiplicative group $\mathbb{C}^\times$ of nonzero complex numbers.

(i) Show that γ is differentiable and satisfies the differential equation

$$\gamma'(t) = \gamma(0)\gamma(t), \ t \in \mathbb{R}.$$

(Hint: There exists $c > 0$ such that $\int_0^c \gamma(s)ds \neq 0$, and then

$$\gamma(t) = \left(\int_0^c \gamma(s)ds\right)^{-1} \int_0^{c+t} \gamma(s)ds$$

for all $t \in \mathbb{R}$.)

(ii) Deduce that there exists $z \in \mathbb{C}$ such that $\gamma(t) = e^{zt}$ for all $t \in \mathbb{R}$.

Exercise 2.12.30. Let G be a compact Abelian group and let α and β be distinct characters of G. Show the *orthogonality relation* $\int_G \alpha(x)\overline{\beta(x)}dx = 0$. (Hint: For $\gamma \in \widehat{G} \setminus \{1_G\}$, choose $x_0 \in G$ such that $\gamma(x_0) \neq 1$ and observe that $\int_G \gamma(x)dx = \gamma(x_0) \int_G \gamma(x)dx$.)

Exercise 2.12.31. Let G be a compact Abelian group with normalized Haar measure and let $1 \leq p < \infty$. With convolution, $L^p(G)$ is a commutative Banach algebra. For $\chi \in \widehat{G}$ and $f \in L^p(G)$, let

$$\varphi_\chi(f) = \widehat{f}(\chi) = \int_G f(x)\overline{\chi(x)}dx.$$

Show that the map $\chi \to \varphi_\chi$ is a bijection between $\widehat{G}$ and $\Delta(L^p(G))$ and that $\Delta(L^p(G))$ is discrete.

Exercise 2.12.32. In Theorem 2.7.12 it was shown that, for a locally compact Abelian group G, the Gelfand transform $f \to \widehat{f}$ from $L^1(G)$ to $C_0(\widehat{G})$ is onto (equivalently, the norms $f \to \|f\|_1$ and $f \to \|\widehat{f}\|_\infty$ are equivalent) only when G is finite. For the real line, one can explicitly construct a sequence of functions $f_n \in L^1(\mathbb{R})$, $n \in \mathbb{N}$, such that $\|f_n\|_1 = 1$ for all n, whereas $\|\widehat{f_n}\|_\infty \to 0$ as $n \to \infty$. In fact, normalizing Lebesgue measure so that $[0, 1]$ has measure one, define f_n by

$$f_n(t) = \frac{1}{\sqrt{\pi}} \exp(-(1 + ni)t^2),$$

$t \in \mathbb{R}$, and show that this sequence has the stated properties. (Hint: Use the formula

$$\int_{\mathbb{R}} \exp(-ist - zt^2)dt = \left(\frac{\pi}{z}\right)^{1/2} \exp\left(-\frac{s^2}{4z}\right),$$

which holds for all $s \in \mathbb{R}$ and all $z \in \mathbb{C}$ with $\text{Re} z > 0$.)

Exercise 2.12.33. Let $f \in C(\mathbb{T}) \subseteq L^1(\mathbb{T})$. Show that the following conditions are equivalent.
  (i) $f \in P(\mathbb{T})$.
  (ii) There exists $g \in A(\mathbb{D})$ such that $g|_\mathbb{T} = f$.
  (iii) $\widehat{f}(-n) = 0$ for all $n \in \mathbb{N}$.

Exercise 2.12.34. Let μ denote Lebesgue measure on the unit interval $[0,1]$ and let $L^\infty(\mu)$ be the space of equivalence classes modulo sets of measure zero of complex valued essentially bounded measurable functions on $[0,1]$. With the essential supremum norm, pointwise multiplication and $f \to \bar{f}$, $L^\infty(\mu)$ is a unital commutative C^*-algebra. Let $\Delta = \Delta(L^\infty(\mu))$ and $L^\infty(\mu) \to C(\Delta)$, $f \to \widehat{f}$ the Gelfand isomorphism. Observe that $\widehat{f} \to \int_0^1 f(t)d\mu(t)$ is a bounded linear functional of norm one on $C(\Delta)$. By the Riesz representation theorem there is a regular probability measure $\widehat{\mu}$ on Δ satisfying

$$\int_\Delta \widehat{f}(\varphi)d\widehat{\mu}(\varphi) = \int_0^1 f(t)d\mu(t)$$

for all $f \in L^\infty(\mu)$.
  (i) Show that $\widehat{\mu}(U) > 0$ for every nonempty open subset U of Δ.
  (ii) Show that for every nonempty open subset U of Δ there exists $f_U \in L^\infty(\mu)$ such that $\widehat{f_U} = 1_U$ $\widehat{\mu}$-almost everywhere.

Exercise 2.12.35. Retain the setting and notation of Exercise 2.12.34. Prove that Δ is extremally disconnected, that is, the closure of every open subset of Δ is open.
(Hint: Let U be an open subset of Δ and $f \in L^\infty(\mu)$ such that $\widehat{f} = 1_U$ $\widehat{\mu}$-almost everywhere. Deduce from continuity of $\widehat{f}$ that $\widehat{f}$ takes only the values 0 and 1.)

Exercise 2.12.36. Let A be a commutative Banach $*$-algebra. Then A is called symmetric if $\varphi(x^*) = \overline{\varphi(x)}$ for all $x \in A$ and $\varphi \in \Delta(A)$. Prove that the following two conditions are equivalent.

(i) A is symmetric.

(ii) $-1 \notin \sigma_A(x^*x)$ for every $x \in A$.

(Hint: Without loss of generality, assume that A has an identity e. For (ii) $\Rightarrow$ (i), use (ii) to show that if x is a selfadjoint element of A and α and β are real numbers with $\beta \neq 0$, then $(\alpha + i\beta)e - x$ is invertible.)

Exercise 2.12.37. Let $\mathbb{D} = \{z \in \mathbb{C} : |z| \leq 1\}$, the closed unit disc, and $A = P(\mathbb{D})$. For $f \in A$, define f^* by $f^*(z) = \overline{f(\overline{z})}$.

(i) Show that $f \to f^*$ is an involution on A.

(ii) Does this involution turn A into a C^*-algebra?

(iii) Is $\sigma_A(f^*f) \subseteq \mathbb{R}$ for all $f \in A$?

(iv) Which $\varphi \in \Delta(A)$ satisfy $\varphi(f^*f) \geq 0$ for all $f \in A$?

(v) Does there exist an involution $f \to \widetilde{f}$ on A such that $\varphi(\widetilde{f}) = \overline{\varphi(f)}$ for all $f \in A$ and $\varphi \in \Delta(A)$?

Exercise 2.12.38. Let G be a locally compact Abelian group such that $G \neq \{e\}$. Construct a function $f \in L^1(G)$ such that $\|f^* * f\| \neq \|f\|^2$, thereby showing that $\|\cdot\|_1$ fails to be a C^*-norm.

(Hint: In case G has at least three elements e, a and b, choose a compact symmetric neighbourhood V of e with the property that the three sets V, aV and bV are pairwise disjoint and consider the function $f = 1_V + i1_{aV} + 1_{bV}$.)

Exercise 2.12.39. Let A be C^*-algebra with identity e. Then e is an extreme point of the unit ball $A_1 = \{a \in A : \|a\|_1 \leq 1\}$. To prove this, proceed as follows.

(i) Suppose that $e = \frac{1}{2}(a + b)$, $a, b \in A_1$. Show that there exist selfadjoint elements x and y of A_1 such that $e = \frac{1}{2}(x + y)$ and $xy = yx$.

(ii) Let B be the closed subalgebra of A generated by x, y and e. Apply Theorem 2.4.5 and show that $\widehat{x}(\varphi) = \widehat{y}(\varphi)$ for all $\varphi \in \Delta(B)$.

Exercise 2.12.40. Let $S^n = \{x \in \mathbb{R}^{n+1} : \|x\| = 1\}$, the unit sphere in $\mathbb{R}^{n+1}$, $n \geq 1$. Use the Stone–Weierstrass theorem to show that $C(S^n)$ admits a system of $n + 1$ generators.

(Remark: Using cohomology theory, one can prove that $C(S^n)$ cannot admit a system of less than $n + 1$ generators.)

Exercise 2.12.41. Let X be a compact subset of $\mathbb{C}$ and suppose that $\mathbb{C} \setminus X$ has infinitely many connected components. Prove that $R(X)$ cannot be generated by finitely many rational functions.

Exercise 2.12.42. For $0 < r < R < \infty$ let $K(r, R)$ denote the compact annulus

$$K(r, R) = \{z \in \mathbb{C} : r \leq |z| \leq R\}.$$

Prove that the uniform algebra $A(K(r, R))$ is generated by the two functions $z \to z$ and $z \to 1/z$.

Exercise 2.12.43. Let A be a commutative Banach algebra and let $\varphi_1, \ldots, \varphi_n$ be distinct elements of $\Delta(A)$. Show that the mapping

$$x \to (\varphi_1(x), \ldots, \varphi_n(x))$$

maps A onto $\mathbb{C}^n$.

Exercise 2.12.44. Let K be a compact subset of $\mathbb{C}^n$, $n \in \mathbb{N}$. Show that there exist a unital commutative Banach algebra A and elements $x_1, \ldots, x_n$ of A such that

$$K = \sigma_A(x_1, \ldots, x_n).$$

Why does this not contradict Theorem 2.3.6?

Exercise 2.12.45. Let A be a commutative Banach algebra with identity e and let $x_1, \ldots, x_n \in A$. Let Λ denote the set of all $\lambda = (\lambda_1, \ldots, \lambda_n) \in \mathbb{C}^n$ with the property that for any $y_1, \ldots, y_n \in A$, the element $\sum_{j=1}^n y_j(\lambda e - x_j)$ is not invertible in A. Prove that

$$\Lambda = \sigma_A(x_1, \ldots, x_n).$$

(Hint: Let $\lambda = (\lambda_1, \ldots, \lambda_n) \in \Lambda$. To show that $\lambda \in \sigma_A(x_1, \ldots, x_n)$, observe that the set of all elements $\sum_{j=1}^n y_j(\lambda e - x_j)$, $y_j \in A$, either equals A or is a proper ideal of A and hence is contained in a maximal ideal.)

Exercise 2.12.46. Prove that the two-dimensional torus

$$T = \{(z, w) \in \mathbb{C}^2 : |z| = |w| = 1\}$$

in $\mathbb{C}^2$ has as its polynomially convex hull the 4-dimensional bicylinder $\mathbb{D} \times \mathbb{D}$. (Remark: The question of whether there is any relation between the topological dimension of a compact subset of $\mathbb{C}^n$ and the topological dimension of its polynomially convex hull has been a matter of some interest.)

Exercise 2.12.47. Consider the following subset

$$Y = \left\{(z, w) \in \mathbb{C}^2 : z \neq 0, w = \frac{1}{z}\right\}$$

of $\mathbb{C}^2$. Show that for every compact subset X of Y, the polynomially convex hull $\widehat{X}_p$ of X is contained in Y.

Exercise 2.12.48. In Proposition 2.8.8, consider the following choices of w:
  (i) $w(n) = 2^n$ for all $n \in \mathbb{Z}$;
  (ii) $w(n) = 2^n$ for $n \geq 0$ and $w(n) = 1$ for $n < 0$;
  (iii) $w(n) = 1 + 2^n$ for all $n \in \mathbb{Z}$;
  (iv) $w(n) = 1 + 2^n$ for $n \geq 0$ and $w(n) = 1$ for $n < 0$.
For which of these choices is $K(R_-, R_+)$ a circle? For which of them is $\widehat{l^1(\mathbb{Z}, w)}$ closed under complex conjugation?

Exercise 2.12.49. Determine the structure space of the Beurling algebra $l^1(\mathbb{Z}, \omega)$ for the following weights ω:

(i) $\omega(n) = e^{|n|}$, $n \in \mathbb{Z}$.

(ii) $\omega_\alpha(n) = (1 + |n|)^\alpha$, $n \in \mathbb{Z}$, $0 < \alpha < \infty$.

Exercise 2.12.50. Let ω be a continuous weight function on $\mathbb{R}^+$. Then the limit $\lim_{t \to \infty} \omega(t)^{1/t}$ exists and is equal to $\rho = \inf\{\omega(t)^{1/t} : t > 0\}$. Suppose that $\rho > 0$ and let

$$S = \{z \in \mathbb{C} : \mathrm{Re}\, z \geq -\ln\rho\}.$$

The purpose of this exercise is to determine, by analogy with Beurling algebras on $\mathbb{R}$ (Proposition 2.8.7), the structure space of the convolution algebra $L^1(\mathbb{R}^+, \omega)$.

(i) For $z \in S$, show that

$$\varphi_z(f) = \int_0^\infty f(t)e^{-zt}dt, \ f \in L^1(\mathbb{R}^+, \omega),$$

defines an element of $\Delta(L^1(\mathbb{R}^+, \omega))$.

(ii) Prove that every element of $\Delta(L^1(\mathbb{R}^+, \omega))$ is of the form φ_z for some $z \in S$.

(iii) Deduce that $L^1(\mathbb{R}^+, \omega)$ is semisimple.

(iv) Show that the map $z \to \varphi_z$ is a homeomorphism from the halfplane S onto $\Delta(L^1(\mathbb{R}^+, \omega))$.

Exercise 2.12.51. Let ω and ρ be as in the preceding exercise and assume that $\rho = 0$. Show that then $L^1(\mathbb{R}^+, \omega)$ is radical. An example of such a radical weight is $\omega(t) = \exp(-t^2), t \in \mathbb{R}^+$.

Exercise 2.12.52. Let $\mathbb{T}$ be the multiplicative group of complex numbers of absolute value one with normalized Haar measure. Recall that the Fourier transform of $f \in C(\mathbb{T})$ on $\widehat{\mathbb{T}} = \mathbb{Z}$ is defined by $\widehat{f}(n) = \int_\mathbb{T} f(z)z^{-n}dz$, $n \in \mathbb{Z}$. Prove that $f \to \widehat{f}$ furnishes an isometric isomorphism of the Fourier algebra $A(\mathbb{T})$ to $l^1(\mathbb{Z})$.

Exercise 2.12.53. Show that the Fourier algebra $A(\mathbb{Z})$ of the group of integers is isometrically isomorphic to $L^1(\mathbb{T})$.

Exercise 2.12.54. Let G be a locally compact group and $A(G)$ the Fourier algebra of G as studied in Section 2.9. Exploit Lemmas 2.9.3 and 2.9.5 (with $K = \{e\}$) to establish the existence of a net $(u_\alpha)_\alpha$ in $A(G)$ with the following properties:

(1) $\|u_\alpha\|_{A(G)} = u_\alpha(e) = 1$ for all α;

(2) $\|vu_\alpha\|_{A(G)} \to 0$ for every $v \in A(G)$ with $v(e) = 0$.

Exercise 2.12.55. Let G be a noncompact locally compact group. Show that there exists a bounded continuous function on G which fails to be uniformly continuous (and hence is not almost periodic).

Exercise 2.12.56. Let G be a locally compact group and

$$N = \{x \in G : f(x) = f(e) \text{ for all } f \in AP(G)\}.$$

Using only the fact that $f \in AP(G)$ implies that $L_a R_b f \in AP(G)$ for all $a, b \in G$, give a direct proof that N is a normal subgroup of G and that, for $x, y \in G$, $f(x) = f(y)$ for all $f \in AP(G)$ if and only if $y^{-1}x \in N$.

Exercise 2.12.57. Let G be a locally compact Abelian group and let $P : L^\infty(G) \to L^\infty(G)$ be a norm-bounded projection such that $P(L_x f) = L_x(P(f))$ for all $f \in L^\infty(G)$ and $x \in G$. Show that P maps $AP(G)$ into $AP(G)$ and that there exists a finite measure μ on the Bohr compactification $b(G)$ such that $P(f) = f * \mu$ for all $f \in AP(G) = C(b(G))$.

Exercise 2.12.58. Let A be a commutative Banach algebra and let D be a continuous derivation of A. The *Singer–Wermer theorem* states that $Dx \in \text{rad}(A)$ for every $x \in A$. In particular, there are no nonzero continuous derivations on a semisimple commutative Banach algebra.

Prove the Singer–Wermer theorem as follows. For $\varphi \in \Delta(A)$ and $x \in A$, consider the function $z \to \varphi(\exp(zD)x)$. Show that this is a bounded holomorphic function in the entire complex plane (note that $x \to \varphi(\exp(zD)x)$ is a multiplicative linear functional on A). Conclude that $\varphi(Dx) = 0$.

Let A be a commutative Banach algebra and $\varphi \in \Delta(A)$. A linear functional D on A is called a *point derivation at* φ if $D(ab) = \varphi(a)D(b) + \varphi(b)D(a)$ for all $a, b \in A$.

Exercise 2.12.59. Show that there is a nonzero continuous point derivation on $\text{Lip}_\alpha[0,1]$ at every $t \in [0,1]$.
(Hint: Let $(t_n)_n \subseteq [0,1]$ be a sequence such that $t_n \to t$ and $t_n \neq t$ for all n. Define $l_n \in (\text{Lip}_\alpha[0,1])^*$ by

$$l_n(f) = \frac{f(t_n) - f(t)}{|t_n - t|^\alpha},$$

and let l be a w^*-accumulation point of the sequence $(l_n)_n$ in $(\text{Lip}_\alpha[0,1])^*$.)

Exercise 2.12.60. Let $t \in [0,1]$ and let I and J be the closed ideals in $C^n[0,1]$ defined by

$$I = \{f \in C^n[0,1] : f(t) = 0\} \text{ and } J = \{f \in C^n[0,1] : f(t) = f'(t) = 0\}.$$

It follows from Taylor's formula that I^2 is dense in J. Let D be a continuous point derivation of $C^n[0,1]$ at t, that is,

$$D(fg) = f(t)D(g) + g(t)D(f)$$

for all $f, g \in C^n[0,1]$. Show that $D(J) = \{0\}$ and hence D is of the form $D(f) = \alpha f(t) + \beta f'(t)$ for some $\alpha, \beta \in \mathbb{C}$. Conclude that $D(f) = \beta f'(t)$ for all $f \in C^n[0,1]$.

Exercise 2.12.61. Let $A \subseteq C(X)$ and $B \subseteq C(Y)$ be uniform algebras. Let

$$A \otimes_u B = \{f \in C(X \times Y) : f(\cdot, y) \in A \text{ for all } y \in Y$$
$$\text{and } f(x, \cdot) \in B \text{ for all } x \in X\}.$$

Show that $A \otimes_u B$ is a uniform algebra on $X \times Y$ and that $\Delta(A \otimes_u B)$ can be canonically identified with $\Delta(A) \times \Delta(B)$. $A \otimes_u B$ is called the *uniform tensor product* (slice product) of A and B.

Exercise 2.12.62. Let A and B be commutative Banach algebras such that A is semisimple and B is finite dimensional. Prove, without using the fact that B has the approximation property, that $A \widehat{\otimes}_\pi B$ is semisimple. (Hint: Let $\Delta(B) = \{\psi_1, \ldots, \psi_m\}$ and choose $b_j \in \cap\{\ker \psi_k : k \neq j\}$ such that $\psi_j(b_j) = 1$. Then $b_1, \ldots, b_m$ form a basis of B.)

Exercise 2.12.63. Let G and H be discrete Abelian groups with dual groups $\widehat{G}$ and $\widehat{H}$. Prove that the Gelfand homomorphism maps $l^1(G \times H)$ into the projective tensor product $C(\widehat{G}) \widehat{\otimes}_\pi C(\widehat{H})$.

2.13 Notes and references

Theorem 2.1.2, characterizing multiplicative linear functionals on (not necessarily commutative) Banach algebras, has been established independently by Gleason [44] and Kahane and Zelazko [64] using analytic tools. The fairly elementary algebraic proof given here was found by Roitman and Sternfeld [109] and the preliminary Lemma 2.1.1 is due to Zelazko [141]. There exist an extensive theory and a wealth of interesting examples of radical commutative Banach algebras. These play a fundamental role in the investigation of automatic continuity problems (see [25] for a comprehensive account). We have confined ourselves to including just two illustrative examples. The continuity results Corollaries 2.1.10 and 2.1.12 and the uniqueness of norm property, Corollary 2.1.11, trace back to Rickart [106]. Corollaries 2.1.11 and 2.1.12 hold as well for non-commutative semisimple Banach algebras. This follows from Johnson's theorem [61] stating that if A and B are Banach algebras with B semisimple, then every homomorphism from A onto B is continuous. For a short proof of Johnson's theorem, see [101].

The Gelfand representation is the pioneering work of Gelfand. All the basic results presented in Section 2.2 appeared first in [38] and [40] and are nowadays part of any book on Banach algebras. Also, the examples and immediate applications of Gelfand's theory given in Section 2.2 are standard.

Many commutative Banach algebras are generated by finitely many elements. If $a_1, \ldots, a_n$ generate A, then $\Delta(A)$ is canonically homeomorphic to the joint spectrum of $a_1, \ldots, a_n$, which is a compact subset of $\mathbb{C}^n$. It is therefore an important issue to identify the compact subsets of $\mathbb{C}^n$ arising in this manner as joint spectra. Theorem 2.3.6, which states that these are exactly the polynomial convex subsets of $\mathbb{C}^n$, was shown by Shilov, as was Theorem 2.3.7, which says that a compact subset of $\mathbb{C}$ is polynomially convex if and only if its complement is connected [121, 123]. The problem of a topological characterisation of polynomial convex subsets of $\mathbb{C}^n$ for $n \geq 2$ is open. For more details and partial results we refer the reader to [126].

C^*-algebras were first studied by Gelfand and Naimark in their fundamental paper [39]. Theorem 2.4.5, which is usually referred to as the commutative Gelfand-Naimark theorem and which identifies the commutative C^*-algebras as precisely the uniform algebras $C_0(X)$, where X is a locally compact Hausdorff space, as well as the continuous functional calculus (Theorem 2.4.9) can be found in [39]. Let X be a completely regular topological space. The introduction of the Stone–Čech compactification $\beta(X)$ as the structure space of the commutative C^*-algebra $C^b(X)$ (Theorem 2.4.12) is for instance given in [126] and [36].

There is a vast literature on uniform algebras, in particular on $P(X)$, $R(X)$ and $A(X)$, where X is a compact subset of $\mathbb{C}^n$. We refer the reader to the monographs by Stout [126], Gamelin [36] and Leibowitz [78] concerning much more detailed material. Equality to hold at any position in the chain of inclusions $P(X) \subseteq R(X) \subseteq A(X) \subseteq C(X)$ can be interpreted as a result in qualitative approximation theory and is therefore of interest beyond Banach algebra theory. Samples of such results are Theorem 2.5.8 and Theorem 2.5.12, the former being a major step towards Mergelyan's theorem which asserts that if X is a compact subset of $\mathbb{C}$, then $P(X) = A(X)$ precisely when $\mathbb{C} \setminus X$ is connected. Except for $n = 1$, there are no topological characterisations of those compact subsets of $\mathbb{C}^n$ which arise as structure spaces of algebras $P(X)$ and $R(X)$ (Theorem 2.5.7). Examples of compact subsets X of $\mathbb{C}$ with empty interior for which $R(X) \neq C(X)$ have been given by several authors. The example we have presented in Section 2.5 is basically due to Mergelyan [87], somewhat modified by McKissick [85] (see also [73]). The maximality theorem, Theorem 2.5.15, was found by Wermer [136]. Lemma 2.5.14 and the simple proof of Theorem 2.5.15 based on it was discovered by Cohen [22]. The related result displayed in Exercise 2.12.25 was shown by Rudin [112].

Theorem 2.6.6 is due to Arens [4] and can also be found in [126] and [36]. It is worth pointing out that when X is a compact subset of $\mathbb{C}^n$ for some $n > 1$, then $\Delta(A(X))$ need not be homeomorphic with a subset of $\mathbb{C}^n$ [126].

The convolution algebras $L^1(G)$ of locally compact Abelian groups, which are the central object of study in commutative harmonic analysis, form a large and extremely important class of commutative Banach algebras. The fact that the structure space of $L^1(G)$ identifies canonically with the dual group $\widehat{G}$ of G, endowed with the topology of uniform convergence of characters on compact subsets of G (Theorems 2.7.2 and 2.7.5) is classical. We refer to [54], [105], and [113]. Note that for G the group of real numbers, the Gelfand transform is nothing but the Fourier transform. To show semisimplicity of $L^1(G)$, we have exploited the left regular representation of $L^1(G)$ on $L^2(G)$ and the semisimplicity of commutative C^*-algebras. A highly non-trivial fact is that the Gelfand homomorphism of $L^1(G)$ into $C_0(\widehat{G})$ is surjective only when G is finite (Theorem 2.7.12). There are approaches to Theorem 2.7.12 different from the one chosen here, either using the Pontryagin duality theorem or some other tools none of which we want to employ in this context (see [28, Theorem B.4.6], [34], and [45]).

Beurling algebras behave in many respects similarly to L^1-algebras. For instance, Theorems 2.8.2 and 2.8.5 exposing the Gelfand representation of $L^1(G,\omega)$, parallel Theorems 2.7.3 and 2.7.5. Some technical complications, however, arise from the facts that weights are only locally bounded and that the set of ω-bounded generalized characters is less handy than the dual group $\widehat{G}$. The concrete realizations of $\Delta(L^1(\mathbb{R},\omega))$ and $\Delta(l^1(\mathbb{Z},\omega))$ by means of a vertical strip and an annulus in the complex plane, respectively, are classical [41, Chapter III]. The elementary proof of semisimplicity of $L^1(G,\omega)$ given here (Theorem 2.8.10) is due to Bhatt and Dedania [16].

The Fourier algebra $A(G)$ of a locally compact group G was introduced by Eymard [32] as the predual of the group von Neumann algebra $VN(G)$. The realization of $A(G)$ which we have taken as the definition and all the basic results, such as Theorem 2.9.4 and Lemma 2.9.5, are contained in [32]. Our presentation follows the one in [25]. Eymard has also shown that $A(G)$ is isometrically isomorphic to $L^1(\widehat{G})$ when G is Abelian. This is one of the reasons why the large class of Fourier algebras currently attracts a lot of attention within the theory of commutative Banach algebras. A result of Leptin [79] says that $A(G)$ has a bounded approximate identity if and only if G is a so-called amenable group. One of the many open questions is whether $A(G)$ always possesses an (unbounded) approximate identity.

The Bohr or almost periodic compactification $b(G)$ of a locally compact Abelian group G originated from a paper by Bohr [18] who was the first to study almost periodic functions on the real line. Discussions of the subject under various different aspects can be found in the monographs by Hewitt and Ross [54, 55], Loomis [81], and Weil [134]. In particular, the fairly elementary proof showing that one-sided almost periodic functions are necessarily two-sided almost periodic is due to Loomis. In Section 2.10 we have established the existence and properties of $b(G)$ by applying Gelfand's theory to the commutative C^*-algebra of almost periodic functions.

Tensor products of commutative Banach algebras have been investigated by several authors. Theorem 2.11.2, which canonically identifies $\Delta(A\widehat{\otimes}_\gamma B)$ with the product space $\Delta(A)\times\Delta(B)$, was independently shown by Tomiyama [128] and Gelbaum [37], following earlier work of Hausner [48, 49] and G.P. Johnson [62] on $L^1(G,A)$ and $C(X,A)$. The more subtle question of when the projective tensor product $A\widehat{\otimes}_\pi B$ is semisimple was addressed in [128], where Theorem 2.11.6 can be found. The fact that condition (ii) of Theorem 2.11.6 need not be satisfied and consequently the projective tensor product of two semisimple commutative Banach algebras need not be semisimple, was discovered by Milne [89] by exploiting the existence of Banach spaces which don't share the approximation property [31]. Note in this context that Theorem 2.11.6 contradicts Corollary 1 of [77].

3

Functional Calculus, Shilov Boundary, and Applications

Let A be a commutative Banach algebra. This chapter focuses on several important problems which evolve from the Gelfand representation theory and concern the structure space $\Delta(A)$ and the structure of A itself. The most significant new tool to be used in this context are so-called holomorphic functional calculi for Banach algebra elements. The single-variable holomorphic functional calculus associates with a complex-valued function f, which is defined and holomorphic in a neighbourhood of the spectrum of an element x of A, an element $f(x)$ of A. This calculus and the properties of the assignment $f \to f(x)$ are developed in Section 3.1. There is a generalisation to several-variable holomorphic functions a weaker form of which we discuss at the end of Section 3.1.

A first application of the single-variable functional calculus concerns the topological group $G(A)$ of invertible elements of a unital commutative Banach algebra A, especially the description of its connected component of the identity (Section 3.2). A very intricate problem is the identification of those elements of $\Delta(A)$ that extend to elements of $\Delta(B)$ whenever B is any commutative Banach algebra containing A as a closed subalgebra. This question is dealt with at several places in this chapter and it has been the motivation for Shilov to introduce the boundary $\partial(A)$ carrying his name. The Shilov boundary is the smallest closed subset of $\Delta(A)$ on which every function $|\hat{a}|$, $a \in A$, attains its maximum (Section 3.3). Not only does the Shilov boundary play an important role in the extension problem, it is also linked to the concept of (joint) topological zero divisors which we study thoroughly in Section 3.4.

The structure space of a unital commutative Banach algebra is compact. The converse, a deep fact, holds for semisimple algebras. Although some special cases are easier to obtain, the decisive result is based on Shilov's idempotent theorem, the proof of which requires (at least so far) the multivariable holomorphic functional calculus. Shilov's idempotent theorem states that the characteristic function of a compact open subset of $\Delta(A)$ is the Gelfand transform of an idempotent in A. This theorem is unquestionably one of the

E. Kaniuth, *A Course in Commutative Banach Algebras*, Graduate Texts in Mathematics, DOI 10.1007/978-0-387-72476-8_3, © Springer Science+Business Media, LLC 2009

highlights in commutative Banach algebra theory and is presented in Section 3.5, followed by the proof that a semisimple commutative Banach algebra with compact structure space has to be unital. We also study the impact of Shilov's idempotent theorem on decomposing A into a direct sum of ideals.

3.1 The holomorphic functional calculus

In Theorem 2.4.9 we have described a continuous functional calculus for elements of a commutative C^*-algebra A. In this section we replace A by an arbitrary unital Banach algebra and develop the so-called *holomorphic functional calculus* which provides an efficient method to construct from a given algebra element new elements with specified properties.

Let A be a unital Banach algebra and $x \in A$. Suppose that U is an open set containing $\sigma_A(x)$, and denote by $R(U)$ the set of all rational functions on U. That is, $f \in R(U)$ if and only if $f = (p/q)|_U$, where p and q are polynomials with $q(z) \neq 0$ for all $z \in U$. Since $\sigma_A(q(x)) = q(\sigma_A(x))$ (Lemma 1.2.10), we have that $0 \notin \sigma_A(q(x))$ and therefore $q(x)$ is invertible in A. We define $f(x) \in A$ by
$$f(x) = p(x)q(x)^{-1}.$$
As U is a nonempty open set, the representation p/q of f is unique apart from common factors of numerator and denominator. Moreover, polynomials in x and the inverses of such polynomials commute with each other. It follows that $f(x)$ is independent of the choice of p and q. Let
$$R(x) = \bigcup \{R(U) : U \text{ open, } U \supseteq \sigma_A(x)\}.$$

Then $R(x)$ is an algebra and $f(x) \in A$ is well defined for every $f \in R(x)$. The proof of the following lemma, which is left to the reader as an exercise, is straightforward using Lemma 1.2.10 and the fact that $\varphi(q(x)^{-1}) = 1/\varphi(q(x))$ for all $\varphi \in \Delta(A)$.

Lemma 3.1.1. *The mapping $f \to f(x)$ is a homomorphism from $R(x)$ into A and satisfies $\varphi(f(x)) = f(\varphi(x))$ for all $\varphi \in \Delta(A)$ and $\sigma_A(f(x)) = f(\sigma_A(x))$.*

For an open subset U of $\mathbb{C}$ let $H(U)$ denote the algebra of all holomorphic functions on U. For $x \in A$, let $H(x)$ be the algebra of all functions that are holomorphic in some neighbourhood of $\sigma_A(x)$; that is,
$$H(x) = \bigcup \{H(U) : U \text{ open, } U \supseteq \sigma_A(x)\}.$$

With pointwise operations, $H(x)$ is an algebra. We wish to extend the homomorphism from $R(x)$ into A to a homomorphism from $H(x)$ into A. Clearly, in what follows we need some basic tools from one variable complex analysis. Proofs of the following two lemmas can, for instance, be found in [23, Chapter VIII, Propositions 1.1 and 1.7]. In the sequel, for a rectifiable closed curve

$\gamma : [a, b] \to \mathbb{C}$ and $z \in \mathbb{C} \setminus \gamma[a, b]$, $w(\gamma, z)$ denotes the winding number of γ relative to the point z.

Lemma 3.1.2. *Let U be an open subset of $\mathbb{C}$ and K a compact subset of U. Then there are closed, piecewise smooth curves $\gamma_1, \ldots, \gamma_m$ in $U \setminus K$ such that for any holomorphic function f on U and $z \in K$,*

$$f(z) = \frac{1}{2\pi i} \sum_{j=1}^{m} \int_{\gamma_j} \frac{f(w)}{w - z} \, dw.$$

In particular, $\sum_{j=1}^{m} w(\gamma_j, z) = 1$.

Lemma 3.1.3. *If $f : U \to \mathbb{C}$ is a holomorphic function, then for every compact subset K of U and $\epsilon > 0$ there exists a rational function r, the poles of which are contained in $\mathbb{C} \setminus K$, such that*

$$\|f|_K - r|_K\|_\infty \leq \epsilon.$$

Remark 3.1.4. Let A be a unital Banach algebra, $\gamma : [a, b] \to \mathbb{C}$ a rectifiable curve, and $F : \gamma[a, b] \to A$ a continuous mapping. For any partition

$$\mathcal{Z} = \{a = t_0 < t_1 < \ldots < t_n = b\}$$

of $[a, b]$ let $\delta(\mathcal{Z}) = \max\{t_j - t_{j-1} : 1 \leq j \leq n\}$. Now, using that F is uniformly continuous, precisely the same arguments as those showing the existence of the Riemann integral along such a curve, in the present situation yield that the limit

$$\lim_{\delta(\mathcal{Z}) \to 0} \sum_{j=1}^{n} (\gamma(t_j) - \gamma(t_{j-1})) F(\gamma(t_j))$$

exists in A. This element in A is denoted $\int_\gamma F(z) dz$. It is immediate from the definition that

$$\varphi \left(\int_\gamma F(z) dz \right) = \int_\gamma \varphi(F(z)) dz$$

for every $\varphi \in A^*$. Conversely, by the Hahn–Banach theorem, $\int_\gamma F(z) dz$ is uniquely determined by this equation. Moreover, for each $x \in A$,

$$x \cdot \int_\gamma F(z) dz = \int_\gamma x F(z) dz.$$

In fact, denoting by $L_x : A \to A$ the mapping $y \to xy$, we have $\varphi \circ L_x \in A^*$ for every $\varphi \in A^*$ and hence

$$\varphi\left(x\int_\gamma F(z)dz\right) = \varphi \circ L_x\left(\int_\gamma F(z)dz\right)$$

$$= \int_\gamma \varphi \circ L_x(F(z))dz = \int_\gamma \varphi(xF(z))dz$$

$$= \varphi\left(\int_\gamma xF(z)dz\right).$$

Proposition 3.1.5. *Let A be a Banach algebra with identity e, $x \in A$, and U an open neighbourhood of $\sigma_A(x)$. Suppose that $\gamma_1, \ldots, \gamma_n$ are closed, piecewise smooth curves in $U \setminus \sigma_A(x)$ having the properties of Lemma 3.1.2. Then for any rational function f on U,*

$$f(x) = \frac{1}{2\pi i}\sum_{k=1}^n \int_{\gamma_k} f(z)(ze - x)^{-1}dz.$$

Proof. First, we show that

$$\frac{1}{2\pi i}\sum_{k=1}^n \int_{\gamma_k}(ze - x)^{-1}dz = e.$$

For that, by the Hahn–Banach theorem, it suffices to verify that

$$\varphi(e) = \varphi\left(\frac{1}{2\pi i}\sum_{k=1}^n \int_{\gamma_k}(ze - x)^{-1}dz\right)$$

for all $\varphi \in A^*$. Thus, fix $\varphi \in A^*$ and define a function g on $\mathbb{C} \setminus \sigma_A(x)$ by

$$g(z) = \varphi((ze - x)^{-1}).$$

Then g is holomorphic (compare the proof of Theorem 1.2.8), and by Remark 3.1.4,

$$\varphi\left(\frac{1}{2\pi i}\sum_{k=1}^n \int_{\gamma_k}(ze - x)^{-1}dz\right) = \frac{1}{2\pi i}\sum_{k=1}^n \int_{\gamma_k}\varphi((ze - x)^{-1})dz$$

$$= \frac{1}{2\pi i}\sum_{k=1}^n \int_{\gamma_k} g(z)dz.$$

Choose $R > \|x\|$, and let $\gamma(t) = Re^{2\pi it}$, $t \in [0, 1]$, and let γ^{-1} denote the inverse of the curve γ. Then $w(\gamma^{-1}, z) = -1$ for every $z \in \sigma_A(x)$. Applying Lemma 3.1.2 with $f = 1$, we get $\sum_{k=1}^n w(\gamma_k, z) = 0$ for all $z \in \mathbb{C}\setminus U = \sigma_A(x)$.

A variant of Cauchy's integral formula (see [23, p. 206 and p. 220]), applied to $\mathbb{C} \setminus \sigma_A(x)$, the holomorphic function g, and the curves $\gamma, \gamma_1, \ldots, \gamma_n$, now yields

$$\sum_{k=1}^n \int_{\gamma_k} g(z)dz = \int_\gamma g(z)dz.$$

On the other hand, for $z \in \gamma[0,1]$,

$$(ze - x)^{-1} = \frac{1}{z}\left(e - \frac{1}{z}x\right)^{-1} = \sum_{j=0}^\infty z^{-(j+1)}x^j,$$

the series being uniformly convergent on $\gamma[0,1]$. Because $\int_\gamma z^{-1}dz = 2\pi i$ and $\int_\gamma z^m dz = 0$ for $m \in \mathbb{Z}$, $m \neq -1$, it follows that

$$\int_\gamma g(z)dz = \int_\gamma \varphi\left(\sum_{j=0}^\infty z^{-(j+1)}x^j\right)dz$$

$$= \int_\gamma \sum_{j=0}^\infty z^{-(j+1)}\varphi(x^j)dz$$

$$= \sum_{j=0}^\infty \varphi(x^j)\int_\gamma z^{-(j+1)}dz$$

$$= 2\pi i\varphi(e).$$

Combining the above equations we obtain

$$\varphi(e) = \frac{1}{2\pi i}\int_\gamma g(z)dz = \frac{1}{2\pi i}\sum_{k=1}^n \int_{\gamma_k} g(z)dz = \varphi\left(\frac{1}{2\pi i}\sum_{k=1}^n \int_{\gamma_k}(ze - x)^{-1}dz\right).$$

Now, let $f(z) = p(z)/q(z)$, $z \in U$, where p and q are polynomials and q is nonzero on U. Let, say,

$$p(z) = \sum_{\nu=0}^s a_\nu z^\nu \quad \text{and} \quad q(z) = \sum_{\mu=0}^t b_\mu z^\mu.$$

Then, since $p(x)$ and $q(x)$ commute,

$$p(z)q(x) - q(z)p(x) = (p(z)e - p(x))q(x) - (q(z)e - q(x))p(x)$$

$$= \left(\sum_{\nu=0}^s a_\nu(z^\nu e - x^\nu)\right)q(x) - \left(\sum_{\mu=0}^t b_\mu(z^\mu e - x^\mu)\right)p(x)$$

$$= (ze - x)\big(p_1(z,x)q(x) - p_2(z,x)p(x)\big),$$

where p_1 and p_2 are polynomials in two variables. Hence

$$(f(z)e - f(x))(ze - x)^{-1} = (p(z)q(x) - q(z)p(x))q(z)^{-1}q(x)^{-1}(ze - x)^{-1}$$
$$= q(z)^{-1}q(x)^{-1}\left(p_1(z,x)q(x) - p_2(z,x)p(x)\right).$$

Therefore, there exist elements $a_1, \ldots, a_m$ of A and rational functions $g_1, \ldots, g_m$ defined on U such that

$$(f(z)e - f(x))(ze - x)^{-1} = \sum_{j=1}^{m} g_j(z)a_j.$$

Since $\int_{\gamma_k} g_j(z)dz = 0$ for all j and k, it follows from the first part of the proof that

$$\frac{1}{2\pi i} \sum_{k=1}^{n} \int_{\gamma_k} f(z)(ze - x)^{-1}dz = \frac{1}{2\pi i} \sum_{k=1}^{n} \int_{\gamma_k} (f(z)e - f(x))(ze - x)^{-1}dz$$

$$+ \frac{1}{2\pi i} \sum_{k=1}^{n} \int_{\gamma_k} f(x)(ze - x)^{-1}dz$$

$$= \frac{1}{2\pi i} \sum_{k=1}^{n} \left(\sum_{j=1}^{m} \int_{\gamma_k} g_j(z)dz\right)a_k$$

$$+ f(x)\frac{1}{2\pi i} \sum_{k=1}^{n} \int_{\gamma_k} (ze - x)^{-1}dz$$

$$= f(x),$$

as was to be shown. $\qquad\square$

We intend to define, for $f \in H(x)$, an element $f(x)$ of A by setting

$$f(x) = \frac{1}{2\pi i} \sum_{k=1}^{n} \int_{\gamma_k} f(z)(ze - x)^{-1}dz,$$

where $\gamma_1, \ldots, \gamma_n$ are as in Lemma 3.1.2. For this definition to make sense, we have to verify that it does not depend on the choice of the curves involved. Once this has been done, Lemma 3.1.5 shows that this definition of $f(x)$ extends the one for rational functions f.

Lemma 3.1.6. *Let U be an open neighbourhood of $\sigma_A(x)$ and let f be a holomorphic function on U. Moreover, let $\gamma_1, \ldots, \gamma_n$ and $\delta_1, \ldots, \delta_m$ be systems of closed, piecewise smooth curves in $U \setminus \sigma_A(x)$ with the properties of Lemma 3.1.2. Then*

$$\sum_{k=1}^{n} \int_{\gamma_k} f(z)(ze - x)^{-1}dz = \sum_{j=1}^{m} \int_{\delta} f(z)(ze - x)^{-1}dz.$$

Proof. Let C_k and D_j denote the trace of γ_k and δ_j, respectively, and set

$$K = \left(\bigcup_{k=1}^{n} C_k \right) \cup \left(\bigcup_{j=1}^{m} D_j \right).$$

Notice that, because the function $z \to \|(ze - x)^{-1}\|$ is continuous on $\mathbb{C} \setminus \sigma_A(x)$ and because K is compact and $K \cap \sigma_A(x) = \emptyset$,

$$M = \left(\sum_{k=1}^{n} L(\gamma_k) + \sum_{j=1}^{m} L(\delta_j) \right) \sup_{z \in K} \|(ze - x)^{-1}\| < \infty.$$

Here, $L(\gamma)$ of course denotes the length of a rectifiable curve γ. By Lemma 3.1.3 there exist rational functions $f_n, n \in \mathbb{N}$, having poles outside $K \cup \sigma_A(x)$ such that $f_n(z) \to f(z)$ uniformly on K. Using Lemma 3.1.5, we can now estimate the norm of the element

$$a = \sum_{k=1}^{n} \int_{\gamma_k} f(z)(ze - x)^{-1} \, dz - \sum_{j=1}^{m} \int_{\delta_j} f(z)(ze - x)^{-1} \, dz$$

of A as follows:

$$\|a\| \leq \left\| 2\pi i f_n(x) - \sum_{k=1}^{n} \int_{\gamma_k} f(z)(ze - x)^{-1} dz \right\|$$

$$+ \left\| 2\pi i f_n(x) - \sum_{j=1}^{m} \int_{\delta_j} f(z)(ze - x)^{-1} dz \right\|$$

$$= \left\| \sum_{k=1}^{n} \int_{\gamma_k} [f_n(z) - f(z)](ze - x)^{-1} dz \right\|$$

$$+ \left\| \sum_{j=1}^{m} \int_{\delta_j} [f_n(z) - f(z)](ze - x)^{-1} dz \right\|$$

$$\leq M \cdot \sup_{z \in K} |f_n(z) - f(z)|.$$

Since the sequence $(f_n)_n$ converges to f uniformly on K, the statement of the lemma follows. □

Definition 3.1.7. Let A be a unital Banach algebra. For $x \in A$ and $f \in H(x)$ we define $f(x) \in A$ as follows. Suppose that f is a holomorphic function on the open set U containing $\sigma_A(x)$, and choose closed, piecewise smooth curves $\gamma_1, \ldots, \gamma_n$ in $U \setminus \sigma_A(x)$ with the properties of Lemma 3.1.2. Then, define $f(x) \in A$ by

$$f(x) = \frac{1}{2\pi i} \sum_{k=1}^{n} \int_{\gamma_k} f(z)(ze - x)^{-1} dz.$$

It follows from Lemma 3.1.6 that this definition does not depend on the choice of U and of the curves $\gamma_1, \ldots, \gamma_n$. Also, by Lemma 3.1.5, it extends the definition of $f(x)$ for rational functions f. The set of mappings $H(x) \to A, f \to f(x), x \in A$, is referred to as the *single-variable holomorphic functional calculus*.

The basic properties of the holomorphic functional calculus are listed in the next theorem.

Theorem 3.1.8. *Let A be a commutative unital Banach algebra. For $x \in A$ the following assertions hold.*

(i) *$f \to f(x)$ is a homomorphism from $H(x)$ into A.*
(ii) *If f is an entire function and $f(z) = \sum_{k=0}^{\infty} a_k z^k$, then*

$$f(x) = \sum_{k=0}^{\infty} a_k x^k,$$

the series being absolutely convergent.
(iii) *Suppose that f and $f_n, n \in \mathbb{N}$, are holomorphic functions on some open set U containing $\sigma_A(x)$ and that f_n converges uniformly to f on every compact subset of U. Then*

$$\|f_n(x) - f(x)\| \to 0.$$

(iv) *For $f \in H(x)$ we have $\varphi(f(x)) = f(\varphi(x))$ for all $\varphi \in \Delta(A)$, and hence*

$$\sigma_A(f(x)) = f(\sigma_A(x)).$$

Proof. (i) Recall first that the definition of $f(x)$ for $f \in H(x)$ does not depend on the choice of curves $\gamma_1, \ldots, \gamma_n$ in $U \setminus \sigma_A(x)$ as long as these have the properties in Lemma 3.1.2. Moreover, if γ is any rectifiable curve with image Γ, then the mapping $g \to \int_{\gamma} g(z) dz$ from $C(\Gamma, A)$ into A is linear. These two facts at once imply that the mapping $f \to f(x)$ is linear.

To verify that this mapping is multiplicative, let f and g be holomorphic on some open neighbourhood U of $\sigma_A(x)$ and choose curves $\gamma_1, \ldots, \gamma_m : [0, 1] \to U \setminus \sigma_A(x)$ as in Lemma 3.1.2, and let $\Gamma_j = \gamma_j[0, 1], 1 \leq j \leq m$. Then there are sequences $(f_n)_n$ and $(g_n)_n$ of rational functions, each of which has its poles outside of $\sigma_A(x) \cup \left(\bigcup_{j=1}^{m} \Gamma_j\right)$ such that $f_n \to f$ and $g_n \to g$ uniformly on $\bigcup_{j=1}^{m} \Gamma_j$. It follows that $f_n g_n \to fg$ uniformly on $\bigcup_{j=1}^{\infty} \Gamma_j$, and since the mapping $r \to r(x)$ from $R(x)$ into A is a homomorphism, we conclude that

$$\|f(x)g(x) - (fg)(x)\| \leq \|f(x) - f_n(x)\| \cdot \|g(x)\|$$
$$+ \|f_n(x)\| \cdot \|g(x) - g_n(x)\|$$
$$+ \|(f_n g_n)(x) - (fg)(x)\|,$$

which converges to 0 as $n \to \infty$.

(ii) Let $R > \|x\|$ and $\gamma(t) = Re^{2\pi it}, t \in [0, 1]$. Then γ has the properties of Lemma 3.1.2, and the series $\sum_{k=0}^{\infty} z^{-(k+1)} x^k$ converges uniformly on $\gamma[0, 1]$. This implies

$$f(x) = \frac{1}{2\pi i} \int_\gamma f(z)(ze - x)^{-1} dz$$

$$= \sum_{k=0}^{\infty} \frac{1}{2\pi i} \int_\gamma \frac{f(z)}{z^{k+1}} x^k dz = \sum_{k=0}^{\infty} \frac{f^{(k)}(0)}{k!} x^k$$

$$= \sum_{k=0}^{\infty} a_k x^k.$$

(iii) follows from the estimate

$$\left\| \int_\gamma g(z)(ze - x)^{-1} dz \right\| \leq L(\gamma) \|g|_{\gamma[0,1]}\| \cdot \sup_{z \in \gamma[0,1]} \|(ze - x)^{-1}\|.$$

(iv) For $\varphi \in \Delta(A)$ and $z \in \mathbb{C} \setminus \sigma_A(x)$,

$$1 = \varphi((ze - x)(ze - x)^{-1}) = (z - \varphi(x))\varphi((ze - x)^{-1}).$$

Thus $\varphi((ze - x)^{-1}) = (z - \varphi(x))^{-1}$, and since $\Delta(A) \subseteq A^*$, we get

$$\varphi(f(x)) = \frac{1}{2\pi i} \sum_{k=1}^{n} \int_{\gamma_k} \varphi(f(z)(ze - x)^{-1}) dz$$

$$= \frac{1}{2\pi i} \sum_{k=1}^{n} \int_{\gamma_k} f(z)(z - \varphi(x))^{-1} dz$$

$$= f(\varphi(x)).$$

Finally, this equation yields that $\sigma_A(f(x)) = \widehat{f(x)}(\Delta(A)) = f(\widehat{x}(\Delta(A))) = f(\sigma_A(x))$. $\qquad \square$

Suppose A is a unital commutative C^*-algebra. Then, for each $x \in A$, we have two functional calculi of A, the holomorphic functional calculus $H(x) \to A$ and the continuous functional calculus $C(\sigma_A(x)) \to A$ (Theorem 2.4.9). It is worth pointing out that these two functional calculi coincide in the following sense. If $f \in H(x)$, then $f(x) = (f|_{\sigma_A(x)})(x)$. Indeed, this follows because $\Delta(A)$ separates the elements of A and

$$\varphi(f(x)) = f(\varphi(x)) = (f|_{\sigma_A(x)})(\varphi(x)) = \varphi((f|_{\sigma_A(x)})(x)).$$

for every $\varphi \in \Delta(A)$. Some applications of the holomorphic functional calculus are presented in the next section. In passing we mention the straightforward extension of the holomorphic functional calculus to nonunital algebras.

Theorem 3.1.9. *Let A be a commutative Banach algebra without identity. For $x \in A$ let*
$$H_0(x) = \{f \in H(x) : f(0) = 0\}.$$
Then the functional calculus $H(x) \to A_e$ (note that $\sigma_{A_e}(x) = \sigma_A(x)$) maps $H_0(x)$ into A, and the mapping $f \to f(x)$ from $H_0(x)$ into A satisfies (i) to (iv) in Theorem 3.1.8.

Proof. Recall that $\Delta(A_e) = \widehat{\Delta(A)} \cup \{\varphi_\infty\}$, where $\widetilde{\varphi}(y + \lambda e) = \varphi(y) + \lambda$ for $\varphi \in \Delta(A), y \in A, \lambda \in \mathbb{C}$. By Theorem 3.1.8, $\psi(f(x)) = f(\psi(x))$ for all $\psi \in \Delta(A_e)$. In particular,
$$\varphi_\infty(f(x)) = f(\varphi_\infty(x)) = f(0) = 0.$$
Thus, if $f(x) = a + \lambda e$ with $a \in A$ and $\lambda \in \mathbb{C}$, then $0 = \varphi_\infty(f(x)) = \lambda$. This shows $f(x) \in A$, and hence, for all $\varphi \in \Delta(A)$,
$$\varphi(f(x)) = \widetilde{\varphi}(f(x)) = f(\widetilde{\varphi}(x)) = f(\varphi(x)).$$
Hence (iv) holds, and (i) to (iii) follow immediately from Theorem 3.1.8.   □

The single-variable holomorphic functional calculus admits a generalisation to functions of n variables, $n \geq 2$, which involves the techniques of the theory of holomorphic functions of n complex variables. To prove Shilov's idempotent theorem in Section 3.5, we need the following weaker version.

Theorem 3.1.10. *Let A be a unital commutative Banach algebra and let $x_1, \ldots, x_n \in A$. Let f be a complex-valued function of n variables which is defined and holomorphic on some open set containing the joint spectrum $\sigma_A(x_1, \ldots, x_n)$ of $x_1, \ldots, x_n$. Then there exists $x \in A$ such that*
$$\widehat{x}(\varphi) = f(\widehat{x_1}(\varphi), \ldots, \widehat{x_n}(\varphi))$$
for all $\varphi \in \Delta(A)$.

To prove Theorem 3.1.10, the following result, which is due to Oka and usually referred to as *Oka's extension theorem*, is employed (see Section 3.7 for references).

Let $n, m \in \mathbb{N}$. Let $p_1, \ldots, p_m$ be polynomials in n complex variables and let $\pi : \mathbb{C}^n \to \mathbb{C}^{n+m}$ denote the mapping defined by
$$\pi(z) = (z, p_1(z), \ldots, p_m(z)).$$
If f is holomorphic on an open neighbourhood of $\pi^{-1}(\mathbb{D}^{n+m})$, then there exists a holomorphic function F, defined on some open neighbourhood of $\mathbb{D}^{n+m}$ such that $F(\pi(z)) = f(z)$ for all $z \in \pi^{-1}(\mathbb{D}^{n+m})$.

In the sequel, as in the case $n = 1$, for an open subset U of $\mathbb{C}^n$, $H(U)$ denotes the algebra of holomorphic functions on U.

Proposition 3.1.11. *Let* $n, m \in \mathbb{N}$ *and* $c_j > 0$ *for* $1 \leq j \leq n + m$, *and let* $p_1, \ldots, p_m$ *be polynomials in* n *variables. Let*

$$D = \{z \in \mathbb{C}^{n+m} : |z_j| \leq c_j \text{ for } j = 1, \ldots, n+m\},$$

and define $\pi : \mathbb{C}^n \to \mathbb{C}^{n+m}$ *as above. If* f *is a function holomorphic on an open neighbourhood of* $\pi^{-1}(D)$, *then there exists a function* F *holomorphic on an open neighbourhood of* D *such that* $F(\pi(z)) = f(z)$ *for all* $z \in \pi^{-1}(D)$.

Proof. Define three mappings $\rho : \mathbb{C}^n \to \mathbb{C}^n$, $\sigma : \mathbb{C}^{n+m} \to \mathbb{C}^{n+m}$, and $\tau : \mathbb{C}^n \to \mathbb{C}^{n+m}$, respectively, by

$$\rho(w_1, \ldots, w_n) = (c_1 w_1, \ldots, c_n w_n),$$

$$\sigma(w_1, \ldots, w_{n+m}) = \left(\frac{w_1}{c_1}, \ldots, \frac{w_{n+m}}{c_{n+m}} \right),$$

$$\tau(w_1, \ldots, w_n) = \left(w_1, \ldots, w_n, \frac{p_1(\rho(w))}{c_{n+1}}, \ldots, \frac{p_m(\rho(w))}{c_{n+m}} \right).$$

Then $\tau(w) = \sigma(\pi(\rho(w)))$ for all $w \in \mathbb{C}^n$, $\sigma(D) = \mathbb{D}^{n+m}$, and $\tau^{-1}(\mathbb{D}^{n+m}) = \rho^{-1}(\pi^{-1}(D))$.

Let U be an open neighbourhood of $\pi^{-1}(D)$ in $\mathbb{C}^n$ and let $f \in H(U)$. Then $\rho^{-1}(U)$ is an open neighbourhood of $\tau^{-1}(\mathbb{D}^{n+m})$ and $f \circ \rho \in H(\rho^{-1}(U))$. By the Oka extension theorem, there exists a holomorphic function G defined on some open neighbourhood V of $\mathbb{D}^{n+m}$ satisfying $G(\tau(w)) = f \circ \rho(w)$ for all $w \in \tau^{-1}(\mathbb{D}^{n+m})$. Now let $F = G \circ \sigma$. Then F is holomorphic on $\sigma^{-1}(V)$ which is an open neighbourhood of D. If now $z \in \pi^{-1}(D)$, then $z = \rho(w)$ with $w \in \tau^{-1}(\mathbb{D}^{n+m})$ and hence

$$f(z) = f(\rho(w)) = G(\tau(w)) = G(\sigma(\pi(\rho(w)))) = F(\pi(z)),$$

as required. $\qquad\square$

In what follows A is always a commutative Banach algebra with identity e and A^n denotes the Cartesian product of n copies of A.

Lemma 3.1.12. *Let* $x = (x_1, \ldots, x_n) \in A^n$ *and let* U *be an open neighbourhood of* $\sigma_A(x)$ *in* $\mathbb{C}^n$. *Then there exists a finitely generated closed subalgebra* B *of* A *containing* $e, x_1, \ldots, x_n$ *such that* $\sigma_B(x) \subseteq U$.

Proof. Let $z = (z_1, \ldots, z_n) \in \mathbb{C}^n \setminus \sigma_A(x)$. Then the ideal generated by the elements $z_j e - x_j$, $1 \leq j \leq n$, is not contained in any maximal ideal of A, and hence there exists $y = (y_1, \ldots, y_n) \in A^n$ such that

$$\sum_{j=1}^{n} (z_j e - x_j) y_j = e.$$

Fix such a y and let $B(z)$ denote the subalgebra of A generated by the elements $e, x_1, \ldots, x_n, y_1, \ldots, y_n$. Then $z \notin \sigma_{B(z)}(x)$ because otherwise $\psi(e) = 0$ for some $\psi \in \Delta(B(z))$. Choose an open neighbourhood $U(z)$ of z in $\mathbb{C}^n$ with $U(z) \cap \sigma_{B(z)}(x) = \emptyset$ and let

$$C = \{z \in \mathbb{C}^n : |z_j| \leq \|x_j\| \text{ for } j = 1, \ldots, n\}.$$

Because $C \setminus U$ is compact and $\sigma_A(x) \subseteq U$, by the preceding paragraph there exist $z^{(1)}, \ldots, z^{(m)} \in C \setminus U$ such that $C \setminus U \subseteq \bigcup_{k=1}^m U(z^{(k)})$. Each of the algebras $B(z^{(k)})$ is finitely generated, and so there exists a finitely generated closed subalgebra B of A with $B(z^{(k)}) \subseteq B$ for $k = 1, \ldots, m$. Now

$$\sigma_B(x) \cap U(z^{(k)}) \subseteq \sigma_{B(z^{(k)})} \cap U(z^{(k)}) = \emptyset$$

for all k and hence $\sigma_B(x) \cap (C \setminus U) = \emptyset$. Since $\sigma_B(x) \subseteq C$, it follows that $\sigma_B(x) \subseteq U$. □

The proof of the following lemma is a simple modification of the proof of Theorem 2.3.6, (i) $\Rightarrow$ (ii). However, we include the argument for the reader's convenience.

Lemma 3.1.13. *Let $\{x_1, \ldots, x_n\}$ be a set of generators for A and $(\lambda_1, \ldots, \lambda_n)$ $\in \mathbb{C}^n \setminus \sigma_A(x_1, \ldots, x_n)$. Then there exists a polynomial p such that*

$$|p(\lambda_1, \ldots, \lambda_n)| > 1 + \|p(x_1, \ldots, x_n)\|.$$

Proof. Because $(\lambda_1, \ldots, \lambda_n) \notin \sigma_A(x_1, \ldots, x_n)$, there exist $y_1, \ldots, y_n \in A$ so that $\sum_{j=1}^n (\lambda_j e - x_j) y_j = e$. Choose $\delta > 0$ such that

$$\delta \cdot \sum_{j=1}^n \|\lambda_j e - x_j\| < \frac{1}{2}.$$

Since $x_1, \ldots, x_n$ generate A, each element of A can be approximated arbitrarily closely by elements of the form $r(x_1, \ldots, x_n)$, where r is a polynomial. So there exist polynomials $q_1, \ldots, q_n$ such that $\|q_j(x_1, \ldots, x_n) - y_j\| < \delta$, $1 \leq j \leq n$. Define a polynomial q by

$$q(z_1, \ldots, z_n) = 1 - \sum_{j=1}^n (z_j - \lambda_j) q_j(z_1, \ldots, z_n).$$

Then $q(\lambda) = 1$, and by the choice of δ and the polynomials q_j,

$$\|q(x_1, \ldots, x_n)\| \leq \sum_{j=1}^n \|x_j - \lambda_j e\| \cdot \|y_j - q_j(x_1, \ldots, x_n)\| < \frac{1}{2}.$$

Then the polynomial $p = \|q(x_1, \ldots, x_n)\|^{-1} q$ satisfies

$$|p(\lambda_1, \ldots, \lambda_n)| > 2 = 1 + \|p(x_1, \ldots, x_n)\|,$$

as is easily verified. □

Proposition 3.1.14. *Let $x = (x_1, \ldots, x_n) \in A^n$ and let U be an open neighbourhood of $\sigma_A(x)$ in $\mathbb{C}^n$. Then there exist $x_{n+1}, \ldots, x_N \in A$ with the following property. Given $f \in H(U)$, there exists a function F, holomorphic on some open neighbourhood of the polydisc $\{z \in \mathbb{C}^N : |z_j| \leq 1 + \|x_j\|, 1 \leq j \leq N\}$, such that*

$$f(\varphi(x_1), \ldots, \varphi(x_n)) = F(\varphi(x_1), \ldots, \varphi(x_N))$$

for all $\varphi \in \Delta(A)$.

Proof. By Lemma 3.1.12 there exists a finitely generated closed subalgebra B of A containing $e, x_1, \ldots, x_n$ such that $\sigma_B(x) \subseteq U$. Choose elements $x_{n+1}, \ldots, x_k$ of B so that $x_1 \ldots, x_k$ generate B.

Given $z \in \mathbb{C}^k \setminus \sigma_A(x_1, \ldots, x_k)$, by Lemma 3.1.13 there exists a polynomial p in k variables such that $|p(z)| > 1 + \|p(x)\|$ and hence $|p(w)| > 1 + \|p(x)\|$ for all w in a neighbourhood of z in $\mathbb{C}^k$. Let P denote the projection $(z_1, \ldots, z_k) \rightarrow (z_1, \ldots, z_n)$ of $\mathbb{C}^k$ onto $\mathbb{C}^n$ and let

$$D = \{z \in \mathbb{C}^k : |z_j| \leq 1 + \|x_j\| \text{ for } j = 1, \ldots, k\}.$$

Then $D \setminus P^{-1}(U)$ is compact and contained in $\mathbb{C}^k \setminus \sigma_A(x_1, \ldots, x_k)$ since

$$P(\sigma_A(x_1, \ldots, x_k)) = \sigma_A(x_1, \ldots, x_n).$$

Therefore there exist finitely many polynomials $p_1, \ldots, p_m$ in k variables such that for each $z \in D \setminus P^{-1}(U)$, $|p_j(z)| > 1 + \|p_j(x)\|$ for at least one $j \in \{1, \ldots, m\}$.

Now, let $N = k + m$, $x_{k+j} = p_j(x)$ for $j = 1, \ldots, m$, and

$$C = \{z \in \mathbb{C}^N : |z_j| \leq 1 + \|x_j\| \text{ for } j = 1, \ldots, N\}.$$

Moreover, as before, define $\pi : \mathbb{C}^k \rightarrow \mathbb{C}^N$ by

$$\pi(z) = (z, p_1(z), \ldots, p_m(z))$$

for $z \in \mathbb{C}^k$. Then $\pi^{-1}(C) \subseteq P^{-1}(D)$ since $\pi(z) \in C$ implies $z \in D$ and $|p_j(z)| \leq 1 + \|p_j(x)\|$ for $j = 1, \ldots, m$. Since $f \circ P$ is holomorphic on the open neighbourhood $P^{-1}(U)$ of $\pi^{-1}(C)$, by Proposition 3.1.11 there exists F, holomorphic on some open neighbourhood of C, such that $F(\pi(z)) = f(P(z))$ for all $z \in \pi^{-1}(C)$.

For $\varphi \in \Delta(A)$ and $j = 1, \ldots, m$, we have

$$p_j(\varphi(x_1), \ldots, \varphi(x_k)) = \varphi(p_j(x_1, \ldots, x_k)) = \varphi(x_{k+j})$$

and hence $\pi(\varphi(x_1), \ldots, \varphi(x_k)) = (\varphi(x_1), \ldots, \varphi(x_N))$. Because $|\varphi(x_j)| \leq \|x_j\|$ for $j = 1, \ldots, k$ and $|p_j(\varphi(x_1), \ldots, \varphi(x_k))| \leq \|p_j(x_1, \ldots, x_k)\|$ for $j = 1, \ldots, m$, we have $\pi(\varphi(x_1), \ldots, \varphi(x_k)) \in C$ and hence $(\varphi(x_1), \ldots, \varphi(x_k)) \in \pi^{-1}(C)$. Thus

$$f(\varphi(x_1), \ldots, \varphi(x_n)) = f(P(\varphi(x_1), \ldots, \varphi(x_k)))$$
$$= F(\pi(\varphi(x_1), \ldots, \varphi(x_k)))$$
$$= F(\varphi(x_1), \ldots, \varphi(x_N)),$$

which is the desired formula. $\qquad\qquad\qquad\qquad\qquad\qquad\qquad\qquad$ □

Now we are in a position to prove Theorem 3.1.10. Thus let A be a unital commutative Banach algebra and $x_1, \ldots, x_n \in A$, and let f be holomorphic on some open neighbourhood of $\sigma_A(x_1, \ldots, x_n)$.

By Proposition 3.1.14 there exist $x_{n+1}, \ldots, x_N \in A$ and a function F defined and holomorphic on an open neighbourhood of the polydisc

$$D = \{z = (z_1, \ldots, z_N) \in \mathbb{C}^N : |z_j| \leq 1 + \|x_j\| \text{ for } j = 1, \ldots, N\}$$

such that, for all $\varphi \in \Delta(A)$,

$$f(\varphi(x_1), \ldots, \varphi(x_n)) = F(\varphi(x_1), \ldots, \varphi(x_N)).$$

The function F admits a power series expansion

$$F(z_1, \ldots, z_N) = \sum_{k \in (\mathbb{N}_0)^N} \lambda_k \, z_1^{k_1} \cdot \ldots \cdot z_N^{k_N},$$

where $k = (k_1, \ldots, k_N)$, which converges in a neighbourhood of D and hence converges absolutely on D. Therefore the series

$$\sum_{k \in (\mathbb{N}_0)^N} |\lambda_k| \cdot \|z_1\|^{k_1} \cdot \ldots \cdot \|z_N\|^{k_N}$$

converges. Consequently, the series

$$\sum_{k \in (\mathbb{N}_0)^N} \lambda_k \, z_1^{k_1} \cdot \ldots \cdot z_N^{k_N}$$

converges in norm to an element y of A. It follows that

$$\widehat{y}(\varphi) = \sum_{k \in (\mathbb{N}_0)^N} \lambda_k \varphi(x_1)^{k_1} \cdot \ldots \cdot \varphi(x_N)^{k_N}$$
$$= F(\widehat{x_1}(\varphi), \ldots, \widehat{x_N}(\varphi))$$
$$= f(\widehat{x_1}(\varphi), \ldots, \widehat{x_n}(\varphi))$$

for all $\varphi \in \Delta(A)$.

3.2 Some applications of the functional calculus

The one-variable holomorphic functional calculus, as explored in the preceding section, is a powerful tool in the investigation of commutative Banach algebras. Its applications concern, among others, the structure of the group $G(A)$ of invertible elements of A, the existence of idempotents, approximation theory, and the question of whether compactness of $\Delta(A)$ forces A to be unital. We start with the latter problem.

Theorem 3.2.1. *Let A be a commutative Banach algebra and suppose that $\Delta(A)$ is compact. Let $x \in A$ be such that $\widehat{x}(\varphi) \neq 0$ for all $\varphi \in \Delta(A)$, and let f be a holomorphic function on some open neighbourhood of $\widehat{x}(\Delta(A))$. Then there exists $y \in A$ so that $\widehat{y} = f \circ \widehat{x}$.*

Proof. By hypothesis, $\widehat{x}(\Delta(A))$ is a compact subset of $\mathbb{C} \setminus \{0\}$. Choose disjoint open sets U and V in $\mathbb{C}$ such that $\widehat{x}(\Delta(A)) \subseteq U$, $0 \in V$, and f is holomorphic on U. Define $g : U \cup V \to \mathbb{C}$ by $g|_U = f$ and $g|_V = 0$ and let $y = g(x) \in A$ (see Theorems 3.1.8 and 3.1.9). Then

$$\widehat{y}(\varphi) = \varphi(y) = \varphi(g(x)) = g(\varphi(x)) = f(\varphi(x)) = f \circ \widehat{x}(\varphi)$$

for all $\varphi \in \Delta(A)$ since $\widehat{x}(\Delta(A)) \subseteq U$. $\qquad\square$

Corollary 3.2.2. *Let A be a semisimple commutative Banach algebra. Suppose that $\Delta(A)$ is compact and that there exists $x \in A$ such that $\widehat{x}(\varphi) \neq 0$ for all $\varphi \in \Delta(A)$. Then A has an identity.*

Proof. Let f be the function $f(z) = z^{-1}$ on $\mathbb{C} \setminus \{0\}$. Since $\widehat{x}(\Delta(A)) \subseteq \mathbb{C} \setminus \{0\}$, by Theorem 3.2.1 there exists $y \in A$ such that

$$\varphi(y) = f(\varphi(x)) = \frac{1}{\varphi(x)}$$

for all $\varphi \in \Delta(A)$. Then the element $u = xy \in A$ satisfies

$$\varphi(ua) = \varphi(x)\varphi(y)\varphi(a) = \varphi(a)$$

for all $a \in A$ and $\varphi \in \Delta(A)$. Semisimplicity of A implies that $ua = a$ for all $a \in A$. So u is an identity for A. $\qquad\square$

Corollary 3.2.3. *Let A be a semisimple commutative Banach $*$-algebra. If $\Delta(A)$ is compact and the Gelfand homomorphism of A is a $*$-homomorphism, then A is unital.*

Proof. According to Corollary 3.2.2 it suffices to show the existence of some $x \in A$ such that $\widehat{x}(\varphi) \neq 0$ for all $\varphi \in \Delta(A)$. For each $\varphi \in \Delta(A)$, there exists $y_\varphi \in A$ with $\varphi(y_\varphi) \neq 0$. By continuity, $\psi(y_\varphi) \neq 0$ for all ψ in some neighbourhood V_φ of φ in $\Delta(A)$. Since $\Delta(A)$ is compact, there exist $\varphi_1, \ldots, \varphi_n \in \Delta(A)$ such that

$$\Delta(A) = \bigcup_{j=1}^{n} V_{\varphi_j}.$$

Now consider the element

$$x = \sum_{j=1}^{n} y_{\varphi_j} y_{\varphi_j}^*$$

of A. If $\varphi \in \Delta(A)$, then $\varphi \in V_{\varphi_k}$ for some $1 \le k \le n$. Since the Gelfand homomorphism of A is a *-homomorphism, it follows that

$$\hat{x}(\varphi) = \sum_{j=1}^{n} \widehat{y_{\varphi_j} y_{\varphi_j}^*}(\varphi) = \sum_{j=1}^{n} \left| \hat{y}_{\varphi_j}(\varphi) \right|^2 \ge |\varphi(y_{\varphi_k})|^2 > 0.$$

Thus $\hat{x}(\varphi) \neq 0$ for every $\varphi \in \Delta(A)$. $\square$

It is true in general that a semisimple commutative Banach algebra with compact structure space is unital. The proof, however, requires Shilov's idempotent theorem, which in turn builds on the multivariable holomorphic functional calculus, and is presented in Section 3.5. We now proceed to study $G(A)$ when A is unital.

In the sequel, we denote by log the usual branch of the logarithm with domain $\mathbb{C} \setminus (-\infty, 0]$.

Lemma 3.2.4. *Let $x \in A$ be such that $\sigma_A(x) \subseteq \mathbb{C} \setminus (-\infty, 0]$. Then*

$$\exp(\log x) = x.$$

Proof. Let $g \in H(x)$ and let $p(z) = \sum_{k=0}^{m} \alpha_k z^k$ be any polynomial. Then, because the mapping from $H(x)$ into A is a homomorphism and the definition of $f(x)$ for $f \in H(x)$ does not depend on the choice of the curves γ_j in Definition 3.1.7, we have

$$
p(g(x)) = \sum_{k=0}^{m} \alpha_k g(x)^k = \frac{1}{2\pi i} \sum_{k=0}^{m} \alpha_k \left(\sum_{j=1}^{n} \int_{\gamma_j} g(z)^k (ze - x)^{-1} dz \right)
$$

$$
= \frac{1}{2\pi i} \sum_{j=1}^{n} \int_{\gamma_j} (p \circ g)(z)(ze - x)^{-1} dz
$$

$$
= (p \circ g)(x).
$$

Now let $p_m(z) = \sum_{k=0}^{m} z^k / k!$, $m \in \mathbb{N}$. The sequence $(p_m)_m$ converges to the exponential function uniformly on compact subsets of $\mathbb{C}$, and hence $\exp(\log z)$ is the uniform limit of $p_m(\log z)$ on compact subsets of $\mathbb{C} \setminus (-\infty, 0]$. This implies

$$p_m(\log x) \longrightarrow \exp(\log x) \quad \text{and} \quad (p_m \circ \log)(x) \longrightarrow (\exp \circ \log)(x) = x.$$

Since $p_m(\log x) = (p_m \circ \log)(x)$ by the above calculation, we conclude that $\exp(\log x) = x$. $\square$

Corollary 3.2.5. *If $x \in A$ is such that $\|e - x\| < 1$, then $x = \exp y$ for some $y \in A$.*

Proof. If $\|e - x\| < 1$, then $\sigma_A(x) \subseteq \{z \in \mathbb{C} : |z - 1| < 1\}$ and hence $x = \exp(\log x) \in \exp A$ by Lemma 3.2.4.   □

We are now able to identify $\exp(A)$ within $G(A)$.

Theorem 3.2.6. *Let A be a commutative Banach algebra with identity e. Then $\exp A$ equals the connected component of e in $G(A)$.*

Proof. Note first that Theorem 3.1.8(ii) and the functional equation of the exponential function imply that

$$\exp x \cdot \exp y = \exp(x + y)$$

for all $x, y \in A$. It follows that $\exp A$ is a group and $\exp A \subseteq G(A)$. For $x \in A$, the map $\gamma : t \to \exp(tx)$ from $[0, 1]$ into A is continuous because

$$\|\exp(tx) - \exp(sx)\| \le |t - s| \sum_{j=1}^{\infty} \frac{1}{j!} |t - s|^{j-1} \|x\|^j.$$

Since $\gamma(0) = e$ and $\gamma(1) = \exp x$, we deduce that $\exp A$ is connected.

To prove that $\exp A$ actually equals the connected component of e in $G(A)$, it suffices to show that $\exp A$ is both open and closed in $G(A)$. Let $y = \exp x$ and consider any $z \in A$ with $\|z - y\| < \|y^{-1}\|^{-1}$. Then

$$\|e - y^{-1}z\| \le \|y^{-1}\| \cdot \|y - z\| < 1,$$

and hence $y^{-1}z = \exp a$ for some $a \in A$ by Corollary 3.2.5. Thus

$$z = y \log a = \exp x \exp a = \exp(x + a) \in \exp A.$$

This shows that $\exp A$ is open. To see that $\exp A$ is also closed in $G(A)$, let $y \in G(A)$ be in the closure of $\exp A$ and choose $z \in \exp A$ so that $\|z - y\| < \|y^{-1}\|^{-1}$. Arguing as above, we conclude that $y^{-1}z \in \exp A$ and therefore $y \in \exp A$ since $z \in \exp A$ and $\exp A$ is a group.   □

Theorem 3.2.6 and the next lemma are employed to prove that $G(A)$ is either connected or has infinitely many connected components.

Lemma 3.2.7. *Let a be an element of finite order in $G(A)$. Then a belongs to the connected component of e in $G(A)$.*

Proof. Choose $n \in \mathbb{N}$ such that $a^n = e$, and for each $\lambda \in \mathbb{C}$ define an element $a(\lambda) \in A$ by

$$a(\lambda) = \sum_{j=0}^{n-1} (\lambda - 1)^j (\lambda a)^{n-1-j}.$$

Then, by definition of $a(\lambda)$,

$$(\lambda^n - (\lambda - 1)^n)e = (\lambda a)^n - (\lambda - 1)^n e = (\lambda a - (\lambda - 1)e)a(\lambda).$$

It follows that $\lambda a - (\lambda - 1)e \in G(A)$ whenever $\lambda^n \neq (\lambda - 1)^n$. The map $\gamma : \lambda \to \lambda a - (\lambda - 1)e$ from $\mathbb{C}$ into A is continuous and satisfies $\gamma(0) = e$ and $\gamma(1) = a$. Now, it is not difficult to find a continuous function $t \to \lambda(t)$ from $[0, 1]$ into $\mathbb{C}$ such that $\lambda(0) = 0$, $\lambda(1) = 1$, and $\lambda(t)^n \neq (\lambda(t) - 1)^n$ for all t. Then $t \to \gamma(\lambda(t))$, $t \in [0, 1]$, is a path in $G(A)$ connecting $e = \gamma(\lambda(0))$ and $a = \gamma(\lambda(1))$. Thus a lies in the connected component of the identity in $G(A)$. $\square$

Theorem 3.2.8. *Let A be a unital commutative Banach algebra. Then either $G(A)$ is connected or $G(A)$ has infinitely many connected components.*

Proof. Let C_e denote the connected component of the identity e in $G(A)$. Since elements of finite order of $G(A)$ lie in C_e, it suffices to show that if $x \in G(A)$ is such that $x \notin C_e$, then no two of the elements x^n, $n \in \mathbb{Z}$, belong to the same connected component of $G(A)$. Towards a contradiction, assume that, for some connected component C of $G(A)$, there exist $k, l \in \mathbb{Z}$ with $k > l$ and $x^k, x^l \in C$. Then $Cx^{-l} = C_e$ and $x^{k-l} \in Cx^{-l} = C_e$. Thus $x^n \in C_e$ for some $n \in \mathbb{N}$.

By Theorem 3.2.6, there exists $y \in A$ such that $x^n = \exp y$. Let $u = \exp(-(1/n)y)$, so that $u \in C_e$ (Theorem 3.2.6) and

$$(ux)^n = u^n x^n = \exp(-y) \exp y = e.$$

So ux is an element of $G(A)$ of finite order. By Lemma 3.2.7, $ux \in C_e$. Since C_e is a group, it follows that $x \in C_e$. This contradiction finishes the proof. $\square$

Our next application of the holomorphic functional calculus concerns the existence of idempotents.

Theorem 3.2.9. *Let A be a commutative Banach algebra with identity e. Let $x \in A$ and suppose that $\sigma(x) = \bigcup_{j=1}^m C_j$, where the sets $C_j, 1 \leq j \leq m$, are nonempty, pairwise disjoint, and open and closed in $\sigma(x)$. Then there exist idempotents $e_1, \ldots, e_m$ in A with the following properties.*

(i) $e = \sum_{j=1}^m e_j, e_j \neq 0$ and $e_j e_k = 0$ for $1 \leq j, k \leq m, k \neq j$.
(ii) *Each e_j is contained in the closed linear span of all elements of the form $(\lambda e - x)^{-1}, \lambda \in \rho(x)$.*

Proof. Because $C_1, \ldots, C_m$ are compact, there exist pairwise disjoint open subsets $V_1, \ldots, V_m$ of $\mathbb{C}$ such that $C_j \subseteq V_j$. For each j, choose an open subset W_j of $\mathbb{C}$ such that $W_j \cap \sigma(x) = C_j$. Then the sets $U_j = V_j \cap W_j, 1 \leq j \leq m$, are pairwise disjoint and open and satisfy $U_j \cap \sigma(x) = C_j$. Let $U = \bigcup_{j=1}^m U_j$ and, for each j, define a function f_j on U by

$$f_j(z) = \begin{cases} 1 & \text{if } z \in U_j, \\ 0 & \text{if } z \in U \setminus U_j. \end{cases}$$

Clearly, the holomorphic functions f_j satisfy $f_j^2 = f_j$ and $\sum_{j=1}^m f_j(z) = 1$ for all $z \in U$. Let $e_j = f_j(x) \in A$, $1 \le j \le m$. Then $e_j^2 = e_j$, $e_j e_k = 0$ for $j \ne k$ and

$$e = 1_U(x) = \sum_{j=1}^m f_j(x) = \sum_{j=1}^m e_j.$$

Moreover, for each j, $e_j \ne 0$ since $1 \in f_j(\sigma(x)) = \sigma(e_j)$. Finally, by the definition of $f_j(x)$, this element is a norm limit of finite linear combinations of elements of the form $(\lambda e - x)^{-1}$, where $\lambda \in \mathbb{C} \setminus \sigma(x)$. $\square$

Corollary 3.2.10. *Let A and e be as in Theorem 3.2.9. If, for some $x \in A$, the spectrum $\sigma(x)$ is not connected, then there exists an idempotent e' in A such that $e' \ne 0$ and $e' \ne e$.*

Runge's classical approximation theorem asserts that if K is a compact subset of $\mathbb{C}$ and Λ is a subset of $\mathbb{C}$ such that $\Lambda \cap C \ne \emptyset$ for each bounded connected component C of $\mathbb{C} \setminus K$, then every function f, which is holomorphic in a neighbourhood of K, can be approximated uniformly on K by rational functions with poles only among the points of Λ and at infinity. The next theorem is therefore justifiably often referred to as the *abstract Runge theorem.*

Theorem 3.2.11. *Let A be a commutative Banach algebra with identity e and let $x \in A$. Suppose that Λ is a subset of $\mathbb{C} \setminus \sigma(x)$ such that $\Lambda \cap C \ne \emptyset$ for every bounded connected component C of $\mathbb{C} \setminus \sigma(x)$. Let B denote the smallest closed subalgebra of A containing e, x, and all the elements of the form $(\lambda e - x)^{-1}$, $\lambda \in \Lambda$. Then $(\mu e - x)^{-1} \in B$ for every $\mu \in \mathbb{C} \setminus \sigma(x)$.*

Proof. By the Hahn–Banach theorem, it suffices to show that if $l \in A^*$ is such that $l|_B = 0$, then $l((\mu e - x)^{-1}) = 0$ for all $\mu \in \rho(x)$. As we have shown in the proof of Theorem 1.2.8, the function f on $\rho(x)$ defined by $f(\mu) = l((\mu e - x)^{-1})$ is holomorphic. Now, if $|\mu| > \|x\|$, then the series

$$(\mu e - x)^{-1} = \sum_{n=0}^{\infty} \mu^{-(n+1)} x^n$$

converges absolutely in A. Thus $(\mu e - x)^{-1} \in B$ for all such μ and this implies that f vanishes on the unbounded connected component of $\rho(x)$.

It remains to prove that $f = 0$ on each bounded component C of $\rho(x)$. Since f is holomorphic and $C \cap \Lambda \ne \emptyset$ by hypothesis, it is enough to show that if $\lambda \in \Lambda$ then f vanishes in some neighbourhood of λ. Fix such a λ and consider any $\mu \in \mathbb{C}$ such that $|\mu - \lambda| < \|(\lambda e - x)^{-1}\|^{-1}$. Then $\|(\lambda - \mu)(\lambda e - x)^{-1}\| < 1$ and hence $y = e - (\lambda - \mu)(\lambda e - x)^{-1}$ is invertible with inverse given by

$$y^{-1} = \sum_{n=0}^{\infty} (\lambda - \mu)^n (\lambda e - x)^{-n}$$

(Lemma 1.2.6). It follows that $(\lambda e - x)y = \mu e - x$ is invertible and

$$(\mu e - x)^{-1} = \sum_{n=0}^{\infty} (\lambda - \mu)^n (\lambda e - x)^{-(n+1)}.$$

Because l is continuous and $l|_B = 0$, we obtain that

$$f(\mu) = l((\mu e - x)^{-1}) = \sum_{n=0}^{\infty} (\lambda - \mu)^n l((\lambda e - x)^{-(n+1)}) = 0.$$

Thus f vanishes in a neighbourhood of λ. $\square$

3.3 The Shilov boundary

Let A be a commutative Banach algebra. In this section we prove the existence of a unique smallest closed subset $\partial(A)$ of $\Delta(A)$ on which each of the functions $|\hat{a}|$, $a \in A$, attains its maximum, and we establish a number of interesting results concerning $\partial(A)$ which are used later in the book. In addition, we present several illustrative examples.

Definition 3.3.1. Let X be a set and F a family of bounded complex valued functions on X. A subset R of X is called a *boundary* for F if for each $f \in F$ there exists $y \in R$ such that

$$|f(y)| = \sup_{x \in X} |f(x)|.$$

As a motivating example for introducing this notion of a boundary consider $\mathbb{D}$, the closed unit disc, and $F = P(\mathbb{D})$. Then the maximum modulus principle tells us that $\mathbb{T} = \{z \in \mathbb{C} : |z| = 1\}$, the topological boundary of $\mathbb{D}$ in $\mathbb{C}$, is a boundary for $P(\mathbb{D})$ in the above sense. Conversely, let $z_0 \in \mathbb{T}$ and let $f \in P(\mathbb{D})$ be the function defined by

$$f(z) = \frac{1}{2}(1 + \bar{z}_0 z).$$

Then $f(z_0) = 1$ and $|f(z)| < 1$ for every $z \in \mathbb{D}$, $z \neq z_0$. This means that $\mathbb{T}$ is contained in every boundary for $P(\mathbb{D})$.

It turns out that under very natural assumptions on X and F there always exists a unique smallest closed boundary for F. This is the content of the following theorem.

Theorem 3.3.2. *Let X be a locally compact Hausdorff space and suppose that A is a subalgebra of $C_0(X)$ which strongly separates the points of X. Then the intersection of all closed boundaries for A is a boundary for A.*

Proof. Let $\mathcal{R}$ denote the set of all closed boundaries for A. Then $\mathcal{R}$ is nonempty since $X \in \mathcal{R}$. We introduce a partial ordering on $\mathcal{R}$ by setting $R_1 \geq R_2$ if and only if $R_1 \subseteq R_2$.

We are going to show that $\mathcal{R}$ satisfies the hypothesis of Zorn's lemma. To that end, let $\{R_\lambda : \lambda \in \Lambda\}$ be a linearly ordered subset of $\mathcal{R}$ and let

$$R = \bigcap \{R_\lambda : \lambda \in \Lambda\}.$$

In order to prove that R is a boundary for A, consider any $f \in A, f \neq 0$, and set

$$X_f = \{x \in X : |f(x)| = \|f\|_\infty\}.$$

Then X_f is nonempty and compact because $f \neq 0$ and f vanishes at infinity. Now, for each $\lambda \in \Lambda, R_\lambda \cap X_f \neq \emptyset$ since R_λ is a boundary. The collection $\{R_\lambda \cap X_f : \lambda \in \Lambda\}$ of compact subsets of X is linearly ordered and hence has the finite intersection property. It follows that

$$R \cap X_f = \bigcap_{\lambda \in \Lambda} (R_\lambda \cap X_f) \neq \emptyset.$$

In particular, there exists $x \in R$ such that $|f(x)| = \|f\|_\infty$. This shows that R is a boundary for A.

We have thus seen that every linearly ordered subset of $\mathcal{R}$ has an upper bound. By virtue of Zorn's lemma there exists a maximal element R_0 in $\mathcal{R}$. It remains to show that $R_0 \subseteq R$ for each $R \in \mathcal{R}$. Suppose that $R_0 \not\subseteq R$ for some $R \in \mathcal{R}$. Fix $x_0 \in R_0 \setminus R$ and choose an open neighbourhood U of x_0 such that $U \cap R = \emptyset$. Because A strongly separates the points of X, by Proposition 2.2.14 X carries the weak topology defined by the functions in A. Therefore, we can assume that U is of the form

$$U(x_0, f_1, \ldots, f_m, \epsilon) = \{x \in X : |f_j(x) - f_j(x_0)| < \epsilon \text{ for } 1 \leq j \leq m\},$$

where $0 < \epsilon < 1$ and $f_1, \ldots, f_m \in A$. Moreover, we can assume that

$$|f_j(x) - f_j(x_0)| \leq 1 \quad \text{for all } x \in X \text{ and } 1 \leq j \leq m.$$

Indeed, since $A \subseteq C_0(X)$,

$$M = \sup\{|f_j(x) - f_j(x_0)| : x \in X, 1 \leq j \leq m\} < \infty,$$

and replacing f_j by $h_j = (1/M)f_j$ in case $M > 1$, the functions h_j satisfy the above condition and also

$$U(x_0, h_1, \ldots, h_m, \tfrac{\epsilon}{M}) \subseteq U(x_0, f_1, \ldots, f_m, \epsilon).$$

Now, R_0 is a maximal element in $\mathcal{R}$, and therefore the set $R_0 \setminus U$ cannot be a boundary. Hence there exists $f \in A$ with $|f(y)| < \|f\|_\infty$ for all $y \in R_0 \setminus U$. Of course, replacing f by $\|f\|_\infty^{-1}f$, we can assume that $\|f\|_\infty = 1$. Then $|f(y)| < 1$ for all $y \in R_0 \setminus U$ and hence, since $f \in C_0(X)$,

$$\sup \{|f(y)| : y \in R_0 \setminus U\} < 1.$$

Choose $k \in \mathbb{N}$ such that $|f(y)|^k < \epsilon$ for all $y \in R_0 \setminus U$ and set $g = f^k$. Then $\|g\|_\infty = 1$ and, for every $x \in U$,

$$|g(x)f_j(x) - g(x)f_j(x_0)| \leq \|g\|_\infty |f_j(x) - f_j(x_0)| < \epsilon.$$

Also, for $y \in R_0 \setminus U$,

$$|g(y)f_j(y) - g(y)f_j(x_0)| = |g(y)| \cdot |f_j(y) - f_j(x_0)| < \epsilon.$$

Since R_0 is a boundary we conclude that

$$\|gf_j - f_j(x_0)g\|_\infty < \epsilon$$

for $1 \leq j \leq m$. On the other hand, R is a boundary and hence

$$1 = \|g\|_\infty = |g(x)|$$

for some $x \in R$. It follows that

$$|f_j(x) - f_j(x_0)| = |g(x)f_j(x) - g(x)f_j(x_0)| < \epsilon$$

for all j. Thus $x \in U$, which contradicts $U \cap R = \emptyset$. So $R_0 \subseteq R$ for all $R \in \mathcal{R}$, and this finishes the proof. $\square$

Definition 3.3.3. Let X be a locally compact Hausdorff space and A a subalgebra of $C_0(X)$ which strongly separates the points of X. The intersection of all closed boundaries for A, which is a boundary by Theorem 3.3.2, is called the *Shilov boundary* of A and denoted by $\partial(A)$.

Clearly, $\partial(A)$ is the unique minimal closed boundary for A. Theorem 3.3.2 yields the following characterisation of points in $\partial(A)$ by means of a peak point property which can very effectively be used to compute the Shilov boundary of concrete examples.

Corollary 3.3.4. *A point $x \in X$ belongs to the Shilov boundary of A if and only if given any open neighbourhood U of x, there exists $f \in A$ such that*

$$\|f|_{X \setminus U}\|_\infty < \|f|_U\|_\infty.$$

Proof. First, let $x \in X \setminus \partial(A)$. Then $U = X \setminus \partial(A)$ is an open neighbourhood of x and because $\partial(A)$ is a boundary, we have for all $f \in A$,

$$\|f|_U\|_\infty \leq \|f\|_\infty = \|f|_{\partial(A)}\|_\infty = \|f|_{X \setminus U}\|_\infty.$$

Conversely, let $x \in \partial(A)$ and suppose there exists an open neighbourhood U of x such that

$$\|f|_U\|_\infty \leq \|f|_{X \setminus U}\|_\infty$$

for all $f \in A$. Then $X \setminus U$ is a boundary for A, so that $\partial(A) \subseteq X \setminus U$. This contradicts $x \in \partial(A)$. $\square$

We now examine a number of examples.

Example 3.3.5. (1) Let X be a locally compact Hausdorff space and let A be a subalgebra of $C_0(X)$ with the property that given any closed subset E of X and $x \in X \setminus E$, there exists $f \in A$ such that $f(x) \neq 0$ and $f|_E = 0$. It is obvious that then $\partial(A) = X$. Hence, in particular, $\partial(C_0(X)) = X$.

(2) Let X be a compact subset of $\mathbb{C}$. We claim that $\partial(R(X))$ coincides with $\partial(X)$, the topological boundary of X. Notice first that since each $f \in R(X)$ is holomorphic on the interior X° of X, it follows from the maximum modulus principle that $\partial(X) = X \setminus X^\circ$ is a boundary for $R(X)$.

To see that conversely $\partial(R(X))$ contains $\partial(X)$, we have to verify that every point in $\partial(X)$ fulfills the condition in Corollary 3.3.4. Thus, let $z_0 \in \partial(X)$, and let U be an open neighbourhood of z_0 in X. Choose an open disc V of radius $r > 0$ around z_0 so that $V \cap X \subseteq U$ and pick $z_1 \in V \setminus X$ with $|z_1 - z_0| < r/2$. Let $f \in R(X)$ be the function defined by

$$f(z) = \frac{1}{z - z_1}, \quad z \in X.$$

Then, for $x \in X \setminus U, |z - z_0| \geq r$ and hence

$$|z - z_1| \geq |z - z_0| - |z_1 - z_0| > \frac{r}{2}.$$

It follows that $\|f|_{X \setminus U}\|_\infty \leq 2/r$. On the other hand,

$$\|f|_U\|_\infty \geq \frac{1}{|z_1 - z_0|} > \frac{2}{r}.$$

So z_0 satisfies the hypothesis in Corollary 3.3.4.

(3) Continue to let X be a compact subset of $\mathbb{C}$. Then $\partial(P(X))$ equals the topological boundary of the unbounded component of $\mathbb{C} \setminus X$. To show this, assume first that X is polynomially convex. Then $\mathbb{C} \setminus X$ is connected (Theorem 2.3.7) and $P(X) = R(X)$ by Theorem 2.5.8. Therefore, example (2) yields $\partial(P(X)) = \partial(R(X)) = \partial(X) = \partial(\mathbb{C} \setminus X)$.

Now, for arbitrary X, $P(X)$ is isometrically isomorphic to $P(\widehat{X}_p)$ (Theorem 2.5.7). Hence every boundary for $P(X)$ is a boundary for $P(\widehat{X}_p)$. By the preceding paragraph we obtain

$$\partial(P(X)) = \partial(P(\widehat{X}_p)) = \partial(\mathbb{C} \setminus \widehat{X}_p).$$

However, $\mathbb{C} \setminus \widehat{X}_p$ coincides with the unbounded component C of $\mathbb{C} \setminus X$. Indeed, $\mathbb{C} \setminus C$ is a compact subset of $\mathbb{C}$ with connected complement and hence is polynomially convex. As $X \subseteq \mathbb{C} \setminus C \subseteq \widehat{X}_p$, we get $\mathbb{C} \setminus C = \widehat{X}_p$.

(4) The description of $\partial(P(X))$ in (3) does not remain true for compact subsets X of $\mathbb{C}^n$ when $n \geq 2$. To demonstrate this we show that $\partial(P(\mathbb{D}^n)) = \mathbb{T}^n$ for $n \geq 2$. First, let $w = (e^{it_1}, \ldots, e^{it_n}) \in \mathbb{T}^n$, $t_1, \ldots, t_n \in \mathbb{R}$. Then the polynomial function f defined by

$$f(z_1, \ldots, z_n) = \frac{1}{2^n} \prod_{j=1}^{n} (1 + z_j e^{-it_j})$$

satisfies $f(w) = 1$ and $|f(z)| < 1$ for all $z \in \mathbb{D}^n, z \neq w$. This proves $\mathbb{T}^n \subseteq \partial(P(\mathbb{D}^n))$. It remains to verify that $\mathbb{T}^n$ is a boundary for $P(\mathbb{D}^n)$. To see this, let $f \in P(\mathbb{D}^n)$ and $z = (z_1, \ldots, z_n) \in \mathbb{D}^n$ such that $\|f\|_\infty = |f(z)|$. Then the function

$$w \to f(w, z_2, \ldots, z_n),$$

$w \in \mathbb{D}$, belongs to $P(\mathbb{D})$, and hence

$$|f(z_1, \ldots, z_n)| \leq |f(e^{it_1}, z_2, \ldots, z_n)|$$

for some $t_1 \in \mathbb{R}$. Next, the function

$$w \to f(e^{it_1}, w, z_3, \ldots, z_n)$$

is in $P(\mathbb{D})$. As before, it follows that, for some $t_2 \in \mathbb{R}$,

$$|f(e^{it_1}, z_2, \ldots, z_n)| \leq |f(e^{it_1}, e^{it_2}, z_3, \ldots, z_n)|.$$

Continuing in this manner, we find $t_1, \ldots, t_n \in \mathbb{R}$ such that

$$|f(z_1, \ldots, z_n)| \leq |f(e^{it_1}, \ldots, e^{it_n})|.$$

This shows $\|f\|_\infty = \|f|_{\mathbb{T}^n}\|_\infty$. Thus $\mathbb{T}^n$ is a boundary for $P(\mathbb{D}^n)$ and, since $\mathbb{T}^n \subseteq \partial(P(\mathbb{D}^n))$, we get that $\mathbb{T}^n = \partial(P(\mathbb{D}^n))$. However, $\mathbb{C}^n \setminus \mathbb{D}^n$ is connected and

$$\partial(\mathbb{C}^n \setminus \mathbb{D}^n) = \{z \in \mathbb{D}^n : z_j \in \mathbb{T} \text{ for at least one } j\}$$

does not equal $\mathbb{T}^n$ when $n \geq 2$.

We now introduce the notion of a boundary for an arbitrary commutative Banach algebra.

Definition 3.3.6. Let A be a commutative Banach algebra and $\Gamma : A \to C_0(\Delta(A))$ the Gelfand representation of A. A subset R of $\Delta(A)$ is called a *boundary* for A if R is a boundary for $\Gamma(A)$, the range of the Gelfand homomorphism. In particular, $\partial(\Gamma(A))$ is called the *Shilov boundary* of A and denoted $\partial(A)$.

Let X be a locally compact Hausdorff space and A a closed subalgebra of $C_0(X)$. Then, according to Definitions 3.3.1 and 3.3.6, we have to distinguish between boundaries for the family A of functions on X and boundaries for the commutative Banach algebra A, the latter being the boundaries of $\Gamma(A) \subseteq C_0(\Delta(A))$. However, as explained in the following remark, the two Shilov boundaries are canonically homeomorphic provided that A satisfies some natural conditions.

Remark 3.3.7. Suppose that A strongly separates the points of X. Then the mapping $\phi : x \to \varphi_x$, where $\varphi_x(f) = f(x)$ for $f \in A$, is a homeomorphism from X onto $\phi(X) \subseteq \Delta(A)$ because, by Proposition 2.2.14, X carries the weak topology defined by the functions $f \in A$. Moreover, for every subset Y of X,

$$\|f|_Y\| = \sup_{y \in Y} |f(y)| = \sup_{y \in Y} |\varphi_y(f)| = \|\widehat{f}|_{\phi(Y)}\|_\infty.$$

Therefore every boundary for $A \subseteq C_0(X)$ is a boundary for $\Gamma(A)$. In particular, if $\phi(X)$ is closed $\Delta(A)$, then $\phi(\partial(A)) = \partial(\Gamma(A))$.

In the next remark we draw attention to the question of how the Shilov boundaries of A and of its unitisation are related.

Remark 3.3.8. Let A_e be the commutative Banach algebra obtained from A by adjoining an identity e to A. As we always do, regard $\Delta(A)$ as a subset of $\Delta(A_e)$. Let $\varphi \in \partial(A)$, and let U be an open neighbourhood of φ in $\Delta(A)$. By Corollary 3.3.4 there exists $x \in A$ such that

$$\|\widehat{x}|_U\|_\infty > \|\widehat{x}|_{\Delta(A)\setminus U}\|_\infty.$$

Since $\widehat{x}(\varphi_\infty) = 0$ and U is an open neighbourhood of φ in $\Delta(A_e)$, it follows from Corollary 3.3.4 that φ is an element of the Shilov boundary of A_e. Conversely, since $\widehat{x}(\varphi_\infty) = 0$ for all $x \in A$, it is clear that $\partial(A_e) \cap \Delta(A) \subseteq \partial(A)$. Thus $\partial(A) = \partial(A_e) \cap \Delta(A)$. For a generalisation of this equation see Exercise 3.6.10.

The conclusion of the next theorem was the originally reason for introducing the Shilov boundary.

Theorem 3.3.9. *Let A be a commutative Banach algebra and B a closed subalgebra of A. Then every $\varphi \in \partial(B)$ extends to an element of $\Delta(A)$.*

Proof. Considering A_e and its subalgebra B_e and having in mind that $\partial(B) \subseteq \partial(B_e)$ shows that we can assume that A has an identity e and that $e \in B$. Let $\varphi \in \partial(B)$ and suppose first that $\ker \varphi \subseteq \ker \psi$ for some $\psi \in \Delta(A)$. Then $B \cap \ker \psi = \ker \varphi$ since $\ker \varphi$ is of codimension one in B and $e \in B \setminus \ker \psi$. Thus $\ker(\psi|_B) = \ker \varphi$, and since $\psi(e) = 1 = \varphi(e)$, this gives $\psi|_B = \varphi$.

Thus we are left with the more difficult task of showing that such a ψ always exists. Towards a contradiction, assume that $\ker \varphi \not\subseteq M$ for each $M \in \mathrm{Max}(A)$. Let I denote the ideal of A generated by $\ker \varphi$, that is,

$$I = \left\{ \sum_{i=1}^n a_i b_i : a_i \in A, b_i \in \ker \varphi, n \in \mathbb{N} \right\}.$$

Then $I = A$, because otherwise $\ker \varphi \subseteq I \subseteq M$ for some $M \in \mathrm{Max}(A)$. Let

$$e = \sum_{i=1}^n a_i b_i,$$

where $a_i \in A$ and $b_i \in \ker \varphi$. Replacing each a_i by a suitable scalar multiple of itself, we can assume that $\|\widehat{b}_i\|_\infty \leq 1$ for all i. Now, choose a positive number R such that

$$R \geq \max_{1 \leq i \leq n} \|\widehat{a}_i\|_\infty,$$

and define an open neighbourhood U of φ in $\Delta(B)$ by

$$U = \left\{ \psi \in \Delta(B) : |\psi(b_i) - \varphi(b_i)| < \frac{1}{2nR} \text{ for } 1 \leq i \leq n \right\}$$

$$= \left\{ \psi \in \Delta(B) : |\psi(b_i)| < \frac{1}{2nR} \text{ for } 1 \leq i \leq n \right\}.$$

Since $\varphi \in \partial(B)$, by Corollary 3.3.4 there exists $x \in B$ such that

$$\|\widehat{x}|_{\Delta(B) \backslash U}\|_\infty < \|\widehat{x}|_U\|_\infty.$$

Thus we find $m \in \mathbb{N}$ such that the element $y = (\|\widehat{x}\|_\infty^{-1} x)^m$ of A satisfies

$$\|\widehat{y}\|_\infty = 1 \text{ and } |\psi(y)| < \frac{1}{2nR}$$

for all $\psi \in \Delta(B) \setminus U$. This implies, for $\psi \in \Delta(B) \setminus U$ and each $i = 1, \ldots, n$,

$$|\psi(y)\psi(b_i)| \leq \|\widehat{b}_i\|_\infty |\psi(y)| < \frac{1}{2nR}.$$

Also, for $\psi \in U$,

$$|\psi(y)\psi(b_i)| \leq \|\widehat{y}\|_\infty |\psi(b_i)| < \frac{1}{2nR}.$$

Combining these two inequalities and using that $y = y \sum_{i=1}^n a_i b_i$, we obtain

$$1 = \|\widehat{y}\|_\infty \leq \sum_{i=1}^n \|\widehat{a}_i\|_\infty \sup_{\rho \in \Delta(A)} |\rho(y)\rho(b_i)|$$

$$\leq \sum_{i=1}^n \|\widehat{a}_i\|_\infty \sup_{\psi \in \Delta(B)} |\psi(y)\psi(b_i)|$$

$$\leq R \cdot \sum_{i=1}^n \sup_{\psi \in \Delta(B)} |\psi(y)\psi(b_i)|$$

$$< \frac{1}{2}.$$

This contradiction shows that $\ker \varphi \subseteq \ker \psi$ for some $\psi \in \Delta(A)$, and this finishes the proof of the theorem. $\square$

Theorem 3.3.9 proves very useful when we take up the problem of extending elements of $\Delta(B)$ to elements of $\Delta(A)$ again in Section 3.4 and in Chapter 4. For instance, applying the theory of topological divisors of zero, we show in Theorem 3.4.13 that if A is unital, then each $\varphi \in \partial(B)$ actually extends to some element of $\partial(A)$.

In passing we give an example which shows that it may well happen that the elements in $\partial(B)$ are the only elements of $\Delta(B)$ extending to all of A.

Example 3.3.10. Let $A = C(\mathbb{T})$ and $B = P(\mathbb{T})$. Then

$$\partial(B) = \mathbb{T} = \Delta(A) \text{ and } \Delta(B) = \mathbb{D},$$

and obviously no $\varphi_z \in \Delta(B)$, $|z| < 1$, extends to an element of $\Delta(C(\mathbb{T}))$.

The next theorem generalises, in the commutative case, the fact that if A is a unital Banach algebra and $x, y \in A$ are such that x is invertible and $\|y - x\| < \|x^{-1}\|^{-1}$, then y is invertible too (compare the proof of Lemma 1.2.7(ii)).

Theorem 3.3.11. *Let A be a unital commutative Banach algebra and suppose that x and y are elements of A satisfying*

$$|\widehat{x}(\varphi) - \widehat{y}(\varphi)| < |\widehat{x}(\varphi)|$$

for all $\varphi \in \partial(A)$. Then x is invertible if and only if y is invertible.

Proof. Because $\partial(A)$ is compact and the function

$$\varphi \to |\widehat{x}(\varphi)| - |\widehat{x}(\varphi) - \widehat{y}(\varphi)|$$

is continuous, the hypothesis implies that

$$c = \inf_{\varphi \in \partial(A)} \{|\widehat{x}(\varphi)| - |\widehat{x}(\varphi) - \widehat{y}(\varphi)|\} > 0.$$

Choose $n \in \mathbb{N}$ such that $nc > r(x - y)$, and consider the sequence

$$nx, (n-1)x + y, \ldots, (n-k)x + ky, \ldots, ny$$

of elements of A. Suppose the statement of the theorem is wrong, so that either $nx \in G(A)$ and $ny \notin G(A)$, or $nx \notin G(A)$ and $ny \in G(A)$. In addition, assume that for each invertible element in the above sequence, both its predecessor and its successor (as long as they exist) are invertible too. Apparently, then $nx \in G(A)$ implies $ny \in G(A)$ and conversely. This contradiction shows that there exists $0 \leq k \leq n$ such that

$$(n-k)x + ky$$

is invertible, but one of its immediate neighbours is not invertible. Let $l = k-1$ if $0 < k \leq n$ and

$$(n - (k-1))x + (k-1)y \notin G(A),$$

and in the remaining cases put $l = k + 1$. Since $(n - l)x + ly$ is not invertible, there exists $\varphi_0 \in \Delta(A)$ such that

$$\varphi_0((n - l)x + ly) = 0.$$

Let $z = ((n - k)x + ky)^{-1}$. It follows that

$$
\begin{aligned}
r(x - y) < nc &= \inf_{\varphi \in \partial(A)} \{n|\widehat{x}(\varphi)| - n|\widehat{x}(\varphi) - \widehat{y}(\varphi)|\} \\
&\leq \inf_{\varphi \in \partial(A)} \{n|\widehat{x}(\varphi)| - k|\widehat{x}(\varphi) - \widehat{y}(\varphi)|\} \\
&\leq \inf_{\varphi \in \partial(A)} |n\widehat{x}(\varphi) - k(\widehat{x}(\varphi) - \widehat{y}(\varphi))| \\
&= \inf_{\varphi \in \partial(A)} |\widehat{z}(\varphi)|^{-1} \\
&= \left(\sup_{\varphi \in \partial(A)} |\widehat{z}(\varphi)| \right)^{-1} = \left(\sup_{\varphi \in \Delta(A)} |\widehat{z}(\varphi)| \right)^{-1} \\
&= \inf_{\varphi \in \Delta(A)} |\widehat{z}(\varphi)|^{-1} \leq |n\widehat{x}(\varphi_0) - k(\widehat{x}(\varphi_0) - \widehat{y}(\varphi_0))| \\
&= |n\widehat{x}(\varphi_0) - k(\widehat{x}(\varphi_0) - \widehat{y}(\varphi_0)) - (n - l)\widehat{x}(\varphi_0) - l\widehat{y}(\varphi_0)| \\
&= |(l - k)\widehat{x}(\varphi_0) + (k - l)\widehat{y}(\varphi_0)| = |\widehat{x}(\varphi_0) - \widehat{y}(\varphi_0)| \\
&\leq r(x - y).
\end{aligned}
$$

So we have reached a contradiction, and this shows that nx is invertible if and only if ny is.   □

We now determine the Shilov boundary of tensor products. Recall that if γ is an algebra cross-norm on $A \otimes B$ which dominates the injective norm ϵ, then $\Delta(A \widehat{\otimes}_\gamma B)$ identifies with the product space $\Delta(A) \times \Delta(B)$ by means of the homeomorphism $(\varphi, \psi) \to \varphi \widehat{\otimes}_\gamma \psi$ (Theorem 2.11.2).

Proposition 3.3.12. *Let A and B be commutative Banach algebras and let γ be an algebra cross-norm on $A \otimes B$ dominating ϵ. Then the Shilov boundary of $A \widehat{\otimes}_\gamma B$ equals $\partial(A) \times \partial(B) \subseteq \Delta(A) \times \Delta(B) = \Delta(A \widehat{\otimes}_\gamma B)$.*

Proof. The Shilov boundary of $A \widehat{\otimes}_\gamma B$ is the same as that of $A \otimes B$, because a closed boundary for a dense subalgebra is always a boundary for the whole algebra.

We show first that $\partial(A) \times \partial(B)$ is a boundary for $A \widehat{\otimes}_\alpha B$. Let

$$z = \sum_{j=1}^{n} x_j \otimes y_j \in A \otimes B, \quad x_j \in A, \quad y_j \in B, \quad 1 \leq j \leq n.$$

There exist $\varphi \in \Delta(A)$ and $\psi \in \Delta(B)$ such that

$$\|\widehat{z}\|_\infty = |(\varphi \otimes \psi)(z)| = \left| \sum_{j=1}^{n} \varphi(x_j)\psi(y_j) \right|.$$

Consider the element $x = \sum_{j=1}^{n} \psi(y_j) x_j$ of A. There exists $\varphi_0 \in \partial(A)$ such that $|\varphi_0(x)| = \|\widehat{x}\|_\infty$. Similarly, for $y = \sum_{j=1}^{n} \varphi_0(x_j) y_j \in B$, there exists $\psi_0 \in \partial(B)$ such that $|\psi_0(y)| = \|\widehat{y}\|_\infty$. Combining all this, we obtain

$$\|\widehat{z}\|_\infty = \left| \sum_{j=1}^{n} \psi(y_j) \widehat{x}_j(\varphi) \right| = |\widehat{x}(\varphi)| \leq \|\widehat{x}\|_\infty$$
$$= |\varphi_0(x)| = |\psi(y)| \leq \|\widehat{y}\|_\infty = |\psi_0(y)|$$
$$= |(\varphi_0 \otimes \psi_0)(z)|.$$

So the function $|\widehat{z}|$ attains its maximum at $\varphi_0 \otimes \psi_0 \in \partial(A) \times \partial(B)$. Thus $\partial(A) \times \partial(B)$ is a boundary for $A \widehat{\otimes}_\gamma B$, whence $\partial(A) \times \partial(B) \supseteq \partial(A \widehat{\otimes}_\gamma B)$.

Conversely, let $\varphi_0 \in \partial(A)$ and $\psi_0 \in \partial(B)$. Let U be any neighbourhood of (φ_0, ψ_0) in $\Delta(A) \times \Delta(B)$, and choose open neighbourhoods V of φ_0 in $\Delta(A)$ and W of ψ_0 in $\Delta(B)$ such that $V \times W \subseteq U$. Since $\varphi_0 \in \partial(A)$, by Corollary 3.3.4 there exists $x \in A$ such that $\|\widehat{x}\|_\infty = 1$ and $|\widehat{x}(\varphi)| < 1$ for all $\varphi \in \Delta(A) \setminus V$. Similarly, there exists $y \in B$ such that $\|\widehat{y}\|_\infty = 1$ and $|\widehat{y}(\psi)| < 1$ for all $\psi \in \Delta(B) \setminus W$. Then $\|\widehat{x \otimes y}\|_\infty = 1$ and $|\widehat{x \otimes y}(\omega)| < 1$ for all $\omega \in (\Delta(A) \times \Delta(B)) \setminus U$. Since U was arbitrary, it follows from Corollary 3.3.4 that $(\varphi_0, \psi_0) \in \partial(A \otimes B) = \partial(A \widehat{\otimes}_\gamma B)$. $\qquad \square$

We remind the reader that $\sigma_A(x) \subseteq \widehat{x}(\Delta(A)) \cup \{0\}$ for every element x of a commutative Banach algebra A and that $\sigma_A(x) = \widehat{x}(\Delta(A))$ when A is unital (Theorem 2.2.5). The next theorem shows that analogous assertions hold when replacing $\Delta(A)$ with the Shilov boundary of A and $\sigma_A(x)$ with its topological boundary.

Theorem 3.3.13. *Let A be a commutative Banach algebra and $x \in A$. Then*

$$\partial(\sigma_A(x)) \subseteq \widehat{x}(\partial(A)) \cup \{0\}.$$

If A is unital, then $\partial(\sigma_A(x)) \subseteq \widehat{x}(\partial(A))$.

Proof. We first verify that $\widehat{x}(\partial(A)) \cup \{0\}$ is closed in $\mathbb{C}$. For that, let $(\lambda_n)_n$ be a sequence in $\widehat{x}(\partial(A))$ converging to some $\lambda_0 \neq 0$. We can assume that $|\lambda_n| \geq \frac{1}{2}|\lambda_0|$ for all $n \in \mathbb{N}$. Let $\varphi_n \in \partial(A)$ such $\widehat{x}(\varphi_n) = \lambda_n, n \in \mathbb{N}$. Because $\partial(A)$ is closed in $\Delta(A)$ and $\widehat{x} \in C_0(\Delta(A))$, the set

$$C = \left\{ \varphi \in \partial(A) : |\widehat{x}(\varphi)| \geq \frac{1}{2}|\lambda_0| \right\}$$

is compact. Since $\varphi_n \in C$, after passing to a subnet if necessary, we can assume that $\varphi_n \to \varphi$ for some $\varphi \in \partial(A)$. It follows that

$$\lambda_0 = \lim_{n \to \infty} \lambda_n = \lim_{n \to \infty} \varphi_n(x) = \varphi(x) \in \widehat{x}(\partial(A)),$$

as claimed.

Towards a contradiction, suppose now that there exists $\lambda \in \partial(\sigma_A(x))$ such that $\lambda \notin \widehat{x}(\partial(A)) \cup \{0\}$. Then there exists $0 < \delta < 1$ such that $|\widehat{x}(\varphi) - \lambda| > \delta r_A(x)$ for all $\varphi \in \partial(A)$. Since $\lambda \in \partial(\sigma_A(x))$, we find $\mu \in \mathbb{C} \setminus \sigma_A(x)$ such that $|\mu - \lambda| < \delta|\lambda|/2$. Temporarily, let $B = A$ if A has an identity e and $B = A_e$ otherwise. Then $\mu e - x$ is invertible in B and $(\mu e - x)^{-1} = (1/\mu)e - y$ for some $y \in A$. It follows that

$$\widehat{y}(\varphi) = \frac{1}{\mu} \cdot \frac{\widehat{x}(\varphi)}{\widehat{x}(\varphi) - \mu}$$

for all $\varphi \in \Delta(A)$. Now, for $\varphi \in \partial(A)$,

$$|\widehat{x}(\varphi) - \mu| \geq |\widehat{x}(\varphi) - \lambda| - |\mu - \lambda| \geq \delta r_A(x) - \frac{\delta}{2}|\lambda| \geq \frac{\delta}{2} r_A(x),$$

and hence

$$\sup_{\varphi \in \partial(A)} |\widehat{y}(\varphi)| \leq \frac{2}{\delta|\mu| r_A(x)} \sup_{\varphi \in \partial(A)} |\widehat{x}(\varphi)| \leq \frac{2}{\delta|\mu|}.$$

On the other hand, there exists $\psi \in \Delta(A)$ such that $\lambda = \widehat{x}(\psi)$. Then

$$\sup_{\varphi \in \partial(A)} |\widehat{y}(\varphi)| = \sup_{\varphi \in \Delta(A)} |\widehat{y}(\varphi)| \geq |\widehat{y}(\psi)| = \frac{|\lambda|}{|\mu| \cdot |\lambda - \mu|} > \frac{2}{\delta|\mu|}.$$

This contradiction shows that $\partial(\sigma_A(x)) \subseteq \widehat{x}(\partial(A)) \cup \{0\}$.

Finally, suppose that A has an identity e. By the first part of the proof, it only remains to show that if $0 \in \partial(\sigma_A(x))$, then $0 \in \widehat{x}(\partial(A))$. To that end, assume that $0 \notin \widehat{x}(\partial(A))$. Since $\widehat{x}(\partial(A))$ is compact, $|\lambda| \geq \delta$ for some $\delta > 0$ and all $\lambda \in \widehat{x}(\partial(A))$. Since $0 \in \partial(\sigma_A(x))$, there exists $\mu \in \rho_A(x)$ with $|\mu| < \delta/2$. Then $y = (\mu e - x)^{-1}$ satisfies

$$r_A(y) = \sup_{\varphi \in \partial(A)} |\widehat{y}(\varphi)| = \inf_{\varphi \in \partial(A)} \frac{1}{|\widehat{x}(\varphi) - \mu|} \leq \frac{2}{\delta}.$$

However, since $0 \in \sigma_A(x)$ and A is unital, there exists $\varphi \in \Delta(A)$ with $\widehat{x}(\varphi) = 0$. Thus

$$r_A(y) \geq |\widehat{y}(\varphi)| = \frac{1}{|\mu|} > \frac{2}{\delta},$$

a contradiction. So $0 \in \widehat{x}(\partial(A))$, as required. $\square$

In the preceding theorem, it is not true in general that $\partial(\sigma_A(x)) = \widehat{x}(\partial(A))$ if A is unital (Exercise 3.6.13). However, this equality holds when A is generated by x (Exercise 3.6.6). In this context, also compare Exercise 3.6.8.

Our final result in this section shows that $\partial(A)$ can be finite only if it equals $\Delta(A)$.

Theorem 3.3.14. *Let A be a commutative Banach algebra and suppose that $\partial(A) \neq \Delta(A)$. Then $\partial(A)$ contains an infinite number of points.*

Proof. Let φ and ψ be any two distinct elements of $\Delta(A)$. We prove first that there exists $a \in A$ such that $\hat{a}(\varphi) \neq 0$ and $\hat{a}(\psi) = 0$. This is evident if A is unital, but requires some argument in the general case. Let x be an element of A such that $\hat{x}(\varphi) \neq \hat{x}(\psi)$. If $\hat{x}(\psi) = 0$, we are done. If $\hat{x}(\psi) \neq 0$, we can assume that $\hat{x}(\psi) = 1$, and hence $\hat{x}(\varphi) \neq 1$. If, in addition, $\hat{x}(\varphi) \neq 0$, then taking $a = x - x^2$ we have $\hat{a}(\psi) = 0$ and $\hat{a}(\varphi) = \hat{x}(\varphi)(1 - \hat{x}(\varphi)) \neq 0$. This leaves the case $\hat{x}(\varphi) = 0$ and $\hat{x}(\psi) = 1$. Choose any $y \in A$ with $\hat{y}(\varphi) \neq 0$. If $\hat{y}(\psi) = 0$, take $a = y$. If $\hat{y}(\psi) \neq 0$, let $a = x - \psi(y)^{-1}y$. Then $\hat{a}(\psi) = 0$ whereas

$$\hat{a}(\varphi) = \hat{x}(\varphi) - \hat{y}(\psi)^{-1}\hat{y}(\varphi) = -\hat{y}(\psi)^{-1}\hat{y}(\varphi) \neq 0.$$

Hence an element a with the desired properties exists.

Now, let $\varphi_1, \ldots, \varphi_n$ be any finite set of elements of $\partial(A)$ and choose $\varphi \in \Delta(A) \setminus \partial(A)$. For each $i = 1, \ldots, n$, by what we have seen above there exists $a_i \in A$ such that $\hat{a}_i(\varphi) \neq 0$ and $\hat{a}_i(\varphi_i) = 0$. If $a = a_1 \cdots a_n$, then $\hat{a}(\varphi) \neq 0$ whereas $\hat{a}(\varphi_i) = 0$ for all i. This shows that the points $\varphi_1, \ldots, \varphi_n$ cannot exhaust $\partial(A)$ and hence $\partial(A)$ must be infinite. □

Corollary 3.3.15. *If $\partial(A)$ is finite, then so is $\Delta(A)$.*

We close this section by mentioning a class of Banach algebras for which the Shilov boundary always equals the whole structure space.

Example 3.3.16. Let A be a commutative symmetric Banach $*$-algebra. Then $\partial(A) = \Delta(A)$. Since A is symmetric, $\hat{\bar{x}}(\varphi) = \widehat{x^*}(\varphi)$ for all $x \in A$ and $\varphi \in \Delta(A)$. Thus $\Gamma(A)$ is a subalgebra of $C_0(\Delta(A))$ which strongly separates the points of $\Delta(A)$ and is closed under complex conjugation. Then, by the Stone–Weierstrass theorem, $\Gamma(A)$ is dense in $C_0(\Delta(A))$. This readily implies that the Shilov boundary of A is the whole of $\Delta(A)$.

3.4 Topological divisors of zero

The theme of this section is to consider a concept which naturally extends that of zero divisors in the purely algebraic setting to normed algebras. In keeping with the main focus of this book, we confine ourselves to commutative Banach algebras. As it turns out, this concept has several interesting applications.

Definition 3.4.1. Let A be a commutative normed algebra. A nonzero element x of A is called a *topological divisor of zero* or *topological zero divisor* if there exists a sequence $(x_n)_n$ in A such that $\|x_n\| = 1$ for all n and $x_n x \to 0$ as $n \to \infty$.

For any $x \in A$, let

$$d(x) = \inf\{\|xy\| : y \in A, \|y\| = 1\} = \inf\left\{\frac{\|xy\|}{\|y\|} : y \in A, y \neq 0\right\}.$$

It is obvious that $x \in A$ is a topological divisor of zero if and only if $d(x) = 0$. This observation leads to the following generalisation. For any finitely many elements $x_1, \ldots, x_n$ of A, let

$$d(x_n, \ldots, x_n) = \inf \left\{ \sum_{j=1}^{n} \|x_j y\| : y \in A, \|y\| = 1 \right\}$$

$$= \inf \left\{ \sum_{j=1}^{n} \frac{\|x_j y\|}{\|y\|} : y \in A, y \neq 0 \right\}.$$

Definition 3.4.2. A subset S of A consists of *joint topological zero divisors* if $d(x_1, \ldots, x_n) = 0$ for any finitely many $x_1, \ldots, x_n \in S$.

Remark 3.4.3. Suppose that A has an identity e and let $x \in A$ be a topological divisor of zero. Then x is not invertible in A. Indeed, if $x_n \in A, n \in \mathbb{N}$, are such that $\|x_n\| = 1$ and $xx_n \to 0$ and if $yx = e$ for some $y \in A$, then

$$1 = \|x_n\| = \|yxx_n\| \leq \|y\| \cdot \|xx_n\| \to 0,$$

which is impossible.

The converse fails to hold in general. For example, let $A = P(\mathbb{D})$ and $f(z) = z$ for all $z \in \mathbb{D}$. Then f is not invertible. But f fails to be a topological zero divisor because if $(g_n)_n \subseteq A$ is a sequence satisfying $\|fg_n\|_\infty \to 0$, then $\|g_n|_{\mathbb{T}}\|_\infty \to 0$ and hence $\|g_n\|_\infty \to 0$ by the maximum modulus principle.

We now first identify the topological zero divisors in certain algebras of continuous functions on topological spaces (Corollary 3.4.5).

Proposition 3.4.4. *Let X be a locally compact Hausdorff space and let A be a subalgebra of $C_0(X)$ which strongly separates the points of X. Then, for any $f \in A$,*
$$d(f) = \inf\{|f(x)| : x \in \partial(A)\}.$$

Proof. Let $g \in A, g \neq 0$, and choose $x \in \partial(A)$ such that $|g(x)| = \|g\|_\infty$. Then

$$\inf\{|f(y)| : y \in \partial(A)\} \leq |f(x)| = \frac{|fg(x)|}{|g(x)|} \leq \frac{\|fg\|_\infty}{\|g\|_\infty}.$$

Thus $\inf\{|f(y)| : y \in \partial(A)\} \leq d(f)$. To show that actually equality holds, we can assume without loss of generality that $d(f) > 0$. Let

$$R = \{x \in X : |f(x)| \geq d(f)\}.$$

Then R is compact since $f \in C_0(X)$. We claim that R is a boundary for A. Towards a contradiction, assume that there exists $g \in A$ such that $\|g\|_\infty > |g(y)|$ for all $y \in R$. Then $\|g\|_\infty > \|g|_R\|_\infty$ since R is compact. Of course, we can assume that $\|g\|_\infty = 1$. For every $n \in \mathbb{N}$, there exists $x_n \in \partial(A)$ so that

$$|f(x_n)g^n(x_n)| = \|fg^n\|_\infty.$$

Since $\|g^n\|_\infty = 1$, it follows that

$$d(f) \le \|fg^n\|_\infty = |f(x_n)g^n(x_n)| \le |f(x_n)|.$$

Hence $x_n \in R$ and therefore

$$|g^n(x_n)| \le \|g^n|_R\|_\infty = \|g|_R\|_\infty^n$$

for each n. Now $\|g|_R\|_\infty^n \to 0$ as $n \to \infty$ since $\|g|_R\|_\infty < 1$. Thus $d(f) = 0$, contradicting our assumption.

Finally, as A strongly separates the points of X and R has been seen to be a boundary, Theorem 3.3.2 shows that $\partial(A) \subseteq R$. This in turn implies that

$$d(f) \le \inf\{|f(x)| : x \in R\} \le \inf\{|f(x)| : x \in \partial(A)\},$$

as required. $\qquad\square$

The following corollary is an immediate consequence of Proposition 3.4.4.

Corollary 3.4.5. *Let X be a locally compact Hausdorff space and let A be a subalgebra of $C_0(X)$ which strongly separates the points of X. If $f \in A$, then f is a topological divisor of zero if and only if*

$$\inf\{|f(x)| : x \in \partial(A)\} = 0.$$

In particular, if $\partial(A)$ is compact then f is a topological divisor of zero if and only $f(x) = 0$ for some $x \in \partial(A)$.

Corollary 3.4.6. *Let A be a unital commutative Banach algebra and suppose that $\|x\|^2 \le k\|x^2\|$ for some $k > 0$ and all $x \in A$. Then an element x of A is a topological divisor of zero if and only if $\varphi(x) = 0$ for some $\varphi \in \partial(A)$.*

Proof. The hypothesis on A implies that the Gelfand homomorphism $\Gamma : A \to C(\Delta(A))$ is injective and that the two norms $y \to \|y\|$ and $y \to \|\widehat{y}\|_\infty$ on A are equivalent (compare Exercise 2.12.9). Thus there are positive constants c and d such that $c\|y\| \le \|\widehat{y}\|_\infty \le d\|y\|$ for all $y \in A$. Now, if $y_n \in A$, $n \in \mathbb{N}$, are such that $\|y_n\| = 1$ and $xy_n \to 0$, then

$$d(\widehat{x}) \le \frac{\|\widehat{x}\,\widehat{y}_n\|_\infty}{\|\widehat{y}_n\|_\infty} \le \frac{d\|xy_n\|}{c\|y_n\|} \to 0,$$

and conversely, if $\|\widehat{y}_n\|_\infty = 1$ and $\widehat{x}\,\widehat{y}_n \to 0$, then

$$d(x) \le \frac{\|xy_n\|}{\|y_n\|} \le \frac{\|\widehat{x}\,\widehat{y}_n\|_\infty}{c\|\widehat{y}_n\|_\infty} = \frac{1}{c}\|\widehat{x}\,\widehat{y}_n\|_\infty \to 0.$$

Consequently, $x \in A$ is a topological divisor of zero in A if and only if $\widehat{x}$ is a topological divisor of zero in $\Gamma(A)$. Now, by Corollary 3.4.5, $\widehat{x}$ is a topological divisor of zero in $\Gamma(A)$ precisely when $\varphi(x) = \widehat{x}(\varphi) = 0$ for some $\varphi \in \partial(\widehat{A}) = \partial(A)$. $\qquad\square$

The next theorem characterises topological divisors of zero in terms of noninvertibility in all superalgebras.

Theorem 3.4.7. *Let A be a commutative Banach algebra with identity e. For $x \in A$ the following conditions are equivalent.*

(i) *x is a topological divisor of zero.*
(ii) *If B is any unital commutative Banach algebra and j is an isometric isomorphism from A into B with $j(e) = e_B$, where e_B denotes the identity of B, then $j(x)$ is not invertible in B.*

Proof. The implication (i) $\Rightarrow$ (ii) is simple. In fact, let x be a topological zero divisor in A. Then, since j is isometric, $j(x)$ is a topological divisor of zero in B and hence $j(x)$ is not invertible in B by Remark 3.4.3.

Now, suppose that (ii) holds and that nevertheless x fails to be a topological divisor of zero in A. Then $d(x) > 0$, and replacing x with $d(x)^{-1}x$, we can assume that $d(x) \geq 1$ and hence $\|xz\| \geq \|z\|$ for all $z \in A$. Let C denote the linear space of all formal power series

$$\tilde{x}(t) = \sum_{n=0}^{\infty} x_n t^n, \quad x_n \in A, \quad \sum_{n=0}^{n} \|x_n\| < \infty.$$

Then C is an algebra because

$$\left(\sum_{n=0}^{\infty} x_n t^n \right) \left(\sum_{n=0}^{\infty} y_n t^n \right) = \sum_{n=0}^{\infty} \left(\sum_{k+l=n} x_k y_l \right) t^n$$

and

$$\sum_{n=0}^{\infty} \left\| \sum_{k+l=n} x_k y_l \right\| \leq \sum_{k=0}^{\infty} \|x_k\| \cdot \sum_{l=0}^{\infty} \|y_l\| < \infty.$$

Thus the norm

$$\left\| \sum_{n=0}^{\infty} x_n t^n \right\| = \sum_{n=0}^{\infty} \|x_n\|$$

turns C into a commutative normed algebra. It can be easily verified that C is complete. Actually, as a linear space, C is isometrically isomorphic to the Banach space $l^1(\mathbb{N}_0, A)$.

Let J be the ideal $J = (e - xt)C$ of C, I the closure of J in C, and $B = C/I$. We now identify an element y of A with the constant function $t \to y$ and let $j : A \to B$ denote the mapping $y \to y + I$. It is clear that j is an algebra homomorphism. Moreover, $j(e) = e + I$, the identity of B, and j is norm decreasing. To show that j is actually isometric, let $y \in A$ and $\sum_{n=0}^{\infty} z_n t^n \in C$. Then, using that $\|xz\| \geq \|z\|$ for all $z \in A$, we get

$$\left\| y - (e - xt) \sum_{n=0}^{\infty} z_n t^n \right\| = \left\| y - z_0 - \sum_{n=1}^{\infty} z_n t^n + x \sum_{n=0}^{\infty} z_n t^{n+1} \right\|$$

$$= \left\| y - z_0 - \sum_{n=1}^{\infty} (z_n - x z_{n-1}) t^n \right\|$$

$$= \| y - z_0 \| + \sum_{n=1}^{\infty} \| z_n - x z_{n-1} \|$$

$$\geq \| y - z_0 \| + \sum_{n=1}^{\infty} (\| x z_{n-1} \| - \| z_n \|)$$

$$\geq \| y - z_0 \| + \sum_{n=1}^{\infty} (\| z_{n-1} \| - \| z_n \|)$$

$$= \| y - z_0 \| + \| z_0 \|$$

$$\geq \| y \|.$$

This proves that $\| j(y) \| \geq \| y \|$.

Finally, $j(x)$ is invertible in B since $e - xt \in I$ and hence

$$(x + I)(et + I) = xt + I = e + I.$$

This contradicts hypothesis (ii). Thus (ii) $\Rightarrow$ (i). $\qquad\square$

Corollary 3.4.8. *Let A be a commutative unital Banach algebra and let $\varphi \in \partial(A)$. Then every element of $\ker \varphi$ is a topological divisor of zero.*

Proof. We show that every $x \in \ker \varphi$ satisfies condition (ii) of Theorem 3.4.7. Let B be any unital commutative Banach algebra such that there exists an isometric isomorphism j from A into B with $j(e) = e_B$. Since $\varphi \in \partial(A)$, by Theorem 3.3.9 there exists $\psi \in \Delta(B)$ such that $\psi(j(x)) = \varphi(x)$ for all $x \in A$. Thus $j(x) \in \ker \psi$ and hence $j(x)$ is not invertible in B. By Theorem 3.4.7, x is a topological divisor of zero in A. $\qquad\square$

Corollary 3.4.8 is considerably improved later (Theorem 3.4.11) to the effect that $\ker \varphi$ even consists of joint topological zero divisors.

Remark 3.4.9. Let A be a commutative Banach algebra with identity e and let $\varphi \in \Delta(A)$. Suppose that $\ker \varphi$ consists of joint topological zero divisors. Let B be a commutative unital Banach algebra containing A. Then φ extends to some $\psi \in \Delta(B)$. Towards a contradiction, assume that no such extension ψ exists. Then it can be shown exactly as in the proof of Theorem 3.3.9 that there exist $a_1, \ldots, a_n \in \ker \varphi$ and $b_1, \ldots, b_n \in B$ such that $e = \sum_{j=1}^{n} a_j b_j$. Because $d(a_1, \ldots, a_n) = 0$, it follows that

$$d(a_1 b_1, \ldots, a_n b_n) = \inf \left\{ \sum_{j=1}^{n} \| a_j b_j y \| : y \in B, \| y \| = 1 \right\}$$

$$\leq d(a_1, \ldots, a_n) \max_{1 \leq j \leq n} \|b_j\|$$
$$= 0.$$

On the other hand, since $\sum_{j=1}^{n} a_j b_j = e$,

$$1 = \inf \left\{ \left\| \sum_{j=1}^{n} a_j b_j y \right\| : y \in A, \|y\| = 1 \right\} \leq d(a_1 b_1, \ldots, a_n b_n).$$

This contradiction shows that φ extends to some $\psi \in \Delta(B)$.

For uniform algebras elements of the Shilov boundary can be characterised in terms of topological divisors of zero as follows.

Theorem 3.4.10. *Let X be a compact Hausdorff space, A a uniform algebra on X, and $\varphi \in \Delta(A)$. Then φ belongs to $\partial(A)$ if and only if $\ker \varphi$ consists of joint topological zero divisors.*

Proof. Let $\varphi \in \partial(A)$ and $f_1, \ldots, f_n \in \ker \varphi$. Since $\varphi \in \partial(A)$, $\varphi = \varphi_x$ for some $x \in X$. Given $\epsilon > 0$, there exists an open neighbourhood V of x such that $|f_j(y)| < \epsilon$ for all $y \in V$ and $1 \leq j \leq n$. Since $x \in \partial(A)$, there exists $g \in A$ such that $\|g\|_\infty > \|g|_{X \setminus V}\|_\infty$ (Corollary 3.3.4). Of course, we can assume that $\|g\|_\infty = 1$. Then $\|g^m|_{X \setminus V}\|_\infty \leq \epsilon$ for all sufficiently large $m \in \mathbb{N}$. It follows that, for $1 \leq j \leq n$,

$$|f_j(y) g^m(y)| \leq \epsilon \|f_j\|_\infty$$

for all $y \in X \setminus V$ and $|f_j(y) g^m(y)| \leq \epsilon$ for all $y \in V$. Because $\|g^m\|_\infty = 1$, these estimates together show that

$$d(f_1, \ldots, f_n) \leq \epsilon \cdot \max\{1, \|f_1\|_\infty, \ldots, \|f_n\|_\infty\}.$$

So $d(f_1, \ldots, f_n) = 0$ since $\epsilon > 0$ was arbitrary.

Conversely, suppose that $\ker \varphi$ consist of joint topological divisors of zero, and let ϕ denote the isometric isomorphism $f \to f|_{\partial(A)}$ from A into $C(\partial(A))$. Then $\varphi \circ \phi^{-1} \in \Delta(\phi(A))$ and $\ker(\varphi \circ \phi^{-1})$ consists of joint topological zero divisors. As shown in Remark 3.4.9, there exists $\psi \in \Delta(C(\partial(A)))$ with $\psi|_{\phi(A)} = \varphi \circ \phi^{-1}$. Now $\psi(g) = g(x)$ for some $x \in \partial(A)$ and all $g \in C(\partial(A))$. It follows that

$$\varphi(f) = (\varphi \circ \phi^{-1})(\phi(f)) = \psi(f|_{\partial(A)}) = f(x) = \varphi_x(f)$$

for all $f \in A$. So $\varphi \in \partial(A)$. $\qquad\qquad\qquad\qquad\qquad\qquad\qquad$ $\square$

Theorem 3.4.10 can be used to show that the 'only if' part of the assertion of Theorem 3.4.10 remains true for arbitrary unital commutative Banach algebras.

Theorem 3.4.11. *Let A be a unital commutative Banach algebra and let $\varphi \in \partial(A)$. Then $\ker \varphi$ consists of joint topological zero divisors.*

Proof. It suffices to show that given $a_1, \ldots, a_q \in A$ such that $d(a_1, \ldots, a_q) > 0$, there is no maximal ideal of A containing all of $a_1, \ldots, a_q$ and corresponding to some point in $\partial(A)$. Of course, we can assume $d(a_1, \ldots, a_q) \geq 1$; that is, $\sum_{j=1}^{q} \|a_j y\| \geq \|y\|$ for all $y \in A$. We claim that

$$\sum_{j=1}^{q} r(a_j y) \geq r(y)$$

for all $y \in A$. To verify this turns out to be quite intricate.

Let B denote the algebra of all formal power series

$$\widetilde{x}(t_1, \ldots, t_q) = \sum x_{n_1, \ldots, n_q} t_1^{n_1} \cdot \ldots \cdot t_q^{n_q}$$

in q variables $t_1, \ldots, t_q$, where summation extends over all $(n_1, \ldots, n_q) \in \mathbb{N}_0^q$, $x_{n_1, \ldots, n_q} \in A$ and $\sum \|x_{n_1, \ldots, n_q}\| < \infty$. Recall that multiplication in B is given by

$$\sum x_{n_1, \ldots, n_q} t_1^{n_1} \cdot \ldots \cdot t_q^{n_q} \cdot \sum y_{m_1, \ldots, m_q} t_1^{m_1} \cdot \ldots \cdot t_q^{m_q} = \sum z_{p_1, \ldots, p_q} t_1^{p_1} \cdot \ldots \cdot t_q^{p_q},$$

where, for $(p_1, \ldots, p_q) \in \mathbb{N}_0^q$,

$$z_{p_1, \ldots, p_q} = \sum_{\substack{n_j + m_j = p_j \\ 1 \leq j \leq q}} x_{n_1, \ldots, n_q} y_{m_1, \ldots, m_q}.$$

It is not difficult to check (compare the proof of Theorem 3.4.7 in the case $q = 1$) that B, equipped with the norm

$$\|\widetilde{x}\| = \sum \|x_{n_1, \ldots, n_q}\|,$$

becomes a commutative Banach algebra. The map

$$\phi : x \to \widetilde{x}(t_1, \ldots, t_q) = x$$

is an isometric isomorphism from A into B. Using ϕ, we identify A with $\phi(A)$. Let

$$z = \sum_{j=1}^{q} a_j t_j \in B.$$

We prove by induction that $\|z^k y\| \geq \|y\|$ for all $y \in A$ and $k \in \mathbb{N}$. Clearly,

$$\|zy\| = \left\| \sum_{j=1}^{q} (a_j y) t_j \right\| = \sum_{j=1}^{q} \|a_j y\| \geq \|y\|$$

by assumption. For the inductive step, suppose that $\|z^{k-1} y\| \geq \|y\|$ for all y and note that

$$z^{k-1} = \sum_{n_1+\ldots+n_q=k-1} c_{n_1,\ldots,n_q} a_1^{n_1} \cdot \ldots \cdot a_q^{n_q} t_1^{n_1} \cdot \ldots \cdot t_q^{n_q},$$

where $c_{n_1,\ldots,n_q} > 0$. For any $y \in A$,

$$\|z^{k-1}y\| = \sum_{n_1+\ldots+n_q=k-1} c_{n_1,\ldots,n_q} \|a_1^{n_1} \cdot \ldots \cdot a_q^{n_q} y\|.$$

This implies

$$\|z^k y\| = \left\| z \cdot \sum_{n_1+\ldots+n_q=k-1} c_{n_1,\ldots,n_q} a_1^{n_1} \cdot \ldots \cdot a_q^{n_q} t_1^{n_1} \cdot \ldots \cdot t_q^{n_q} \right\|$$

$$= \left\| \sum_{j=1}^{q} \sum_{n_1+\ldots+n_q=k-1} c_{n_1,\ldots,n_q} a_1^{n_1} \cdot \ldots \cdot a_j^{n_j+1} \cdot \ldots \cdot a_q^{n_q} y \, t_1^{n_1} \cdot \ldots \cdot t_q^{n_q} \right\|$$

$$= \sum_{n_1+\ldots+n_q=k-1} c_{n_1,\ldots,n_q} \left(\sum_{j=1}^{q} \|a_j a_1^{n_1} \cdot \ldots \cdot a_q^{n_q} y\| \right)$$

$$\geq \sum_{n_1+\ldots+n_q=k-1} c_{n_1,\ldots,n_q} \|a_1^{n_1} \cdot \ldots \cdot a_q^{n_q} y\|$$

$$= \|z^{k-1}y\|.$$

Thus, the inductive hypothesis shows that $\|z^k y\| \geq \|y\|$. Replacing y with y^k, we get

$$\|(zy)^k\|^{1/k} \geq \|y^k\|^{1/k}$$

for all $y \in A$ and $k \in \mathbb{N}$ and hence $r(zy) \geq r(y)$ for all $y \in A$. Using that the spectral radius is subadditive and submultiplicative and that $r(t_j) = 1$, $1 \leq j \leq q$, we obtain

$$r(y) \leq r(zy) = r\left(\sum_{j=1}^{q} a_j y t_j \right)$$

$$\leq \sum_{j=1}^{q} r(a_j y t_j) \leq \sum_{j=1}^{q} r(a_j y) r(t_j)$$

$$= \sum_{j=1}^{q} r(a_j y)$$

for all $y \in A$. This establishes the above claim.

With $\Gamma : A \to C(\Delta(A)), y \to \widehat{y}$ denoting the Gelfand homomorphism, we can reformulate what we have shown so far by

$$\sum_{j=1}^{q} \|\widehat{a_j}\widehat{y}\|_\infty = \sum_{j=1}^{q} r(a_j y) \geq r(y) = \|\widehat{y}\|_\infty$$

for all $y \in A$; that is, $d(\widehat{a}_1, \ldots, \widehat{a}_q) \geq 1$.

Let C denote the closure of $\Gamma(A)$ in $C(\Delta(A))$. Then, by Theorem 3.4.10, the functions $\widehat{a}_1, \ldots, \widehat{a}_q$ cannot simultaneously belong to $\ker \psi$ for any $\psi \in \partial(C)$. Now, there is a continuous bijection $\varphi \to \widetilde{\varphi}$ between $\Delta(A)$ and $\Delta(C)$ satisfying $\widetilde{\varphi}(\widehat{a}) = \varphi(a)$ for all $a \in A$. In addition, since $\Delta(A)$ is compact, this bijection is a homeomorphism and maps $\partial(A)$ onto $\partial(C)$ (Remark 3.3.7). It follows that none of the ideals $\ker \varphi, \varphi \in \partial(A)$, can contain all of $a_1, \ldots, a_q$.

This finishes the proof of the theorem. $\hfill\square$

From Theorem 3.3.9 we know that if B is a commutative Banach algebra with identity e and A is a closed subalgebra of B containing e, then every $\varphi \in \partial(A)$ extends to some $\widetilde{\varphi} \in \Delta(B)$.

We conclude this section by showing that combining Theorem 3.3.9 and Theorem 3.4.10 leads to a major improvement in that actually such an extension $\widetilde{\varphi}$ can be found in $\partial(B)$. In preparation for this, we need a technical lemma.

Lemma 3.4.12. *Let A be a commutative normed algebra and let M be a subset of A consisting of joint topological divisors of zero. Then the closed ideal of A generated by M also consists of joint topological divisors of zero.*

Proof. Let I be the ideal generated by M, so that

$$I = \left\{ \sum_{j=1}^{n} x_j y_j : x_j \in M, y_j \in A, n \in \mathbb{N} \right\}.$$

Let $a_1, \ldots, a_m \in I$ and, for $i = 1, \ldots, m$, write $a_i = \sum_{j=1}^{n} x_{ij} y_{ij}$, where $x_{ij} \in M$ and $y_{ij} \in A$. Let

$$d = d(x_{11}, \ldots, x_{1n}, x_{21}, \ldots, x_{2n}, \ldots, x_{m1}, \ldots x_{mn}).$$

Then

$$d(a_1, \ldots, a_m) = \inf_{\|y\|=1} \sum_{i=1}^{m} \left\| \sum_{j=1}^{n} x_{ij} y_{ij} y \right\| \leq \inf_{\|y\|=1} \sum_{i=1}^{m} \sum_{j=1}^{n} \|x_{ij} y_{ij} y\|$$
$$\leq d \cdot \max\{\|y_{ij}y\| : 1 \leq i \leq m, 1 \leq j \leq n, \|y\| = 1\}$$
$$= 0.$$

Now, given $a_1, \ldots, a_m \in \overline{I}$ and $\epsilon > 0$, there exist $b_1, \ldots, b_m \in I$ such that $\|a_j - b_j\| \leq \epsilon$ for $j = 1, \ldots, m$ and an element y of A with $\|y\| = 1$ such that $\sum_{j=1}^{m} \|b_j y\| \leq \epsilon$. It follows that

$$\sum_{j=1}^{m} \|a_j y\| \leq \sum_{j=1}^{m} (\|a_j - b_j\| + \|b_j y\|) \leq (m + 1)\epsilon.$$

Because $\epsilon > 0$ was arbitrary, we conclude that $d(a_1, \ldots, a_m) = 0$. $\hfill\square$

Theorem 3.4.13. *Let B be a commutative Banach algebra with identity e and let A be a closed subalgebra of B containing e. Then every $\varphi \in \partial(A)$ extends to some $\widetilde{\varphi} \in \partial(B)$.*

Proof. Let $\Gamma : B \to C(\Delta(B)), x \to \widehat{x}$ denote the Gelfand homomorphism of B and let C_A and C_B be the closure of $\Gamma(A)$ and of $\Gamma(B)$ in $C(\Delta(B))$, respectively.

For $\varphi \in \partial(A)$, define $\widehat{\varphi} : \Gamma(A) \to \mathbb{C}$ by $\widehat{\varphi}(\widehat{x}) = \varphi(x)$ for $x \in A$. By Theorem 3.3.9 there exists $\psi \in \Delta(B)$ such that $\psi|_A = \varphi$. Thus

$$|\widehat{\varphi}(\widehat{x})| = |\varphi(x)| = |\psi(x)| \leq \|\widehat{x}\|_\infty$$

for all $x \in A$. Hence $\widehat{\varphi}$ extends uniquely to some $\epsilon(\varphi) \in \Delta(C_A)$, and the map $\epsilon : \varphi \to \epsilon(\varphi)$ is an embedding of $\partial(A)$ into $\Delta(C_A)$. Now, every boundary of $\Gamma(A)$ is a boundary for C_A. So $\epsilon(\partial(A))$ is a boundary for C_A and hence equals $\partial(C_A)$.

Fix $\varphi \in \partial(A)$. By Theorem 3.4.10, $\ker \epsilon(\varphi)$ consists of joint topological zero divisors. Let I be the ideal of C_B generated by $\ker \epsilon(\varphi) \subseteq C_A \subseteq C_B$. By Lemma 3.4.12, I consists of joint topological zero divisors. Let $\psi \to \widehat{\psi}$ be the bijection between $\Delta(B)$ and $\Delta(C_B)$ given by $\widehat{\psi}(\widehat{y}) = \psi(y)$ for all $y \in B$. Recall that this maps $\partial(B)$ onto $\partial(C_B)$.

Finally, let $\psi \in \Delta(B)$ such that $\psi|_A = \varphi$. Then $I \subseteq \ker \widehat{\psi}$ because otherwise there exist $x \in \ker \varphi$ and $y \in B$ such that $\widehat{x}\widehat{y} \notin \ker \widehat{\psi}$, which contradicts

$$\widehat{\psi}(\widehat{x}\,\widehat{y}) = \psi(x)\psi(y) = \varphi(x)\psi(y) = 0.$$

Thus $\ker \widehat{\psi}$ consists of joint topological zero divisors. Since C_B is a uniform algebra, by Theorem 3.4.10 $\widehat{\psi}$ belongs to the Shilov boundary of C_B, and this implies that $\psi \in \partial(B)$, as desired. $\square$

3.5 Shilov's idempotent theorem and applications

Our concern in this section is the following celebrated idempotent theorem due to Shilov. Recall that an element a of an algebra A is called an *idempotent* if it satisfies $a^2 = a$.

Theorem 3.5.1. *Let A be a commutative Banach algebra and let C be a compact open subset of $\Delta(A)$. Then there exists an idempotent a in A such that $\widehat{a}$ equals the characteristic function of C.*

Shilov's idempotent theorem is not only a beautiful result on its own but in turn admits a variety of important applications. Most notably, it allows us to show that a semisimple commutative Banach algebra with compact structure space is unital. Recall that we have already seen special cases in earlier sections. Unfortunately, no proof of Theorem 3.5.1 is known which does not require the use of some kind of multivariable holomorphic functional calculus.

Lemma 3.5.2. *Let A be a commutative Banach algebra and let φ_1 and φ_2 be distinct elements of $\Delta(A)$. Then there exists $x \in A$ such that $\varphi_1(x) = 1$ and $\varphi_2(x) = 0$.*

Proof. What follows is a standard argument which is also used to prove the classical Stone–Weierstrass theorem. However, we include the proof for the reader's convenience.

The set of Gelfand transforms strongly separates the points of $\Delta(A)$. Therefore there exist elements a_1, a_2, and b of A such that $\varphi_1(a_1) \neq 0, \varphi_2(a_2) \neq 0$, and $\varphi_1(b) \neq \varphi_2(b)$. Let $c_j = (1/\varphi_j(a_j))a_j$ for $j = 1, 2$, and let $c = c_1 + c_2 - c_1 c_2 \in A$. Then

$$\varphi_j(c) = \varphi_j(c_1) + \varphi_j(c_2) - \varphi_j(c_1)\varphi_j(c_2) = 1,$$

$j = 1, 2$. Let

$$x = \frac{1}{\varphi_1(b) - \varphi_2(b)} \left(b - \varphi_2(b)c\right) \in A.$$

Then, because $\varphi_1(c) = \varphi_2(c) = 1$,

$$\varphi_1(x) = \frac{1}{\varphi_1(b) - \varphi_2(b)} \left(\varphi_1(b) - \varphi_2(b)\varphi_1(c)\right) = 1$$

and

$$\varphi_2(x) = \frac{1}{\varphi_1(b) - \varphi_2(b)} \left(\varphi_2(b) - \varphi_2(b)\varphi_2(c)\right) = 0.$$

So x has the required properties. $\qquad\square$

Proposition 3.5.3. *Let A be a unital commutative Banach algebra and let U_1 and U_2 be disjoint open subsets of $\Delta(A)$ such that $\Delta(A) = U_1 \cup U_2$. Then there exists $x \in A$ such that $\widehat{x}|_{U_1} = 0$ and $\widehat{x}|_{U_2} = 1$.*

Proof. Given $\varphi \in U_1$ and $\psi \in U_2$, by Lemma 3.5.2 there exists $a_{\varphi,\psi} \in A$ such that $\varphi(a_{\varphi,\psi}) = 0$ and $\psi(a_{\varphi,\psi}) = 1$. Let

$$V_{\varphi,\psi} = \left\{\sigma \in U_1 : |\sigma(a_{\varphi,\psi})| < \frac{1}{2}\right\} \text{ and } W_{\varphi,\psi} = \left\{\tau \in U_2 : |\tau(a_{\varphi,\psi})| > \frac{1}{2}\right\}.$$

These sets are open neighbourhoods of φ and ψ, respectively, and $V_{\varphi,\psi} \subseteq U_1$ and $W_{\varphi,\psi} \subseteq U_2$. Now, fix $\psi \in U_2$. Then, since U_1 is open and compact, there exists a finite subset E_ψ of U_1 such that

$$U_1 = \bigcup_{\varphi \in E_\psi} V_{\varphi,\psi}.$$

Thus, if $\sigma \in U_1$, then $|\sigma(a_{\varphi,\psi})| < \frac{1}{2}$ for at least one $\varphi \in E_\psi$. Let

$$W_\psi = \bigcap_{\varphi \in E_\psi} W_{\varphi,\psi},$$

which is an open neighbourhood of ψ in U_2. Since U_2 is compact, there exist $\psi_1, \ldots, \psi_m \in U_2$ such that $U_2 = \bigcup_{j=1}^m W_{\psi_j}$. Now, consider the finite subset

$$M = \{a_{\varphi,\psi_j} : 1 \leq j \leq m, \varphi \in E_{\psi_j}\}$$

of A and enumerate M, say $M = \{x_1, \ldots, x_r\}$. Let

$$C_j = \{(\varphi(x_1), \ldots, \varphi(x_r)) : \varphi \in U_j\},$$

$j = 1, 2$. Then C_1 and C_2 are compact and, since $U_1 \cup U_2 = \Delta(A)$,

$$C_1 \cup C_2 = \sigma_A(x_1, \ldots, x_r).$$

Assume $C_1 \cap C_2 \neq \emptyset$. Then there exist $\sigma \in U_1$ and $\tau \in U_2$ such that

$$\sigma(a_{\varphi,\psi_j}) = \tau(a_{\varphi,\psi_j})$$

for each $1 \leq j \leq m$ and all $\varphi \in E_{\psi_j}$. Now, $\tau \in W_{\psi_j}$ for some j and then $\sigma \in V_{\varphi,\psi_j}$ for some $\varphi \in E_{\psi_j}$. By the definition of $W_{\psi_j}, \tau \in W_{\varphi,\psi_j}$ and hence

$$|\sigma(a_{\varphi,\psi_j})| < \frac{1}{2} \text{ and } |\tau(a_{\varphi,\psi_j})| > \frac{1}{2},$$

which is impossible. Thus C_1 and C_2 are disjoint compact subsets of $\mathbb{C}^r$, and hence we can find disjoint open neighbourhoods W_1 and W_2 of C_1 and C_2 in $\mathbb{C}^r$, respectively.

Define $f : W_1 \cup W_2 \to \mathbb{C}$ by $f|_{W_1} = 0$ and $f|_{W_2} = 1$. Then f is a holomorphic function on the neighbourhood $W_1 \cup W_2$ of $\sigma_A(x_1, \ldots, x_r)$. By Theorem 3.1.10, there exists $x \in A$ such that

$$\widehat{x}(\varphi) = f(\varphi(x_1), \ldots, \varphi(x_r))$$

for all $\varphi \in \Delta(A)$. It follows that $\widehat{x}(\varphi) = 0$ for all $\varphi \in U_1$ and $\widehat{x}(\varphi) = 1$ for all $\varphi \in U_2$. $\qquad \square$

When A is semisimple, the element x of A in the preceding proposition is an idempotent because $\widehat{x^2} = \widehat{x}$. Thus Shilov's idempotent theorem has been established so far for semisimple unital commutative Banach algebras. We now continue with a construction which allows us to drop the hypothesis that A be semisimple.

Lemma 3.5.4. *Let A be a commutative Banach algebra with identity e and let $b \in A$ be such that $\widehat{b^2} = \widehat{b}$. Then there exists $a \in A$ such that $\widehat{a} = \widehat{b}$ and $a^2 = a$.*

Proof. Recall that for any $x \in A$ the geometric series $\sum_{n=0}^\infty x^n$ converges in A whenever $r(x) < 1$. Let $x = 4(b^2 - b)$. Then $\widehat{x} = 0$ by hypothesis and hence x is in the radical of A and therefore $r(x) = 0$. Since $\left|\binom{-1/2}{n}\right| \leq 1$ for all $n \in \mathbb{N}_0$, the series

$$\sum_{n=0}^{\infty} \binom{-1/2}{n} x^n$$

converges in A. For the obvious reason, we denote this element by $(e+x)^{-1/2}$. We claim that

$$(e+x)^{-1/2}(e+x)^{-1/2}(e+x) = e,$$

a formula the reader will expect. Indeed, using the well-known equation

$$\sum_{k+l=m} \binom{-1/2}{k}\binom{-1/2}{l} = (-1)^m$$

for $m \in \mathbb{N}_0$, we get

$$\left((e+x)^{-1/2}\right)^2 (e+x) = (e+x)\sum_{n=0}^{\infty}\binom{-1/2}{n}x^n \cdot \sum_{m=0}^{\infty}\binom{-1/2}{m}x^m$$

$$= (e+x)\sum_{m=0}^{\infty}\left(\sum_{k+l=m}\binom{-1/2}{k}\binom{-1/2}{l}\right)x^m$$

$$= (e+x)\sum_{m=0}^{\infty}(-1)^m x^m$$

$$= e + \sum_{m=1}^{\infty}(-1)^m x^m + \sum_{m=0}^{\infty}(-1)^m x^{m+1}$$

$$= e + \sum_{m=0}^{\infty}\left((-1)^{m+1} + (-1)^m\right)x^{m+1}$$

$$= e.$$

Thus $\left((e+x)^{-1/2}\right)^2 = (e+x)^{-1}$, and setting

$$a = \left(b - \frac{1}{2}e\right)(e+x)^{-1/2} + \frac{1}{2}e,$$

it follows that

$$a(a-e) = \left(b - \frac{1}{2}e\right)^2 \left((e+x)^{-1/2}\right)^2 - \frac{1}{4}e$$

$$= (e+x)^{-1}\left(b^2 - b + \frac{1}{4}e\right) - \frac{1}{4}e$$

$$= 0.$$

Hence a is an idempotent, and it only remains to verify that $\hat{a} = \hat{b}$. To that end, let $y = \left(b - \frac{1}{2}e\right)\sum_{n=1}^{\infty}\binom{-1/2}{n}x^n$ and note that

$$a = \left(b - \frac{1}{2}e\right)\left(e + \sum_{n=1}^{\infty}\binom{-1/2}{n}x^n\right) + \frac{1}{2}e = b + y.$$

Now, $y \in \mathrm{rad}(A)$ because $x \in \mathrm{rad}(A)$ and $\mathrm{rad}(A)$ is a closed ideal. Thus $\widehat{y} = 0$ and hence $\widehat{a} = \widehat{b}$. $\square$

Lemma 3.5.4 completes the proof of Theorem 3.5.1 when A is unital. Now, assume that A does not have an identity and consider the unitisation A_e of A. Embedding $\Delta(A)$ into $\Delta(A_e)$ as usual, C is still an open and closed set. Thus, there is an idempotent u in A_e such that $\widehat{u}|_C = 1$ and $\widehat{u}|_{\Delta(A) \setminus C} = 0$. Because $\widehat{u}(\varphi_\infty) = 0$, u is in A.

We next present a number of applications of Shilov's idempotent theorem. The first one has been announced several times.

Theorem 3.5.5. *Let A be a semisimple commutative Banach algebra. If $\Delta(A)$ is compact, then A has an identity.*

Proof. Since $\Delta(A)$ is compact, by Theorem 3.5.1 there exists $e \in A$ such that $\widehat{e} = 1$ on $\Delta(A)$. It follows that $\widehat{xe - x}(\varphi) = \widehat{x}(\varphi)\widehat{e}(\varphi) - \widehat{x}(\varphi) = 0$ for all $x \in A$ and $\varphi \in \Delta(A)$. Since A is semisimple, we conclude that $xe - x = 0$ for all $x \in A$, whence e is an identity for A. $\square$

Later (Corollary 4.2.11) we give a much simpler proof (one which does not require Shilov's idempotent theorem) of Theorem 3.5.5 for so-called regular semisimple commutative Banach algebras.

Corollary 3.5.6. *Let A be a commutative Banach algebra and suppose that $\Delta(A)$ is totally disconnected. Then $\widehat{A} = \{\widehat{a} : a \in A\}$ is dense in $C_0(\Delta(A))$.*

Proof. Let $f \in C_0(\Delta(A))$ and $\epsilon > 0$ be given. Because f vanishes at infinity and every point of $\Delta(A)$ has a neighbourhood basis of compact open sets, there exists a compact open subset K of $\Delta(A)$ such that $|f(\varphi)| < \epsilon$ for all $\varphi \in \Delta(A) \setminus K$. Now, K can be written as a disjoint union of compact open sets $E_1, \ldots, E_r$ such that $|f(\varphi) - f(\psi)| < \epsilon$ for all $\varphi, \psi \in E_j, 1 \leq j \leq r$. So there exist $c_1, \ldots, c_r \in \mathbb{C}$ with the property that the function $g = \sum_{j=1}^{r} c_j 1_{E_j}$ satisfies $|f(\varphi) - g(\varphi)| < \epsilon$ for all $\varphi \in K$. By Shilov's idempotent theorem, there exist $a_j \in A, 1 \leq j \leq r$, so that $\widehat{a}_j = 1_{E_j}$. For the element $a = \sum_{j=1}^{r} c_j a_j$ of A it follows that $\|\widehat{a} - f\|_\infty < \epsilon$. Hence $\widehat{A}$ is dense in $C_0(\Delta(A))$. $\square$

As a final application of the Shilov idempotent theorem we now investigate the relation between coverings of $\Delta(A)$ through disjoint open subsets and decomposition of A into the direct sum of ideals. We start with the more general situation of not necessarily finite coverings.

Theorem 3.5.7. *Let A be a nonunital commutative Banach algebra and let $\Delta(A) = \bigcup_{\lambda \in \Lambda} F_\lambda$ be a decomposition of $\Delta(A)$ into open and compact subsets $F_\lambda, \lambda \in \Lambda$. Then there exists a family of closed ideals $I_\lambda, \lambda \in \Lambda$, with the following properties.*

(i) $\Delta(I_\lambda) = F_\lambda$ *for each λ.*

(ii) $I_\lambda \cap \left(\sum_{\substack{\mu \in \Lambda \\ \mu \neq \lambda}} I_\mu \right) \subseteq \operatorname{rad}(A)$ for each λ.

(iii) $\sum_{\lambda \in \Lambda} I_\lambda$ is dense in A provided that every proper closed ideal of A is contained in a maximal modular ideal.

Proof. By the idempotent theorem, for each $\lambda \in \Lambda$, there exists an idempotent u_λ in A such that
$$\widehat{u}_\lambda|_{F_\lambda} = 1 \quad \text{and} \quad \widehat{u}_\lambda|_{\Delta(A) \setminus F_\lambda} = 0.$$

Let $I_\lambda = Au_\lambda$. Then I_λ is an ideal, and I_λ is closed in A because if $(x_n)_n \subseteq I_\lambda$ and $x_n \to x$ in A, then $x_n = x_n u_\lambda \to x u_\lambda$ and hence $x = x u_\lambda \in I_\lambda$.

To show (i), recall that
$$\Delta(I_\lambda) = \Delta(A) \setminus h(I_\lambda) = \{\varphi \in \Delta(A) : \varphi(x) \neq 0 \text{ for some } x \in I_\lambda\}.$$

Now, if $\varphi(x) \neq 0$ and $x = yu_\lambda, y \in A$, then $\varphi(u_\lambda) \neq 0$. Thus $\Delta(I_\lambda) \subseteq F_\lambda$. Conversely, if $\varphi \in F_\lambda$ then $\varphi(u_\lambda) \neq 0$ and hence $\varphi \notin h(I_\lambda)$.

For (ii), fix λ and let $J = \sum_{\mu \neq \lambda} I_\mu$ and $D = I_\lambda \cap J$. We have to verify that $D \subseteq \operatorname{rad}(A)$, equivalently, that $h(D) = \Delta(A)$. Towards a contradiction, suppose that there exists $\varphi \in \Delta(A) \setminus h(D)$ and choose $x \in D$ with $\varphi(x) \neq 0$. Since $x \in I_\lambda, x = x u_\lambda$ and hence $\varphi(u_\lambda) \neq 0$. Thus $\varphi \in F_\lambda$ by (i). We show that also $\varphi \in F_\mu$ for some $\mu \in \Lambda, \mu \neq \lambda$. Since $x \in J$, x can be written as a sum $x = \sum_{j=1}^n c_j x_j$, where $x_j \in I_{\mu_j}, \mu_1, \ldots, \mu_n \in \Lambda$ and $\mu_j \neq \lambda$ for all j. It follows that, since $x_j = x_j u_{\lambda_j}$,

$$0 \neq \varphi(x) = \sum_{j=1}^n c_j \varphi(x_j) = \sum_{j=1}^n c_j \varphi(x_j) \varphi(u_{\lambda_j}),$$

and therefore $\varphi(u_{\lambda_j}) \neq 0$ for some j, so that $\varphi \in F_{\lambda_j}$. This contradicts the fact that $F_\lambda \cap F_\mu = \emptyset$ for $\mu \neq \lambda$.

(iii) Because $\Delta(I_\lambda) = F_\lambda$ for all λ and $\Delta(A) = \bigcup_{\lambda \in \Lambda} F_\lambda$, no element of $\Delta(A)$ annihilates $\sum_{\lambda \in \Lambda} I_\lambda$ and hence this ideal is dense in A by hypothesis. $\square$

We continue with a converse to Theorem 3.5.7.

Theorem 3.5.8. *Let A be a nonunital commutative Banach algebra and let $\{I_\lambda : \lambda \in \Lambda\}$ be a family of unital closed ideals of A satisfying property (ii) of Theorem 3.5.7 and such that the ideal $\sum_{\lambda \in \Lambda} I_\lambda$ is dense in A. Then $\Delta(A) = \bigcup_{\lambda \in \Lambda} \Delta(I_\lambda)$ and the sets $\Delta(I_\lambda), \lambda \in \Lambda$, are open and disjoint.*

Proof. Of course, each $\Delta(I_\lambda) = \Delta(A) \setminus h(I_\lambda)$ is open in $\Delta(A)$. Let $\lambda, \mu \in \Lambda$ such that $\lambda \neq \mu$, and suppose that there exists $\varphi \in \Delta(I_\lambda) \cap \Delta(I_\mu)$. Choose $a \in I_\lambda$ and $b \in I_\mu$ such that $\varphi(a) \neq 0$ and $\varphi(b) \neq 0$. Then $ab \in I_\lambda \cap I_\mu$ and $I_\lambda \cap I_\mu \subseteq \operatorname{rad}(A)$ by hypothesis. This contradicts $\varphi(\operatorname{rad}(A)) = \{0\}$.

It remains to show that $\Delta(A) = \bigcup_{\lambda \in \Lambda} \Delta(I_\lambda)$. If $\varphi \in \Delta(A)$ and φ annihilates all I_λ, then $\varphi(\sum_{\lambda \in \Lambda} I_\lambda) = \{0\}$. However, this is impossible because φ is continuous and $\sum_{\lambda \in \Lambda} I_\lambda$ is dense in A. $\square$

Now we turn to finite coverings.

Theorem 3.5.9. *Let A be a unital commutative Banach algebra.*

(i) *If $\Delta(A)$ is a disjoint union $\Delta(A) = \bigcup_{j=1}^{m} F_j$ of open (and closed) subsets F_j, then there exist unital closed ideals $I_1, \ldots, I_m$ of A such that $A = \oplus_{j=1}^{m} I_j$ and $\Delta(I_j) = F_j$ for $j = 1, \ldots, m$.*
(ii) *Conversely, if A is the direct sum of closed ideals $I_1, \ldots, I_m$, then the sets $\Delta(I_j)$ are open and closed in $\Delta(A)$ and $\Delta(A)$ is the disjoint union of the sets $\Delta(I_j), 1 \leq j \leq m$.*

Proof. A straightforward induction argument shows that for both (i) and (ii), it suffices to consider the case $m = 2$.

(i) Since $\Delta(A)$ is compact, F_1 and F_2 are compact. By Shilov's idempotent theorem, there exists an idempotent $e_1 \in A$ such that $\widehat{e}_1 = 1_{F_1}$. Let e denote the identity of A and set $e_2 = e - e_1$. Then e_2 is an idempotent and $\widehat{e}_2 = 1_{F_2}$. Let $I_j = e_j A$ for $j = 1, 2$. Then I_1 and I_2 are closed ideals of A and $\Delta(I_j) = F_j, j = 1, 2$ (compare Theorem 3.5.7). Note that $e_1 e_2 = 0$ since

$$e_1 + e_2 = e = e^2 = e_1^2 + e_2^2 + 2e_1 e_1 = e_1 + e_2 + 2e_1 e_2.$$

Thus, if $x \in I_1 \cap I_2$, then $x = xe_2 = xe_1 e_2 = 0$. Hence $I_1 + I_2$ is the direct sum of I_1 and I_2. Finally, for $x \in A$,

$$x = xe = xe_1 + xe_2 \in I_1 + I_2.$$

This finishes the proof of (i).

For (ii), as in the proof Theorem 3.5.8, it follows that $\Delta(A) = \Delta(I_1) \cup \Delta(I_2)$ and $\Delta(I_1) \cap \Delta(I_2) = \emptyset$. Of course, $\Delta(I_1)$ and $\Delta(I_2)$ are open (and hence closed) in $\Delta(A)$. □

Corollary 3.5.10. *Let A be a commutative Banach algebra.*

(i) *Suppose that $\Delta(A)$ is a disjoint union $\Delta(A) = \bigcup_{j=1}^{m} F_j$, where F_1 is closed and $F_2, \ldots, F_m$ are compact. Then there exist closed ideals $I_1, \ldots, I_m$ of A such that $A = \oplus_{j=1}^{m} I_j, \Delta(I_j) = F_j$ for $j = 1, \ldots, m$, and $I_2, \ldots, I_m$ are unital.*
(ii) *Conversely, let $I_1, I_2, \ldots, I_m$ be closed ideals of A such that $A = \oplus_{j=1}^{m} I_j$ and $I_2, \ldots, I_m$ are unital. Then $\Delta(A)$ is the disjoint union of the closed set $\Delta(I_1)$ and the compact sets $\Delta(I_2), \ldots, \Delta(I_m)$.*

Proof. To prove (i), in view of Theorem 3.5.9 we can assume that A does not have an identity. Let A_e be the algebra obtained by adjoining an identity e to A. Let $E_1 = F_1 \cup \{\varphi_\infty\}$ and $E_j = F_j$ for $j = 2, \ldots, m$. Then

$$\Delta(A_e) = \Delta(A) \cup \{\varphi_\infty\} = \bigcup_{j=1}^{m} E_j,$$

a disjoint union of open and closed subsets. By Theorem 3.5.9, there exist closed unital ideals $J_1, \ldots, J_m$ of A_e, such that $A_e = \oplus_{j=1}^m J_j$ and $\Delta(J_j) = E_j$, $1 \le j \le m$.

Notice first that $J_j \subseteq A$ for $2 \le j \le m$. Indeed, otherwise $\varphi_\infty(x) \ne 0$ for some $x \in J_j$ and hence $\varphi_\infty \in \Delta(J_j)$, which is impossible since $\Delta(J_j) \subseteq \Delta(A)$. Define closed ideals $I_1, \ldots, I_m$ of A by $I_1 = J_1 \cap A$ and $I_j = J$, for $j = 2, \ldots, m$.

We claim that $\Delta(I_1) = F_1$. Since $I_1 \subseteq J_1$, we have $\Delta(I_1) \subseteq \Delta(J_1) = E_1$. But $\varphi_\infty(I_1) = \{0\}$, whence $\Delta(I_1) \subseteq E_1 \setminus \{\varphi_\infty\} = F_1$. Conversely, let $\varphi \in F_1 \subseteq \Delta(J_1)$ and choose $x \in J_1$ with $\varphi(x) \ne 0$. Since $\varphi \in \Delta(A)$, there exists $y \in A$ such that $\varphi(y) \ne 0$. It follows that $xy \in J_1 \cap A = I_1$ and $\varphi(xy) \ne 0$, and hence $\varphi \in \Delta(I_1)$.

It remains to show that $A = \oplus_{j=1}^m I_j$. Given $x \in A$, there exist elements $x_1 \in J_1, \ldots, x_m \in J_m$ such that $x = x_1 + \ldots + x_m$. Since $x, x_2, \ldots, x_m \in A$, it follows that $x_1 \in A \cap J_1 = I_1$. Thus $A = I_1 + \ldots + I_m$. However, this sum is direct. $\square$

3.6 Exercises

Exercise 3.6.1. Let $u : \mathbb{T} \to \mathbb{C}$ be a function of the form $u(z) = \sum_{n=-\infty}^\infty \alpha_n z^n$, where $\sum_{n=-\infty}^\infty |\alpha_n| < \infty$. Suppose that f is a function which is holomorphic in a neighbourhood of the compact subset $u(\mathbb{T})$ of $\mathbb{C}$. Show that $f \circ u$ has a representation of the form $f \circ u(z) = \sum_{n=-\infty}^\infty \beta_n z^n$, where $\sum_{n=-\infty}^\infty |\beta_n| < \infty$.

Exercise 3.6.2. Let A be a commutative Banach algebra with identity e. Apply the functional calculus for entire functions to establish the following assertions.

(i) There exists no nonzero element x in $\mathrm{rad}(A)$, the radical of A, such that $\exp x = e$.

(ii) $\exp(2\pi i k e) = e$ for all $k \in \mathbb{Z}$.

(iii) If $\Delta(A)$ is connected and $x \in A$ is such that $\exp x = e$, then $x = 2\pi i e$ for some $k \in \mathbb{Z}$.

(iv) If $\Delta(A)$ is not connected, then there exist $x, y \in A$ such that $\exp x = \exp y$ and $x - y \notin (2\pi i \mathbb{Z})e$.

Exercise 3.6.3. Let A be a unital commutative Banach algebra and $x \in A$. Suppose that f is holomorphic in a neighbourhood of $\sigma_A(x)$ and g is holomorphic in some neighbourhood of $\sigma_A(f(x))$. Show that $(g \circ f)(x) = g(f(x))$.

Exercise 3.6.4. Let A be a commutative Banach algebra with identity e and let $a, b \in A$. Suppose that $\exp a = \exp b$ and $\hat{a} = \hat{b}$. To prove that $a = b$, proceed as follows.

(i) Let $c = a - b \in \mathrm{rad}(A)$ and show that

$$c\left(e + \sum_{j=1}^{\infty} \frac{1}{(j+1)!} c^j\right) = 0.$$

(ii) Show that the element

$$e + \sum_{j=1}^{\infty} \frac{1}{(j+1)!} c^j$$

is invertible and conclude that $c = 0$.

Exercise 3.6.5. Let A be a commutative Banach algebra with identity e and let $\Lambda(A)$ denote the set of all $\varphi \in \Delta(A)$ with the property that given any extension B of A, φ admits an extension to some element of $\Delta(B)$. Prove that $\Lambda(A)$ is closed in $\Delta(A)$.
(Hint: For each extension B of A, consider the restriction map $\psi \to \psi|_A$ from the set $\{\psi \in \Delta(B) : \psi(e) = 1\}$ to $\Delta(A)$.)

Exercise 3.6.6. Let A be a unital commutative Banach algebra which is generated by some element a. Show that the homeomorphism $\varphi \to \varphi(a)$ between $\Delta(A)$ and $\sigma_A(a)$ maps $\partial(A)$ onto the topological boundary of $\sigma_A(a)$.

Exercise 3.6.7. Let A be a commutative Banach algebra and C a compact and open subset of $\Delta(A)$. Prove that $C \cap \partial(A) \neq \emptyset$.

Exercise 3.6.8. Let A be a unital commutative Banach algebra and $a \in A$. Prove that $\hat{a}(\Delta(A)) = \hat{a}(\partial(A))$.
(Hint: Suppose there exists $\varphi_0 \in \Delta(A)$ such that $\hat{a}(\varphi_0) \notin \hat{a}(\partial(A))$. Choose a polynomial p such that $|p(\hat{a}(\varphi_0))| > |p(\hat{a}(\varphi))|$ for all $\varphi \in \partial(A)$ and consider the element $b = p(a) \in A$.)

Exercise 3.6.9. Let A be a commutative Banach algebra and $A_e = A \oplus \mathbb{C}e$ the unitisation of A. Show that $\partial(A_e) = \partial(A) \cup \{\varphi_\infty\}$, where $\varphi_\infty(a + \lambda e) = \lambda$ for $a \in A$ and $\lambda \in \mathbb{C}$.

Exercise 3.6.10. Let A be a commutative Banach algebra and I a closed ideal of A. As usual, identify $\Delta(I)$ with $\Delta(A) \setminus h(I)$. Then

$$\partial(I) = \partial(A) \cap \Delta(I).$$

To verify this, observe first that $\partial(A) \cap \Delta(I)$ is a (closed) boundary for the algebra $\hat{I} = \{\hat{y} : y \in I\}$, so that $\partial(I) \subseteq \partial(A) \cap \Delta(I)$. To establish the converse inclusion, exploit Corollary 3.3.4. Let $\varphi \in \partial(A) \cap \Delta(I)$ and let V be an open neighbourhood of φ in $\Delta(I)$. Then there exists $x \in A$ such that

$$1 = \|\hat{x}|_V\|_\infty < \|\hat{x}|_{\Delta(A)\setminus V}\|_\infty.$$

Choose $y \in I$ with $\hat{y}(\varphi) \neq 0$ and show that for sufficiently large $n \in \mathbb{N}$, the element yx^n of I satisfies

$$\|\widehat{yx^n}|_V\|_\infty > \|\widehat{yx^n}|_{\Delta(I)\setminus V}\|_\infty.$$

Exercise 3.6.11. Let I be a closed ideal of a commutative Banach algebra A. In general, there is no relation between $\partial(A/I)$ and $\partial(A) \cap h(I)$. Give an example of a unital A and an ideal I such that $\partial(A) \cap h(I) = \emptyset$.

Exercise 3.6.12. Let A be a commutative Banach algebra and let $\varphi \in \partial(A)$. Suppose that φ is an isolated point of $\partial(A)$. Use Exercise 3.6.10 to show that φ is an isolated point in $\Delta(A)$.

Exercise 3.6.13. Let X denote the solid cylinder $\mathbb{D} \times [0,1] \subseteq \mathbb{C} \times \mathbb{R}$, and let

$$A = \{f \in C(X) : z \to f(z,t) \text{ is holomorphic on } \mathbb{D}^\circ \text{ for each } t \in [0,1]\}.$$

Show that $\partial(A) = \{(z,t) : |z| = 1, 0 \le t \le 1\}$. Find $f \in A$ such that $f(\partial(A)) \not\subseteq \partial(\sigma_A(f))$.

Exercise 3.6.14. Let X be as in the preceding exercise and

$$B = \{f \in C(X) : z \to f(z,1) \text{ is holomorphic on } \mathbb{D}^\circ\}.$$

Then $\partial(B) = X$.

Exercise 3.6.15. Let A be as in Exercise 3.6.13 and let $f \in A$ be the function $f(z,t) = tz, z \in \mathbb{D}, t \in [0,1]$. Prove that $\partial(\sigma_A(f))$ is a proper subset of $\widehat{f}(\partial(A))$. Thus the inclusion in Theorem 3.3.13 may well be proper.

Exercise 3.6.16. Let G be a bounded region in $\mathbb{C}$ whose boundary consists of finitely many simply closed curves. Show that $\partial(A(\overline{G}))$ equals the topological boundary of G.

Exercise 3.6.17. Let A be the closed subalgebra of $C(\mathbb{D})$ generated by $P(\mathbb{D})$ and the function $z \to \overline{z}$. Show that $\partial(A) = \mathbb{D}$, but $\Delta(A) \ne \mathbb{D}$.

Exercise 3.6.18. Show that $\partial(A) = \Delta(A)$ when A is either of the algebras $\text{Lip}_\alpha[0,1]$ or $C^n[0,1]$.

Exercise 3.6.19. As in Exercise 2.12.61, let $A \otimes_u B$ be the uniform tensor product of two uniform algebras A and B. Show that $\partial(A \otimes_u B) = \partial(A) \times \partial(B)$.

Exercise 3.6.20. Let A be a uniform algebra. The set

$$S(A) = \{l \in A^* : \|l\| = l(1) = 1\}$$

is called the set of *states* of A. Then $S(A)$ is a w^*-compact convex subset of A^* and hence, by the Krein-Milman theorem, it is the closed convex hull of the set $\text{ex}(S(A))$ of its extreme points. The elements of $\text{ex}(S(A))$ are called *pure states*. Show that the closure $\overline{\text{ex}(S(A))}$ contains the Shilov boundary $\partial(A)$. Actually, one can prove that $\partial(A) = \overline{\text{ex}(S(A))}$.

Let X be a locally compact Hausdorff space and A a subalgebra of $C_0(X)$ which strongly separates the points of X. A point $x \in X$ is said to be a *peak point* for A if there exists some $f \in A$ such that $|f(x)| = \|f\|_\infty = 1$ and $|f(y)| < 1$ for all $y \neq x$. The set of all peak points for A is called the *Bishop boundary* for A and denoted $\rho(A)$.

Exercise 3.6.21. Let $w \in \mathbb{D}$ and consider the function $f(z) = \frac{1}{2}(z + w), z \in \mathbb{D}$, to show that w is a peak point for the disc algebra $A(\mathbb{D})$ if and only if $|w| = 1$.

Exercise 3.6.22. Let $X = \mathbb{D} \times \mathbb{D} \subseteq \mathbb{C}^2$ and $A = P(X)$. The following assertions (i) and (ii) show that

$$\partial(A) = \rho(A) = \{(z, w) \in X : |z| = |w| = 1\}.$$

(i) Let $f \in A$. For $z \in \mathbb{D}$ define f^z on $\mathbb{D}$ by $f^z(w) = f(z, w)$. Similarly, for $w \in \mathbb{D}$, define f_w on $\mathbb{D}$ by $f_w(z) = f(z, w)$. Observe that $f^z, f_w \in A(\mathbb{D})$ and conclude that $|f|$ attains its maximum on $\mathbb{T} \times \mathbb{T}$.

(ii) Let $(z_0, w_0) \in \mathbb{T} \times \mathbb{T}$. Show that (z_0, w_0) is a peak point for A by considering the function $f(z, w) = \frac{1}{4}(z + z_0)(w + w_0)$, $(z, w) \in X$.

It is worth pointing out that in the example of Exercise 3.6.22 the Shilov boundary is much smaller than the topological boundary of $\Delta(A) = \mathbb{D} \times \mathbb{D}$, which equals $\partial(\mathbb{D}) = (\mathbb{D} \times \mathbb{T}) \cup (\mathbb{T} \times \mathbb{D})$.

Exercise 3.6.23. Let $X = \{(z, t) : z \in \mathbb{D}, t \in [-1, 1]\}$ and let A be the algebra consisting of all continuous functions f on X with the property that $z \to f(z, 0)$ is holomorphic on the open unit disc. Prove that a point $(z, t) \in X$ is a peak point for A if and only if either $t \neq 0$ or $t = 0$ and $|z| = 1$. Note that the Shilov boundary of A equals X.

Exercise 3.6.24. Let $X = \{z = (z_1, \ldots, z_n) \in \mathbb{C}^n : \|z\|^2 = \sum_{j=1}^n |z_j|^2 \leq 1\}$, the closed unit ball in $\mathbb{C}^n$ ($n \in \mathbb{N}$). Show that

$$\rho(P(X)) = \partial(P(X)) = \{z \in X : \|z\| = 1\}.$$

Exercise 3.6.25. Let X be a compact Hausdorff space and $x_0 \in X$, and suppose that x_0 has a countable neighbourhood basis. Prove that x_0 is a peak point for $C(X)$ by finding a continuous function $f : X \to [0, 1]$ such that $f^{-1}(1) = \{x_0\}$.

Exercise 3.6.26. Let $A = \{f \in A(\mathbb{D}) : f(0) = f(1)\} \subseteq A(\mathbb{D})$.

(i) Show that $\Delta(A)$ is homeomorphic to $\mathbb{D} \setminus \{0\}$, where $\mathbb{D} \setminus \{0\}$ is identified, as a set and topologically, with the quotient space $\mathbb{D}/\sim$ of $\mathbb{D}$ which is obtained by the equivalence relation $z \sim w$ if and only if $z = 0$ and $w = 1$ or $z = 1$ and $w = 0$.

(ii) Show that $\partial(A) = \mathbb{T}$ and $\rho(A) = \mathbb{T} \setminus \{1\}$.

Exercise 3.6.27. Consider functions of the form $\delta_0 + z\delta_1 \in l^1(\mathbb{Z})$, $z \in \mathbb{T}$, to show that every point of $\mathbb{T} = \Delta(l^1(\mathbb{Z}))$ is a peak point for $l^1(\mathbb{Z})$. Thus $\partial(l^1(\mathbb{Z})) = \Delta(l^1(\mathbb{Z}))$.
(Remark: Later it follows from regularity of $L^1(G)$ (Theorem 4.4.14) that $\partial(L^1(G)) = \Delta(L^1(G))$ for every locally compact Abelian group.)

Exercise 3.6.28. Prove that in the definition of a topological divisor of zero (Definition 3.4.1) the condition $\|x_n\| = 1$ may be replaced by $\|x_n\| \geq \delta$ for some $\delta > 0$.

Exercise 3.6.29. Let A be a unital, not necessarily commutative Banach algebra which is not isomorphic to the complex number field. Conclude from Corollary 3.4.8 and the Gelfand–Mazur theorem (Theorem 1.2.9) that A has topological divisors of zero.

Exercise 3.6.30. Use Corollary 3.4.8 to show that a nonzero element of the radical of a commutative Banach algebra is a topological divisor of zero.

Exercise 3.6.31. Let A be a commutative Banach algebra and let x be a nonzero element of A which is not a divisor of zero. Prove that x is a topological divisor of zero if and only if $xA \neq \overline{xA}$.
(Hint: Consider the linear mapping $L_x : y \to xy$ of A.)

Exercise 3.6.32. Show that in Exercise 3.6.31, the assumption that x is not a divisor of zero is essential. That is, find an example of a commutative Banach algebra A and a zero divisor x in A such that $xA \neq \overline{xA}$.

Exercise 3.6.33. For $k \in \mathbb{N}$, let $g_k \in L^1(\mathbb{T})$ denote the function $g_k(z) = z^k$. Let $f \in L^1(\mathbb{T})$ and compute $\|g_k * f\|_1$ to conclude that f is a topological zero divisor.

Exercise 3.6.34. Let $f : [0,1] \to \mathbb{C}$ be a continuous function. Show directly, without appealing to Corollary 3.4.5 or Corollary 3.4.6, that f is a topological zero divisor in $C[0,1]$ if and only if f vanishes at some point of $[0,1]$.
(Hint: Suppose that $0 < t_0 < 1$ and that $f(t_0) = 0$. For $k \in \mathbb{N}$, consider the tent function g_k defined by

$$g_k(t) = \begin{cases} 1 + k(t - t_0) & \text{for } t_0 - 1/k \leq t \leq t_0, \\ 1 - k(t - t_0) & \text{for } t_0 \leq t \leq t_0 + 1/k, \\ 0 & \text{for } |t - t_0| \geq 1/k. \end{cases}$$

Show that $\|g_k f\|_\infty \to 0$ as $k \to \infty$.)

Exercise 3.6.35. Let A be a commutative Banach algebra such that $\Delta(A)$ is finite. Show that there exist idempotents $e_1, \ldots, e_n \in A$ with the following properties:
(1) $e_j \neq e_k$ for $j \neq k$, $1 \leq j, k \leq n$.
(2) Every $x \in A$ admits a representation $x = y + \sum_{j=1}^n \lambda_j e_j$, where $\lambda_1, \ldots, \lambda_n \in \mathbb{C}$ and $y \in \text{rad}(A)$.

Exercise 3.6.36. Let A be a semisimple commutative Banach algebra. An idempotent u of A is called minimal if for any other idempotent v of A, either $uv = 0$ or $uv = u$. Show that the following conditions are equivalent.

(i) There exists a minimal idempotent in A.

(ii) There exists a compact and open connected component in $\Delta(A)$.

Exercise 3.6.37. Let A be a semisimple commutative Banach algebra and suppose that $\Delta(A)$ is discrete and that the set of all $a \in A$ such that $\hat{a}$ has finite support is dense in A. Show that the set of all finite linear combinations of idempotents in A is dense in A.

Exercise 3.6.38. Let X be a compact subset of $\mathbb{C}^n$, $n \in \mathbb{N}$. Use the Shilov idempotent theorem to show that if X is connected, then so is the polynomial convex hull of X. Does the converse also hold?

Exercise 3.6.39. Let X be a compact subset of $\mathbb{C}^n$. Suppose that X is polynomially convex and totally disconnected. Use Theorem 3.5.6 to show that $P(X) = C(X)$.

3.7 Notes and references

Most of the material covered in this chapter is the work of Shilov or at least originated from it. The single-variable holomorphic functional calculus, as presented in Section 3.1, essentially amounts to the exploitation of Cauchy's integral formula in one variable to define functions of Banach algebra elements and is due to Gelfand [38]. The much more sophisticated several-variable functional calculus, which of course requires Cauchy's theory for holomorphic functions in several-variables, was developed by Shilov [123] for finitely generated algebras and in the general case by Arens and Calderon [7]. We refer the reader to, for example, [108, Chapter III], [126, Section 8], [99, Section 3.5], and [56]. The beautiful proof of Theorem 3.1.10 given here is taken from [19, Section 20]. Proofs of Oka's extension theorem can be found in [47] and [137].

In Section 3.2 we have put together some of the more immediate applications of the one-variable functional calculus, notably those concerning $G(A)$, the group of invertible elements of a unital commutative Banach algebra A. The realisation of the connected component of the identity of $G(A)$ as $\exp A$ (Theorem 3.2.6) is standard. The fact that $G(A)$ is either connected or has infinitely many connected components (Theorem 3.2.8) was shown by Lorch [82]. Actually, Theorem 3.2.8 can be deduced from the following theorem. The quotient group $G(A)/\exp A$ is isomorphic to $H^1(\Delta(A), \mathbb{Z})$, the first Čech cohomology group of $\Delta(A)$ with integer coefficients. This result, which is one of the most important ones on the subject, is referred to as the Arens-Royden theorem as it was proved, independently, by Arens [6] and Royden [111].

The existence of a smallest closed boundary $\partial(A)$ (Theorem 3.3.2) was established by Shilov (see [41, Section 24]). A proof which does not utilize Zorn's lemma is also available [126, Section 7]. It is worth pointing out that Shilov introduced $\partial(A)$, which is called the Shilov boundary, to determine which elements of $\Delta(A)$ extend to elements of $\Delta(B)$ whenever B is any commutative Banach algebra containing A as a closed subalgebra. Temporarily, call such an element of $\Delta(A)$ extensible. Then Theorem 3.3.9, which is also due to Shilov, combined with Lemma 4.4.5 [88] shows that the Shilov boundary is exactly the set of all extensible elements of $\Delta(A)$.

The notion of a topological divisor of zero was introduced by Shilov [120] who used the terminology generalised divisor of zero and was aware of the connection with the extension problem. However, topological and joint topological divisors of zero have been studied by several authors and appear in various contexts. Theorem 3.4.7, for instance, which identifies the topological divisors of zero of a unital commutative Banach algebra as precisely those elements of A that are not invertible in any superalgebra of A, was shown, among many other results, by Arens [4, 5]. If A is a uniform algebra on a compact Hausdorff space, then $\varphi \in \Delta(A)$ belongs to $\partial(A)$ if and only if $\ker \varphi$ consists of joint topological divisors of zero (Theorem 3.4.10), and we have seen in Theorem 3.4.11 that the 'only if' part holds for arbitrary unital commutative Banach algebras [142].

The Shilov idempotent theorem and its various important consequences presented in Section 3.5, such as Theorem 3.5.5 and Theorem 3.5.9, are all due to Shilov [123]. Unfortunately, it is not known how to prove the idempotent theorem without recourse to the several-variable holomorphic functional calculus, or at least some variant of it. An interesting approach was chosen in [99, Section 3.5], utilising an implicit function theorem which in turn is also based on the heavy machinery of several-variable complex analysis. The reader who is especially interested in the topology of $\Delta(A)$, will appreciate Theorem 3.5.5, which is the ultimate solution to the question of whether compactness of $\Delta(A)$ forces A to be unital.

4

Regularity and Related Properties

The main theme of this chapter is the concept of regularity, which plays a central role in the study of the ideal structure of a commutative Banach algebra. This concept originates from regularity of algebras of functions on locally compact Hausdorff spaces, applied to the range of the Gelfand homomorphism. The relevance of regularity of a commutative Banach algebra A for the ideal theory is mainly due to the fact that it is equivalent to coincidence of the Gelfand topology and the hull-kernel topology on $\Delta(A)$.

Accordingly, we start by introducing the hull-kernel topology in Section 4.1. In Section 4.2 we relate regularity to the hull-kernel topology and present fundamental properties of regular commutative Banach algebras, such as normality and the existence of partitions of unity. In addition, we prove that regularity is inherited by ideals and quotients, by the unitisation and by tensor products. Every commutative Banach algebra A possesses a greatest closed regular subalgebra, $\mathrm{reg}(A)$ (Section 4.3). This is used to show that if I is a closed ideal of A, then A is regular if both I and A/I are regular. As an example, $\mathrm{reg}(C_0(X, A))$ is determined.

In Section 4.4 we establish regularity of $L^1(G)$ for a locally compact Abelian group G. This is one of the most profound results in commutative harmonic analysis, and, as usual, the proof is based on the Plancherel theorem which in turn is a consequence of the inversion formula. To keep our treatment as self-contained as possible, we have included a proof of the inversion formula which utilises the Gelfand theory of commutative C^*-algebras (Section 2.4).

Recently, certain properties weaker than regularity have been investigated. These properties concern questions such as when, for a semisimple commutative Banach algebra A, spectral radii, or spectra of elements of A remain unchanged when embedding A into a larger algebra B, and also the problem of extending elements of $\Delta(A)$ to elements of $\Delta(B)$. In Section 4.5 we discuss most of the relevant results in this context that have been obtained. A related property is the so-called unique uniform norm property which we address in Section 4.6.

E. Kaniuth, *A Course in Commutative Banach Algebras*, Graduate Texts in Mathematics,
DOI 10.1007/978-0-387-72476-8_4, © Springer Science+Business Media, LLC 2009

The final section of this chapter is devoted to the study of Beurling algebras $L^1(G, \omega)$. In contrast to $L^1(G)$, Beurling algebras in general fail to be regular. Somewhat surprisingly, $L^1(G, \omega)$ turns out to be regular if it has the unique uniform norm property. Our main objective, however, is to establish Domar's theorem which asserts that $L^1(G, \omega)$ is regular whenever the weight ω is non-quasianalytic.

4.1 The hull-kernel topology

Let A be a commutative Banach algebra. Recall that there is a bijection between $\Delta(A)$, the set of all homomorphisms of A onto $\mathbb{C}$, and $\mathrm{Max}(A)$, the set of all maximal modular ideals in A, given by $\varphi \to \ker \varphi$. In this way we always identify $\Delta(A)$ and $\mathrm{Max}(A)$. So far we only considered the Gelfand topology on $\Delta(A)$. We now introduce a new topology on $\Delta(A) = \mathrm{Max}(A)$, the so-called hull-kernel topology, which is much more appropriate for studying the ideal structure of A. In general, the hull-kernel topology is weaker than the Gelfand topology, and we show soon (Theorem 4.2.3) that the two topologies coincide if and only if $\Gamma(A) = \{\widehat{x} : x \in A\}$ is a regular algebra of functions on $\Delta(A)$.

Definition 4.1.1. For $E \subseteq \Delta(A) = \mathrm{Max}(A)$ the *kernel* of E, denoted by $k(E)$, is defined as

$$k(E) = \{x \in A : \varphi(x) = 0 \text{ for all } \varphi \in E\} = \bigcap\{M \in \mathrm{Max}(A) : M \in E\}$$

if $E \neq \emptyset$, whereas $k(\emptyset) = A$. For $\varphi \in \Delta(A)$ we write $k(\varphi)$ instead of $k(\{\varphi\}) = \ker \varphi$. If $B \subseteq A$, then the *hull* $h(B)$ of B is defined by

$$h(B) = \{\varphi \in \Delta(A) : B \subseteq k(\varphi)\} = \{M \in \mathrm{Max}(A) : B \subseteq M\}.$$

Also, for $x \in A$, we simply write $h(x)$ instead of $h(\{x\})$.

It is clear that $k(E)$ is a closed ideal in A, and that $h(B)$ is a closed subset of $\Delta(A)$ since the functions $\widehat{x}, x \in A$, are continuous on $\Delta(A)$. We next list some elementary properties of the formation of hulls and kernels.

Lemma 4.1.2. *Let $B, B_1,$ and B_2 be subsets of A and let $E, E_1,$ and E_2 be subsets of $\Delta(A)$. Then*

(i) $B_1 \subseteq B_2 \Longrightarrow h(B_1) \supseteq h(B_2)$.
(ii) $h(\overline{B}) = h(B)$ and $\overline{B} \subseteq k(h(B))$.
(iii) $h(B) = h(k(h(B)))$.
(iv) $E_1 \subseteq E_2 \Longrightarrow k(E_1) \supseteq k(E_2)$.
(v) $E \subseteq h(k(E))$ and $k(E) = k(h(k(E)))$.
(vi) $h(k(E_1 \cup E_2)) = h(k(E_1)) \cup h(k(E_2))$.

Proof. (i), (ii), and (iv) are obvious from the definitions. We show the remaining assertions.

(iii) If $M \in h(k(h(B)))$ then $M \supseteq k(h(B)) \supseteq B$, so that $M \in h(B)$. Conversely, if $\varphi \in \Delta(A)$ is such that $k(\varphi) \not\supseteq h(k(h(B)))$, then $\varphi(a) \neq 0$ for some $a \in k(h(B))$ and hence $\varphi \not\in h(B)$.

(v) $E \subseteq h(k(E))$ is clear, and therefore $k(E) \supseteq k(h(k(E)))$ by (iv). On the other hand, taking $B = k(E)$ in (i), we get $k(E) \subseteq k(h(k(E)))$.

(vi) First, $h(k(E_1)) \cup h(k(E_2)) \subseteq h(k(E_1) \cap k(E_2)) = h(k(E_1 \cup E_2))$. For the converse inclusion, let

$$\varphi \in h(k(E_1 \cup E_2)) = h(k(E_1) \cap k(E_2)) \subseteq h(k(E_1)k(E_2))$$

and assume that $\varphi \not\in h(k(E_2))$. Choose $a \in k(E_2)$ with $\varphi(a) \neq 0$. Then for all $x \in k(E_1)$, $\varphi(x)\varphi(a) = \varphi(xa) = 0$. This implies that $\varphi \in h(k(E_1))$.   □

Definition 4.1.3. Let A be a commutative Banach algebra. For $E \subseteq \Delta(A)$ the *hull-kernel closure* $\overline{E}$ of E is defined to be $\overline{E} = h(k(E))$. The correspondence $E \to \overline{E}$, $E \subseteq \Delta(A)$, is a closure operation, that is, satisfies the following conditions.

(a) $E \subseteq \overline{E}$ and $\overline{\overline{E}} = \overline{E}$.
(b) $\overline{E_1 \cup E_2} = \overline{E_1} \cup \overline{E_2}$.

Indeed, (b) is exactly property (vi) in Lemma 4.1.2, and by (v) $E \subseteq h(k(E)) = \overline{E}$ and

$$\overline{\overline{E}} = h(k(h(k(E)))) = h(k(E)) = \overline{E},$$

so that (a) holds. Thus, there is a unique topology on $\Delta(A)$ such that, for each subset E of $\Delta(A)$, $\overline{E} = h(k(E))$ is the closure of E. This topology is called the *hull-kernel topology* (*hk*-topology).

Example 4.1.4. (1) Let X be a locally compact Hausdorff space. Then the hull-kernel topology on $X = \Delta(C_0(X))$ agrees with the given topology. In fact, given a closed subset E of X and $x_0 \in X \setminus E$, by Urysohn's lemma there exists $f \in C_0(X)$ such that $f(x_0) \neq 0$ and $f(x) = 0$ for all $x \in E$. This shows that $E = h(k(E))$, a hull-kernel closed set.

(2) The hull-kernel topology on $\mathbb{D} = \Delta(A(\mathbb{D}))$ is genuinely weaker than the usual topology on $\mathbb{D}$. This follows from the fact that the set of zeros of a nonzero holomorphic function in a region cannot have an accumulation point within that region. Thus every hull-kernel closed subset of $\mathbb{D}$ has an at most countable intersection with the open unit disc.

We continue with some basic properties of the hull-kernel topology, which are used later.

Lemma 4.1.5. *Let I be a closed ideal of A.*

(i) *Let $q : A \to A/I$ denote the quotient homomorphism. The map $\varphi \to \varphi \circ q$ is a homeomorphism for the hull-kernel topologies between $\Delta(A/I)$ and the closed subset $h(I)$ of $\Delta(A)$.*

(ii) *The map $\varphi \to \varphi|_I$ is a homeomorphism for the hull-kernel topologies between the open subset $\Delta(A) \setminus h(I)$ of $\Delta(A)$ and $\Delta(I)$.*

Proof. (i) The map $\phi : \varphi \to \varphi \circ q$ clearly is a bijection between $\Delta(A/I)$ and $h(I) \subseteq \Delta(A)$. Now, for any subset E of $\Delta(A/I)$ and $\varphi \in \Delta(A/I)$, $\ker \varphi \supseteq \cap \{\ker \psi : \psi \in E\}$ if and only if

$$\ker \phi(\varphi) = q^{-1}(\ker \varphi) \supseteq q^{-1}\left(\bigcap\{\ker \psi : \psi \in E\}\right)$$
$$= \bigcap\{q^{-1}(\ker \psi) : \psi \in E\} = \bigcap\{\ker \phi(\psi) : \psi \in E\}$$
$$= \bigcap\{\ker \tau : \tau \in \phi(E)\}.$$

Thus E is hull-kernel closed in $\Delta(A/I)$ if and only if $\phi(E)$ is hull-kernel closed in $h(I)$.

(ii) We have seen earlier that the map $\phi : \varphi \to \varphi|_I$ is a bijection between $\Delta(A) \setminus h(I)$ and $\Delta(I)$. It is obvious that ϕ is continuous for the hull-kernel topologies. Now let E be a hull-kernel closed subset of $\Delta(A) \setminus h(I)$ and let $\psi \in \Delta(I)$ be such that

$$\ker \psi \supseteq \bigcap\{\ker(\varphi|_I) : \varphi \in E\}.$$

We have to verify that

$$\ker \phi^{-1}(\psi) \supseteq \bigcap\{\ker \phi^{-1}(\varphi|_I) : \varphi \in E\} = \bigcap\{\ker \varphi : \varphi \in E\}.$$

However, if $x \in \ker \varphi$ for all $\varphi \in E$, then $xy \in \ker(\varphi|_I)$ for all $\varphi \in E$ and $y \in I$, whence $\psi(x)\psi(y) = \psi(xy) = 0$ for all $y \in I$. Since $\psi(y) \neq 0$ for some $y \in I$, we get $\psi(x) = 0$. So $x \in \ker \phi^{-1}(\psi)$, and the above inclusion holds. $\square$

Lemma 4.1.6. *Let A be a commutative Banach algebra without identity and let $a \in A$ be such that $\widehat{a}$ is continuous in the hull-kernel topology on $\Delta(A)$. Then $\widehat{a}$ is also continuous on $\Delta(A_e)$ with respect to the hull-kernel topology.*

Proof. Recall that $\Delta(A_e) = \Delta(A) \cup \{\varphi_\infty\}$, where each $\varphi \in \Delta(A)$ is identified with its canonical extension $x + \lambda e \to \varphi(x) + \lambda$, $x \in A$, $\lambda \in \mathbb{C}$. In the sequel we denote by h and k the hull and kernel operations with respect to A and by h_e and k_e those with respect to A_e. For any subset E of $\Delta(A_e)$, it follows immediately from the definitions of hulls and kernels that

$$h_e(k_e(E)) \subseteq h(k(E \cap \Delta(A))) \cup \{\varphi_\infty\}.$$

Let F be a nonempty closed subset of $\mathbb{C}$ and $E = \{\varphi \in \Delta(A_e) : \varphi(a) \in F\}$. By the hypothesis on a, $E \cap \Delta(A)$ is hull-kernel closed in $\Delta(A)$. To show that

E is hull-kernel closed in $\Delta(A_e)$, we have to distinguish the two cases $0 \in F$ and $0 \notin F$. If $0 \in F$ and hence $\varphi_\infty \in E$, the above inclusion gives

$$h_e(k_e(E)) \subseteq (E \cap \Delta(A)) \cup \{\varphi_\infty\} = E,$$

whence E is hull-kernel closed in $\Delta(A_e)$. If $0 \notin F$, then $\varphi_\infty \notin E$ and therefore $E \subseteq \Delta(A)$. Let $\delta = \inf\{|\alpha| : \alpha \in F\}$. Then $\delta > 0$ and $|\varphi(a)| \geq \delta$ for all $\varphi \in E$. In particular, E is compact. Because $E = \Delta(A/k(E))$ and $\widehat{a}(\varphi) \neq 0$ for all $\varphi \in E$, Theorem 3.2.1 applies and yields the existence of some $b \in A$ such that $\varphi(b) = 1/\varphi(a)$ for all $\varphi \in E$. Now let

$$x = e - ab \in A_e.$$

Then x satisfies $\varphi_\infty(x) = 1$ and $\varphi(x) = 0$ for all $\varphi \in E$. So $\varphi_\infty \notin h_e(k_e(E))$, and hence

$$h_e(k_e(E)) \subseteq h(k(E \cap \Delta(A))) = h(k(E)) = E.$$

This shows that E is also hull-kernel closed in $\Delta(A_e)$, as required.   □

Lemma 4.1.7. *Let α be an algebra cross-norm on $A \otimes B$ such that $\alpha \geq \epsilon$. Then the map $\varphi \widehat{\otimes}_\alpha \psi \rightarrow (\varphi, \psi)$ from $\Delta(A \widehat{\otimes}_\alpha B)$ onto $\Delta(A) \times \Delta(B)$ is continuous for the hull-kernel topology on $\Delta(A \widehat{\otimes}_\alpha B)$ and the product of the hull-kernel topologies on $\Delta(A) \times \Delta(B)$.*

Proof. Let E be a hull-kernel closed subset of $\Delta(A)$. We claim that the set $F = \{\varphi \widehat{\otimes}_\alpha \psi : \varphi \in E, \psi \in \Delta(B)\}$ is hull-kernel closed in $\Delta(A \widehat{\otimes}_\alpha B)$. We have $k(F) \supseteq k(E) \widehat{\otimes}_\alpha B$ and hence $\overline{F} = h(k(F)) \subseteq h(k(E) \widehat{\otimes}_\alpha B) \subseteq F$ since $(\varphi \widehat{\otimes}_\alpha \psi)(a \otimes b) = 0$ for all $b \in B$ only when $\varphi \in h(k(E)) = E$.

Thus the projection from $\Delta(A \widehat{\otimes}_\alpha B)$ onto $\Delta(A)$ is hull-kernel continuous, and similarly for $\Delta(A \widehat{\otimes}_\alpha B) \rightarrow \Delta(B)$. The statement of the lemma follows.   □

Conversely, as the following example shows, the map $(\varphi, \psi) \rightarrow \varphi \widehat{\otimes}_\alpha \psi$ is not generally continuous for the hull-kernel topologies.

Example 4.1.8. Let $\mathbb{D}$ denote the closed unit disc, and let $A = A(\mathbb{D})$ and $B = C(\mathbb{D})$. Then the map $\phi : (\varphi, \psi) \rightarrow \varphi \widehat{\otimes}_\pi \psi$ from $\Delta(A) \times \Delta(B)$ onto $\Delta(A \widehat{\otimes}_\pi B)$ fails to be continuous for the hull-kernel topologies. To simplify notation, identify as sets both of $\Delta(A)$ and $\Delta(B)$ with $\mathbb{D}$ and $\Delta(A \widehat{\otimes}_\pi B)$ with $\mathbb{D} \times \mathbb{D}$. Then the diagonal $\Delta = \{(z, z) : z \in \mathbb{D}\}$ is hull-kernel closed in $\Delta(A \widehat{\otimes}_\pi B)$. Indeed, Δ is the zero set of the function $z \otimes 1 - 1 \otimes w \in A \otimes B$, that is, the function $(z, w) \rightarrow z - w$. So $W = \Delta(A \widehat{\otimes}_\pi B) \setminus \Delta$ is hull-kernel open.

Assuming that ϕ is hull-kernel continuous, there exist nonempty hull-kernel open subsets U of $\Delta(A)$ and V of $\Delta(B)$, respectively, such that $U \times V \subseteq \phi^{-1}(W)$. Because the hull-kernel topology on $\Delta(B)$ equals the usual topology on $\mathbb{D}$, V can be taken to be an ordinary open disc contained in $\mathbb{D}$. Now the hull-kernel closed subset $\Delta(A) \setminus U$ of $\Delta(A)$ can contain at most countably many interior points of $\mathbb{D}$ (Example 4.1.4). In particular, $V \cap U$ is a nonempty subset of $\Delta(A)$. It follows that $U \times V$ intersects Δ, a contradiction.

Also, the next two lemmas are used in subsequent sections.

Lemma 4.1.9. *Let I be a closed ideal of the commutative Banach algebra A and let E be an hk-closed subset of $\Delta(A)$ such that $E \cap h(I) = \emptyset$ and $k(E)$ is modular. Then I contains an identity modulo $k(E)$.*

Proof. Because $A/(I + k(E))$ is unital and

$$h(I + k(E)) = h(I) \cap h(k(E)) = h(I) \cap E = \emptyset,$$

it follows that $I + k(E) = A$. Let $u \in A$ be such that $ux - x \in k(E)$ for all $x \in A$. Then $u = v + y$, where $v \in I$ and $y \in k(E)$, and hence $vx - x = ux - x + yx \in k(E)$ for all $x \in A$. $\qquad\square$

Lemma 4.1.10. *Let A be a semisimple commutative Banach algebra with bounded approximate identity and regard A as a closed ideal of its multiplier algebra $M(A)$. Then $\Delta(A)$ is hull-kernel dense in $\Delta(M(A))$.*

Proof. We have to show that $h(k(\Delta(A))) = \Delta(M(A))$. For that, consider an arbitrary $T \in k(\Delta(A))$, so T is a multiplier of A such that $\varphi(T) = 0$ for all $\varphi \in \Delta(A) = \Delta(M(A)) \setminus h(A)$. To prove that $T = 0$, by semisimplicity of A it suffices to show that $\psi(Ta) = 0$ for all $a \in A$ and $\psi \in \Delta(A)$. Let $\varphi \in \Delta(M(A))$ denote the unique extension of ψ. Then

$$\psi(Ta) = \varphi(L_{Ta}) = \varphi(L_a T) = \varphi(L_a)\varphi(T) = 0,$$

as desired. $\qquad\square$

4.2 Regular commutative Banach algebras

Let T be a T_1 topological space and $\mathcal{F}$ a family of complex valued functions on T. Recall from point set topology that $\mathcal{F}$ is said to be *regular* if for any given closed subset E of T and $t \in T \setminus E$, there exists $f \in \mathcal{F}$ with $f(t) \neq 0$ and $f|_E = 0$. This leads to the following definition.

Definition 4.2.1. A commutative Banach algebra A is called *regular* if its algebra of Gelfand transforms is regular in the above sense, that is, given any closed subset E of $\Delta(A)$ and $\varphi_0 \in \Delta(A) \setminus E$, there exists $x \in A$ such that $\varphi_0(x) \neq 0$ and $\varphi(x) = 0$ for all $\varphi \in E$.

Some authors (see [108] and [19], for example) call such Banach algebras *completely regular* rather than *regular*. However, the term regular is more widely used.

Example 4.2.2. (1) Every commutative C^*-algebra A is regular. Indeed, A is isomorphic to $C_0(\Delta(A))$, and Urysohn's lemma ensures that for any locally compact Hausdorff space T, $C_0(T)$ is a regular space of functions.

(2) It is easily seen that $C^n[a,b]$ is regular since, when $\Delta(C^n[a,b])$ is identified with $[a,b]$, the Gelfand homomorphism is nothing but the identity.

(3) The disc algebra $A(\mathbb{D})$ fails to be regular since the Gelfand homomorphism is the identity mapping and a nonzero holomorphic function cannot vanish on, say, a nonempty open set.

It is fairly difficult to prove and is postponed to Section 4.4 that $L^1(G)$, for G a locally compact Abelian group, is regular. In fact, this is one of the most crucial results in commutative harmonic analysis.

We continue to identify $\Delta(A)$ with $\mathrm{Max}(A)$ via the mapping $\varphi \to \ker\varphi$ (Theorem 2.1.7) and proceed with relating regularity of a commutative Banach algebra A to properties of the hull-kernel topology on $\Delta(A)$.

Theorem 4.2.3. *For a commutative Banach algebra A, the following conditions are equivalent.*

(i) *A is regular.*
(ii) *The hull-kernel topology and the Gelfand topology on $\Delta(A)$ coincide.*
(iii) *The hull-kernel topology on $\Delta(A)$ is Hausdorff, and every point in $\Delta(A)$ possesses a hull-kernel neighbourhood with modular kernel.*

Proof. We show the chain of implications (i) $\Rightarrow$ (ii) $\Rightarrow$ (iii) $\Rightarrow$ (i). If I is a closed ideal of A, we consider $\Delta(A/I)$ as embedded into $\Delta(A)$ (Lemma 4.1.5).

Suppose that A is regular and consider a subset E of $\Delta(A)$ that is closed in the Gelfand topology. Then, for every $\varphi \in \Delta(A) \setminus E$, there exists $x_\varphi \in A$ with $\widehat{x_\varphi}|_E = 0$ and $\widehat{x_\varphi}(\varphi) \neq 0$. This means that $k(E) \not\subset \ker\varphi$ for every $\varphi \in \Delta(A) \setminus E$, and hence $E = h(k(E))$, which is an hk-closed set. This proves that the two topologies on $\Delta(A)$ coincide.

To prove (ii) $\Rightarrow$ (iii) we only have to show that every $\varphi_0 \in \Delta(A)$ has a neighbourhood V with modular kernel $k(V)$. Fix $x \in A$ with $\varphi_0(x) \neq 0$ and let

$$V = \left\{ \varphi \in \Delta(A) : |\varphi(x)| > \frac{1}{2}|\varphi_0(x)| \right\}.$$

Then V is open and $\overline{V}$, the closure of V in the Gelfand topology, is contained in the set $\{\varphi \in \Delta(A) : |\varphi(x)| \geq \frac{1}{2}|\varphi_0(x)|\}$. Because $\widehat{x}$ vanishes at infinity, $\overline{V}$ is compact. Now, by hypothesis, $\overline{V} = h(k(V)) = \Delta(A/k(V))$. Thus the semisimple algebra $A/k(V)$ has a compact structure space and $\psi(x+k(V)) \neq 0$ for every $\psi \in \overline{V}$. Corollary 3.2.2 now yields that $A/k(V)$ has an identity.

Finally, suppose that (iii) holds. To show (i), let E be a subset of $\Delta(A)$ which is closed in the Gelfand topology and let $\varphi_0 \in \Delta(A) \setminus E$. Choose an open hull-kernel neighbourhood V of φ_0 with modular kernel $k(V)$. Since $A/k(V)$ has an identity, $h(k(V)) = \Delta(A/k(V))$ is compact with respect to the Gelfand topology, and hence so is $E_0 = E \cap h(k(V))$. Consequently, E_0 is hk-compact. Now, $\varphi_0 \notin E_0$, and the hull-kernel topology is Hausdorff by hypothesis. By the standard covering argument, φ_0 and E_0 can be separated by hk-open sets. Thus, let U be an hk-open set containing E_0 such that $\varphi_0 \notin \overline{U} = h(k(U))$.

Then there exists $y \in A$ such that $\varphi_0(y) \neq 0$, but $\varphi(y) = 0$ for all $\varphi \in \overline{U}$. On the other hand, $\varphi_0 \in V$ and $\Delta(A) \setminus V$ is hk-closed. Hence there exists $z \in k(\Delta(A) \setminus V)$ with $\varphi_0(z) \neq 0$. Let $x = yz$, then $\varphi_0(x) = \varphi_0(y)\varphi_0(z) \neq 0$ and $\varphi(x) = 0$ for all $\varphi \in (\Delta(A) \setminus V) \cup \overline{U}$. Now

$$(\Delta(A) \setminus V) \cup \overline{U} \supseteq [\Delta(A) \setminus h(k(V))] \cup [E \cap h(k(V)) \supseteq E,$$

so that $\widehat{x}|_E = 0$. This shows that A is regular. $\square$

By Lemma 2.2.14, the Gelfand topology on $\Delta(A)$ equals the weak topology with respect to the functions $\widehat{x}$, $x \in A$. Therefore the equivalence of (i) and (ii) in Theorem 4.2.3 can obviously be reformulated as follows.

Corollary 4.2.4. *A is regular if and only if $\widehat{x}$ is hull-kernel continuous on $\Delta(A)$ for each $x \in A$.*

Remark 4.2.5. In [75, Theorem 7.1.2] it is claimed that a commutative Banach algebra is regular provided that the hull-kernel topology on $\Delta(A)$ is Hausdorff. Of course, this is true when A is unital. However, even though we are unaware of a counterexample, this strengthening of the implication (iii) $\Rightarrow$ (i) in Theorem 4.2.3 does not seem to be correct.

In what follows we show that in the definition of regularity the singleton $\{\varphi\}$ can be replaced by any compact subset of $\Delta(A)$ which is disjoint from E, and we investigate how regularity behaves under the standard operations on Banach algebras, such as adjoining an identity and forming closed ideals, quotients, and tensor products.

Theorem 4.2.6. *Let A be a commutative Banach algebra.*

(i) *Let I be closed ideal of A. If A is regular, then so are the algebras I and A/I.*

(ii) *A is regular if and only if A_e, the unitisation of A, is regular.*

Proof. (i) Because A is regular, by Theorem 4.2.3 the Gelfand topology coincides with the hk-topology on $\Delta(A)$. By Lemma 4.1.5(ii), the map $\varphi \to \varphi|_I$ is a homeomorphism for the hk-topologies on $\Delta(A) \setminus h(I)$ and $\Delta(I)$, and the same is true of the Gelfand topologies by Lemma 2.2.15(ii). So the Gelfand topology and the hk-topology on $\Delta(I)$ coincide. Another application of Theorem 4.2.3 now shows that I is regular.

Similarly, using Lemma 4.1.5(i) and Lemma 2.2.15(i), it follows that A/I is regular.

(ii) If A_e is regular, so is A by (i). Conversely, suppose that A is regular. Then, for every $a \in A$, $\widehat{a}$ is hk-continuous on $\Delta(A)$ by Corollary 4.2.4 and hence on $\Delta(A_e)$ by Lemma 4.1.6. This of course implies that $\widehat{x}$ is hk-continuous on $\Delta(A_e)$ for each $x \in A_e$. So A_e is regular by Corollary 4.2.4. $\square$

We show later (Theorem 4.3.8) that conversely A is regular whenever A has a closed ideal I such that both I and A/I are regular. This result is more difficult and involves the existence of a greatest closed regular ideal in a commutative Banach algebra. It is worth mentioning that a closed subalgebra of a regular algebra need not be regular. In fact, $C(\mathbb{D})$ is regular whereas the closed subalgebra $A(\mathbb{D})$ is not.

Lemma 4.2.7. *Let I be an ideal in the regular commutative Banach algebra A. Given any $\varphi_0 \in \Delta(A) \setminus h(I)$, there exists $u \in I$ such that $\widehat{u} = 1$ in some neighbourhood of φ_0.*

Proof. Because A is regular, by Theorem 4.2.3 the hull-kernel topology on $\Delta(A)$ is Hausdorff and φ_0 possesses a neighbourhood with modular kernel. Therefore we can choose a neighbourhood V of φ_0 such that $\overline{V} \cap h(I) = \emptyset$ and $k(V)$ is modular. By Lemma 4.1.9 there exists $u \in I$ such that $\widehat{u}|_V = 1$.   □

The following theorem is one of the most striking results on regular commutative Banach algebras, as becomes apparent in this and several of the subsequent sections.

Theorem 4.2.8. *Let A be a regular commutative Banach algebra, and suppose that I is an ideal in A and K is a compact subset of $\Delta(A)$ with $K \cap h(I) = \emptyset$. Then there exists $x \in I$ such that*

$$\widehat{x}|_K = 1 \ \text{ and } \ \widehat{x} = 0 \ \text{ on some neighbourhood of } h(I).$$

Proof. We first show the existence of some $y \in I$ with $\widehat{y}|_K = 1$. As K is compact, by the preceding lemma there exist open subsets V_i of $\Delta(A)$ and $u_i \in I, 1 \leq i \leq r$, such that $\widehat{u}_i|_{V_i} = 1$ and $K \subseteq \bigcup_{i=1}^{r} V_i$. We inductively define elements y_i of A by $y_1 = u_1$ and

$$y_{i+1} = y_i + u_{i+1} - y_i u_{i+1}, \quad 1 \leq i \leq r-1.$$

It then follows by induction that $y_i \in I$ and $\widehat{y}_j|_{\bigcup_{i=1}^{j} V_i} = 1$. Indeed, if y_j has this latter property, then

$$\varphi(y_{j+1}) = \varphi(y_j) + \varphi(u_{j+1}) - \varphi(y_j)\varphi(u_{j+1})$$
$$= \begin{cases} 1 + \varphi(u_{j+1}) - \varphi(u_{j+1}) & \text{for } \varphi \in \bigcup_{i=1}^{j} V_i \\ \varphi(y_j) + 1 - \varphi(y_j) & \text{for } \varphi \in V_{j+1}, \end{cases}$$

and hence $\widehat{y_{j+1}} = 1$ on $\bigcup_{i=1}^{j+1} V_i$. Now $y = y_r$ has the desired properties.

Choose an open subset V of $\Delta(A)$ with $K \subseteq V$ and $\overline{V} \subseteq \Delta(A) \setminus h(I)$. Because

$$K \cap h(k(\Delta(A) \setminus V)) = K \cap (\Delta(A) \setminus V) = \emptyset,$$

we can apply the arguments of the first part of the proof to K and the ideal $J = k(\Delta(A) \setminus V)$ to obtain $z \in J$ with $\widehat{z}|_K = 1$. By the first part of the proof, there exists $y \in I$ such that $\widehat{y}|_K = 1$. Then the element $x = yz$ of I satisfies $\widehat{x}(\varphi) = 1$ for all $\varphi \in K$ and

$$\operatorname{supp} \widehat{x} \subseteq \operatorname{supp} \widehat{z} \subseteq \overline{V} \subseteq \Delta(A) \setminus h(I),$$

so that $\widehat{x}$ vanishes in a neighbourhood of $h(I)$. $\qquad\square$

We continue with a series of interesting and very useful applications of Theorem 4.2.8.

Corollary 4.2.9. *Every regular commutative Banach algebra A is normal in the sense that whenever $E \subseteq \Delta(A)$ is closed, $K \subseteq \Delta(A)$ is compact and $E \cap K = \emptyset$, then there exists $x \in A$ such that* $\operatorname{supp} \widehat{x} \subseteq \Delta(A) \setminus E$ *and $\widehat{x}|_K = 1$.*

Corollary 4.2.10. *Let A be a regular commutative Banach algebra such that its range under the Gelfand homomorphism $A \to C_0(\Delta(A))$ is closed under complex conjugation. Suppose that K and E are disjoint closed subsets of $\Delta(A)$ with K compact. Then there exists $x \in A$ such that*

$$\widehat{x}|_K = 1, \quad 0 \leq \widehat{x} \leq 1 \quad \text{and} \quad \operatorname{supp} \widehat{x} \subseteq \Delta(A) \setminus E.$$

Proof. By Theorem 4.2.8 there exist $y \in A$ such that $\widehat{y}|_K = 1$ and $\operatorname{supp} \widehat{y} \subseteq \Delta(A) \setminus E$. By hypothesis, there exists $z \in A$ such that $\widehat{z} = \overline{\widehat{y}}$. Let f be the entire function defined by

$$f(w) = \sin^2\left(\frac{\pi}{2}w\right)$$

and let $x = f(yz)$. Then by Theorem 3.1.8,

$$\widehat{x}(\varphi) = \varphi(f(yz)) = f(\varphi(y)\varphi(z)) = \sin^2\left(\frac{\pi}{2}|\varphi(y)|^2\right)$$

for all $\varphi \in \Delta(A)$. Thus

$$\widehat{x}|_K = 1, \ 0 \leq \widehat{x} \leq 1 \text{ and } \operatorname{supp} \widehat{x} \subseteq \Delta(A) \setminus E.$$

$\qquad\square$

Corollary 4.2.11. *Let A be a semisimple regular commutative Banach algebra. If $\Delta(A)$ is compact, then A has an identity.*

Proof. By Theorem 4.2.8 there is $u \in A$ such that $\widehat{u} = 1$ on $\Delta(A)$ and hence $\widehat{x - ux} = 0$ on $\Delta(A)$ for all $x \in A$. A being semisimple, this yields $ux = x$ for all $x \in A$. $\qquad\square$

In Corollary 3.5.5 we have already shown as an application of the Shilov idempotent theorem that the conclusion of Corollary 4.2.11 holds true without assuming that A be regular. However, since the proof of Shilov's idempotent theorem requires a several-variable functional calculus, it appears to be justified to give a simpler proof in the case of a regular semisimple algebra.

In a regular commutative Banach algebra A we can find *partitions of unity* on $\Delta(A)$ subordinate to a given finite open cover of a compact set. Corollary 4.2.9 represents essentially the case $n = 1$ of the following result.

Corollary 4.2.12. *Let A be a regular commutative Banach algebra. Suppose that K is a compact subset of $\Delta(A)$ and $U_1, \ldots, U_n$ are open subsets of $\Delta(A)$ such that $K \subseteq \bigcup_{j=1}^n U_j$. Then there exist $a_1, \ldots, a_n \in A$ with the following properties.*

(i) $(\widehat{a_1} + \ldots + \widehat{a_n})|_K = 1$.
(ii) $\widehat{a_j}|_{\Delta(A) \setminus U_j} = 0$ *for each $j = 1, \ldots, n$.*

Proof. Choose open subsets V_j of $\Delta(A), 1 \leq j \leq n$, such that $\overline{V}_j \subseteq U_j$ and $K \subseteq \bigcup_{j=1}^n V_j$. Let

$$I_j = k(\Delta(A) \setminus V_j), \quad 1 \leq j \leq n, \text{ and } I = I_1 + \ldots + I_n.$$

Then $h(I_j) = \Delta(A) \setminus V_j$ and hence

$$h(I) = \bigcap_{j=1}^n h(I_j) = \bigcap_{j=1}^n (\Delta(A) \setminus V_j) = \Delta(A) \setminus \bigcup_{j=1}^n V_j.$$

Thus $h(I) \cap K = \emptyset$, and Theorem 4.2.8 guarantees the existence of some $a \in I$ with $\widehat{a}|_K = 1$. Write a as $a = a_1 + \ldots + a_n$ where $a_j \in I_j$. Then $a_1, \ldots, a_n$ satisfy (i) and (ii). $\qquad\square$

Corollary 4.2.12 turns out to be a key tool when studying the ideal structure of regular commutative Banach algebras.

The main objective in the remainder of this section is to establish further permanence properties of regularity.

Lemma 4.2.13. *Let A and C be commutative Banach algebras and let $f : \Delta(A) \to \Delta(C)$ be an injective map with the following properties.*

(i) *f is continuous with respect to the hull-kernel topologies.*
(ii) *$f^{-1} : f(\Delta(A)) \to \Delta(A)$ is continuous for the Gelfand topologies.*

Then A is regular whenever C is.

Proof. We remind the reader that a commutative Banach algebra B is regular if and only if the Gelfand topology and the hk-topology on $\Delta(B)$ coincide (Theorem 4.2.3). Let E be a subset of $\Delta(A)$ which is closed in the Gelfand topology. Then $f(E)$ is closed in the Gelfand topology of $f(\Delta(A))$ by condition (ii). So $f(E) = F \cap f(\Delta(A))$ for some subset F of $\Delta(C)$ which is closed in the Gelfand topology of $\Delta(C)$. Because C is regular, F is hk-closed in $\Delta(C)$. It now follows from (i) that $E = f^{-1}(F)$ is hk-closed in $\Delta(A)$. Thus the Gelfand topology on $\Delta(A)$ is coarser than the hk-topology, and hence A is regular. $\qquad\square$

Theorem 4.2.14. *Let $j : A \to B$ be an injective algebra homomorphism between commutative Banach algebras. Suppose that B is regular and that $j(A)$ is an ideal in B. Then A is regular.*

Proof. Let $I = \overline{j(A)}$, which is a closed ideal in B. Since B is regular, so is I by Theorem 4.2.6. Consider the dual mapping

$$j^* : \Delta(I) \to \Delta(A), \quad \psi \to \psi \circ j.$$

Then j^* is injective since $j(A)$ is dense in I. We show that j^* is also surjective. To this end, let $\varphi \in \Delta(A)$ be given and select $a \in A$ such that $\varphi(a) = 1$. Define $\psi : I \to \mathbb{C}$ by

$$\psi(y) = \varphi(j^{-1}(j(a)y)),$$

$y \in I$. Clearly, ψ is linear and $\psi \circ j(x) = \varphi(ax) = \varphi(x)$ for all $x \in A$. Moreover, for $y_1, y_2 \in I$,

$$\begin{aligned}
\psi(y_1 y_2) &= \varphi(a)\varphi(j^{-1}(j(a)y_1 y_2)) \\
&= \varphi(j^{-1}(j(a)y_1 j(a)y_2)) \\
&= \varphi(j^{-1}(j(a)y_1))\varphi(j^{-1}(j(a)y_2)) \\
&= \psi(y_1)\psi(y_2).
\end{aligned}$$

This shows that $\psi \in \Delta(I)$ and $j^*(\psi) = \varphi$.

Thus j^* is a bijection and clearly continuous for the Gelfand topologies. We claim that $(j^*)^{-1}$ is continuous for the hull-kernel topologies. For that we have to show that if F is an hk-closed subset of $\Delta(I)$ and $\varphi \in \Delta(A)$ annihilates $k(j^*(F))$, then $\varphi \in j^*(F)$. Fix $a \in A$ with $\varphi(a) = 1$ and let $y \in k(F) \subseteq I$ be arbitrary. Then

$$\begin{aligned}
(j^*)^{-1}(\varphi)(y) &= (j^*)^{-1}(\varphi)(j(a))(j^*)^{-1}(\varphi)(y) \\
&= (j^*)^{-1}(\varphi)(j(a)y) \\
&= \varphi(j^{-1}(j(a)y)) \\
&= 0,
\end{aligned}$$

because $j^{-1}(j(a)y) \in k(j^*(F))$. Hence $(j^*)^{-1}(\varphi) \in h(k(F)) = F$, whence $\varphi \in j^*(F)$.

An application of Lemma 4.2.13 with $C = I$ and $f = (j^*)^{-1}$ now yields that A is regular.   □

The following lemma is frequently used in Section 4.3.

Lemma 4.2.15. *Let A and B be commutative Banach algebras, and let $\phi : A \to B$ be a homomorphism with dense range. If A is regular, then so is B.*

Proof. We have to show that given a closed subset F of $\Delta(B)$ and $\psi \in \Delta(B) \setminus F$, there exists $b \in B$ such that $\widehat{b} = 0$ on F and $\widehat{b}(\psi) \neq 0$.

Consider the dual mapping $\phi^* : \Delta(B) \to \Delta(A), \psi \to \psi \circ \phi$. We claim that $\phi^*(\psi) \notin \overline{\phi^*(F)}$. Assuming that $\phi^*(\psi) \in \overline{\phi^*(F)}$, we find a net $(\psi_\alpha)_\alpha$ in F such that $\phi^*(\psi_\alpha) \to \phi^*(\psi)$; that is, $\psi_\alpha(\phi(a)) \to \psi(\phi(a))$ for all $a \in A$.

Because $\phi(A)$ is dense in B, this implies that $\psi_\alpha \to \psi$. Thus $\psi \in F$, which is a contradiction.

So $\phi^*(\psi) \notin \overline{\phi^*(F)}$ and since A is regular, there exists $a \in A$ such that $\widehat{a} = 0$ on $\phi^*(F)$ and $\widehat{a}(\phi^*(\psi)) \neq 0$. Now let $b = \phi(a)$. Then $\widehat{b}$ vanishes on F, whereas $\widehat{b}(\psi) \neq 0$. $\qquad\square$

Theorem 4.2.16. *Let A and B be commutative Banach algebras and suppose that A is semisimple and regular. If ϕ is an injective homomorphism from A into B, then*

$$(\Delta(B) \cup \{0\}) \circ \phi = \Delta(A) \cup \{0\}.$$

Proof. Clearly, $(\Delta(B) \cup \{0\}) \circ \phi \subseteq \Delta(A) \cup \{0\}$. Towards a contradiction, assume there exists $\varphi_0 \in \Delta(A) \setminus \Delta(B) \circ \phi$. Then, for any $\psi \in \Delta(B) \cup \{0\}$, there exists $a_\psi \in A$ such that $\psi \circ \phi(a_\psi) \neq \varphi_0(a_\psi)$. Let $\varepsilon_\psi = \frac{1}{2}|\psi \circ \phi(a_\psi) - \varphi_0(a_\psi)|$ and

$$W_\psi = \{\rho \in \Delta(B) \cup \{0\} : |\rho(\phi(a_\psi)) - \varphi_0(a_\psi)| > \varepsilon_\psi\}.$$

Then W_ψ is an open neighbourhood of ψ in $\Delta(B) \cup \{0\} \subseteq B^*$. Since $\Delta(B) \cup \{0\}$ is compact in the w^*-topology, there exist $\psi_1, \ldots, \psi_n \in \Delta(B) \cup \{0\}$ such that

$$\Delta(B) \cup \{0\} = \bigcup_{j=1}^{n} W_{\psi_j}.$$

Let $a_j = a_{\psi_j} (1 \leq j \leq n)$, $\epsilon = \min\{\varepsilon_{\psi_j} : 1 \leq j \leq n\}$, and

$$V = \{\varphi \in \Delta(A) \cup \{0\} : |\varphi(a_j) - \varphi_0(a_j)| < \epsilon \text{ for } 1 \leq j \leq n\}.$$

Then V is an open neighbourhood of φ_0 in $\Delta(A) \cup \{0\}$. Furthermore,

$$V \cap (\Delta(B) \cup \{0\}) \circ \phi = \emptyset.$$

Indeed, for $\varphi \in V$ and $\psi \in W_{\psi_j}$, we have

$$|\psi \circ \phi(a_j) - \varphi(a_j)| \geq |\psi(\phi(a_j)) - \varphi_0(a_j)| - |\varphi_0(a_j) - \varphi(a_j)| > \varepsilon_{\psi_j} - \epsilon \geq 0,$$

so that $V \cap (W_{\psi_j} \circ \phi) = \emptyset$.

Choose an open subset U of $\Delta(A)$ such that $\varphi_0 \in U$, $\overline{U}$ is compact and $\overline{U} \subseteq \Delta(A) \cap V$. Since A is regular, there exists $a_1 \in A$ such that $\varphi_0(a_1) = 1$ and $\varphi(a_1) = 0$ for all $\varphi \in \Delta(A) \setminus U$. Let $I = k(\Delta(A) \setminus V)$. Then $h(I) \cap \overline{U} = \emptyset$ and Theorem 4.2.8 yields the existence of some $a_2 \in A$ such that $\varphi(a_2) = 0$ for all $\varphi \in \Delta(A) \setminus V$ and $\varphi(a_2) = 1$ for all $\varphi \in U$. The element $a_1 a_2$ of A then satisfies $\varphi(a_1 a_2) = \varphi(a_1)$ for all $\varphi \in \Delta(A)$. Because A is semisimple, we conclude that $a_1 a_2 = a_1$.

We next consider $\phi(a_2)$ as an element of B_e and claim that $(1 - \phi(a_2))B_e = B_e$. Assuming that $(1 - \phi(a_2))B_e$ is a proper ideal of B_e, it is contained in some maximal ideal of B_e. Thus there exists $\tau \in \Delta(B_e)$ such that

$$\tau(y) - \tau(y)\tau(\phi(a_2)) = 0$$

for all $y \in B_e$. We have seen above that $V \cap (\Delta(B) \circ \phi) = \emptyset$. Since $\varphi(a_2) = 0$ for all $\varphi \in \Delta(A) \setminus V$, it follows that $\tau(\phi(a_2)) = 0$. This implies $\tau = 0$, which is impossible.

Thus $(1 - \phi(a_2))B_e = B_e$ and hence there exists $b \in B_e$ such that $e = b - b\phi(a_2)$. Since $a_1 a_2 = a_1$, we obtain

$$\phi(a_1) = (b - b\phi(a_2))\phi(a_1) = b\phi(a_1) - b\phi(a_1 a_2) = 0.$$

Since ϕ is injective, $a_1 = 0$. This contradicts $\widehat{a}_1(\varphi_0) = 1$ and hence the existence of some $\varphi_0 \in \Delta(A) \setminus \Delta(B) \circ \phi$.   □

Corollary 4.2.17. *Let B be a commutative Banach algebra and A a subalgebra of B. Suppose that, for some norm, A is a semisimple regular Banach algebra. Then*

(i) *Every element of $\Delta(A)$ extends to some element of $\Delta(B)$.*
(ii) $\sigma_A(x) \cup \{0\} = \sigma_B(x) \cup \{0\}$ *for all $x \in A$.*

Proof. (i) is an immediate consequence of Theorem 4.2.16. To show (ii), we apply Theorem 4.2.16 taking for ϕ the inclusion map $j : A \to B$. It follows that

$$\begin{aligned}
\sigma_A(x) \cup \{0\} &= \widehat{x}(\Delta(A) \cup \{0\}) \\
&= \widehat{x}((\Delta(B) \cup \{0\}) \circ j) \\
&= \widehat{j(x)}(\Delta(B) \cup \{0\}) \\
&= \sigma_B(x) \cup \{0\}
\end{aligned}$$

for every $x \in A$.   □

The preceding corollary is related to concepts which will be studied in Section 4.6. More precisely, let A be any semisimple regular commutative Banach algebra. Then Corollary 4.2.17(i) says that A has the so-called multiplicative Hahn–Banach property.

Corollary 4.2.18. *Let A be a semisimple regular commutative Banach algebra, and let $|\cdot|$ be any algebra norm on A. Then $r_A(x) \leq |x|$ for all $x \in A$.*

Proof. Let B be the completion of A with respect to $|\cdot|$. Then part (ii) of Corollary 4.2.17 implies that

$$r_A(x) = \sup\{|\lambda| : \lambda \in \sigma_A(x)\} = \sup\{|\lambda| : \lambda \in \sigma_B(x)\} \leq |x|$$

for all $x \in A$.   □

We conclude this section by characterizing regularity of tensor products $A \widehat{\otimes}_\alpha B$ through regularity of A and B.

Lemma 4.2.19. *Let A and B be commutative Banach algebras and let α be an algebra cross-norm on $A \otimes B$ such that $\alpha \geq \epsilon$. Then $A \widehat{\otimes}_\alpha B$ is regular whenever both A and B are regular.*

Proof. We identify $\Delta(A) \times \Delta(B)$ and $\Delta(A \widehat{\otimes}_\alpha B)$ as topological spaces by means of the map $(\varphi, \psi) \to \varphi \widehat{\otimes}_\alpha \psi$ (Theorem 2.11.2). Let E be a closed subset of $\Delta(A \widehat{\otimes}_\alpha B)$ and let $\varphi_0 \in \Delta(A)$ and $\psi_0 \in \Delta(B)$ be such that $(\varphi_0, \psi_0) \notin E$. There exist open neighbourhoods U of φ_0 in $\Delta(A)$ and V of ψ_0 in $\Delta(B)$ such that $(U \times V) \cap E = \emptyset$. Because A and B are regular, there exist $a \in A$ and $b \in B$ such that $\widehat{a}(\varphi_0) \neq 0$, $\widehat{a} = 0$ on $\Delta(A) \setminus U$, $\widehat{b}(\psi_0) \neq 0$ and $\widehat{b} = 0$ on $\Delta(B) \setminus V$. Then the element $a \otimes b$ satisfies $\widehat{a \otimes b}(\varphi_0, \psi_0) \neq 0$ and $\widehat{a \otimes b}$ vanishes on

$$((\Delta(A) \setminus U) \times \Delta(B)) \cup (\Delta(A) \times (\Delta(B) \setminus V))$$

and hence on E. □

Theorem 4.2.20. *Let A and B be commutative Banach algebras and let α be a cross-norm on $A \otimes B$ which dominates ϵ. Then the tensor product $A \widehat{\otimes}_\alpha B$ is regular if and only if both A and B are regular.*

Proof. By Lemma 4.2.19, $A \widehat{\otimes}_\alpha B$ is regular whenever A and B are regular. So suppose that conversely $A \widehat{\otimes}_\alpha B$ is regular. To see that A is regular, by Corollary 4.2.4 it suffices to show that, for each $a \in A$, the function $\varphi \to \varphi(a)$ is hk-continuous on $\Delta(A)$.

Select $\psi \in \Delta(B)$ and let $\phi_\psi : A \widehat{\otimes}_\alpha B \to A$ be the continuous homomorphism satisfying $\phi_\psi(a \otimes b) = \psi(b)a$ for all $a \in A$ and $b \in B$ (Lemma 2.11.5). The kernel I of ϕ_ψ is a closed ideal, and the dual map $\phi_\psi^* : \varphi \to \varphi \circ \phi_\psi$ is a bijection between $\Delta(A)$ and $\Delta((A \widehat{\otimes}_\alpha B)/I)$. It actually is a homeomorphism for the hull-kernel topologies by Lemma 4.1.5. Note that $\varphi \circ \phi_\psi = \varphi \widehat{\otimes}_\alpha \psi$ and choose $b \in B$ with $\psi(b) = 1$. Then

$$\varphi(a) = (\varphi \widehat{\otimes}_\alpha \psi)(a \otimes b) = \phi_\psi^*(\varphi)(a \otimes b)$$

for all $a \in A$. Now, $A \widehat{\otimes}_\alpha B$ is regular and therefore the function $\varphi \widehat{\otimes}_\alpha \psi \to (\varphi \widehat{\otimes}_\alpha \psi)(a \otimes b)$ is hk-continuous on $\Delta(A \widehat{\otimes}_\alpha B)$ (Corollary 4.2.4). Since ϕ_ψ^* is hk-continuous, it follows that $\varphi \to \varphi(a)$ is hk-continuous on $\Delta(A)$. □

4.3 The greatest regular subalgebra

The first purpose of this section is to establish the existence of a closed regular subalgebra of a commutative Banach algebra A which contains all closed regular subalgebras of A.

Lemma 4.3.1. *Let A be a commutative Banach algebra and B a closed subalgebra of A. If B is regular, then for every $b \in B$ the Gelfand transform $\widehat{b}$ is continuous on $\Delta(A)$ with respect to the hull-kernel topology.*

Proof. Suppose first that A has an identity e and that $e \in B$. Let $r : \Delta(A) \to \Delta(B)$ denote the restriction map $\varphi \to \varphi|_B$. Because B is regular, the Gelfand transform of $b \in B$ on $\Delta(B)$ is hk-continuous by Corollary 4.2.4. It therefore suffices to show that r is continuous for the hull-kernel topologies on $\Delta(A)$ and $\Delta(B)$. To see this, let F be a hk-closed subset of $\Delta(B)$. Then

$$F = \{\psi \in \Delta(B) : \psi(k(F)) = 0\}$$

and hence

$$r^{-1}(F) = \{\varphi \in \Delta(A) : \varphi(k(F)) = 0\},$$

which is hk-closed in $\Delta(A)$.

In the general situation, consider A_e and the subalgebra $B_e = B + \mathbb{C}e$. Since B is regular, so is B_e (Theorem 4.2.6). By the first paragraph, for $b \in B$, $\hat{b}$ is hull-kernel continuous on $\Delta(A_e)$ and hence on $\Delta(A)$.   □

Theorem 4.3.2. *Let A be a commutative Banach algebra. Then A contains a greatest closed regular subalgebra, denoted* reg(A).

Proof. Let reg(A) be the closed subalgebra of A generated by the collection $\mathcal{B}$ of all closed regular subalgebras B of A. We have to show that reg(A) is regular.

Let $B \in \mathcal{B}$ and $b \in B$. Then, by the preceding lemma, $\hat{b}$ is hk-continuous on $\Delta(\text{reg}(A))$. Thus Gelfand transforms of products $b_1 b_2 \cdot \ldots \cdot b_m$, $b_j \in B_j \in \mathcal{B}$, $1 \leq j \leq m$, and hence of finite linear combinations of such products are also hk-continuous on $\Delta(\text{reg}(A))$. The elements of this form are dense in reg(A). Therefore, for each $a \in \text{reg}(A)$, the Gelfand transform $\hat{a}$ is a uniform limit on $\Delta(\text{reg}(A))$ of hk-continuous functions, hence itself is hk-continuous.

It now follows from Corollary 4.2.4 that reg(A) is regular.   □

Remark 4.3.3. The greatest regular subalgebra of a unital commutative Banach algebra may well be trivial. For instance, this happens with the disc algebra $A(\mathbb{D})$. To see this, let $f \in A(\mathbb{D})$ and suppose that f is hull-kernel continuous. Then, for each $\varepsilon > 0$, the set $C_\varepsilon = \{z \in \mathbb{D} : |f(z) - f(0)| \leq \varepsilon\}$ is hk-closed in $\mathbb{D}$. It follows that $C_\varepsilon \cap \{z \in \mathbb{C} : |z| \leq r\}$ is finite for each $0 < r < 1$. This of course forces f to be constant with value $f(0)$ on $\{z \in \mathbb{C} : |z| \leq r\}$. Thus reg$(A(\mathbb{D}))$ consists only of the constant functions.

The same arguments as in the proof of Theorem 4.3.2 show that A possesses a largest closed regular ideal. More precisely, let regid(A) be the closed subalgebra of A generated by all the closed regular ideals of A. Then regid(A) is regular and it is a closed ideal which contains every closed regular ideal of A.

Lemma 4.3.4. *There exists a largest closed regular ideal* regid(A) *of A, and for every $x \in A$, $\hat{x}$ is hull-kernel continuous on the open subset $\Delta(\text{regid}(A))$ of $\Delta(A)$.*

Proof. It only remains to show the second statement. Let $J = \mathrm{regid}(A)$ and $x \in A$. Let φ_0 be an arbitrary element of $\Delta(J)$ and choose $y \in J$ such that $\varphi_0(y) \neq 0$. Then $\widehat{y} \neq 0$ in a neighbourhood V of φ_0 and $xy \in J$. Since J is regular both $\widehat{xy}$ and $\widehat{y}$ are hk-continuous functions on $\Delta(J)$. Now $\widehat{x}(\varphi) = \widehat{xy}(\varphi)\widehat{y}(\varphi)^{-1}$ for all $\varphi \in V$ and hence $\widehat{x}$ is hk-continuous at φ_0. □

Lemma 4.3.5. *Let A be a commutative Banach algebra, J the greatest regular ideal of A and suppose that A/J is regular. Then the hull $h(J)$ has empty interior in $\Delta(A)$.*

Proof. Assume that there exists a nonempty open subset U of $\Delta(A)$ which is contained in $h(J) = \Delta(A/J)$. Because A/J is regular, U is hk-open in $h(J)$. Let W be an hk-open subset of $\Delta(A)$ such that $W \cap h(J) = U$, and let $K = k(\Delta(A) \setminus W)$. Then $\Delta(K) = W$ and hence $\Delta(K) \cap h(J) = U$, and $V = \Delta(K) \cap \Delta(J)$ is hk-open in $\Delta(K)$. Then $\Delta(K) = U \cup V$ and U and V are both hk-open in $\Delta(K)$.

Now, for every $x \in A$, $\widehat{x}$ is hk-continuous on U since A/J is regular. By Lemma 4.3.4, $\widehat{x}$ is also hk-continuous on V. In particular, $\widehat{x}$ is hk-continuous on $\Delta(K)$ for every $x \in K$. Thus K is a regular ideal of A by Corollary 4.2.4. But K is not contained in J since $U \neq \emptyset$ and $U \subseteq h(J)$. This contradiction shows that $h(J)$ has an empty interior. □

Corollary 4.3.6. *Let A and J be as in Lemma 4.3.5 and let E be a closed subset of $\Delta(A)$ such that $k(E) = \{0\}$. Then $E = \Delta(A)$.*

Proof. By Lemma 4.3.5, $\Delta(J)$ is dense in $\Delta(A)$. It is therefore enough to show that $\Delta(J) \subseteq E$. Assume that $F = E \cap \Delta(J)$ is a proper subset of $\Delta(J)$. Then, since J is regular, there exists a nonzero element x of J such that $\widehat{x} = 0$ on F. It follows that $\widehat{x}$ vanishes on all of E. This contradicts $k(E) = \{0\}$ and shows that $E \supseteq \Delta(J)$. □

Lemma 4.3.7. *Let J be the greatest closed regular ideal of A and suppose that A/J is regular. Let I be an arbitrary closed ideal of A. Then there exists a closed ideal K of A/I such that both K and $(A/I)/K$ are regular.*

Proof. Let $q : A \rightarrow A/I$ denote the quotient homomorphism. Since J is regular, Lemma 4.2.15 implies that $K = \overline{q(J)}$ is a regular ideal of A/I. Yet, $(A/I)/K$ is also regular. Indeed, since

$$(A/I)/K = (A/I)/(q^{-1}(K)/I) = A/q^{-1}(K)$$

and since $q^{-1}(K)$ contains J, $(A/I)/K$ is a quotient algebra of A/J. Because A/J is regular by hypothesis, it follows from Theorem 4.2.6 that $(A/I)/K$ is regular. □

Recall that if A is regular and I is a closed ideal of A, then both I and A/I are regular (Theorem 4.2.6). We are now ready to prove the converse.

Theorem 4.3.8. *Let A be a commutative Banach algebra and suppose that A has a closed ideal I such that both I and A/I are regular. Then A is regular.* +

Proof. Let J be the largest regular closed ideal of A. Then $J \supseteq I$ and since A/I is regular, it follows that A/J is regular as well. So A satisfies the hypotheses of Lemma 4.3.7.

Let E be any closed subset of $\Delta(A)$. We have to show that E is hk-closed in $\Delta(A)$. To that end, let $B = A/k(E)$, let $q : A \to B$ denote the quotient homomorphism and consider the dual mapping

$$q^* : \Delta(B) \to h(k(E)) \subseteq \Delta(A), \quad \psi \to \psi \circ q.$$

Let $F = q^*1 - 1(E) \subseteq \Delta(B)$. We claim that $k(F) = \{0\}$.

For that, let $x \in A$ be such that $q(x) \in k(F)$ and let $\varphi \in E$. Then $\varphi = \psi \circ q$ for some $\psi \in F$ and hence $\varphi(x) = \psi(q(x)) = 0$. Thus $x \in k(E)$ and therefore $q(x) = 0$. Applying Lemma 4.3.7 with $I = k(E)$, we see that $B = A/k(E)$ and $F \subseteq \Delta(B)$ satisfy the hypotheses of Corollary 4.3.6. It follows that $F = \Delta(B)$, and this implies

$$E = q^*(q^{*-1}(E)) = q^*(F) = q^*(\Delta(B)) = h(k(E)),$$

so that E is hk-closed in $\Delta(A)$.   □

Corollary 4.3.9. *Suppose that A possesses a sequence $(I_j)_{j \in \mathbb{N}}$ of closed subalgebras with the following properties.*

(i) *I_j is an ideal in I_{j+1} for each $j \in \mathbb{N}$ and $\bigcup_{j=1}^{\infty} I_j$ is dense in A.*
(ii) *I_1 and I_{j+1}/I_j are regular for each $j \in \mathbb{N}$.*

Then A is regular.

Proof. Applying Theorem 4.3.8 and induction, it follows from the hypotheses that I_j is regular for every j. Thus $\bigcup_{j=1}^{\infty} I_j \subseteq \text{reg}(A)$, and since $\bigcup_{j=1}^{\infty} I_j$ is dense in A, we obtain that $\text{reg}(A) = A$ (Lemma 4.2.15).   □

As an important example, we proceed to determine $\text{reg}(C_0(X, A))$, where X is a locally compact Hausdorff space and A any commutative Banach algebra. Although the result turns out to be what one would expect, it is highly non-trivial to achieve.

Lemma 4.3.10. *Let A be a commutative Banach algebra and let B be a Banach algebra consisting of A-valued functions on a set X with pointwise operations. Let*

$$R = \{f \in B : f(X) \subseteq \text{reg}(A)\},$$

and suppose that R is closed in B and R is regular. Then $\text{reg}(B) = R$.

Proof. For each $x \in X$, consider the algebra homomorphism

$$\phi_x : \mathrm{reg}(B) \to A, \quad f \to f(x).$$

Lemma 4.2.15 yields that $\overline{\phi_x(\mathrm{reg}(B))}$ is a regular subalgebra of A and hence contained in $\mathrm{reg}(A)$. Thus, for every $f \in \mathrm{reg}(B)$, we have $f(x) \in \mathrm{reg}(A)$ for all $x \in X$. Consequently, $\mathrm{reg}(B) \subseteq R$. By hypothesis, R is closed and regular. Thus it follows that $\mathrm{reg}(B) = R$. $\qquad\square$

Theorem 4.3.11. *Let X be a locally compact Hausdorff space and A a commutative Banach algebra. Then*

$$\mathrm{reg}(C_0(X, A)) = C_0(X, \mathrm{reg}(A)).$$

In particular, $C_0(X, A)$ is regular if and only if A is regular.

Proof. Because $C_0(X, \mathrm{reg}(A))$ is a closed subalgebra of $C_0(X, A)$, by Lemma 4.3.10 it suffices to show that $C_0(X, \mathrm{reg}(A))$ is regular. We know from Lemma 4.2.19 that the projective tensor product of the two regular algebras $C_0(X)$ and $\mathrm{reg}(A)$ is regular. Therefore, we establish a homomorphism

$$\phi : C_0(X) \widehat{\otimes}_\pi \mathrm{reg}(A) \to C_0(X, \mathrm{reg}(A))$$

with dense range. Regularity of $C_0(X, \mathrm{reg}(A))$ then follows from Lemma 4.2.15.

For $f \in C_0(X)$ and $a \in A$, let $fa \in C_0(X, A)$ be defined by $fa(x) = f(x)a$ for all $x \in X$. Then the map $(f, a) \to fa$ is bilinear and maps $C_0(X) \times \mathrm{reg}(A)$ into $C_0(X, \mathrm{reg}(A))$. Hence there is a unique linear map

$$\phi : C_0(X) \otimes \mathrm{reg}(A) \to C_0(X, \mathrm{reg}(A))$$

such that $\phi(f \otimes a) = fa$ for all $f \in C_0(X)$ and $a \in \mathrm{reg}(A)$. For $f, g \in C_0(X)$ and $a, b \in A$, we have

$$\phi((f \otimes a)(g \otimes b)) = \phi(fg \otimes ab) = (fg)(ab) = (fa)(gb) = \phi(f \otimes a)\phi(g \otimes b).$$

So ϕ is a homomorphism.

We verify next that ϕ is continuous with respect to the projective tensor norm on $C_0(X) \otimes \mathrm{reg}(A)$. For $f_1, \ldots, f_n \in C_0(X)$ and $a_1, \ldots, a_n \in \mathrm{reg}(A)$, we have

$$\left\| \phi\left(\sum_{j=1}^n f_j \otimes a_j \right) \right\| = \left\| \sum_{j=1}^n f_j a_j \right\| \le \sum_{j=1}^n \|f_j\|_\infty \|a_j\|.$$

It follows that $\|\phi(u)\| \le \pi(u)$ for every $u \in C_0(X) \otimes \mathrm{reg}(A)$. Hence ϕ is continuous and extends uniquely to a homomorphism

$$\widetilde{\phi} : C_0(X) \widehat{\otimes}_\pi \mathrm{reg}(A) \to C_0(X, \mathrm{reg}(A)).$$

It remains to show that the range of $\tilde{\phi}$ is dense in $C_0(X, \mathrm{reg}(A))$. To that end, let $F \in C_0(X, \mathrm{reg}(A))$ and $\epsilon > 0$ be given. There exists a compact subset K of X such that $\|F(x)\| \leq \epsilon/2$ for every $x \in X \setminus K$. For each $x \in K$, let

$$U_x = \{y \in X : \|F(y) - F(x)\| < \epsilon/2\}.$$

The sets $U_x, x \in K$, form an open cover of the compact set K. Hence there exist $x_1, \ldots, x_n \in K$ such that $K \subseteq \bigcup_{j=1}^n U_{x_j}$. Because X is a locally compact Hausdorff space we can find a partition of unity subordinate to this finite open cover of K. This means that there exist non-negative continuous functions $f_1, \ldots, f_n$ on X with compact support such that $\mathrm{supp}\, f_j \subseteq U_{x_j}$ for $j = 1, \ldots, n$ and

$$\left(\sum_{j=1}^n f_j\right)(X) \subseteq [0,1] \quad \text{and} \quad \left(\sum_{j=1}^n f_j\right)\bigg|_K = 1.$$

Now let

$$u = \sum_{j=1}^n f_j \otimes F(x_j) \in C_0(X) \otimes \mathrm{reg}(A).$$

Then, for every $y \in X$,

$$\|F(y) - \phi(u)(y)\| = \left\|F(y) - \sum_{j=1}^n f_j(y) F(x_j)\right\|$$

$$\leq \sum_{j=1}^n f_j(y)\|F(y) - F(x_j)\| + \left(1 - \sum_{j=1}^n f_j(y)\right)\|F(y)\|.$$

It follows that, if $y \notin U_{x_j}$ for all $j = 1, \ldots, n$, then

$$\|F(y) - \phi(u)(y)\| \leq \|F(y)\| < \varepsilon/2,$$

whereas if $y \in U_{x_k}$ for at least one k, then

$$\|F(y) - \phi(u)(y)\| \leq \sum_{\substack{k=1 \\ y \in U_{x_k}}}^n f_k(y)\|F(y) - F(x_k)\|$$

$$+ \left(1 - \sum_{j=1}^n f_j(y)\right)\|F(y)\|$$

$$< \frac{\epsilon}{2} + \left(1 - \sum_{j=1}^n f_j(y)\right)\|F(y)\|.$$

Hence $\|F(y) - \phi(u)(y)\| < \epsilon/2$ for $y \in K$, and if $y \in U_{x_k} \setminus K$ for some k, then

$$\|F(y) - \phi(u)(y)\| < \frac{\epsilon}{2} + \left(1 - \sum_{j=1}^n f_j(y)\right)\frac{\epsilon}{2} \leq \epsilon.$$

Thus $\|F(y) - \phi(u)(y)\| \leq \epsilon$ for all $y \in X$. This shows that $\phi(C_0(X) \otimes \mathrm{reg}(A))$ is dense in $C_0(X, \mathrm{reg}(A))$, which finishes the proof of the theorem. $\square$

4.4 Regularity of $L^1(G)$

The most important examples of regular commutative Banach algebras are the L^1-algebras of locally compact Abelian groups. However, to prove regularity of $L^1(G)$ is rather difficult because one has to appeal to another fundamental theorem in harmonic analysis, the Plancherel theorem. We have chosen an approach to the inversion formula and Plancherel's theorem which utilizes Gelfand's theory of commutative C^*-algebras (Section 2.4) and is therefore much closer to the general theme of this book than are other proofs.

Definition 4.4.1. Let G be a locally compact Abelian group and $C^b(G)$ the space of all bounded continuous functions on G endowed with the supremum norm $\|\cdot\|_\infty$. Let $\lambda : f \to \lambda_f$ denote the regular representation of $L^1(G)$ on $L^2(G)$ as introduced in Section 2.7 and recall that the group C^*-algebra, $C^*(G)$, is the closure of the $\lambda(L^1(G))$ in $\mathcal{B}(L^2(G))$.

Let $C^\infty(G)$ be the set of all $f \in C^b(G)$ such that there exist $T \in C^*(G)$ and a sequence $(f_n)_n$ in $L^1(G) \cap C^b(G)$ with the following properties.

(i) $\lambda_{f_n} \to T$ in $\mathcal{B}(L^2(G))$ as $n \to \infty$.
(ii) $\|f_n - f\|_\infty \to 0$ as $n \to \infty$.

Lemma 4.4.2. *Given $f \in C^\infty(G)$, the operator $T \in C^*(G)$ in Definition 4.4.1 is unique and is denoted T_f.*

Proof. Let $T, S \in C^*(G)$ and suppose that $(f_n)_n$ and $(g_n)_n$ are sequences in $L^1(G) \cap C^b(G)$ such that, as $n \to \infty$,

$$\lambda_{f_n} \to T, \quad \lambda_{g_n} \to S \text{ and } \|f_n - f\|_\infty \to 0, \quad \|g_n - f\|_\infty \to 0.$$

Let $Q = T - S$. Then, for arbitrary $g \in C_c(G)$,

$$\|(f_n - g_n) * g\|_\infty \leq \|f_n - g_n\|_\infty \|g\|_1 \leq \|g\|_1(\|f_n - f\|_\infty + \|g_n - f\|_\infty) \to 0$$

and

$$\|(f_n - g_n) * g - Q(g)\|_2 \leq \|(\lambda_{f_n} - T)(g)\|_2 + \|(\lambda_{g_n} - S)(g)\|_2 \to 0.$$

In particular, for every compact subset K of G, $((f_n - g_n)*g)|_K \to 0$ uniformly and $((f_n - g_n) * g)|_K \to Q(g)|_K$ in $L^2(K)$. Both facts together imply that $Q(g)|_K = 0$ in $L^2(K)$. This holds for all compact subsets K of G. Thus it follows that $Q(g) = 0$. Since $C_c(G)$ is dense in $L^2(G)$, we conclude that $Q = 0$. $\qquad\square$

Remark 4.4.3. Let $f \in L^1(G) \cap C^\infty(G)$. Then, taking $f_n = f$ for all $n \in \mathbb{N}$, we see that $T_f = \lambda_f$. Hence the three Gelfand transforms $\widehat{T}_f$, $\widehat{\lambda}_f$, and $\widehat{f}$ coincide on $\widehat{G}$. Thus, defining $\widehat{f} = \widehat{T}_f$ for $f \in C^\infty(G)$, the assignment $f \to \widehat{f}$ coincides on $L^1(G) \cap C^\infty(G)$ with the Gelfand transformation of $L^1(G)$. In addition, we have $\|\widehat{f}\|_\infty = \|\widehat{T}_f\|_\infty = \|T_f\|$.

In preparation for the inversion formula and the Plancherel theorem we have to provide a series of technical lemmas.

Lemma 4.4.4. *The map $f \to T_f$ from $C^\infty(G)$ into $C^*(G)$ is linear and injective.*

Proof. Linearity of the map is obvious. Thus it remains to show that $f = 0$ whenever $T_f = 0$. So suppose there exists a sequence $(f_n)_n$ in $L^1(G) \cap C^b(G)$ such that $\lambda_{f_n} \to 0$ and $\|f_n - f\|_\infty \to 0$. Then, for any $g \in C_c(G)$, $\|f_n * g\|_2 \to 0$ and

$$\|f_n * g - f * g\|_\infty \leq \|f_n - f\|_\infty \|g\|_1 \to 0.$$

As in the proof of Lemma 4.4.2, it follows that $f * g = 0$ in $L^2(G)$. However, since $f * g$ is continuous, we get that $\int_G f(x)g(x)dx = 0$. Finally, since $C_c(G)$ is dense in $L^1(G)$ and $f \in L^\infty(G) = L^1(G)^*$, it follows that $f = 0$. $\square$

Lemma 4.4.5. *Let $S \in C^*(G)$ and $g, h \in C_c(G)$. Then*

(i) $g * S(h) \in C^\infty(G)$ and $T_{g*S(h)} = S\lambda_{g*h}$.
(ii) $S(g) * S(g)^* \in C^\infty(G)$ and $T_{S(g)*S(g)^*} = SS^*\lambda_{g*g^*}$.

Proof. (i) We know that $g * S(h) \in C_0(G) \subseteq C^b(G)$. Let $(f_n)_n \subseteq C_c(G)$ be such that $\lambda_{f_n} \to S$ in $C^*(G)$. Then, for every $x \in G$,

$$
\begin{aligned}
|(f_n * g * h)(x) - (g * S(h))(x)| &\leq \int_G |g(y)| \cdot |\lambda_{f_n}(h)(y^{-1}x) - S(h)(y^{-1}x)|dy \\
&\leq \|g\|_2 \cdot \|R_x(\lambda_{f_n}(h)) - R_x(S(h))\|_2 \\
&\leq \|\lambda_{f_n} - S\| \cdot \|h\|_2 \|g\|_2,
\end{aligned}
$$

which tends to zero as $n \to \infty$. Moreover, for each $u \in C_c(G)$,

$$
\begin{aligned}
\|\lambda_{f_n*g*h}(u) - S\lambda_{g*h}\|_2 &\leq \|\lambda_{f_n} - S\| \cdot \|g * h * u\|_2 \\
&\leq \|\lambda_{f_n} - S\| \cdot \|u\|_2 \|g * h\|_1,
\end{aligned}
$$

and hence

$$\|\lambda_{f_n*g*h} - S\lambda_{g*h}\| \leq \|\lambda_{f_n} - S\| \cdot \|g * h\|_1 \to 0.$$

This shows that $g * S(h) \in C^\infty(G)$ and $T_{g*S(h)} = S\lambda_{g*h}$.

(ii) Clearly, $S(g) * S(g)^* \in C_0(G) \subseteq C^b(G)$. Let $(f_n)_n$ be a sequence in $C_c(G)$ with $\lambda_{f_n} \to S$. Then

$$\|\lambda_{f_n*f_n^**g*g^*} - SS^*\lambda_{g*g^*}\| = \|\lambda_{f_n}\lambda_{f_n}^*\lambda_{g*g^*} - SS^*\lambda_{g*g^*}\|,$$

which converges to zero as $n \to \infty$. Moreover, we have $f_n * f_n^* * g * g^* \in L^1(G) \cap C^b(G)$ and

$$
\begin{aligned}
\|f_n * f_n^* * g * g^* - S(g) * S(g)^*\|_\infty &= \|\lambda_{f_n}(g) * \lambda_{f_n}(g)^* - S(g) * S(g)^*\|_\infty \\
&\leq \|\lambda_{f_n}(g)\|_2 \|\lambda_{f_n}(g)^* - S(g)^*\|_2 \\
&\quad + \|S(g)^*\|_2 \|\lambda_{f_n}(g) - S(g)\|_2 \\
&\leq \|\lambda_{f_n} - S\| \cdot \|g\|_2(\|\lambda_{f_n}(g)\|_2 + \|S(g)\|_2),
\end{aligned}
$$

which also converges to 0 as $n \to \infty$. This proves (ii). $\square$

Lemma 4.4.6. *Let $f \in C^\infty(G)$, $x \in G$ and $\alpha \in \widehat{G}$. Then*

(i) $f^* \in C^\infty(G)$ *and* $\widehat{f^*} = \overline{\widehat{f}}$.
(ii) $L_x f \in C^\infty(G)$ *and* $\widehat{L_x f}(\alpha) = \overline{\alpha(x)}\widehat{f}(\alpha)$.
(iii) $\alpha f \in C^\infty(G)$ *and* $\widehat{\alpha f} = L_\alpha \widehat{f}$.

Proof. Let $(f_n)_n \subseteq L^1(G) \cap C^b(G)$ and $T \in C^*(G)$ such that $\lambda_{f_n} \to T$ and $\|f_n - f\|_\infty \to 0$.

(i) Because $\|f_n^* - f^*\|_\infty \to 0$ and $\lambda_{f_n^*} - T^* = (\lambda_{f_n} - T)^* \to 0$, it follows that $f^* \in C^\infty(G)$ and $T_{f^*} = T^* = T_f^*$.

(ii) Let $S \in \mathcal{B}(L^2(G))$ be defined by $S(g) = T(L_x g)$, $g \in L^2(G)$. Then, for all $g \in C_c(G)$, $\|L_x f_n - L_x f\|_\infty \to 0$ and

$$\|\lambda_{L_x f_n}(g) - S(g)\|_2 = \|f_n * L_x(g) - T(L_x(g))\| \leq \|\lambda_{f_n} - T\| \cdot \|g\|_2.$$

This shows that $\|\lambda_{L_x f_n} - S\| \to 0$ and hence $S \in C^*(G)$, $L_x f \in C^\infty(G)$ and $S = T_{L_x f}$. Thus $\widehat{L_x f} = \widehat{S}$. On the other hand, $\widehat{L_x h}(\alpha) = \overline{\alpha(x)}\widehat{h}(\alpha)$ for every $h \in L^1(G)$ and hence

$$\widehat{S}(\alpha) = \lim_{n \to \infty} \widehat{L_x f_n}(\alpha) = \overline{\alpha(x)} \lim_{n \to \infty} \widehat{f_n}(\alpha) = \overline{\alpha(x)}\widehat{T}(\alpha).$$

This shows that $\widehat{f}(\alpha)\overline{\alpha(x)} = \widehat{L_x f}(\alpha)$.

(iii) Define $S \in \mathcal{B}(L^2(G))$ by

$$S(g)(x) = \alpha(x)S(\overline{\alpha}g)(x), \quad g \in L^2(G), x \quad \in G.$$

Since $\|\alpha f_n - \alpha f\|_\infty \to 0$ and

$$\lambda_{\alpha h}(g)(x) = ((\alpha h) * g)(x) = \alpha(x)(h * (\overline{\alpha}g))(x) = \alpha(x)\lambda_h(\overline{\alpha}g)(x)$$

for $h \in L^1(G), g \in C_c(G)$ and $x \in G$, we get that $\lambda_{\alpha f_n} \to S$. Consequently, $S \in C^*(G), \alpha f \in C^\infty(G)$, and $S = T_{\alpha f}$. Finally, by Lemma 2.7.3,

$$\widehat{\alpha f} = \widehat{S} = \lim_{n \to \infty} \widehat{\alpha f_n} = \lim_{n \to \infty} L_\alpha \widehat{f_n} = L_\alpha \widehat{T} = L_\alpha \widehat{f},$$

as was to be shown. $\square$

Lemma 4.4.7. *Let $f \in C^\infty(G)$. If $\widehat{f}$ is real valued then $f(e) \in \mathbb{R}$, and if $\widehat{f} \geq 0$ then $f(e) \geq 0$.*

Proof. Suppose that $\widehat{f}$ is real valued and let $T = T_f$. The Gelfand homomorphism of $L^1(G)$ preserves involution and is injective. Hence we have $\widehat{T^*} = \overline{\widehat{T}} = \overline{\widehat{f}} = \widehat{f} = \widehat{T}$ and hence $T^* = T$. Now, let $f_n \in L^1(G) \cap C^b(G)$, $n \in \mathbb{N}$, be such that $\|f_n - f\|_\infty \to 0$ and $\lambda_{f_n} \to T$. Then, setting $g_n = (f_n + f_n^*)/2 \in L^1(G) \cap C^b(G)$,

$$\left\| g_n - \frac{1}{2}(f + f^*) \right\|_\infty \to 0 \text{ and } \lambda_{g_n} \to \frac{1}{2}(T + T^*) = T.$$

This implies $(f + f^*)/2 \in C^\infty(G)$ and $T_{(f+f^*)/2} = T_f$, whence $f = f^*$ and, in particular, $f(e) = \overline{f(e)}$.

Now let $\widehat{f} \geq 0$. Then $f(e) \in \mathbb{R}$ by what we have shown in the preceding paragraph. Towards a contradiction, assume that $f(e) < 0$ and choose symmetric neighbourhoods V and W of e in G such that $W^2 \subseteq V$ and $\mathrm{Re} f(x) < 0$ for all $x \in V$. In addition, choose $g \in C_c^+(G)$, $g \neq 0$, with $\mathrm{supp}\, g \subseteq W$. We want to compute $(f * g * g^*)(e)$. Note that if $\int_G g(y)g^*(y^{-1}x^{-1})dy \neq 0$ for some x, then $y^{-1}x^{-1} \in \mathrm{supp}\, g^* \subseteq W$ for some $y \in \mathrm{supp}\, g \subseteq W$, so that $x \in V$. Thus

$$(f * g * g^*)(e) = \int_V \mathrm{Re} f(x)(g * g^*)(x^{-1})dx + i \int_V \mathrm{Im} f(x)(g * g^*)(x^{-1})dx,$$

which fails to be ≥ 0 since $\mathrm{Re} f(x) < 0$ for all $x \in V$.

To reach a contradiction, we show that $(f * g * g^*)(e) \geq 0$. There exists $S = S^* \in C^*(G)$ such that $S^2 = T$. Indeed, $C^*(G)$ is isomorphic to $C_0(\widehat{G})$ and $\widehat{T}$ is positive, so that there exists $S \in C^*(G)$ such that $\widehat{S} = (\widehat{T})^{1/2}$. Choose $(f_n)_n \subseteq L^1(G) \cap C^b(G)$ such that $\lambda_{f_n} \to T$ and $\|f_n - f\|_\infty \to 0$ and $(g_n)_n \subseteq L^1(G)$ so that $\lambda_{g_n} \to S$. Then

$$\|T - \lambda_{g_n * g_n^*}\| \leq \|S^2 - \lambda_{g_n^*} S\| + \|S\lambda_{g_n^*} - \lambda_{g_n}\lambda_{g_n^*}\|$$
$$\leq \|S\| \cdot \|S - \lambda_{g_n^*}\| + \|\lambda_{g_n}\| \cdot \|\lambda_{g_n} - S\|,$$

which converges to 0 as $n \to \infty$. Since $\lambda_{f_n} \to T$ we get $\lambda_{f_n} - \lambda_{g_n * g_n^*} \to 0$, and hence, using $\|f_n - f\|_\infty \to 0$,

$$(f * g * g^*)(e) = \int_G f(x)(g * g^*)(x^{-1})dx = \lim_{n\to\infty}(f_n * g * g^*)(e)$$
$$= \lim_{n\to\infty}\langle \lambda_{f_n}(g), g \rangle = \lim_{n\to\infty}\langle \lambda_{g_n * g_n^*}(g), g \rangle$$
$$= \lim_{n\to\infty}\int_G (g_n * g)(x)(g_n * g)^*(x^{-1})dx$$
$$= \lim_{n\to\infty}\|\lambda_{g_n}(g)\|_2^2 = \|S(g)\|_2^2$$
$$\geq 0.$$

This contradiction shows that $f(e) \geq 0$. $\square$

Lemma 4.4.8. *Let* $\xi \in C_c^{\mathbb{R}}(\widehat{G})$ *and* $\epsilon > 0$. *Then there exist functions* $f_1, f_2 \in C^\infty(G)$ *such that* $\widehat{f_1}, \widehat{f_2} \in C_c^{\mathbb{R}}(\widehat{G}), \widehat{f_1} \geq \xi \geq \widehat{f_2}$ *and* $f_1(e) - f_2(e) \leq \varepsilon$.

Proof. To start with, let K be any compact subset of $\widehat{G}$ and $\eta > 0$. Then, given $\alpha \in K$, there exist neighbourhoods U_α of α and V_α of e such that $|\beta(x) - 1| < \eta$ for all $x \in V_\alpha$ and $\beta \in U_\alpha$ (Lemma 2.7.4). Since K is compact, the usual covering argument shows that there exists a neighbourhood V of e in G such that $|\alpha(x) - 1| < \eta$ for all $\alpha \in K$ and $x \in V$. Then, if $f \in C_c^+(G)$ is such that $\operatorname{supp} f \subseteq V$ and $\widehat{f}(1_G) = 1$, we have

$$\left| \widehat{f}(\alpha) - 1 \right| = \left| \int_G f(x)(\alpha(x) - 1)dx \right| \leq \sup_{x \in V} |\alpha(x) - 1| \cdot \int_G f(y)dy \leq \eta$$

for all $\alpha \in K$.

Let now $K = \operatorname{supp} \xi$ and $\eta > 0$. Then, by the preceding paragraph, there exists a symmetric neighbourhood V of e such that $|\widehat{f}(\alpha) - 1| \leq \eta$ for all $f \in C_c^+(G)$ with $\|f\|_1 = 1$ and $\operatorname{supp} f \subseteq V$. For all such f, it follows that

$$\left| \widehat{f * f^*}(\alpha) - 1 \right| \leq |\widehat{f}(\alpha)\overline{\widehat{f}(\alpha)} - \widehat{f}(\alpha)| + |\widehat{f}(\alpha) - 1| \leq 2|\widehat{f}(\alpha) - 1| \leq 2\eta.$$

This in turn implies the following facts.

(1) Given $\delta > 0$, there exists a function g_δ of the form $g_\delta = f_\delta * f_\delta^*$ with $f_\delta \in C_c^+(G)$ such that $1 + \delta \geq \widehat{g}_\delta(\alpha) \geq 1 - \delta$ for all $\alpha \in K$.
(2) There exists g of the form $g = f * f^*$, where $f \in C_c^+(G)$, such that $\widehat{g}(\alpha) \geq 1$ for all $\alpha \in K$.

Let $T \in C^*(G)$ with $\widehat{T} = \xi$ and define $T_1, T_2 \in C^*(G)$ by

$$T_1 = T\lambda_{g_\delta + \delta g} \text{ and } T_2 = T\lambda_{g_\delta - \delta g}.$$

From (1) and (2) we get

$$\widehat{T_1} = \xi (g_\delta + \delta g)^\wedge \geq \xi(1 - \delta + \delta \widehat{g}) \geq \xi \geq \widehat{T}(1 + \delta - \delta \widehat{g}) \geq \widehat{T_2}.$$

Let $C_\infty^*(G) = \{T_u : u \in C^\infty(G)\} \subseteq C^*(G)$. Both g and g_δ are of the form $h * h^*$ with $h \in C_c(G)$. So Lemma 4.4.5 shows that

$$T_1, T_2 \in C_\infty^*(G) \text{ and } T\lambda_g \in C_\infty^*(G).$$

Thus there are f, f_1, and f_2 in $C^\infty(G)$ such that $\widehat{f} = \widehat{T\lambda_g}$ and $\widehat{f_j} = \widehat{T_j}, j = 1, 2$. Then $\widehat{f} = \widehat{T}\widehat{\lambda_g} = \xi \widehat{h * h^*}$ is real valued and hence $f(e) \in \mathbb{R}$ by Lemma 4.4.7. Moreover

$$\widehat{f_1} - \widehat{f_2} = 2\delta \widehat{T\lambda_g} = 2\delta \widehat{f}.$$

The injectivity of the map $h \to T_h$ from $C^\infty(G)$ into $C^*(G)$ (Lemma 4.4.4) implies that $f_1 - f_2 = 2\delta f$. In particular $f_1(e) - f_2(e) = 2\delta f(e)$. Because $f(e)$ is real and since the definition of g and hence $T\lambda_g$ and f do not depend on δ, it follows that $f_1(e) - f_2(e) < \epsilon$ for suitably chosen $\delta > 0$. This completes the proof of the lemma.   □

Now we are ready to prove the inversion formula.

Theorem 4.4.9. *(Inversion formula) With suitable normalisation of the Haar measure on $\widehat{G}$, for all $f \in C^\infty(G)$ such that $\widehat{f}$ has compact support and all $x \in G$, we have*

$$f(x) = \int_{\widehat{G}} \widehat{f}(\alpha)\alpha(x)d\alpha.$$

Proof. We are going to define a Haar integral on $C_c(\widehat{G})$. Let $\xi \in C_c^{\mathbb{R}}(\widehat{G})$. Then, by Lemma 4.4.8

$$\sup\{f(e) : f \in C^\infty(G), \widehat{f} \in C_c(\widehat{G}), \widehat{f} \leq \xi\}$$

$$= \inf\{f(e) : f \in C^\infty(G), \widehat{f} \in C_c(\widehat{G}), \widehat{f} \geq \xi\}.$$

Denote this real number by $I(\xi)$. Then $I : \xi \to I(\xi)$ is a real linear functional on $C_c^{\mathbb{R}}(\widehat{G})$. To see this, let $\xi, \eta \in C_c^{\mathbb{R}}(\widehat{G})$ and $\lambda \in \mathbb{R}$. Using the supremum in the above equation gives $I(\xi) + I(\eta) \leq I(\xi + \eta)$, and using the infimum shows $I(\xi + \eta) \leq I(\xi) + I(\eta)$. Clearly, $I(\lambda\xi) = \lambda I(\xi)$ if $\lambda \geq 0$. Since

$$I(-\xi) = \sup\{f(e) : f \in C^\infty(G), \widehat{f} \in C_c(\widehat{G}), -\widehat{f} \geq \xi\}$$

$$= -\inf\{(-f)(e) : f \in C^\infty(G), \widehat{f} \in C_c(\widehat{G}), -\widehat{f} \geq \xi\}$$

$$= -I(\xi),$$

it follows that $I(\lambda\xi) = \lambda I(\xi)$ for all $\lambda \in \mathbb{R}$.

The functional I is positive, since if $\xi \in C_c^+(\widehat{G})$, then

$$I(\xi) = \inf\{f(e) : f \in C^\infty(G), \widehat{f} \in C_c(\widehat{G}), \widehat{f} \geq \xi\}$$

and $\widehat{f} \geq 0$ ensures that $f(e) \geq 0$ by Lemma 4.4.7. Thus I extends uniquely to a positive linear functional on $C_c(\widehat{G})$.

We next observe that I is nontrivial. For that, choose $T \in C^*(G)$ and $g \in C_c(G)$ such that $\widehat{T} \in C_c(\widehat{G})$ and $T(g) \neq 0$. Setting $h = T(g)$, Lemma 4.4.5 implies that $h * h^* \in C^\infty(G)$ and

$$\widehat{h * h^*} = \widehat{TT^*}\widehat{\lambda_{g*g^*}} = |\widehat{T}|^2 |\widehat{g}|^2 \in C_c(\widehat{G}),$$

whence $I(\widehat{h * h^*}) = h * h^*(e) = \|h\|_2^2 > 0$.

It remains to verify that I is translation invariant. Note first that, by Lemma 4.4.6(iii), for $\alpha \in \widehat{G}$ and $f \in C^\infty(G)$ with $\widehat{f} \in C_c(\widehat{G})$,

$$I(L_\alpha\widehat{f}) = I(\widehat{\alpha f}) = (\alpha f)(e) = f(e) = I(\widehat{f}).$$

Using this and Lemma 4.4.6(iii) again, we obtain, for arbitrary $\xi \in C_c^{\mathbb{R}}(\widehat{G})$,

$$I(L_{\overline{\alpha}}\xi) = \sup\{I(\widehat{g}) : g \in C^\infty(G), \widehat{g} \in C_c(\widehat{G}), \widehat{g} \geq L_{\overline{\alpha}}\xi\}$$

$$= \sup\{I(L_\alpha\widehat{g}) : g \in C^\infty(G), \widehat{g} \in C_c(\widehat{G}), L_\alpha\widehat{g} \geq \xi\}$$

$$= \sup\{I(\widehat{\alpha g}) : \alpha g \in C^\infty(G), \widehat{\alpha g} = L_\alpha\widehat{g} \in C_c(\widehat{G}), \widehat{\alpha g} \geq \xi\}$$

$$= \sup\{I(\widehat{f}) : f \in C^\infty(G), \widehat{f} \in C_c(G), \widehat{f} \geq \xi\}$$

$$= I(\xi).$$

Thus, for a suitably normalized Haar measure $d\alpha$ on $\widehat{G}$, we have

$$I(\xi) = \int_{\widehat{G}} \xi(\alpha) d\alpha$$

for all $\xi \in C_c(\widehat{G})$. In particular, if $f \in C^{\infty}(G)$ is such that $\widehat{f} \in C_c(\widehat{G})$, then by Lemma 4.4.6(ii),

$$f(x) = I(\widehat{L_{x^{-1}}f}) = \int_{\widehat{G}} \widehat{L_{x^{-1}}f}(\alpha) d\alpha = \int_{\widehat{G}} \alpha(x)\widehat{f}(\alpha) d\alpha$$

for all $x \in G$.   □

Let $g \in C_c(G)$ and $\xi \in L^2(G)$. Then $g * \xi \in C_0(G) \cap L^2(G)$. The bounded linear transformations $\xi \rightarrow g * \xi$ and $\xi \rightarrow \lambda_g(\xi)$ of $L^2(G)$ agree on $C_c(G)$. Thus, taking $\xi = T(f)$ where $f \in C_c(G)$ and $T \in C^*(G)$, we obtain that

$$g * T(f) = \lambda_g(T(f)) = T(g * f).$$

This simple fact is used when we deduce the Plancherel theorem from the inversion formula and some of the above lemmas.

Theorem 4.4.10. *(Plancherel theorem) Let G be a locally compact Abelian group and let the Haar measure on $\widehat{G}$ be normalised so that the inversion formula holds. Let $E = C^{\infty}(G) \cap L^2(G)$. Then*

(i) *E is a dense linear subspace of $L^2(G)$.*
(ii) *The set $\widehat{E} = \{\widehat{f} : f \in E\}$ is a dense linear subspace of $L^2(\widehat{G})$.*
(iii) *The mapping $f \rightarrow \widehat{f}$ from E to $\widehat{E}$ is isometric and extends uniquely to a Hilbert space isomorphism from $L^2(G)$ onto $L^2(\widehat{G})$.*

Proof. (i) Clearly, E is a linear subspace of $L^2(G)$. To prove that E is dense in $L^2(G)$, let F denote the set of all functions of the form $g * T(h) = T(g * h)$, where $g, h \in C_c(G)$ and $T \in C^*(G)$ is such that $\widehat{T} \in C_c(\widehat{G})$. By Lemma 4.4.5(i), $F \subseteq C^{\infty}(G)$ and hence $F \subseteq E$. It therefore suffices to show that F is dense in $L^2(G)$.

To that end, let $h \in C_c(G), h \neq 0$, and $\epsilon > 0$ be given. Denote by $|X|$ the Haar measure of a measurable subset X of G, and choose an open neighbourhood U of e in G such that $|U \cdot \operatorname{supp} h| \leq 2 |\operatorname{supp} h|$. Since h is uniformly continuous, we find a symmetric open neighbourhood W of e contained in U such that

$$|h(yx) - h(x)| \leq \epsilon (2|\operatorname{supp} h|)^{-1/2}$$

for all $x \in G$ and $y \in W$. For all $x \in G$ and $u \in C_c^+(G)$ such that $\operatorname{supp} u \subseteq W$ and $\|u\|_1 = 1$, it then follows that

$$|(u * h)(x) - h(x)| \leq \int_G |u(y)| \cdot |h(y^{-1}x) - h(x)| dx \leq \epsilon(2|\operatorname{supp} h|)^{-1/2}$$

and hence, by the choice of U, W, and u,

$$\|u * h - h\|_2^2 = \int_{W \cdot \text{supp } h} |(u * h)(x) - h(x)|^2 dx \leq \epsilon^2.$$

Now let V be a symmetric neighbourhood of e such that $V^2 \subseteq W$ and let $f \in C_c^+(G)$ such that $\|f\|_1 = 1$ and $\text{supp } f \subseteq V$. Then $\text{supp}(f * f) \subseteq W$ and $\|f * f\|_1 = 1$. There exists $T \in C^*(G)$ with $\widehat{T} \in C_c(\widehat{G})$ and $\|T(h) - \lambda_f(h)\|_2 \leq \epsilon$. Summarizing, we have $f * T(h) = T(f * h) \in F$ and

$$
\begin{aligned}
\|f * T(h) - h\|_2 &\leq \|f * T(h) - f * \lambda_f(h)\|_2 + \|(f * f) * h - h\|_2 \\
&\leq \|f\|_1 \cdot \|T(h) - \lambda_f(h)\|_2 + \epsilon \\
&\leq 2\epsilon.
\end{aligned}
$$

This proves that F is dense in $L^2(G)$.

To establish (ii), let $\xi \in L^2(\widehat{G})$ and $\epsilon > 0$ be given. Since $C^*(G)^\wedge \supseteq C_c(\widehat{G})$ and $C_c(\widehat{G})$ is dense in $L^2(\widehat{G})$, there exists $T \in C^*(G)$ such that $\widehat{T} \in C_c(\widehat{G})$ and $\|\widehat{T} - \xi\|_2 \leq \epsilon$. Arguing as in the proof of Lemma 4.4.8, there exists $h \in C_c(G)$ such that $f = h * h^*$ satisfies

$$|\widehat{f}(\alpha) - 1| \leq \epsilon (\|\widehat{T}\|_\infty^2 |\text{supp } \widehat{T}|)^{-1/2}$$

for all $\alpha \in \text{supp } \widehat{T}$. Then $T(f) \in E$ and, by Lemma 4.4.5(i),

$$
\begin{aligned}
\widehat{T}\widehat{f} = \widehat{T\lambda_{h*h^*}} &= \widehat{T}\lambda_{h*h^*} = \widehat{T_{h*T(h^*)}} \\
&= \widehat{h * T(h^*)} = \lambda_h(\widehat{T(h^*)}) = \widehat{T(h * h^*)} \\
&= \widehat{T(f)}.
\end{aligned}
$$

It follows that

$$
\begin{aligned}
\|\widehat{T} - \widehat{T(f)}\|_2^2 &= \int_{\widehat{G}} |\widehat{T}(\alpha)|^2 |\widehat{f}(\alpha) - 1|^2 d\alpha \\
&= \int_{\text{supp } \widehat{T}} |\widehat{T}(\alpha)|^2 |\widehat{f}(\alpha) - 1|^2 d\alpha \\
&\leq \|\widehat{T}\|_\infty^2 |\text{supp } \widehat{T}| \epsilon^2 (\|\widehat{T}\|_\infty^2 |\text{supp } \widehat{T}|)^{-1} \\
&= \epsilon^2
\end{aligned}
$$

and hence $\|\xi - \widehat{T(f)}\|_2 \leq \|\xi - \widehat{T}\|_2 + \|\widehat{T} - \widehat{T(f)}\|_2 \leq 2\epsilon$.

Because F and $\widehat{E}$ are dense in $L^2(G)$ and $L^2(\widehat{G})$, respectively, for (iii) it only remains to show that the mapping $f \to \widehat{f}, F \to \widehat{E}$ is isometric. If $f \in F$, say $f = S(g * h)$, where $g, h \in C_c(G)$, and $S \in C^*(G)$ such that $\widehat{S} \in C_c(\widehat{G})$, then $f * f^* \in C^\infty(G)$ and

$$\widehat{f * f^*} = \widehat{S}\,\overline{\widehat{S}}((g * h) * (g * h)^*)^\wedge \in C_c(\widehat{G})$$

by Lemma 4.4.5(ii). So for each $f \in F$, $f * f^*$ satisfies the hypotheses of Theorem 4.4.9. Using Lemma 4.4.5, it follows that

$$\|f\|_2^2 = \int_G f(x)f^*(x^{-1})dx = (f * f^*)(e)$$

$$= \int_{\widehat{G}} \widehat{f * f^*}(\alpha)d\alpha = \int_{\widehat{G}} (S\lambda_{g*h}(S\lambda_{g*h})^*)^\wedge(\alpha)d\alpha$$

$$= \int_{\widehat{G}} |\widehat{T}_{S(g*h)}(\alpha)|^2 d\alpha = \int_{\widehat{G}} |\widehat{f}(\alpha)|^2 d\alpha$$

$$= \|\widehat{f}\|_2^2.$$

Thus $f \to \widehat{f}$ is isometric on F, as required. $\qquad\square$

Definition 4.4.11. The unique extension of the mapping $f \to \widehat{f}$ from E to all of $L^2(G)$, also denoted $f \to \widehat{f}$, is called the *Plancherel transformation* and $\widehat{f}$, for $f \in L^2(G)$, is called the *Plancherel transform* of f. By Theorem 4.4.10, $f \to \widehat{f}$ is a Hilbert space isomorphism from $L^2(G)$ onto $L^2(\widehat{G})$.

The Plancherel formula and linearisation imply

Corollary 4.4.12. *For $f, g \in L^2(G)$ the Parseval identity*

$$\int_G f(x)\overline{g(x)}dx = \int_{\widehat{G}} \widehat{f}(\alpha)\overline{\widehat{g}(\alpha)}d\alpha$$

holds.

Corollary 4.4.13. *For $f \in L^2(G)$ and $\alpha \in \widehat{G}$,*

$$\widehat{f^*} = \overline{\widehat{f}}, \ \widehat{\overline{f}} = (\widehat{f})^* \ \text{and} \ \widehat{\alpha f} = L_\alpha \widehat{f}.$$

Proof. It is sufficient to check all three equations for functions f in a dense linear subspace of $L^2(G)$. Now, the first and the third equations hold in $C^\infty(G)$, and hence in E, by (i) and (iii) of Lemma 4.4.6, respectively. As to the second, it is enough to verify that if $f \in C^\infty(G)$ then $\overline{f} \in C^\infty(G)$ and $\widehat{T}_{\overline{f}}(\beta) = \overline{\widehat{T}_f(\beta^{-1})}$ for all $\beta \in \widehat{G}$. However, this follows immediately from the corresponding equation for functions f in $L^1(G)$. $\qquad\square$

Now regularity of $L^1(G)$, our main goal in this section, follows quickly.

Theorem 4.4.14. *Let G be a locally compact Abelian group. Then $L^1(G)$ is regular.*

Proof. Let E be a closed subset of $\widehat{G}$ and $\alpha \in \widehat{G} \backslash E$. We have to find $f \in L^1(G)$ such that $\widehat{f}(\alpha) \neq 0$ and $\widehat{f}|_E = 0$. Choose a neighbourhood U of α in $\widehat{G}$ and a symmetric neighbourhood V of 1_G in $\widehat{G}$ such that $UV \cap E = \emptyset$.

By the Plancherel theorem we find functions u and v in $L^2(G)$ with the following properties.

(1) $\widehat{u} \in C_c^+(\widehat{G})$, supp $\widehat{u} \subseteq U$ and $\widehat{u}(\alpha) \neq 0$.
(2) $\widehat{v} \in C_c^+(\widehat{G})$, supp $\widehat{v} \subseteq V$ and $\widehat{v}(1_G) \neq 0$.

Then $f = uv \in L^1(G)$ and, by Corollaries 4.4.12 and 4.4.13,

$$
\begin{aligned}
\widehat{f}(\beta) &= \int_G u(x)v(x)\overline{\beta(x)}dx = \int_G u(x)\,\overline{\beta\overline{v}(x)} \\
&= \int_{\widehat{G}} \widehat{u}(\gamma)\,\overline{(\beta\overline{v})(\gamma)}d\gamma = \int_{\widehat{G}} \widehat{u}(\gamma)\,\overline{L_\beta\widehat{\overline{v}}(\gamma)} \\
&= \int_{\widehat{G}} \widehat{u}(\gamma)\,\overline{\widehat{\overline{v}}(\overline{\beta}\gamma)} = \int_{\widehat{G}} \widehat{u}(\gamma)\,\widehat{v}(\overline{\gamma}\beta)d\gamma \\
&= (\widehat{u} * \widehat{v})(\beta)
\end{aligned}
$$

for all $\beta \in \widehat{G}$. Now (1) and (2) imply that $\operatorname{supp}(\widehat{u} * \widehat{v}) \subseteq UV \subseteq \widehat{G} \setminus E$ and $(\widehat{u} * \widehat{v})(\alpha) > 0$. Thus f has the desired properties.   □

In Theorem 4.3.11 we have shown that if X is a locally compact Hausdorff space and A is a commutative Banach algebra, then $\operatorname{reg}(C_0(X, A)) = C_0(X, \operatorname{reg}(A))$. Using regularity of $L^1(G)$, the analogous result can be established for $L^1(G, A)$. The proof, however, is somewhat more technical. The reader is invited to carry out two of the main steps in the exercises (Exercises 4.8.21, 4.8.22 and 4.8.23).

We conclude this section with another consequence of regularity of $L^1(G)$.

Remark 4.4.15. We already know that if G is a compact Abelian group, then $\widehat{G}$ is discrete. Conversely, suppose that G is a locally compact Abelian group and that $\widehat{G}$ is discrete. By Theorem 4.4.14 there exists $f \in L^1(G)$ such that $\widehat{f}(\alpha) = 0$ for $\alpha \neq 1_G$ and $\widehat{f}(1_G) = 1$. Then $f = f * f^*$ and hence $f \in L^2(G)$ and f can be assumed to be continuous. The Plancherel formula implies

$$
\int_G (f(x) - 1)\overline{g(x)}dx = \sum_{\alpha \in \widehat{G}} \widehat{f}(\alpha)\overline{\widehat{g}(\alpha)} - \overline{\widehat{g}(1_G)} = 0
$$

for all $g \in C_c(G)$. This forces $f(x) = 1$ for all $x \in G$ and hence G has to be compact.

4.5 Spectral extension properties

Let A be a commutative Banach algebra. An *extension* of A is a Banach algebra B such that A is algebraically, but not necessarily continuously, embedded in B. We then view A as a subalgebra of B. In this case, $\sigma_B(x) \cup \{0\} \subseteq \sigma_A(x) \cup \{0\}$ and hence $r_B(x) \leq r_A(x)$ for all $x \in A$. These observations suggest the following definition.

Definition 4.5.1. A commutative Banach algebra A has the *spectral extension property* if $r_B(x) = r_A(x)$ for every extension B of A and every $x \in A$.

A is said to have the *strong spectral extension property* if $\sigma_B(x) \cup \{0\} = \sigma_A(x) \cup \{0\}$ holds for every commutative extension B of A and every $x \in A$.

Finally, A has the *multiplicative Hahn–Banach property* if, given any commutative extension B of A, every $\varphi \in \Delta(A)$ extends to some element of $\Delta(B)$.

It is clear from the very definition that the strong spectral extension property implies the spectral extension property. For any commutative Banach algebra C and $y \in C$,

$$\sigma_C(y) \setminus \{0\} \subseteq \{\psi(y) : \psi \in \Delta(C)\} \subseteq \sigma_C(y)$$

(Theorem 2.2.5). Therefore the multiplicative Hahn–Banach property implies the strong spectral extension property.

We have seen in Corollary 4.2.17 that every regular semisimple commutative Banach algebra A has the multiplicative Hahn–Banach property. The purpose of this section is to characterise the semisimple commutative Banach algebras having either of these three properties by a condition similar to, but weaker than, regularity of A and conditions involving the Shilov boundary $\partial(A)$ of A.

Lemma 4.5.2. *For a commutative Banach algebra A, the following conditions are equivalent.*

(i) *A has the spectral extension property.*
(ii) *Every submultiplicative norm $|\cdot|$ on A satisfies $r_A(a) \leq |a|$ for all $a \in A$.*

Proof. (i) $\Rightarrow$ (ii) Let $|\cdot|$ be any submultiplicative norm on A and let $(B, \|\cdot\|)$ be the completion of $(A, |\cdot|)$. By (i), for all $a \in A$,

$$r_A(a) = r_B(a) = \lim_{n \to \infty} \|a^n\|^{1/n} \leq \|a\| = |a|.$$

(ii) $\Rightarrow$ (i) Let $(B, \|\cdot\|)$ be any extension of A. Then r_B is a submultiplicative norm on B and hence on A. Thus (ii) implies that $r_A(a) \leq r_B(a)$ for all $a \in A$, as required. $\square$

For an element $a \in A$, define the *permanent radius* $r_p(a)$ of a to be

$$r_p(a) = \inf\{r_B(a) : B \text{ is an extension of } A\}.$$

Clearly, $r_p(a) \leq r_A(a)$. More precisely,

$$r_p(a) = \inf\{|a| : |\cdot| \text{ is a submultiplicative norm on } A\}.$$

Indeed, if $|\cdot|$ is any submultiplicative norm on A and B is the completion of A with respect to $|\cdot|$, then B is an extension of A and hence $r_p(a) \leq r_B(a) \leq |a|$ for every $a \in A$.

Theorem 4.5.3. *For a semisimple commutative Banach algebra A the following are equivalent.*

(i) *A has the spectral extension property.*
(ii) *If E is a closed subset of $\Delta(A)$ that does not contain the Shilov boundary of A, then there exists an element $a \in A$ such that $\hat{a} = 0$ on E and $r_p(a) > 0$.*
(iii) *Whenever B is a commutative extension of A, every $\varphi \in \partial(A)$ extends to some element of $\Delta(B)$.*

Proof. (i) $\Rightarrow$ (ii) Let E be a closed subset of $\Delta(A)$ that does not contain the Shilov boundary of A. Towards a contradiction, assume that E has the property that for any $a \in A$, $\hat{a}|_E = 0$ implies $a = 0$. Then $|a| = \|\hat{a}|_E\|_\infty$ defines a submultiplicative norm on A. Because A has the spectral extension property, Lemma 4.5.2 shows that $r_A(a) \leq |a|$. Now, since A is semisimple, $r_A(a) = \|\hat{a}\|_\infty$ and hence $\|\hat{a}\|_\infty = \|\hat{a}|_E\|_\infty$ for all $a \in A$. Thus E is a boundary for A and therefore has to contain the Shilov boundary. This contradiction shows the existence of some nonzero element $a \in A$ such that $\hat{a} = 0$ on E. Finally, using once more the facts that A is semisimple and has the spectral extension property, we have $r_p(a) = r_A(a) > 0$.

(ii) $\Rightarrow$ (iii) Let B be a commutative extension of A and consider the restriction map

$$\phi : \Delta(B) \cup \{0\} \to \Delta(A) \cup \{0\}, \quad \psi \to \psi|_A.$$

Then ϕ is continuous with respect to the w^*-topologies. Moreover, $\Delta(B) \cup \{0\}$ is a w^*-closed subset of the unit ball of B^*, because a w^*-limit of elements of $\Delta(B)$ is either 0 or an element of $\Delta(B)$. Thus $\Delta(B) \cup \{0\}$ is w^*-compact and hence so is $\phi(\Delta(B) \cup \{0\}) \subseteq \Delta(A) \cup \{0\}$. Let

$$E = \Delta(A) \cap \phi(\Delta(B) \cup \{0\}),$$

which is a closed subset of $\Delta(A)$. For all $a \in A$,

$$r_B(a) = \sup\{|\psi(a)| : \psi \in \Delta(B)\} = \sup\{|\varphi(a)| : \varphi \in E\},$$

since, by definition of E, $\phi(\Delta(B)) \setminus E$ consists only of the zero functional.

Also, by definition of E, every $\varphi \in E$ extends to some element of $\Delta(B)$. It therefore suffices to show that $\partial(A) \subseteq E$. Assume that E does not contain $\partial(A)$. Then, by hypothesis (ii), there exists $a \in A$ such that $r_p(a) > 0$ and $\hat{a}|_E = 0$. Thus, by the above formula for $r_B(a)$,

$$0 < r_p(a) \leq r_B(a) = \|\hat{a}|_E\|_\infty = 0.$$

This contradiction shows that $\partial(A) \subseteq E$.

(iii) $\Rightarrow$ (i) Let $a \in A$ and let B be any extension of A. To show that $r_B(a) = r_A(a)$, we can assume that B is commutative. In fact, if C is the closure of A in B, then $r_B(a) = r_C(a)$. Choose $\varphi \in \partial(A)$ such that $|\varphi(a)| = r_A(a)$. By (iii), φ extends to some $\psi \in \Delta(B)$. It follows that

$$r_A(a) = |\varphi(a)| = |\psi(a)| \leq r_B(a) \leq r_A(a),$$

and so (i) holds.   □

Before proceeding, we insert a consequence concerning the existence of zero divisors.

Corollary 4.5.4. *Let A be a semisimple commutative Banach algebra and suppose that A has the spectral extension property. If A is not one-dimensional, then A contains zero divisors.*

Proof. Notice that if E is any proper closed subset of $\partial(A)$, then, by Theorem 4.5.3, there exists $a \neq 0$ in A such that $\widehat{a} = 0$ on E. Now $\partial(A)$ contains at least two elements. Indeed, this follows from Theorem 3.3.14 if $\partial(A) \neq \Delta(A)$ and is clear otherwise since A is not one-dimensional and semisimple.

Thus we can find two nonempty disjoint open subsets U and V of $\partial(A)$. Let $E = \partial(A) \setminus U$ and $F = \partial(A) \setminus V$. Then there exist nonzero elements $a, b \in A$ such that $\widehat{a} = 0$ on E and $\widehat{b} = 0$ on F. It follows that $\widehat{ab} = \widehat{a}\,\widehat{b} = 0$ on $\partial(A)$ and hence on all of $\Delta(A)$. By semisimplicity of A, we get $ab = 0$.   □

An obvious method to construct an extension is as follows.

Lemma 4.5.5. *Let A be a semisimple commutative Banach algebra. Then $C_0(\partial(A))$ is an extension of A. Furthermore, if $\varphi \in \Delta(A)$ extends to some element of $\Delta(C_0(\partial(A)))$, then $\varphi \in \partial(A)$.*

Proof. Because A is semisimple, the mapping $a \to \widehat{a}|_{\partial(A)}$ is an injective homomorphism of A into $C_0(\partial(A))$. Now, every element of $\Delta(C_0(\partial(A)))$ equals the evaluation of functions at some point of $\partial(A)$. Hence, if $\varphi \in \Delta(A)$ extends to some element of $\Delta(C_0(\partial(A)))$, then $\varphi(a) = \widehat{a}(\psi) = \psi(a)$ for some $\psi \in \partial(A)$ and all $a \in A$. This proves $\varphi = \psi \in \partial(A)$.   □

Both the strong spectral extension property and the multiplicative Hahn–Banach property are stronger than (but, as Examples 4.5.8 and 4.5.10 show, not equivalent to) the spectral extension property. Therefore the question arises of what distinguishes the spectral extension property from either of the other two properties. The next two theorems provide satisfactory answers.

Theorem 4.5.6. *Let A be a semisimple commutative Banach algebra. Then A has the strong spectral extension property if and only if A has the spectral extension property and the Shilov boundary $\partial(A)$ of A satisfies*

$$\widehat{a}(\partial(A)) \cup \{0\} = \widehat{a}(\Delta(A)) \cup \{0\}$$

for all $a \in A$.

Proof. Suppose first that A has the strong spectral extension property. Then A has the spectral extension property. By Lemma 4.5.5, $B = C_0(\partial(A))$ is an extension of A. For every $f \in B$,

$$\sigma_B(f) \cup \{0\} = f(\partial(A)) \cup \{0\}$$

(Example 1.2.3). It follows that

$$\widehat{a}(\partial(A)) \cup \{0\} = \sigma_B(a) \cup \{0\} = \sigma_A(a) \cup \{0\} = \widehat{a}(\Delta(A)) \cup \{0\}$$

for all $a \in A$.

Now assume that, conversely, A has the spectral extension property and satisfies $\widehat{a}(\partial(A)) \cup \{0\} = \widehat{a}(\Delta(A)) \cup \{0\}$ for all $a \in A$. Let B be any commutative extension of A. If $\lambda \in \sigma_A(a)$ and $\lambda \neq 0$, then $\lambda = \widehat{a}(\varphi)$ for some $\varphi \in \Delta(A)$ (Theorem 2.2.5). By assumption φ can be chosen to be in the Shilov boundary of A. Because A has the spectral extension property, by Theorem 4.5.3, (i) $\Rightarrow$ (iii), φ extends to some element of $\Delta(B)$. It follows that $\lambda = \varphi(a) \in \sigma_B(a)$. This shows that $\sigma_A(a) \cup \{0\} \subseteq \sigma_B(a) \cup \{0\}$ and, since the reverse inclusion holds anyway, equality follows. $\qquad\square$

Theorem 4.5.7. *Let A be a semisimple commutative Banach algebra. Then A has the multiplicative Hahn–Banach property if and only if A has the spectral extension property and the Shilov boundary of A equals $\Delta(A)$.*

Proof. Suppose first that A has the multiplicative Hahn–Banach property. Then $\partial(A) = \Delta(A)$ by Lemma 4.5.5. Let B be any extension of A and let C be the closure of A in B. Given $a \in A$, there exists $\varphi \in \Delta(A)$ such that $|\varphi(a)| = r_A(a)$. Now φ extends to some $\psi \in \Delta(C)$, and this gives

$$r_A(a) = |\varphi(a)| = |\psi(a)| \leq r_C(a) = r_B(a),$$

since C is a closed subalgebra of B. Thus $r_A(a) = r_B(a)$.

Conversely, if A has the spectral extension property and $\partial(A)$ equals $\Delta(A)$, then the implication (i) $\Rightarrow$ (ii) of Theorem 4.5.3 shows that A has the multiplicative Hahn–Banach property. $\qquad\square$

Recall that regularity implies the multiplicative Hahn–Banach property, this in turn implies the strong spectral extension property and, finally, the strong spectral extension property implies the spectral extension property. We now present three examples showing that none of these implications can by reversed. All these examples are constructed in a similar manner in that they possess an ideal isomorphic to $C_0(Z)$ with quotient isomorphic to $A(Y)$ for properly chosen spaces Z and Y.

Example 4.5.8. Suppose that $0 < r < R$ and let $X = \{z \in \mathbb{C} : |z| \leq R\}$ and $U = \{z \in \mathbb{C} : |z| < r\}$. Let

$$A = \{f \in C(X) : f \text{ is holomorphic on } U\},$$

endowed with the uniform norm. Because a uniform limit of holomorphic functions is holomorphic, A is closed in $C(X)$. Of course, the mapping $x \to \varphi_x$, where $\varphi_x(f) = f(x)$ for all $f \in A$, is an embedding of X into $\Delta(A)$. To see that every $\varphi \in \Delta(A)$ is of this form, let $Y = \{z \in \mathbb{C} : |z| \le r\}$ and consider the closed ideal $I = \{f \in A : f|_Y = 0\}$ of A. Then the mapping $f \to f|_{X \setminus Y}$ is an isometric isomorphism from I onto $C_0(X \setminus Y)$, and it follows from Tietze's extension theorem that A/I is isometrically isomorphic to $A(Y)$ (Exercise 4.8.24). Let $\varphi \in \Delta(A)$. If $\varphi \in h(I)$ then $\varphi = \varphi_x$ for some $x \in X \setminus Y$, whereas if $\varphi(I) \ne \{0\}$, then $\varphi(g) = g(y)$ for all $g \in I$ and some $y \in Y$ (Theorem 2.6.6). In the latter case, choosing $g \in I$ such that $\varphi(g) \ne 0$, we obtain $\varphi(f) = \varphi(fg)\varphi(g)^{-1} = f(y)$ for all $f \in A$. Thus $\Delta(A)$ can be identified with X and, with this identification, the Gelfand transform $\hat{f}$ coincides with f for every $f \in A$.

It follows from the maximum modulus principle that the Shilov boundary of A equals the annulus $\{z \in \mathbb{C} : r \le |z| \le R\}$. Now the function $f(z) = z$ does not attain all its values on the Shilov boundary, and so A does not have the strong spectral extension property by Theorem 4.5.6.

Yet, A has the spectral extension property. To verify this, we apply Theorem 4.5.3. Thus, let E be a proper closed subset of X that does not contain the Shilov boundary. Then there exists a nonempty open subset of $X \setminus Y$ which is disjoint from E. Hence there exists $f \in A$ such that $f = 0$ on E and $f = 1$ on some nonempty open subset W of X. On the other hand, A contains a nonzero function g with $\operatorname{supp} g \subseteq W$. It follows that $fg = g$ and hence

$$0 < r_p(g) = r_p(fg) \le r_p(f)r_p(g),$$

whence $r_p(f) \ge 1$. An application of Theorem 4.5.3, (ii) $\Rightarrow$ (i), now shows that A has the spectral extension property.

The next example shares the multiplicative Hahn–Banach property but fails to be regular.

Example 4.5.9. Let $X = \mathbb{D} \times [0, 1]$ and let A be the algebra of all continuous complex-valued functions f on X with the property that $z \to f(z, 0)$ is holomorphic on $\mathbb{D}^\circ$. Endowed with the supremum norm, A is a commutative Banach algebra. Arguing as in the previous example and taking the closed ideal $I = C_0(\mathbb{D} \times (0, 1])$ with quotient isomorphic to $A(\mathbb{D})$, it is easily seen that $\Delta(A)$ can be identified with X in such a way that $\hat{f}$ equals f for all $f \in A$.

In this case, however, the Shilov boundary of A is all of X. Indeed, since A contains every continuous function on X which is zero on $\mathbb{D} \times \{0\}$, it follows that $\partial(A) \supseteq \mathbb{D} \times (0, 1]$ and hence $\partial(A) = X$. Of course, A fails to be regular because $A/I = A(\mathbb{D})$ is not regular.

However, A has the multiplicative Hahn–Banach property. Since $\partial(A) = X$, this follows from Theorem 4.5.7 once we have shown that A has the spectral extension property. This is done by again applying Theorem 4.5.3, (ii) $\Rightarrow$ (i). Thus let E be a closed subset of X that does not contain the Shilov boundary, which simply means that $E \neq X$. Precisely as in Example 4.5.8, it can be shown that there exist functions $f, g \in A$ such that $f = 0$ on E, $g \neq 0$, and $fg = g$. Then $r_p(f) > 0$ and condition (ii) of Theorem 4.5.3 is fulfilled.

Finally, we give an example of a commutative Banach algebra A which has the strong spectral extension property, but does not have the multiplicative Hahn–Banach property. As one might expect, such an example is not so easy to discover. The one that follows involves the theory of holomorphic functions of two complex variables.

Example 4.5.10. Let X denote the closed ball of radius 2 around zero in $\mathbb{C}^2$ and Y the open ball of radius 1 around zero in $\mathbb{C}^2$. Let

$$A = \{f \in C(X) : f|_Y \text{ is holomorphic}\}.$$

Endowed with the supremum norm, A becomes a commutative Banach algebra. As usual, for each $x \in X$, let $\varphi_x : A \to \mathbb{C}$ be the evaluation at x. It is then shown in the same way as in Example 4.5.9 that the mapping $x \to \varphi_x$ is a homeomorphism between X and $\Delta(A)$ (Exercise 4.8.25). Identifying X and $\Delta(A)$ in this way, the Gelfand homomorphism of A is the identity. Notice next that

$$\partial(A) = \{(z, w) \in \mathbb{C}^2 : 1 \leq |z|^2 + |w|^2 \leq 4\}.$$

Indeed, by the maximum modulus principle for holomorphic functions of two variables, $\|f|_{\overline{Y}}\|_\infty = \|f|_{\partial(Y)}\|$ for every $f \in A$. Moreover, the ideal

$$I = \{f \in A : f|_Y = 0\} = C_0(X \setminus \overline{Y})$$

is regular, whence $X \setminus \overline{Y} = \Delta(I) = \partial(I)$. So $\partial(A) = X \setminus Y$ and, in particular, $\partial(A) \neq \Delta(A)$. Therefore A cannot have the multiplicative Hahn–Banach property (Theorem 4.5.7).

We observe next that A has the spectral extension property. To see this, we once more apply Theorem 4.5.3, (ii) $\Rightarrow$ (i). Thus, let E be a closed subset of X not containing $\partial(A)$. Because E does not contain $X \setminus \overline{Y}$, we can find an open subset U of X such that $U \subseteq \partial(A)$ and $\overline{U} \cap E = \emptyset$. There exists $f \in A$ such that $f = 0$ on E and $f = 1$ on U. Also, there exists $g \in A, g \neq 0$, with supp $g \subseteq U$. Then $fg = g$ and hence, as in the previous two examples, $r_p(f) \geq 1$. Theorem 4.5.3 now shows that A has the spectral extension property.

To conclude that A even has the strong spectral extension property, by Theorem 4.5.6 we have to verify that $f(\partial(A)) \cup \{0\} = f(\Delta(A)) \cup \{0\}$ for every $f \in A$. It is this point where use is made of the fact that the functions $f|_Y, f \in A$, are holomorphic functions of two variables. From a well-known property of holomorphic functions of several-variables [76, Theorem 1.2.6], it follows that if f attains a value c at some point $(z, w) \in Y$, then there exists $(z', w') \in \partial(Y)$ such that $f(z', w') = c$. Therefore

$$f(\Delta(A)) = f(X) = f(X \setminus Y) = f(\partial A),$$

as required.

4.6 The unique uniform norm property

The unique uniform norm property, which is the subject of this section, is intimately related to the properties studied in Section 4.5.

Definition 4.6.1. Let A be an algebra. A submultiplicative (not necessarily complete) norm $|\cdot|$ on A is called a *uniform norm* if it satisfies the square property $|x^2| = |x|^2$ for all $x \in A$.

Lemma 4.6.2. *Let A be a commutative Banach algebra and let $|\cdot|$ be a uniform norm on A. Then $|x| \leq r_A(x)$ for all $x \in A$. Let*

$$E = \{\varphi \in \Delta(A) : |\varphi(x)| \leq |x| \text{ for all } x \in A\}.$$

Then E is a closed subset of $\Delta(A)$ and

$$|x| = \sup\{|\varphi(x)| : \varphi \in E\}$$

for all $x \in A$.

Proof. Let $(B, |\cdot|)$ be the completion of A with respect to $|\cdot|$. Since A is dense in B and elements of $\Delta(B)$ are continuous, $\psi|_A \in \Delta(A)$ for each $\psi \in \Delta(B)$ and the map $\psi \to \psi|_A$ is a bijection from $\Delta(B)$ onto the set of all elements of $\Delta(A)$ which are continuous with respect to the norm $|\cdot|$ on A. By definition, this set is nothing but the set E.

Because $|\cdot|$ is a uniform norm on B, for $x \in B$ we have $|x^{2^n}| = |x|^{2^n}$ for all $n \in \mathbb{N}$ and hence $|x| = r_B(x)$. For $x \in A$, it follows that

$$|x| = r_B(x) = \sup\{|\varphi(x)| : \varphi \in E\},$$

which is the second statement of the lemma, and also

$$|x| = r_B(x) = \sup\{|\psi(x)| : \psi \in \Delta(B)\}$$
$$\leq \sup\{|\varphi(x)| : \varphi \in \Delta(A)\}$$
$$= r_A(x),$$

which is the first statement of the lemma.   □

Lemma 4.6.2 leads to the following criterion for semisimplicity.

Corollary 4.6.3. *For a commutative Banach algebra A the following conditions are equivalent.*

(i) *A is semisimple.*
(ii) *The spectral radius is a uniform norm on A.*
(iii) *A admits a uniform norm.*

Proof. We have already observed in Section 2.1 that A is semisimple if and only if r_A is a (uniform) norm on A. Thus it suffices to show (iii) $\Rightarrow$ (i). If $\|\cdot\|$ is the original norm on A and $|\cdot|$ is a uniform norm on A, then $|x| \le r_A(x) = \|\widehat{x}\|_\infty$ for all $x \in A$ by Lemma 4.6.2. Hence $\widehat{x} = 0$ implies $x = 0$. □

Definition 4.6.4. A commutative Banach algebra A is said to have the *unique uniform norm property* if r_A is the only uniform norm on A. A closed subset E of $\Delta(A)$ is called a *set of uniqueness* for A if the assignment

$$x \to \sup\{|\widehat{x}(\varphi)| : \varphi \in E\}$$

defines a norm on A.

Note that if A has the spectral extension property then it has the unique uniform norm property. In fact, let $|\cdot|$ be a uniform norm on A and let B be the completion of $(A, |\cdot|)$. Then

$$r_A(x) = r_B(x) = \lim_{n \to \infty} |x^{2^n}|^{1/2^n} = |x|$$

for every $x \in A$.

Of course, the Shilov boundary is a set of uniqueness by definition. The theorem below characterizes the unique uniform norm property in several different ways.

Theorem 4.6.5. *Let A be a semisimple commutative Banach algebra. Then the following four conditions are equivalent.*

(i) *A has the unique uniform norm property.*
(ii) *The Shilov boundary $\partial(A)$ of A is the smallest set of uniqueness.*
(iii) *If F is a closed subset of $\Delta(A)$ that does not contain $\partial(A)$, then there exists $x \in A$ such that $x \ne 0$ and $\widehat{x}|_F = 0$.*
(iv) *If B is any semisimple commutative extension of A, then every element of $\partial(A)$ extends to a multiplicative linear functional on B.*

Proof. (i) $\Rightarrow$ (ii) If $F \subseteq \Delta(A)$ is a closed set of uniqueness, then

$$|x| = \sup\{|\varphi(x)| : \varphi \in F\}$$

defines a uniform norm on A. By (i), $r_A(x) = |x|$ for all $x \in A$. Thus F is a closed boundary, whence $F \supseteq \partial(A)$.

(ii) $\Rightarrow$ (iii) is obvious.

Suppose that (iii) holds and let B be any semisimple commutative extension of A. As in the proof of the implication (ii) $\Rightarrow$ (iii) of Theorem 4.5.3, consider the map

$$\phi : \Delta(B) \cup \{0\} \to \Delta(A) \cup \{0\}, \quad \psi \to \psi|_A.$$

Then (compare the proof of Theorem 4.5.3) the set

$$F = \Delta(A) \cap \phi(\Delta(B) \cup \{0\})$$

is closed in $\Delta(A)$. Suppose that F does not contain $\partial(A)$. Then, by (iii), there exists $x \in A, x \neq 0$, such that $\hat{x} = 0$ on F. Since $\varphi(x) = 0$ for every $\varphi \in \phi(\Delta(B)) \setminus F$ and since B is semisimple, it follows that

$$0 = \sup\{|\varphi(x)| : \varphi \in F\} = \sup\{|\psi(x)| : \psi \in \Delta(B)\} = r_B(x).$$

This contradicts $x \neq 0$ and shows that $\partial(A) \subseteq F$. Hence, by definition of F, each $\varphi \in \partial(A)$ extends to some element of $\Delta(B)$.

(iv) $\Rightarrow$ (i) Let $|\cdot|$ be a uniform norm on A and let B be the completion of $(A, |\cdot|)$. Since $r_B(x) = |x|$ for all $x \in B$, B is a semisimple commutative extension of A. For each $x \in A$, there exists $\varphi_x \in \partial(A)$ such that $r_A(x) = |\varphi_x(x)|$. By hypothesis (iv), there exists $\psi_x \in \Delta(B)$ extending φ_x. It follows that $r_A(x) = |\psi_x(x)| \leq |x|$ and hence $|x| = r_A(x)$ for every $x \in A$. $\qquad\square$

A further interesting notion that has been introduced in this context is that of weak regularity.

Definition 4.6.6. A semisimple commutative Banach algebra A is called *weakly regular* if given any proper closed subset E of $\Delta(A)$, there exists a nonzero element a of A with $\hat{a}|_E = 0$.

The algebra studied in Example 4.5.9 is, as we have seen there, weakly regular but not regular. An immediate consequence of the equivalence of conditions (i) and (iii) in Theorem 4.6.5 is the following

Corollary 4.6.7. *For a semisimple commutative Banach algebra A, the following conditions are equivalent.*

(i) A *is weakly regular.*
(ii) A *has the unique uniform norm property and satisfies $\partial(A) = \Delta(A)$.*

From Theorem 4.6.5 we can deduce that the unique uniform norm property is inherited by certain ideals.

Corollary 4.6.8. *Let A be a semisimple commutative Banach algebra having the unique uniform norm property. Let I be a spectral synthesis ideal of A, that is, an ideal with the property that $I = k(h(I))$. Then I also has the unique uniform norm property.*

Proof. By Theorem 4.6.5, (iii) $\Rightarrow$ (i), it is sufficient to show that if F is a closed subset of $\Delta(I)$ not containing the Shilov boundary $\partial(I)$, then F is not a set of uniqueness for I. Let

$$E = (\Delta(A) \setminus \Delta(I)) \cup F = h(I) \cup F.$$

Then E is a closed subset of $\Delta(A)$ which does not contain $\partial(A)$ as $\partial(I) \subseteq \partial(A)$. Because A has the unique uniform norm property, by (i) $\Rightarrow$ (iii) of Theorem 4.6.5 there exists a nonzero element $a \in A$ such that $\hat{a}|_E = 0$. Then $a \in k(h(I))$ since $h(I) \subseteq E$. Thus $a \in I$ and $\hat{a}|_F = 0$, and hence F fails to be a set of uniqueness for I. $\qquad\Box$

It is not known whether in the preceding corollary the hypothesis that I be a spectral synthesis ideal can be dropped. We continue with an example showing that the unique uniform norm property is in general not inherited by quotient algebras A/I even when I is a spectral synthesis ideal. This is in strong contrast to the hereditary properties of regularity.

Example 4.6.9. Let A be the commutative Banach algebra considered in Example 4.5.8 where we now choose $r = 1$ and $R = 2$. Then A has the spectral extension property and hence the unique uniform norm property. Let

$$I = \{f \in A : f(z) = 0 \text{ for all } z \in \mathbb{D}\}.$$

Then I is a spectral synthesis ideal and A/I is isometrically isomorphic to the disc algebra $A(\mathbb{D})$ which does not have the unique uniform norm property. Indeed, every closed subset of $\mathbb{D}$ which has an accumulation point in $\mathbb{D}^\circ$ is a set of uniqueness.

The construction in the following proposition is a generalisation of the one performed in Example 4.5.9. The subsequent Theorem 4.6.11 shows that this construction provides a useful method to construct examples.

Proposition 4.6.10. *Let $(B, \|\cdot\|_B)$ be a unital semisimple commutative Banach algebra. Let $X = \Delta(B) \times [0, 1]$ and*

$$A = \{f \in C(X) : f(\cdot, 0) = \hat{b} \text{ for some } b \in B\}.$$

Then, with the norm

$$\|f\| = \max(\|f\|_\infty, \|f(\cdot, 0)\|_B),$$

A is a semisimple commutative Banach algebra satisfying

$$\Delta(A) = X \quad and \quad \partial(A) = \Delta(A).$$

Proof. It is clear that A is a semisimple commutative Banach algebra. For $\psi \in \Delta(B)$ and $t \in [0,1]$, define $\varphi_{\psi,t} \in \Delta(A)$ by $\varphi_{\psi,t}(f) = f(\psi,t), f \in A$. Then $\phi : (\psi,t) \to \varphi_{\psi,t}$ maps $\Delta(B) \times [0,1]$ continuously into $\Delta(A)$.

To show that ϕ is injective, let (ψ_1,t_1) and (ψ_2,t_2) be two distinct elements of X. If $t_1 = t_2 = 0$, then $\psi_1 \neq \psi_2$ and since B is semisimple, there exists $b \in B$ such that $\psi_1(b) \neq \psi_2(b)$ and $f \in C(X)$, defined by $f(\psi,t) = \psi(b)$, belongs to A and satisfies

$$\varphi_{\psi_1,t_1}(f) \neq \varphi_{\psi_2,t_2}(f).$$

Now, if one of t_1, t_2 is non-zero, say $t_1 \neq 0$, then there exists a continuous function $f : X \to [0,1]$ such that $f(\psi_1,t_1) = 1$ and

$$f|_{(\Delta(B)\times\{0\})\cup\{(\psi_2,t_2)\}} = 0.$$

Then also $\varphi_{\psi_1,t_1}(f) \neq \varphi_{\psi_2,t_2}(f)$.

Because X is compact, ϕ is a homeomorphism onto its range. Therefore, it remains to show that every $\varphi \in \Delta(A)$ is of the form $\varphi = \varphi_{\psi,t}$ for some $\psi \in \Delta(B)$ and $t \in [0,1]$. To that end, define a closed ideal I of A by

$$I = \{f \in A : f(\Delta(B) \times \{0\}) = \{0\}\}$$

and a closed subalgebra J of A by

$$J = \{f \in A : f(\psi,t) = f(\psi,0) \text{ for all } \psi \in \Delta(B), t \in [0,1]\}.$$

Notice that J is isomorphic to B and I is isomorphic to $C_0(\Delta(B) \times (0,1])$. If $f \in A$ and $b \in B$ such that $\hat{b} = f$ on $\Delta(B) \times \{0\}$, then

$$f = (f - \hat{b}) + \hat{b} \in I + J.$$

Hence A is the vector space direct sum of I and J. Since J contains the identity, φ is nonzero on J, and therefore there exists $\psi \in \Delta(B)$ such that $\varphi(f) = f(\psi,0)$ for all $f \in J$. If $\varphi(I) = \{0\}$, then clearly $\varphi = \varphi_{\psi,0}$. If $\varphi(I) \neq \{0\}$, then there exists $(\rho,t) \in \Delta(B) \times (0,1]$ such that $\varphi(f) = f(\rho,t)$ for all $f \in I$. For any $f \in I$ and $g \in J$, we then have, since I is an ideal,

$$f(\rho,t)g(\rho,t) = fg(\rho,t) = \varphi(fg) = \varphi(f)\varphi(g) = f(\rho,t)g(\psi,0).$$

Choosing f such that $f(\rho,t) \neq 0$, we obtain that $g(\rho,t) = g(\psi,0)$ for all $g \in J$. Now, let $f = f_1 + f_2$, where $f_1 \in I$ and $f_2 \in J$, be an arbitrary element of A. It follows that

$$\begin{aligned}
\varphi(f) &= \varphi(f_1) + \varphi(f_2) = f_1(\rho,t) + f_2(\psi,0) \\
&= f_1(\rho,t) + f_2(\rho,t) \\
&= f(\rho,t).
\end{aligned}$$

This proves that $\varphi = \varphi_{s,t}$. Finally, it is now obvious that $\partial(A) = \Delta(A)$. $\square$

Theorem 4.6.11. *Let B and A be as in Proposition 4.6.10. Then*

(i) *A has the spectral extension property and hence the unique uniform norm property.*

(ii) *A is weakly regular.*

(iii) *A is regular if and only if B is regular.*

Proof. (i) It is shown exactly as in Example 4.5.9 that A has the spectral extension property.

(ii) Because A has the unique uniform norm property and $\partial(A) = \Delta(A)$ by Proposition 4.6.10, Corollary 4.6.7 shows that A is weakly regular.

(iii) Suppose first that B is not regular. Then there exists a closed subset F of $\Delta(B)$ and $\psi \in \Delta(B) \setminus F$ such that $\widehat{b}(F) \neq \{0\}$ for every $b \in B$ with $\widehat{b}(\psi) = 1$. Let $E = F \times \{0\}$ and $\varphi = \varphi_{\psi,0}$. Then E is a closed subset of $\Delta(A)$ and $\varphi \notin E$. Since $f = \widehat{b}$ on $\Delta(B) \times \{0\}$ for every $f \in A$, it follows that there exists no $f \in A$ such that $\varphi(f) = 1$ and $f(E) = \{0\}$. Hence A is not regular.

Conversely, suppose that B is regular. Let E be a closed subset of $\Delta(A)$ and $(\psi, t) \in \Delta(A) \setminus E$. If $t \neq 0$, then $F = E \cap (\Delta(B) \times (0,1])$ is closed in $\Delta(B) \times (0,1] = \Delta(I)$ and $(\psi, t) \in \Delta(I) \setminus F$. Since $I = C_0(\Delta(B) \times (0,1])$, there exists $f \in I$ such that $f(\psi, t) = 1$ and $f(F) = \{0\}$. But then $f(E) = \{0\}$ since $f \in C_0(\Delta(B) \times (0,1])$.

Now let $t = 0$. There exists an open neighbourhood U of ψ in $\Delta(B)$ such that $(U \times [0, \varepsilon]) \cap E = \emptyset$ for some $\varepsilon > 0$. Let $F = \Delta(B) \setminus U$. Then $\psi \notin F$ and since B is regular, there exists $b \in B$ such that $\widehat{b}(\psi) = 1$ and $\widehat{b}(F) = \{0\}$. Define $h \in A$ by $h(\rho, s) = \widehat{b}(\rho)$ for $(\rho, s) \in \Delta(A)$. Then $h(\psi, 0) = 1$ and $h(E) = \{0\}$. Let $G = \Delta(B) \times [\varepsilon/2, 1] \subseteq \Delta(A)$. There exists $g \in C(\Delta(B) \times [0,1])$ such that $g(G) = \{0\}$ and $g(\Delta(B) \times \{0\}) = \{1\}$. Since B is unital, $g \in A$. Let $f = gh \in A$. Then $f(\psi, 0) = 1$. For $(\rho, s) \in E$, we have to distinguish two cases. If $0 \leq s \leq \varepsilon/2$, then $\rho \notin U$ and hence $h(\rho, s) = \widehat{b}(\rho) = 0$. If $s \geq \varepsilon/2$, then $(\rho, s) \in G$ and hence $g(\rho, s) = 0$. Thus $f(E) = \{0\}$. This shows that A is regular. $\square$

The following simple lemma is analogous to the corresponding statement for regularity (Lemma 4.2.15).

Lemma 4.6.12. *Let A be a semisimple commutative Banach algebra and B be a dense subalgebra of A. If B has the unique uniform norm property, then so does A.*

Proof. Let $|\cdot|$ be a uniform norm on A. Then $|\cdot| \leq r_A(\cdot) \leq \|\cdot\|$ on A. Let $a \in A$ and choose a sequence $(b_n)_n$ in B such that $\|b_n - a\| \to 0$. Because B has the unique uniform norm property, $|b_n| = r_A(b_n)$ for all n. Now, since the spectral radius is subadditive,

$$|r_A(a) - |a|| = r_A(a) - |a| \leq r_A(a - b_n) + r_A(b_n) - |a|$$
$$= r_A(a - b_n) + |b_n| - |a|$$
$$\leq \|a - b_n\| + |b_n - a|,$$

which converges to 0 as $n \to \infty$. Thus $|\cdot| = r_A(\cdot)$, as required. $\square$

As an application of Theorem 4.6.5 we finally show that the unique uniform norm property behaves well with respect to forming projective tensor products.

Theorem 4.6.13. *Let A and B be commutative Banach algebras and suppose that their projective tensor product $A \widehat{\otimes}_\pi B$ is semisimple. Then $A \widehat{\otimes}_\pi B$ has the unique uniform norm property if and only if both A and B have the unique uniform norm property.*

Proof. Suppose first that both A and B have the unique uniform norm property and let $|\cdot|$ be any uniform norm on $A \widehat{\otimes}_\pi B$. Let

$$S = \{\sigma \in \Delta(A \widehat{\otimes}_\pi B) : |\sigma(x)| \le |x| \text{ for all } x \in A \widehat{\otimes}_\pi B\}.$$

Then, by Lemma 4.6.2, for every $x \in A \widehat{\otimes}_\pi B$,

$$|x| = \sup\{|\sigma(x)| : \sigma \in S\}.$$

We claim that $\|\widehat{x}\|_\infty = |x|$ for all $x \in A \widehat{\otimes}_\pi B$. Clearly, $|x| \le \|\widehat{x}\|_\infty$ and for the reverse inequality it suffices to show that $\partial(A \widehat{\otimes}_\pi B) \subseteq S$. Suppose that this is false. Since $\partial(A \widehat{\otimes}_\pi B) = \partial(A) \times \partial(B)$ (Proposition 3.3.12), there exist $\varphi \in \partial(A)$ and $\psi \in \partial(B)$ such that $\varphi \widehat{\otimes}_\pi \psi \notin S$. Thus we can find open neighbourhoods U of φ in $\Delta(A)$ and V of ψ in $\Delta(B)$ such that $S \cap (U \times V) = \emptyset$. Let $E = \Delta(A) \setminus U$ and $F = \Delta(B) \setminus V$. Then E and F do not contain $\partial(A)$ and $\partial(B)$, respectively.

Because A and B have the unique uniform norm property, by Theorem 4.6.5, (i) $\Rightarrow$ (iii), there exist nonzero elements a of A and b of B such that $\widehat{a} = 0$ on E and $\widehat{b} = 0$ on F. Since

$$S \subseteq (E \times \Delta(B)) \cup (\Delta(A) \times F),$$

it follows that $\sigma(a \otimes b) = 0$ for all $\sigma \in S$. Thus $a \otimes b = 0$. This contradicts the fact that $a \ne 0$ and $b \ne 0$. Hence $\partial(A \widehat{\otimes}_\pi B) \subseteq S$, as was to be shown.

Conversely, let $A \widehat{\otimes}_\pi B$ have the unique uniform norm property. To show that then A has this property, we apply the equivalence of conditions (i) and (iii) in Theorem 4.6.5.

Let E be a closed subset of $\Delta(A)$ which does not contain $\partial(A)$. Then $E \times \Delta(B)$ does not contain $\partial(A \widehat{\otimes}_\pi B)$. Therefore there exists $u \in A \widehat{\otimes}_\pi B$ such that $u \ne 0$ and $\widehat{u}|_{E \times \Delta(B)} = 0$. Select $\varphi \in \Delta(A)$ and $\psi \in \Delta(B)$ with $(\varphi \widehat{\otimes}_\pi \psi)(u) \ne 0$ and let $a = \phi_\psi(u)$, where ϕ_ψ is the homomorphism from $A \widehat{\otimes}_\pi B$ to A satisfying $\phi_\psi(x \otimes y) = \psi(y)x$ for all $x \in A$ and $y \in B$ (Lemma 2.11.5). Then, for any representation $u = \sum_{j=1}^\infty a_j \otimes b_j$, $a_j \in A$, $b_j \in B$, of u,

$$\varphi(a) = \varphi\left(\sum_{j=1}^\infty \psi(b_j)a_j\right) = (\varphi \widehat{\otimes}_\pi \psi)(u).$$

On the other hand, for every $\rho \in E$, $\rho(a) = (\rho \widehat{\otimes}_\pi \psi)(u) = 0$. Thus E satisfies condition (iii) of Theorem 4.6.5. It follows that A has the unique uniform norm property. □

4.7 Regularity of Beurling algebras

In Section 4.4 we have shown that L^1-algebras of locally compact Abelian groups are always regular, and hence they have the unique uniform norm property. We begin this section with a simple example which shows in a very explicit manner how the unique uniform norm property may fail for Beurling algebras. However, perhaps unexpectedly, it turns out that for Beurling algebras satisfaction of the unique uniform norm property already implies regularity.

Example 4.7.1. Let ω be a weight function on the group $\mathbb{Z}$ of integers and let $A = l^1(\mathbb{Z}, \omega)$. Let, as in Proposition 2.8.8,

$$R_+ = \inf\{\omega_n^{1/n} : n \in \mathbb{N}\} \text{ and } R_- = \sup\{\omega_m^{1/m} : m \in -\mathbb{N}\}.$$

Then $R_- \leq R_+$ and we know from Proposition 2.8.8 that $\Delta(A)$ can be identified with the annulus

$$\{z \in \mathbb{C} : R_- \leq |z| \leq R_+\}$$

by means of the map $z \to \varphi_z$, where $\varphi_z(f) = \sum_{n \in \mathbb{Z}} f(n)z^n$. For $R_- \leq r \leq R_+$, let $K_r = \{z \in \mathbb{C} : |z| = r\}$. Suppose that $f \in A$ satisfies $\varphi_z(f) = 0$ for all $z \in K_r$. Then $\varphi_z(\delta_m * f) = 0$ for all $m \in \mathbb{Z}$ and $z \in K_r$, and hence

$$0 = \int_0^{2\pi} \varphi_{re^{it}}(\delta_m * f)dt$$

$$= \sum_{n \in \mathbb{Z}} (\delta_m * f)(n)r^n \int_0^{2\pi} e^{-int}dt = (\delta_m * f)(0)$$

$$= f(m).$$

This shows that $f = 0$. Thus every such circle K_r is a set of uniqueness and therefore $A = l^1(\mathbb{Z}, \omega)$ does not have the unique uniform norm property whenever $R_- < R_+$.

Recall from Section 2.8 (Theorems 2.8.2 and 2.8.5) that $\Delta(L^1(G, \omega))$ is homeomorphic to the set $\widehat{G}(\omega)$ of ω-bounded generalised characters on G by the map $\gamma \to \varphi_\gamma$, where

$$\varphi_\gamma(f) = \int_G f(x)\overline{\gamma(x)}dx \quad (f \in L^1(G, \omega)),$$

and $\widehat{G}(\omega)$ is equipped with the topology of uniform convergence on compact subsets of G. In the sequel we make use of the fact that if $\delta \in \widehat{G}(\omega)$ and $\alpha \in \widehat{G}$, then $\delta\alpha \in \widehat{G}(\omega)$.

Proposition 4.7.2. *Suppose that $L^1(G,\omega)$ has the unique uniform norm property and let $\gamma \in \widehat{G}(\omega)$. Then*

(i) *The map $\alpha \to \gamma\alpha$ is a homeomorphism from $\widehat{G}$ onto $\widehat{G}(\omega)$.*
(ii) *$\partial(L^1(G,\omega))$, the Shilov boundary of $L^1(G,\omega)$, coincides with $\widehat{G}(\omega)$.*

Proof. Note first that for each $\delta \in \widehat{G}(\omega)$, the assignment

$$f \to |f|_\delta = \sup\{|\varphi_{\delta\alpha}(f)| : \alpha \in \widehat{G}\}$$

defines a uniform norm on $L^1(G,\omega)$ since $L^1(G)$ is semisimple, $\delta(x) \neq 0$ for all $x \in G$ and $\varphi_{\delta\alpha}(f) = \varphi_\alpha(\overline{\delta} \cdot f)$. The map $\alpha \to \gamma\alpha$ from $\widehat{G}$ into $\widehat{G}(\omega)$ is continuous and injective. Moreover, the set $\gamma \cdot \widehat{G}$ is closed in $\widehat{G}(\omega)$. Indeed, if $\beta \in \widehat{G}(\omega)$ and $(\alpha_\lambda)_\lambda$ is a net in $\widehat{G}$ such that $\gamma\alpha_\lambda \to \beta$ in $\widehat{G}(\omega)$, then $\alpha_\lambda = \gamma^{-1}(\gamma\alpha_\lambda) \to \gamma^{-1}\beta$ uniformly on compact subsets of G, and this implies that $\gamma^{-1}\beta \in \widehat{G}$.

Assume that $\gamma\widehat{G} \neq \widehat{G}(\omega)$ and choose $\delta \in \widehat{G}(\omega) \setminus \gamma\widehat{G}$. Then, since $\widehat{G}$ is a group, $\gamma\widehat{G} \cap \delta\widehat{G} = \emptyset$. Because $\gamma\widehat{G}$ and $\delta\widehat{G}$ are both closed sets of uniqueness for $L^1(G,\omega)$ and $L^1(G,\omega)$ has the unique uniform norm property, Theorem 4.6.5 yields that $\partial(L^1(G,\omega)) \subseteq \gamma\widehat{G} \cap \delta\widehat{G}$, which is empty. This contradiction shows that $\gamma\widehat{G} = \widehat{G}(\omega)$. Finally, since the function $|\gamma|$ is bounded away from zero on every compact subset of G, it is easy to see that the map $\beta \to \gamma^{-1}\beta$ from $\widehat{G}(\omega)$ onto $\widehat{G}$ is continuous. Thus $\alpha \to \gamma\alpha$ is a homeomorphism.

To prove (ii), let $E \subseteq \widehat{G}(\omega) = \gamma\widehat{G}$ be a closed boundary for $L^1(G,\omega)$. Then $F = \{\alpha \in \widehat{G} : \gamma\alpha \in E\}$ is closed in $\widehat{G}$ and for every $f \in L^1(G)$ we have $\gamma^{-1}f \in L^1(G,\omega)$ and

$$\|\widehat{f}\|_\infty = \sup\left\{|\widehat{\gamma^{-1}f}(\gamma\alpha)| : \alpha \in \widehat{G}\right\} = \|\widehat{\gamma^{-1}f}\|_\infty$$
$$= \|\widehat{\gamma^{-1}f}|_E\|_\infty = \sup\{|\widehat{f}(\alpha)| : \alpha \in F\}$$
$$= \|\widehat{f}|_F\|_\infty.$$

Since $L^1(G)$ is regular, it follows that $F = \widehat{G}$ and hence $E = \widehat{G}(\omega)$.   □

Theorem 4.7.3. *Let G be a locally compact Abelian group and ω a weight on G. Then $L^1(G,\omega)$ has the unique uniform norm property if and only if $L^1(G,\omega)$ is regular.*

Proof. Since for every semisimple commutative Banach algebra regularity implies the unique uniform norm property, we only have to show the 'only if' part.

Let E be a proper closed subset of $\widehat{G}(\omega)$ and let $\gamma \in \widehat{G}(\omega) \setminus E$. By the preceding lemma, the map $\alpha \to \gamma\alpha$ is a homeomorphism between $\widehat{G}$ and $\widehat{G}(\omega)$. Thus $F = \{\alpha \in \widehat{G} : \gamma\alpha \in E\}$ is a closed subset of $\widehat{G}$ which does not contain the trivial character 1_G. Choose a symmetric open neighbourhood U of 1_G in $\widehat{G}$ such that $F \cap U^2 = \emptyset$. Then $E \cap \gamma U^2 = \emptyset$.

Because $L^1(G,\omega)$ is semisimple (Theorem 2.8.10), has the unique uniform norm property, and satisfies $\partial(L^1(G,\omega)) = \widehat{G}(\omega)$ (Proposition 4.7.2(ii)), $L^1(G,\omega)$ is weakly regular by Corollary 4.6.7. Thus there exists a nonzero function $g \in L^1(G,\omega)$ such that $\widehat{g}(\delta) = 0$ for all $\delta \in \widehat{G}(\omega) \setminus \gamma U$. Since g is nonzero, $\widehat{g}(\gamma\beta) \neq 0$ for some $\beta \in U$. Now, let $f = \overline{\beta}g$. Then $f \in L^1(G,\omega)$ and $\widehat{f}(\gamma) = \widehat{g}(\gamma\beta) \neq 0$. Moreover, $\widehat{f}(\delta) = \widehat{g}(\delta\beta) = 0$ for all $\delta \in E$ since

$$\beta E \cap \gamma U = \beta(E \cap \overline{\beta}\gamma U) \subseteq \beta(E \cap \gamma U^2) = \emptyset.$$

This shows that $L^1(G,\omega)$ is regular.   □

Passing to the problem of when $L^1(G,\omega)$ is regular, we now introduce a condition on the weight ω that turns out to entail regularity of $L^1(G,\omega)$.

Definition 4.7.4. Let G be a locally compact Abelian group. A weight ω on G such that $\omega(x) \geq 1$ for all $x \in G$ is said to be *nonquasianalytic* if

$$\sum_{n=-\infty}^{\infty} \frac{\ln \omega(x^n)}{1+n^2} < \infty$$

for all $x \in G$.

Lemma 4.7.5. *Let G be a nonquasianalytic weight on G. Then*

$$\lim_{n\to\infty} \omega(x^n)^{1/n} = \lim_{n\to\infty} \omega(x^{-n})^{-1/n} = 1$$

for all $x \in G$. In particular, $\widehat{G}(\omega) = \widehat{G}$.

Proof. Suppose the statement is false, so that $\lim_{n\to\infty} \omega(x^n)^{1/n} > 1$ for some $x \in G$. Then there exist some $\delta > 0$ and $N \in \mathbb{N}$ such that $\omega(x^n)^{1/n} \geq 1 + \delta$ for all $n \geq N$. For all such n,

$$\frac{\ln \omega(x^n)}{1+n^2} = \frac{\ln \omega(x^n)^{1/n}}{\frac{1}{n} + n} \geq \frac{\ln(1 + \delta)}{\frac{1}{n} + n}.$$

However, this is impossible since ω is nonquasianalytic. Thus $\lim_{n\to\infty} \omega(x^n)^{1/n} \leq 1$ for all $x \in G$.

Since $\omega(y) \geq 1$ for all $y \in G$, it follows that $\lim_{n\to\infty} \omega(x^n)^{1/n} = 1$ for all $x \in G$ and, as noted in Remark 2.8.3, this implies that $\widehat{G}(\omega) = \widehat{G}$. Finally, we also get

$$\lim_{n\to\infty} \omega(x^{-n})^{-1/n} = \left(\lim_{n\to\infty} \omega((x^{-1})^n)^{1/n}\right)^{-1} = 1$$

for every $x \in G$.   □

Clearly, each bounded weight is nonquasianalytic, and in this case $L^1(G, \omega)$ $= L^1(G)$ as algebras. Using the next two lemmas it is very easy to construct unbounded weights which are nonquasianalytic.

Lemma 4.7.6. *A weight ω on $\mathbb{Z}$ is nonquasianalytic if and only if*

$$\sum_{n=-\infty}^{\infty} \frac{\ln \omega(n)}{1 + n^2} < \infty.$$

Proof. For $x \in \mathbb{N}$, this condition implies that

$$\sum_{n=-\infty}^{\infty} \frac{\ln \omega(nx)}{1 + n^2} \leq \sum_{n=-\infty}^{\infty} \frac{\ln(\omega(n)^x)}{1 + n^2} = x \cdot \sum_{n=-\infty}^{\infty} \frac{\ln \omega(n)}{1 + n^2} < \infty.$$

If $x \in -\mathbb{N}$, apply the preceding inequality with $-x$ in place of x and use that

$$\sum_{n=-\infty}^{\infty} \frac{\ln(\omega(-n))}{1 + n^2} = \sum_{n=-\infty}^{\infty} \frac{\ln \omega(n)}{1 + n^2}.$$

Conversely, if ω is nonquasianalytic, taking $x = 1$ in Definition 4.7.4 shows that $\sum_{n=-\infty}^{\infty} (1 + n^2)^{-1} \ln \omega(n) < \infty$. □

In the following, for each $t \in \mathbb{R}$, $\lfloor t \rfloor$ will denote the greatest integer $\leq t$.

Lemma 4.7.7. *A weight ω on $\mathbb{R}$ is nonquasianalytic if and only if*

$$\int_{-\infty}^{\infty} \frac{\ln \omega(t)}{1 + t^2} \, dt < \infty.$$

Proof. Let $C = \sup\{\omega(s) : 0 \leq s \leq 1\}$. Then $C < \infty$ and

$$\omega(t) \leq \omega(\lfloor t \rfloor)\omega(t - \lfloor t \rfloor) \leq C\omega(\lfloor t \rfloor)$$

for all t. This implies

$$\int_{-\infty}^{\infty} \frac{\ln \omega(t)}{1 + t^2} dt = \sum_{k=-\infty}^{\infty} \int_k^{k+1} \frac{\ln w(t)}{1 + t^2} dt$$

$$\leq C \sum_{k=-\infty}^{\infty} \ln \omega(k) \int_k^{k+1} \frac{1}{1 + t^2} dt$$

$$\leq C \sum_{k=-\infty}^{\infty} \frac{\ln \omega(k)}{1 + k^2}.$$

Thus nonquasianalyticity implies that $\int_{-\infty}^{\infty} (1 + n^2)^{-1} \ln \omega(n) dt < \infty$.

Conversely, suppose that ω satisfies this latter condition. It suffices to show that $\sum_{n=-\infty}^{\infty} (1 + n^2)^{-1} \ln \omega(n) < \infty$ for every $x > 0$. Let $m \in \mathbb{N}_0$ such that $m < x \leq m + 1$. Then

$$\sum_{n=-\infty}^{\infty} \frac{\ln \omega(nx)}{1+n^2} \le \sum_{n=-\infty}^{\infty} \frac{\ln(\omega(nm)\omega(n(x-m)))}{1+n^2}$$

$$= \sum_{n=-\infty}^{\infty} \frac{\ln \omega(nm)}{1+n^2} + \sum_{n=-\infty}^{\infty} \frac{\ln \omega(n(x-m))}{1+n^2}$$

$$\le \sum_{n=-\infty}^{\infty} \frac{\ln(\omega(n)^m)}{1+n^2} + \sum_{n=-\infty}^{\infty} \frac{\ln \omega(n(x-m))}{1+n^2}$$

$$= m \sum_{n=-\infty}^{\infty} \frac{\ln \omega(n)}{1+n^2} + \sum_{n=-\infty}^{\infty} \frac{\ln \omega(n(x-m))}{1+n^2}.$$

Thus we can even assume that $0 < x \le 1$. Then, since

$$\omega(nx) \le \omega(tx)\omega((n-t)x) \le C\omega(tx)$$

for all $t \in [n-1, n]$, it follows that

$$\sum_{n=-\infty}^{\infty} \frac{\ln \omega(nx)}{1+n^2} \le C \sum_{n=-\infty}^{\infty} \int_{n-1}^{n} \frac{\ln \omega(tx)}{1+t^2} dt$$

$$= \frac{C}{x} \int_{-\infty}^{\infty} \frac{\ln \omega(s)}{1+(s/x)^2} ds = Cx \int_{-\infty}^{\infty} \frac{\ln \omega(s)}{s^2 + 1/x^2} ds$$

$$\le Cx \int_{-\infty}^{\infty} \frac{\ln \omega(s)}{1+s^2} ds < \infty.$$

This completes the proof that ω is nonquasianalytic.   □

With regard to the proofs of the following three lemmas and of Theorem 4.7.11, we point out that if ω is a nonquasianalytic weight on G and hence $\widehat{G}(\omega) = \widehat{G}$, in order to establish regularity of $L^1(G, \omega)$ it suffices to show that, given any neighbourhood U of 1_G in $\widehat{G}$, there exists $f \in L^1(G, \omega)$ such that $\widehat{f}(1_G) \ne 0$ and $\widehat{f}(\alpha) = 0$ for all $\alpha \in \widehat{G} \setminus U$.

Lemma 4.7.8. *Let ω be a nonquasianalytic weight function on $\mathbb{R}$. Then $L^1(\mathbb{R}, \omega)$ is regular.*

Proof. The main tool in the proof will be the well-known Paley–Wiener theorem. Define a function h on $\mathbb{R}$ by

$$h(t) = \frac{1}{(1+t^2)\omega(t)} \quad (t \in \mathbb{R}).$$

Then $h \in L^1(\mathbb{R}) \cap L^2(\mathbb{R})$ since $\omega(t) \ge 1$. Moreover, since

$$\frac{\ln(1+t^2)}{1+t^2} \le \frac{t^{2/3}}{1+t^2} \le \frac{1}{t^{4/3}}$$

for large t and since ω is nonquasianalytic, Lemma 4.7.7 shows that

$$\int_{-\infty}^{\infty} \frac{|\ln h(t)|}{1+t^2} dt = \int_{-\infty}^{\infty} \frac{\ln(1+t^2)}{1+t^2} dt + \int_{-\infty}^{\infty} \frac{\ln \omega(t)}{1+t^2} dt < \infty.$$

Then, by the Paley–Wiener theorem, there exists a function $g \in L^2(\mathbb{R})$ such that $|g(t)| = h(t)$ almost everywhere and the Plancherel transform of g vanishes almost everywhere on $(-\infty, 0)$. Since $g \in L^1(\mathbb{R}) \cap L^2(\mathbb{R})$, the Plancherel transform of g equals the Fourier transform $\widehat{g}$ of g and $\widehat{g}$ is continuous. So $\widehat{g}(s) = 0$ for all $s \leq 0$. Let

$$s_0 = \inf\{s \in \mathbb{R} : \widehat{g}(s) \neq 0\}.$$

Replacing g by the function $t \to g(t)e^{is_0 t}$ if necessary, we can assume that $s_0 = 0$. It suffices now to show that given any $\epsilon > 0$, there exists $f \in L^1(\mathbb{R}, \omega)$ such that $\widehat{f}(0) \neq 0$ and $\widehat{f}(s) = 0$ when $|s| \geq \epsilon$. Since $s_0 = 0$, there exists $0 < \delta \leq \epsilon$ such that $\widehat{g}(\delta) \neq 0$. Define functions f_1 and f_2 on $\mathbb{R}$ by

$$f_1(t) = g(t)e^{i\delta t} \quad \text{and} \quad f_2(t) = \overline{f_1(t)} = \overline{g(t)}e^{-i\delta t}.$$

Then the function $f = f_1 * f_2 \in L^1(\mathbb{R})$ satisfies $\widehat{f}(0) \neq 0$ and $\widehat{f}(s) = 0$ whenever $|s| \geq \epsilon$. Indeed,

$$\widehat{f}(s) = \widehat{f_1}(s)\widehat{f_2}(s) = \widehat{g}(\delta + s)\overline{\widehat{g}(\delta - s)}$$

and so $\widehat{f}(s) = 0$ whenever $|s| \geq \delta$. Moreover, $\widehat{f}(0) = |\widehat{g}(\delta)|^2 \neq 0$.
Finally, we show that $f \in L^1(\mathbb{R}, \omega)$. We have

$$|f(t)| \leq (|f_1| * |f_2|)(t) = (|g| * |g|)(t) = (h * h)(t)$$

for all $t \in \mathbb{R}$, and hence

$$\begin{aligned}
|f(t)| &\leq \int_{-\infty}^{\infty} \frac{1}{(1+u^2)(1+(t-u)^2)\omega(u)\omega(t-u)} du \\
&\leq \frac{1}{\omega(t)} \int_{-\infty}^{\infty} \frac{1}{(1+u^2)(1+(t-u)^2)} du \\
&= \frac{2\pi}{\omega(t)(t^2+4)}.
\end{aligned}$$

Thus $f\omega \in L^1(\mathbb{R})$, as required.   □

Lemma 4.7.9. *Let ω be a nonquasianalytic weight on the integer group $\mathbb{Z}$. Then $l^1(\mathbb{Z}, \omega)$ is regular.*

Proof. We reduce this case to that of the real line. To this end, we first extend ω to $\mathbb{R}$ by setting

$$\omega_1(t) = \omega(\lfloor t \rfloor) + (t - \lfloor t \rfloor)(\omega(\lfloor t \rfloor + 1) - \omega(\lfloor t \rfloor)),$$

Since we will have lost submultiplicativity through interpolation, we now put .

$$\widetilde{\omega}(t) = \sup\left\{\frac{\omega_1(t+s)\omega_1(0)}{\omega_1(s)} : s \in \mathbb{R}\right\}.$$

Then $\widetilde{\omega}(t) \geq \omega_1(t) \geq 1$ for all $t \in \mathbb{R}$ and $\widetilde{\omega}(k) \geq \omega_1(k) = \omega(k)$ for $k \in \mathbb{Z}$. We show next that

$$\widetilde{\omega}(t_1 + t_2) \leq \widetilde{\omega}(t_1)\widetilde{\omega}(t_2)$$

for all $t_1, t_2 \in \mathbb{R}$. By definition of $\widetilde{\omega}$, it suffices to verify that

$$\widetilde{\omega}(t_1)\widetilde{\omega}(t_2) \geq \frac{\omega_1(t_1 + t_2 + s)\omega_1(0)}{\omega_1(s)}$$

for all $s \in \mathbb{R}$. Now, given s, let $s_1 = t_2 + s$ and $s_2 = s$. Then

$$\begin{aligned}
\widetilde{\omega}(t_1)\widetilde{\omega}(t_2) &\geq \frac{\omega_1(t_1 + s_1)\omega_1(0)}{\omega_1(s_1)} \cdot \frac{\omega_1(t_2 + s_2)\omega_1(0)}{\omega_1(s_2)} \\
&= \frac{\omega_1(t_1 + t_2 + s)}{\omega_1(s)} \cdot \omega_1(0)^2 \\
&\geq \frac{\omega_1(t_1 + t_2 + s)}{\omega_1(s)} \cdot \omega_1(0).
\end{aligned}$$

Thus $\widetilde{\omega}(t_1)\widetilde{\omega}(t_2) \geq \widetilde{\omega}(t_1 + t_2)$. Observe next that, with $C = \max\{\omega(0), \omega(1)\}$, we have

$$\widetilde{\omega}(t) \leq C\omega(\lfloor t \rfloor)$$

for all $t \in \mathbb{R}$. This implies

$$\begin{aligned}
\int_{-\infty}^{\infty} \frac{\ln\widetilde{\omega}(t)}{1+t^2} dt &\leq C \sum_{k=-\infty}^{\infty} \int_k^{k+1} \frac{\ln\omega(\lfloor t \rfloor)}{1+t^2} dt \\
&\leq C \sum_{k=-\infty}^{\infty} \int_k^{k+1} \frac{\ln\omega(k)}{1+k^2} dt \\
&= C \sum_{k=-\infty}^{\infty} \frac{\ln\omega(k)}{1+k^2} < \infty.
\end{aligned}$$

Thus, by Lemma 4.7.7, $\widetilde{\omega}$ is a nonquasianalytic weight on $\mathbb{R}$ and therefore, since $L^1(\mathbb{R}, \widetilde{\omega})$ is regular by Lemma 4.7.8, given any $0 < \epsilon < \pi$, there exists $g \in L^1(\mathbb{R}, \widetilde{\omega})$ such that $\hat{g}(0) \neq 0$ and $\hat{g}(y) = 0$ for all $y \in \mathbb{R}$ with $|y| \geq \epsilon$. Actually, as shown in the proof of Lemma 4.7.8, g can be constructed in such a way that

$$|g(t)|\widetilde{\omega}(t) \leq \frac{1}{1+t^2}$$

for all $t \in \mathbb{R}$. Because

$$|g(k)| \cdot \omega(k) \leq |g(k)|\tilde{\omega}(k) \leq \frac{1}{1+k^2}$$

for all $k \in \mathbb{Z}$, we get that $g|_{\mathbb{Z}} \in l^1(\mathbb{Z}, \omega)$. The Fourier transform of $g|_{\mathbb{Z}}$ is given by the Fourier series

$$\widehat{g|_{\mathbb{Z}}}(z) = \sum_{k=-\infty}^{\infty} g(k)z^{-k}.$$

For each $l \in \mathbb{Z}$, it follows that

$$\int_{-\pi}^{\pi} \widehat{g|_{\mathbb{Z}}}(e^{it})e^{ilt}\,dt = \sum_{k=-\infty}^{\infty} g(k) \int_{-\pi}^{\pi} e^{it(l-k)}\,dt$$
$$= 2\pi g|_{\mathbb{Z}}(l) = 2\pi g(l)$$
$$= \int_{-\infty}^{\infty} \widehat{g}(t)e^{ilt}\,dt$$
$$= \int_{-\pi}^{\pi} \widehat{g}(t)e^{ilt}\,dt.$$

Since the finite linear combinations of functions $t \to a^{ilt}, l \in \mathbb{Z}$, are dense in $L^2[-\pi, \pi]$, we obtain that

$$\widehat{g|_{\mathbb{Z}}}(e^{it}) = \widehat{g}(t)$$

for all $|t| \leq \pi$. Hence $\widehat{g|_{\mathbb{Z}}}$ vanishes on the complement of the arc determined by $e^{-i\epsilon}$ and $e^{i\epsilon}$. Finally, $\widehat{g|_{\mathbb{Z}}}(1_G) = \widehat{g}(0) \neq 0$. □

At the current stage we know that $L^1(G, \omega)$ is regular for any non-quasianalytic weight whenever G equals $\mathbb{R}$ of $\mathbb{Z}$ or when G is compact. The following lemma provides a natural tool to enlarge the class of groups for which regularity can be shown.

Lemma 4.7.10. *Let G_1 and G_2 be locally compact Abelian groups and let $G = G_1 \times G_2$. Suppose that $L^1(G_1, \omega_1)$ and $L^1(G_2, \omega_2)$ are regular for all nonquasianalytic weights ω_1 and ω_2, respectively. Then $L^1(G, \omega)$ is regular for every nonquasianalytic weight ω on G.*

Proof. Let ω be given and define ω_1 and ω_2 by $\omega_1(x_1) = \omega(x_1, e_2), x_1 \in G_1$, and $\omega_2(x_2) = \omega(e_1, x_2), x_2 \in G_2$, where e_1 and e_2 are the identities of G_1 and G_2, respectively. Then clearly both ω_1 and ω_2 are non-quasianalytic weights. For any neighbourhood V_1 of 1_{G_1} in $\widehat{G}_1$ and V_2 of 1_{G_2} in $\widehat{G}_2$, there exist $f_1 \in L^1(G_1, \omega_1)$ and $f_2 \in L^1(G_2, \omega_2)$ satisfying $\widehat{f}_1(1_{G_1}) \neq 0, \widehat{f}_1 = 0$ on $\widehat{G}_1 \setminus V_1$, and similarly for f_2. Define f on $G_1 \times G_2$ by $f(x_1, x_2) = f_1(x_1)f_2(x_2)$. Then $f \in L^1(G, \omega)$ since

$$\omega(x_1, x_2) \leq \omega(x_1, e_2)\omega(e_1, x_2) = \omega_1(x_1)\omega_2(x_2).$$

Furthermore, $\widehat{f}(1_G) = \widehat{f}_1(1_{G_1})\widehat{f}_2(1_{G_2}) \neq 0$ and $\widehat{f}$ vanishes outside of $V_1 \times V_2$. These sets form a neighbourhood basis of 1_G in $\widehat{G}$. Hence it follows that $L^1(G, \omega)$ is regular. □

Using Lemmas 4.7.8 through 4.7.10 and employing the structure theorem for compactly generated locally compact Abelian groups, we can now prove the result at which we are aiming.

Theorem 4.7.11. *Let G be any locally compact Abelian group and ω a non-quasianalytic weight on G. Then $L^1(G, \omega)$ is regular.*

Proof. Let V be any relatively compact neighbourhood of 1_G in $\widehat{G}$. Let φ be a continuous function on $\widehat{G}$ such that $\varphi(1_G) = 1$ and $\varphi = 0$ on $\widehat{G} \setminus V$. Since the image of $C_c(G)$ under the Gelfand homomorphism is dense in $C_0(\widehat{G})$, there exists $k \in C_c(G)$ such that $\|\varphi - \widehat{k}\|_\infty \leq \frac{1}{3}$. Let

$$h = \widehat{k}(1_G)^{-1}k \in C_c(G).$$

Then $\widehat{h}(1_G) = 1$ and, for all $\alpha \in \widehat{G} \setminus V$,

$$|\widehat{h}(\alpha)| = |\widehat{k}(1_G)|^{-1}|\widehat{k}(\alpha)| \leq |\widehat{k}(1_G)|^{-1}\left(\|\widehat{k} - \varphi\|_\infty + |\varphi(\alpha)|\right)$$

$$\leq \frac{1}{3}|\widehat{k}(1_G)|^{-1} \leq \frac{1/3}{2/3}$$

$$= \frac{1}{2}.$$

Now, choose a compactly generated open subgroup H of G containing the support of h. By the structure theorem for compactly generated locally compact Abelian groups (Theorem A.5.5),

$$H = K \times \mathbb{R}^p \times \mathbb{Z}^q,$$

where $p, q \in \mathbb{N}_0$ and K is a compact group. Since $\omega|_H$ is nonquasianalytic, combining Lemmas 4.7.8 through 4.7.10, we conclude that $L^1(H, \omega|_H)$ is regular. Let the Haar measure of H be the Haar measure of G restricted to H, and let

$$U = \left\{\gamma \in \widehat{H} : |\widehat{h}(\gamma)| > \frac{1}{2}\right\}.$$

Then U is an open neighbourhood of 1_H in $\widehat{H}$ since $\widehat{h|_H}(1_H) = \widehat{h}(1_G) = 1$. Because $L^1(H, \omega|_H)$ is regular, there exists $g \in L^1(H, \omega|_H)$ satisfying $\widehat{g}(1_H) \neq 0$ and $\widehat{g}(\gamma) = 0$ for all $\gamma \in \widehat{H} \setminus U$. Let $f : G \to \mathbb{C}$ be defined by $f(x) = g(x)$ for $x \in H$ and $f(x) = 0$ for $x \notin H$. Then $f \in L^1(G, \omega)$ and $\widehat{f}(\alpha) = \widehat{g}(\alpha|_H)$ for all $\alpha \in \widehat{G}$. If $\alpha \notin V$, then

$$\left|\int_H h(x)\overline{\alpha|_H(x)}dx\right| = \left|\int_G h(y)\overline{\alpha(y)}dy\right| = \left|\widehat{h}(\alpha)\right| \leq \frac{1}{2},$$

whence $\alpha|_H \notin U$. This implies that $\widehat{f}(\alpha) = \widehat{g}(\alpha|_H) = 0$ for all $\alpha \in \widehat{G} \setminus V$. Finally, $\widehat{f}(1_G) = \widehat{g}(1_H) \neq 0$, and this completes the proof of the theorem. $\square$

We finish this section by just mentioning that the converse to Theorem 4.7.11 also holds. That is, regularity of $L^1(G, \omega)$ implies that ω is non-quasianalytic.

4.8 Exercises

Exercise 4.8.1. Prove that the only hk-continuous functions in the disc algebra $A(\mathbb{D})$ are the constant functions.

Exercise 4.8.2. Let A be a commutative Banach algebra and I a closed ideal of finite codimension of A. Show that $h(I)$ is finite. Why does the converse not hold?

Exercise 4.8.3. Let A be a commutative Banach algebra and U an open subset of $\Delta(A)$. Suppose that the Gelfand and the hull-kernel topologies agree on U. Let C be a subset of U which is compact in the Gelfand topology. Prove that C is hk-closed in $\Delta(A)$.
(Hint: Assuming that there exists $\varphi \in h(k(C)) \setminus C$, find an element $a \in k(\Delta(A) \setminus U)$ such that $\hat{a} = 1$ on C and $\hat{a}(\varphi) = 0$.)

Exercise 4.8.4. Prove that $\mathrm{Lip}_\alpha[0,1]$, the algebra of Lipschitz functions of order α (Exercise 1.6.11), is regular.

Exercise 4.8.5. Let A be a commutative Banach algebra and suppose that $\Delta(A)$ is totally disconnected. Apply Shilov's idempotent theorem to show that A is regular.
(Hint: Given a closed subset E of $\Delta(A)$ and $\varphi \in \Delta(A) \setminus E$, there exists an open and closed neighbourhood V of φ such that $V \cap E = \emptyset$).

Exercise 4.8.6. Let A be a semisimple commutative Banach algebra with the property that the product of any two nonzero elements of A is nonzero. Prove that if A is regular, then A is at most one-dimensional.

Exercise 4.8.7. Use the conclusion of Exercise 4.8.6 to show that the convolution algebras $L^1(\mathbb{R}^+)$ and $l^1(\mathbb{Z}^+)$ are not regular. For generalisations, see Exercises 4.8.39 and 4.8.40.

Exercise 4.8.8. Let G be a nondiscrete locally compact Abelian group. Then the so-called Wiener–Pitt phenomenon asserts that there exists a noninvertible measure $\mu \in M(G)$ with the property that the Fourier–Stieltjes transform $\hat{\mu}$ of μ satisfies $|\hat{\mu}(\alpha)| \geq 1$ for all $\alpha \in \hat{G}$. Deduce that $\hat{G}$ is not dense in $\Delta(M(G))$ with respect to the Gelfand topology.

Exercise 4.8.9. Show that the measure algebra $M(G)$ of a nondiscrete locally compact Abelian group G fails to be regular.
(Hint: Use Exercise 4.8.8 and Lemma 4.1.10.)

Exercise 4.8.10. Let A be a commutative Banach algebra with bounded approximate identity and let

$$M_{00}(A) = \{T \in M(A) : \hat{T}|_{\Delta(M(A))\setminus\Delta(A)} = 0\}.$$

Show that $M_{00}(A)$ is regular if and only if A is regular.

Exercise 4.8.11. Let A be a semisimple (and hence faithful) commutative Banach algebra and let $T \in M(A)$. Then

(i) $\sigma_{M(A)}(T) = \widehat{T}(\Delta(M(A)))$.

(ii) $\sigma_{M(A)}(T) = \overline{\widehat{T}(\Delta(A))}$ if $\widehat{T}$ is hk-continuous on $\Delta(M(A))$.

Exercise 4.8.12. Let A and B be unital commutative Banach algebras with identities e_A and e_B, respectively, and suppose that A is semisimple and regular. Let $\phi : A \to B$ be an injective homomorphism such that $\phi(e_A) = e_B$. Use results of Section 4.2 to show that $\sigma_B(\phi(x)) = \widehat{x}(\Delta(A))$ for every $x \in A$.

Exercise 4.8.13. Let A and B be commutative Banach algebras such that B is unital and A is nonunital. Let $\phi : A \to B$ be an injective homomorphism. Show that $\sigma_B(\phi(x)) = \widehat{x}(\Delta(A))$ for every $x \in A$.
(Hint: Extend ϕ to an injective homomorphism from the unitisation A_e into B.)

Exercise 4.8.14. Let A be a semisimple and symmetric commutative Banach $*$-algebra. View $\Delta(A)$ as a subset of $\Delta(M(A))$. For $T \in M(A)$, define f_T on $\Delta(A)$ by $f_T(\varphi) = \widehat{T}(\varphi)$.

(i) Show that $f_T \cdot \widehat{A} \subseteq \widehat{A}$. Hence there exists a unique element T^* of $M(A)$ such that $\widehat{T^*(x)}(\varphi) = \overline{f_T(\varphi)}\widehat{x}(\varphi)$ for all $\varphi \in \Delta(A)$ and $x \in A$ (Proposition 2.2.16).

(ii) Assume that $M(A)$ is regular. Show that the involution $T \to T^*$ turns $M(A)$ into a symmetric Banach $*$-algebra.

Exercise 4.8.15. Let A be a symmetric commutative Banach $*$-algebra. Then both the greatest regular subalgebra and the greatest regular ideal of A are $*$-algebras.

Exercise 4.8.16. Let U be a connected open subset of the complex plane and A a unital closed subalgebra of $H^\infty(U)$. Show that $\mathrm{reg}(A)$ consists only of the constant functions.

Exercise 4.8.17. Let A be a commutative Banach algebra and U an open subset of $\Delta(A)$. Suppose that U is hk-dense in $\Delta(A)$ and that the Gelfand and the hull-kernel topologies agree on U. Let

$$A_0(U) = \{a \in A : \widehat{a}|_U \in C_0(U)\},$$

which is a closed ideal of A. Show that

$$k(\Delta(A) \setminus U) \subseteq \mathrm{reg}(A_0(U)).$$

(Hint: Use Theorem 4.2.3 to conclude that $k(\Delta(A) \setminus U)$ is regular.)

Exercise 4.8.18. Let A be a commutative Banach algebra and $n \in \mathbb{N}$. Show that

$$\mathrm{reg}(C^n([0,1], A)) = C^n([0,1], \mathrm{reg}(A)).$$

(Hint: Prove by induction on k that $C^k[0,1] \otimes A$ is dense in $C^k([0,1], A)$, $k \in \mathbb{N}$.)

Exercise 4.8.19. Let A be a regular and semisimple commutative Banach algebra with bounded approximate identity. View A as a closed ideal of the multiplier algebra $M(A)$, and define two subalgebras $M_0(A)$ and $M_{00}(A)$ by

$$M_0(A) = \{T \in M(A) : \widehat{T}|_{\Delta(A)} \in C_0(\Delta(A))\}$$

and

$$M_{00}(A) = \{T \in M(A) : \widehat{T}|_{\Delta(M(A))\backslash\Delta(A)} = 0\}.$$

Prove that $\mathrm{reg}(M_0(A)) = M_{00}(A)$.

Exercise 4.8.20. Let G be a locally compact Abelian group, $M(G)$ the measure algebra of G, and $M_d(G)$ the subalgebra consisting of all discrete measures. Show that $\mathrm{reg}(M(G))$ contains $M_d(G)$. For any closed subgroup H of G, consider $L^1(H)$ as a closed subalgebra of $M(G)$. Then $L^1(H) \subseteq \mathrm{reg}(M(G))$.

Let A be a commutative Banach algebra and G a locally compact Abelian group. It was shown in [69] that $\mathrm{reg}(L^1(G, A)) = L^1(G, \mathrm{reg}(A))$. The conclusions of the next three exercises constitute major steps of the proof.

Exercise 4.8.21. Let G be a locally compact Abelian group and A a commutative Banach algebra. Let $\alpha \in \widetilde{G}$ and define $\phi_\alpha : L^1(G, A) \to A$ by

$$\phi_\alpha(f) = \int_G \alpha(x) f(x) dx \ (f \in L^1(G, A)).$$

Prove that $\phi_\alpha(\mathrm{reg}(L^1(G, A))) \subseteq \mathrm{reg}(A)$.

Exercise 4.8.22. Let B be a regular commutative Banach algebra and let G be an Abelian locally compact group. Let $\phi : L^1(G) \otimes B \to L^1(G, B)$ be the unique linear map such that, for all $f \in L^1(G)$ and $b \in B$,

$$\phi(f \otimes b)(x) = f(x) b$$

for almost all $x \in G$.

(ii) Show that ϕ is a homomorphism and extends to a homomorphism of the projective tensor product $L^1(G) \widehat{\otimes}_\pi B$ into $L^1(G, B)$.

(iii) Adapt the corresponding part of the proof of Theorem 4.3.11 to show that $\phi(C_c(G) \otimes B)$ is dense in $L^1(G, B)$.

(iii) Conclude that $L^1(G, B)$ is regular.

Exercise 4.8.23. Let G and A be as in Exercise 4.8.21. Let $f \in \text{reg}(L^1(G, A))$ and suppose that f is continuous. Show that $\psi(f(x)) = 0$ for all $x \in G$ whenever $\psi \in A^*$ is such that $\psi(\text{reg}(A)) = \{0\}$. Conclude that $f \in L^1(G, \text{reg}(A))$.

Exercise 4.8.24. In Example 4.5.8, fill in the details to show that A/I is isometrically isomorphic to $A(Y)$.

Exercise 4.8.25. Let X and A be as in Example 4.5.10. Carry out the proof of the statement that the point evaluation map $x \to \varphi_x$ is a homeomorphism from X onto $\Delta(A)$.

Exercise 4.8.26. Let X be a locally compact Hausdorff space and A a semisimple commutative Banach algebra. Show that $C_0(X, A)$ has the unique uniform norm property if and only if the same is true of A.

Exercise 4.8.27. Let $|\cdot|_1$ and $|\cdot|_2$ be two equivalent uniform norms on a Banach algebra A. Show that $|\cdot|_1$ and $|\cdot|_2$ are equal.

Exercise 4.8.28. Let $(A, \|\cdot\|)$ be a semisimple commutative Banach algebra and let I be a dense ideal in A which is a Banach algebra under some norm $\|\cdot\|_0$. If A has the unique uniform norm property, then the same is true of I. To verify this, let $|\cdot|$ be a uniform norm on I and prove successively the following assertions.

(i) Since I is an ideal in A, we have $|x| \leq r_I(x) = r_A(x) \leq \|x\|$ for all $x \in I$.

(ii) $|\cdot|$ extends (uniquely) to a uniform seminorm on A, also denoted $|\cdot|$.

(iii) Let $x \in A$ such that $|x| = 0$ and choose a sequence $(x_n)_n$ in I with $\|x_n - x\| \to 0$. Show that $xx_n = 0$ and that this forces $x = 0$.

(iv) Deduce that $|x| = r_I(x)$ for every $x \in I$.

Exercise 4.8.29. Let A be a commutative Banach algebra with approximate identity of norm bound one. Embed A isometrically into $M(A)$ by $x \to L_x$, where $L_x(y) = xy$, $x, y \in A$, and let r denote the spectral radius on $M(A)$. Let $|\cdot|$ be a uniform norm on A and define $|\cdot|_r$ on $M(A)$ by

$$|T|_r = \sup\{|T(x)| : x \in A, r(x) \leq 1\}.$$

(i) Show that $|T|_r < \infty$ and $|T(x)| \leq |T|_r r(x)$ for all $T \in M(A)$ and $x \in A$.

(ii) Let $a \in A$ with $r(a) \leq 1$ and $\epsilon > 0$. Deduce from *Cohen's factorization theorem* (see [19] or [55]) that a can be written as a product $a = bc$ where $r(b) \leq 1$ and $r(c) \leq r(a) + \epsilon$. Use this to prove that $|\cdot|_r$ is submultiplicative and hence an algebra norm on $M(A)$.

(iii) Use the square property of $|\cdot|$ to show that $|T^2|_r = |T|_r^2$ for all $T \in M(A)$.

(iv) Suppose that $M(A)$ has the unique uniform norm property. Prove that $|x| = |L_x|_r = r(x)$ for all $x \in A$. Thus A has the unique uniform norm property.

Exercise 4.8.30. Let A be a unital commutative Banach algebra and let B be a unital commutative extension of A. Prove that the following three conditions are equivalent.

(i) Every $\varphi \in \Delta(A)$ extends to some element of $\Delta(B)$.

(ii) For any finitely many $a_1, \ldots, a_n \in A$,

$$\sigma_A(a_1, \ldots, a_n) \subseteq \sigma_B(a_1, \ldots, a_n).$$

(iii) Every maximal ideal of A is contained in some maximal ideal of B.

Exercise 4.8.31. Let A be a semisimple commutative Banach algebra. Prove that A is weakly regular if and only if A_e, the unitisation of A, is weakly regular.

Exercise 4.8.32. Let A and B be semisimple commutative Banach algebras and let $\phi : A \to B$ be an injective homomorphism with dense range. Show that B is weakly regular whenever A is weakly regular.
(Hint: compare the proof of Lemma 4.2.16.)

Exercise 4.8.33. Let A be a commutative Banach algebra and I a closed ideal of A.

(i) Show that if A is weakly regular, then so is I.

(ii) Give an example showing that A/I need not be weakly regular whenever A is weakly regular.

Exercise 4.8.34. If A is a regular commutative Banach algebra, then $\partial(A) = \Delta(A)$. Conclude from results in Sections 4.5 and 4.6 that the converse statement is not true.

Exercise 4.8.35. Let A be a semisimple commutative Banach algebra with bounded approximate identity and identify A with the closed ideal $\{L_x : x \in A\}$ of $M(A)$. Suppose that A is weakly regular and let $|\cdot|$ be a uniform norm on $M(A)$.

(i) Let E be a closed subset of $\Delta(M(A))$ such that

$$|T| = \sup\{|\varphi(T)| : \varphi \in E\}$$

for all $T \in M(A)$ (Lemma 4.6.2). Prove that $E \supseteq \Delta(A)$.

(ii) Define a seminorm $|\cdot|_\infty$ on $M(A)$ by

$$|T|_\infty = \sup\{|\widehat{T}(\varphi)| : \varphi \in \Delta(A)\}, \ T \in M(A).$$

Conclude from (i) that $|T|_\infty \leq |T|$ for all $T \in M(A)$.

(iii) Suppose that $\Delta(A)$ is a set of uniqueness for $M(A)$. Then $|\cdot|_\infty$ is the smallest uniform norm on $M(A)$.

Exercise 4.8.36. Consider the following weights on $\mathbb{Z}$ and decide for which of them $l^1(\mathbb{Z}, \omega)$ is regular.

(a) $\omega(n) = e^{-n}$ for $n \geq 0$ and $\omega(n) = 1$ for $n \leq 0$.

(b) $\omega(n) = e^{-n^2}$ for $n \geq 0$ and $\omega(n) = 1$ for $n \leq 0$.

Exercise 4.8.37. Let $\alpha > 0$ and let ω be the weight on $\mathbb{R}$ defined by $\omega(t) = (1 + |t|)^\alpha$. Show that ω is nonquasianalytic.

Exercise 4.8.38. Let ω be a weight on the integer group $\mathbb{Z}$ and let R_- and R_+ be as in Proposition 2.8.8. Show that $l^1(\mathbb{Z}, \omega)$ is regular whenever $R_- = R_+$. For the converse statement compare Example 4.7.1.

Exercise 4.8.39. Let ω be a weight on $\mathbb{Z}^+$. Show that $l^1(\mathbb{Z}^+, \omega)$ does not have the unique uniform norm property.
(Hint: Let $R = \inf\{\omega(n)^{1/n} : n \in \mathbb{N}\}$. If $R > 0$, then $l^1(\mathbb{Z}^+, \omega)$ is isomorphic to a subalgebra of $A(\{z \in \mathbb{C} : |z| \le R\})$, whereas if $R = 0$, then $l^1(\mathbb{Z}^+, \omega)$ is not semisimple.)

Exercise 4.8.40. Prove that $L^1(\mathbb{R}^+, \omega)$ does not have the unique uniform norm property for any weight ω on $\mathbb{R}^+$.

Exercise 4.8.41. A weight function ω on a locally compact Abelian group G is said to be *weakly subadditive* if there exists a constant $C \ge 1$ such that

$$\omega(xy) \le C(\omega(x) + \omega(y))$$

for all $x, y \in G$.
   (i) For such a weight ω, show by induction on $k \in \mathbb{N}$ that

$$\omega(x_1 \cdot \ldots \cdot x_n) \le C^k(\omega(x_1) + \ldots + \omega(x_n))$$

for all $n \in \mathbb{N}$ such that $2^{k-1} < n \le 2^k$ and all $x_1, \ldots, x_n \in G$.
   (ii) Deduce from (i) that $\omega(x^n) \le D \cdot \ln n$ for some constant $D > 0$ and all $x \in G$ and $n \in \mathbb{N}$. Conclude that ω is nonquasianalytic.
   (iii) For $\alpha \ge 0$, let ω_α denote the weight on $\mathbb{R}^n$ defined by

$$\omega_\alpha(x) = (1 + \|x\|)^\alpha, \quad x \in \mathbb{R}^n.$$

Show that ω_α is weakly subadditive with $C = 1$ if $\alpha \le 1$ and $C = 2^\alpha$ otherwise.

A commutative Banach algebra A is said to be *boundedly regular* if it has the following property. There exists a constant $C > 0$ such that for each closed subset E of $\Delta(A)$ and any $\varphi \in \Delta(A) \setminus E$, there is an $a \in A$ with $\|a\| \le C$, $\hat{a}(\varphi) = 1$ and $\hat{a} = 0$ on E.

Exercise 4.8.42. Let A be a boundedly regular commutative Banach algebra and I a closed ideal of A.
   (i) Show that A/I is boundedly regular.
   (ii) If A/I is semisimple, then I is boundedly regular.

Exercise 4.8.43. Prove that the projective tensor product $A \widehat{\otimes}_\pi B$ of two boundedly regular commutative Banach algebras A and B is boundedly regular.

4.9 Notes and references

The hull-kernel topology on the maximal ideal space of a commutative Banach algebra was introduced by Gelfand and Shilov [42]. In a more general, purely algebraic setting it is due to Jacobson [60] and is therefore often called the *Jacobson topology*. Proposition 4.1.7 and Example 4.1.8, which shows that the map $\Delta(A) \times \Delta(B) \to \Delta(A \widehat{\otimes}_\pi B)$ need not be continuous for the hull-kernel topologies, can be found in [37]. The notion of regularity and many of the basic results, such as Theorem 4.2.3 and hereditary properties such as Theorem 4.2.6, Corollary 4.2.13, and Lemma 4.2.16, go back to Shilov [122]. Nowadays, they are all standard and can be found in several textbooks. The same is true of the existence of partitions of unity (Corollary 4.2.12) and normality (Corollary 4.2.9) of regular Banach algebras. The converse to Theorem 4.2.6, namely that regularity of a closed ideal I and of A/I together imply regularity of A (Theorem 4.3.8), was shown by Morschel [91] in his diploma thesis. Independently, Tomiyama [128] and Gelbaum [37] show that the projective tensor product $A \widehat{\otimes}_\pi B$ is regular if and only if both A and B are regular (Theorem 4.2.21).

The existence of a greatest regular subalgebra of a semisimple commutative Banach algebra was discovered by Albrecht [1] as an application of the theory of decomposable operators. More elementary proofs were later given by Inoue and Takahasi [58] and Neumann [97], at the same time removing the assumption of semisimplicity. The proof presented here follows [58]. Actually, the concept of regularity and the hull-kernel topology are intimately related to the theory of decomposable multipliers. Concerning this aspect, we refer the interested reader to the monograph by Laursen and Neumann [76]. In [69], Kantrowitz and Neumann have determined the greatest regular subalgebra of several Banach algebras of vector-valued functions, such als $C_0(X, A)$, $L^1(G, A)$, and $C^n([0, 1], A)$. In general, it appears to be very difficult to determine the greatest regular subalgebra of a commutative Banach algebra. For instance, it is unknown whether $\mathrm{reg}(A \widehat{\otimes}_\pi B)$ may strictly contain the projective tensor product of $\mathrm{reg}(A)$ and $\mathrm{reg}(B)$.

Let G be a locally compact Abelian group. Regularity of $L^1(G)$ is one of the most fundamental results in commutative harmonic analysis and the basis of all deeper investigations in the ideal theory of $L^1(G)$ (see Chapter 5). All known proofs of regularity use another fundamental theorem in harmonic analysis, the Plancherel theorem. The most common proof of the Plancherel theorem, in turn, is based on the Pontryagin duality theorem for locally compact Abelian groups which, conversely, can be derived from the Plancherel theorem (see Appendix A.5). We have chosen a more direct, although technical, approach to the Plancherel theorem which goes back to Williamson [139] and utilizes the Gelfand theory of commutative C^*-algebras and therefore meets the intention of the book. It is worth pointing out that the greatest regular subalgebra of $M(G)$, the measure algebra of G, is not known.

We remind the reader that Shilov introduced the boundary carrying his name to decide which elements of $\Delta(A)$ extend to multiplicative linear functionals of every commutative Banach algebra B into which A is embedded algebraically, but not necessarily continuously. Theorem 4.2.17, which is due to Rickart [107], says that this is true for all $\varphi \in \Delta(A)$ whenever A is semisimple and regular. This multiplicative Hahn–Banach property and the two further and weaker spectral extension properties, treated in Section 4.5, were investigated and characterised in a satisfactory manner by Meyer [88]. For example, the condition that $r_B(x) = r_A(x)$ for all $x \in A$ and all commutative extensions B of A turned out to be equivalent to the fact that every $\varphi \in \partial(A)$ extends to an element of $\Delta(B)$. Theorems 4.5.3, 4.5.6, and 4.5.9 are all contained in [88], and so are Examples 4.5.8 and 4.5.9. These examples together with Example 4.5.10, which has been given in [124], show that among the four properties, regularity and the three spectral extension properties, no two are equivalent.

The related and slightly weaker unique uniform norm property was introduced and extensively studied in a number of papers by Bhatt and Dedania. The collection of results presented in Section 4.6 is taken from [14] and [15]. The somewhat unexpected result that for Beurling algebras $L^1(G, \omega)$ the unique uniform norm property is equivalent to regularity (Theorem 4.7.3) was also shown by Bhatt and Dedania [17]. The remarkable and deep Theorem 4.7.11 stating that $L^1(G, \omega)$ is regular whenever the weight ω is nonquasianalytic is due to Domar [26]. Our exposition of Theorem 4.7.11 follows the one in [83]. Unfortunately, no proof seems to be known which avoids the use of the structure theory of locally compact Abelian groups. As a matter of fact, Domar established the even stronger result that $L^1(G, \omega)$ is regular precisely when ω is nonquasianalytic, and that in this case, $L^1(G, \omega)$ is also Tauberian.

5

Spectral Synthesis and Ideal Theory

This final chapter is devoted to ideal theory in commutative Banach algebras with focus on spectral synthesis problems. The proper setting is that of a regular and semisimple commutative Banach algebra A, so that the Gelfand homomorphism $A \to C_0(\Delta(A))$ is injective and the Gelfand topology on $\Delta(A)$ coincides with the hull-kernel topology.

Recall from Chapter 4 that associated to any closed subset E of $\Delta(A)$ is the closed ideal

$$k(E) = \{a \in A : \widehat{a}(\varphi) = 0 \text{ for all } \varphi \in E\}$$

of A and that the hull of a closed ideal I of A is the closed subset

$$h(I) = \{\varphi \in \Delta(A) : \varphi(I) = \{0\}\}$$

of $\Delta(A)$. Then $h(k(E)) = E$, and hence the map $I \to h(I)$ from the collection of all closed ideals in A onto the collection of all closed subsets of $\Delta(A)$ is surjective. The spectral synthesis problem is the question of when the assignment $I \to h(I)$ is injective (in this case, one says that spectral synthesis holds for A) or, more generally, for which closed subsets E of $\Delta(A)$, $k(E)$ is the only closed ideal in A with hull equal to E. All sections of this chapter, except for 5.3 and 5.6, solely concentrate on spectral synthesis problems.

In Section 5.1 we introduce the relevant notions, such as sets of synthesis and Ditkin sets, and develop a key tool, the local membership principle. The genuine interest in producing sets of synthesis and Ditkin sets leads to questions such as how these classes of subsets of $\Delta(A)$ behave under the formation of unions and embedding of $\Delta(A/I)$ into $\Delta(A)$ for a closed ideal I of A. These and similar problems are extensively discussed in Section 5.2.

Spectral synthesis fails for the algebra $C^n[0, 1]$ of n-times continuously differentiable functions on the interval $[0, 1]$. In fact, associated to each point $t \in [0, 1] = \Delta(C^n[0, 1])$ is a chain of $n+1$ distinct ideals with hull the singleton $\{t\}$. Nevertheless, as we show in Section 5.3, every proper closed ideal in $C^n[0, 1]$ is the intersection of such so-called primary ideals. Spectral synthesis

E. Kaniuth, *A Course in Commutative Banach Algebras*, Graduate Texts in Mathematics,
DOI 10.1007/978-0-387-72476-8_5, © Springer Science+Business Media, LLC 2009

need not even hold for A when $\Delta(A)$ is discrete. An example is provided by the Mirkil algebra, which also serves as a counterexample to the union conjecture in that its structure space contains two disjoint sets of synthesis, the union of which fails to be of synthesis (Section 5.4).

A famous theorem of Malliavin, which is beyond the scope of this book, states that spectral synthesis fails for the group algebra $L^1(G)$ of a locally compact Abelian group G whenever G is noncompact (equivalently, $\Delta(L^1(G))$ is nondiscrete). Therefore, in Section 5.5 we utilize the results of Sections 5.2 and 4.4 to study sets of synthesis and Ditkin sets for $L^1(G)$ in detail. Moreover, there is a complete description, which we present in Section 5.6, of the closed ideals in $L^1(G)$ with bounded approximate identities. The hulls of these ideals turn out to be precisely the closed sets in the coset ring of the dual group $\widehat{G}$ of G, and in particular they are Ditkin sets.

Finally, the last section is designated to examine the projective tensor product of two commutative Banach algebras in the context of spectral synthesis.

5.1 Basic notions and local membership

Let A be a regular commutative Banach algebra. We already know that a closed subset E of $\Delta(A)$ is completely determined by its kernel $k(E)$ since $E = h(k(E))$ (Theorem 4.2.3). The following kind of dual question is a very interesting and extremely difficult one. To what extent is a closed ideal I of A determined by its hull $h(I)$? It is not generally true that $I = k(h(I))$. Therefore the question should be rephrased as follows. Given a closed subset E of $\Delta(A)$, what are the different closed ideals of A whose hulls equal E? More specifically, one might ask which closed subsets E of $\Delta(A)$ are the hull of only one closed ideal of A, namely $k(E)$. This is the basic problem of *spectral synthesis*.

To start with, let us consider a commutative C^*-algebra A. Then the Gelfand homomorphism is an isometric isomorphism from A onto $C_0(\Delta(A))$ (Theorem 2.4.5). Now, for any locally compact Hausdorff space X, we have earlier described all the closed ideals of $C_0(X)$. In fact, by Theorem 1.4.6, there is a bijection between closed subsets of X and closed ideals of $C_0(X)$ given by
$$Y \to I(Y) = \{f \in C_0(X) : f(x) = 0 \text{ for all } x \in Y\}.$$

Thus the assignment $E \to k(E) = \{a \in A : \widehat{a}|_E = 0\}$ is a bijection between the collection of closed subsets of $\Delta(A)$ and closed ideals of A. In particular, spectral synthesis holds for any commutative C^*-algebra. Moreover, applying Urysohn's lemma we have seen in the proof of Theorem 1.4.6 that given E, $a \in k(E)$, and $\epsilon > 0$, there exists a continuous function f on $\Delta(A)$ such that $\|f\|_\infty = 1$, $f(\varphi) = 1$ for all φ in the compact set $\{\psi \in \Delta(A) : |\psi(a)| \geq \epsilon\}$ and f has compact support disjoint from E. Now, let $u \in A$ so that $\widehat{u} = f$. Then

$\|ua - a\| = \|(\widehat{u} - 1)\widehat{a}\|_\infty \leq \epsilon$. In the terminology which is introduced soon, this means that every closed subset of $\Delta(A)$ is a Ditkin set.

Throughout this entire section A denotes a commutative Banach algebra. However, all the relevant results require the additional hypotheses that A be semisimple and regular. In studying the ideal theory of A, a certain *local membership principle* (Theorem 5.1.2) turns out to be very useful. This principle is based on the following notion.

Definition 5.1.1. Let M be a subset of A and f a complex-valued function on $\Delta(A)$. Then we say that f *belongs locally to* M *at a point* φ of $\Delta(A)$ if there exist $x \in M$ and a neighbourhood V of φ in $\Delta(A)$ such that $\widehat{x}(\psi) = f(\psi)$ for all $\psi \in V$. Similarly, f *belongs locally to* M *at infinity* if there exist $y \in M$ and a compact subset C of $\Delta(A)$ such that $\widehat{y}(\psi) = f(\psi)$ for all $\psi \in \Delta(A) \setminus C$. Finally, we say that f *belongs locally to* M provided that f belongs locally to M at every $\varphi \in \Delta(A)$ and at infinity.

Theorem 5.1.2. *Let A be regular and let I be an ideal of A and suppose that f is a function on $\Delta(A)$ that belongs locally to I. Then there exists $x \in I$ such that $\widehat{x} = f$. In particular, if A is semisimple and $y \in A$ is such that $\widehat{y}$ belongs locally to I, then $y \in I$.*

Proof. Because f belongs locally to I at infinity, there exist a compact subset C of $\Delta(A)$ and an element x_0 in I such that $\widehat{x}_0(\psi) = f(\psi)$ for all $\psi \in \Delta(A) \setminus C$. Since f belongs locally to I at every point of C, there are open subsets $U_1, \ldots, U_n$ of $\Delta(A)$ and elements $x_1, \ldots, x_n$ of I such that $C \subseteq \bigcup_{j=1}^n U_j$ and $\widehat{x}_j(\varphi) = f(\varphi)$ for all $\varphi \in U_j$, $1 \leq j \leq n$. Because A is regular, by Corollary 4.2.12 we can find elements $u_1, \ldots, u_n \in A$ such that

$$(\widehat{u_1} + \cdots + \widehat{u_n})|_C = 1 \text{ and } \operatorname{supp} \widehat{u}_j \subseteq U_j,$$

$1 \leq j \leq n$. Let $u = \sum_{j=1}^n u_j$ and

$$x = x_0 - x_0 u + \sum_{j=1}^n u_j x_j \in I.$$

Now note that $\widehat{u}_j(\varphi) = 0$ for all j whenever $\varphi \notin \bigcup_{k=1}^n U_k$. On the other hand, if $\varphi \in \bigcup_{k=1}^n U_k$ and J denotes the set of all indices $j \in \{1, \ldots, n\}$ such that $\varphi \in U_j$, then

$$\sum_{j=1}^n \widehat{u}_j(\varphi)\widehat{x}_j(\varphi) = \sum_{j \in J} \widehat{u}_j(\varphi)\widehat{x}_j(\varphi) = f(\varphi) \sum_{j \in J} \widehat{u}_j(\varphi) = f(\varphi) \sum_{j=1}^n \widehat{u}_j(\varphi)$$
$$= f(\varphi)\widehat{u}(\varphi).$$

Since $\widehat{u}(\varphi) = 1$ for $\varphi \in C$ and $\widehat{x}_0(\varphi) = f(\varphi)$ for $\varphi \notin C$, we obtain that, for all $\varphi \in \Delta(A)$,

$$\widehat{x}(\varphi) = \widehat{x}_0(\varphi)(1 - \widehat{u}(\varphi)) + \sum_{j=1}^{n} \widehat{u}_j(\varphi)\widehat{x}_j(\varphi)$$
$$= \widehat{x}_0(\varphi)(1 - \widehat{u}(\varphi)) + f(\varphi)\widehat{u}(\varphi)$$
$$= f(\varphi).$$

This proves the first statement of the theorem. As to the second, we only have to observe that, by the first part, $\widehat{y} = \widehat{x}$ for some $x \in I$. By semisimplicity, this implies $y = x \in I$. $\qquad\square$

In the sequel, for $x \in A$ we simply write $h(x)$ in place of $h(\{x\})$.

Lemma 5.1.3. *Let A be regular, I an ideal of A, and $x \in A$. Then $\widehat{x}$ belongs locally to I at each point of $h(x)^0$, the interior of $h(x)$, and at each point of $\Delta(A) \setminus h(I)$.*

Proof. Because $\widehat{x}(\varphi) = 0$ for all $\varphi \in h(x)$, the first assertion is clear. If $\varphi \notin h(I)$, then by Lemma 4.1.9 there exists $y \in I$ such that $\widehat{y} = 1$ in some neighbourhood V of φ. It follows that $yx \in I$ and

$$\widehat{yx}(\psi) = \widehat{y}(\psi)\widehat{x}(\psi) = \widehat{x}(\psi)$$

for all $\psi \in V$, whence $\widehat{x}$ belongs locally to I at φ. $\qquad\square$

Corollary 5.1.4. *Suppose that A is semisimple and regular. Let $x \in A$ be such that $\widehat{x}$ has compact support and $h(I) \cap \operatorname{supp}\widehat{x} = \emptyset$. Then $x \in I$.*

Proof. Since $\widehat{x}$ has compact support, $\widehat{x}$ belongs locally to I at infinity. By Lemma 5.1.3, $\widehat{x}$ belongs locally to I at every $\varphi \in \Delta(A) \setminus h(I)$ and also at every $\varphi \in h(I)$ since, by hypothesis,

$$h(I) \subseteq \Delta(A) \setminus \operatorname{supp}\widehat{x} = h(x)^0.$$

Theorem 5.1.2 shows that $x \in I$. $\qquad\square$

After these preparations we are able to show that given a closed subset E of $\Delta(A)$, there exists a smallest ideal of A with hull equal to E.

Definition 5.1.5. For any closed subset E of $\Delta(A)$, define an ideal $j(E)$ of A by

$$j(E) = \{x \in A : \widehat{x} \text{ has compact support and } \operatorname{supp}\widehat{x} \cap E = \emptyset\}.$$

If E is a singleton $\{\varphi\}$, we simply write $j(\varphi)$ instead of $j(\{\varphi\})$.

Theorem 5.1.6. *Suppose that A is semisimple and regular and let I be an ideal of A and E a closed subset of $\Delta(A)$. Then $h(I) = E$ if and only if*

$$j(E) \subseteq I \subseteq k(E).$$

In particular, $\overline{j(E)}$ in the smallest closed ideal of A with hull equal to E.

Proof. Suppose first that $j(E) \subseteq I \subseteq k(E)$. Then, since A is regular,

$$E = h(k(E)) \subseteq h(I) \subseteq h(j(E)).$$

To show that actually $h(I) = E$, it therefore suffices to verify that $h(j(E)) \subseteq E$. To that end, let $\varphi \in \Delta(A) \setminus E$ and choose a relatively compact open neighbourhood U of φ such that $\overline{U} \cap E = \emptyset$. Because A is regular, there exists $x \in A$ such that

$$\widehat{x}(\varphi) = 1 \quad \text{and} \quad \widehat{x}|_{\Delta(A) \setminus U} = 0.$$

Thus $\widehat{x}$ has compact support and vanishes on the open neighbourhood $\Delta(A) \setminus \overline{U}$ of E. So $x \in j(E)$, whereas $\varphi(x) \neq 0$. This shows $\varphi \notin h(j(E))$, as required.

Conversely, suppose that $h(I) = E$. Then $I \subseteq k(h(I)) = k(E)$, and if $x \in j(E)$, then $\widehat{x}$ has compact support and $h(I) \cap \operatorname{supp} \widehat{x} = \emptyset$, and this implies $x \in I$ by Corollary 5.1.4.

Finally, this also shows that $h(\overline{j(E)}) = h(j(E)) = E$ and $\overline{j(E)} \subseteq I$ for every closed ideal I of A with $h(I) = E$. $\qquad \square$

We now introduce some further notions that are fundamental to the study of ideal theory in commutative Banach algebras.

Definition 5.1.7. Let A be a commutative Banach algebra and E a closed subset of $\Delta(A)$.

(i) E is called a *spectral set* or *set of synthesis* (some authors also use the term *Wiener set*) if $k(E)$ is the only closed ideal of A with hull equal to E. We say that *spectral synthesis holds for A* or *A admits spectral synthesis* if every closed subset of $\Delta(A)$ is a set of synthesis.

(ii) E is called a *Ditkin set* or *Wiener–Ditkin set* for A if given $x \in k(E)$, there exists a sequence $(y_k)_k$ in $j(E)$ such that $y_k x \to x$ as $k \to \infty$.

(iii) A is called *Tauberian* if the set of all $x \in A$ such that $\widehat{x}$ has compact support is dense in A.

Remark 5.1.8. (1) Suppose that A is semisimple and regular. Then Theorem 5.1.6 shows that a closed subset E of $\Delta(A)$ is a set of synthesis if and only if $k(E) = \overline{j(E)}$, and E is a Ditkin set if and only if E is a set of synthesis and $x \in \overline{xk(E)}$ for every $x \in k(E)$. Furthermore, A is Tauberian precisely when $\emptyset$ is a set of synthesis.

(2) The fact that a singleton $\{\varphi\}$ is a Ditkin set for A is often rephrased by saying that A satisfies *Ditkin's condition at φ*. Similarly, one says that A satisfies *Ditkin's condition at infinity* if $\emptyset$ is a Ditkin set. Moreover, A is said to satisfy *Ditkin's condition* if it satisfies Ditkin's condition at every $\varphi \in \Delta(A)$ and at infinity.

We have already observed that every proper modular ideal of a Banach algebra is contained in some maximal modular ideal (Lemma 1.4.2). It need not generally be the case that every proper closed ideal of a commutative Banach algebra is contained in some maximal modular ideal. However, we have the following

Lemma 5.1.9. *Let A be a regular and semisimple commutative Banach algebra and suppose that A is Tauberian. Then $h(I) \neq \emptyset$ for every proper closed ideal of A. In particular, if $a \in A$ is such that $\hat{a}(\varphi) \neq 0$ for all $\varphi \in \Delta(A)$, then the ideal Aa is dense in A.*

Proof. If I is a proper closed ideal with $h(I) = \emptyset$, then $j(\emptyset) \subseteq I$ by Theorem 5.1.6. However, $j(E)$ is dense in A since A is Tauberian.

The second statement is now obvious. □

In passing we insert a characterisation of sets of synthesis in terms of the dual space A^* of A (Proposition 5.1.13).

Definition 5.1.10. Let V be an open subset of $\Delta(A)$ and let $f \in A^*$. Then f is said to *vanish on V* if $f(x) = 0$ for all $x \in A$ for which $\operatorname{supp} \hat{x}$ is compact and contained in V.

Lemma 5.1.11. *Let A be a semisimple and regular commutative Banach algebra. Given $f \in A^*$, there exists a largest open subset of $\Delta(A)$ on which f vanishes.*

Proof. We first show that if f vanishes on finitely many open subsets $V_1, \ldots, V_n$ of $\Delta(A)$, then f vanishes on $\bigcup_{j=1}^n V_j$. To that end, let $x \in A$ be such that $\operatorname{supp} \hat{x}$ is compact and contained in $\bigcup_{j=1}^n V_j$. Since A is regular, by Corollary 4.2.12 there exist $u_1, \ldots, u_n \in A$ so that $\operatorname{supp} \hat{u}_j \subseteq V_j, 1 \leq j \leq n$, and $\sum_{j=1}^n \hat{u}_j = 1$ on $\operatorname{supp} \hat{x}$. Because A is semisimple, it follows that $x = \sum_{j=1}^n x u_j$, and since $\operatorname{supp} \widehat{xu}_j \subseteq V_j$ for $1 \leq j \leq n$, we conclude that

$$f(x) = \sum_{j=1}^n f(xu_j) = 0,$$

because f vanishes on each V_j.

Now, let $\mathcal{V}$ be the collection of all open subsets of $\Delta(A)$ on which f vanishes and let $U = \bigcup \{V : V \in \mathcal{V}\}$. Then f vanishes on U. Indeed, if $x \in A$ is such that $\operatorname{supp} \hat{x}$ is compact and contained in U, then there exist $V_1, \ldots, V_n \in \mathcal{V}$ with $\operatorname{supp} \hat{x} \subseteq \bigcup_{j=1}^n V_j$, and hence $f(x) = 0$ by the first part of the proof. Thus f vanishes on U and, by definition, U is the largest open subset of $\Delta(A)$ on which f vanishes. □

Definition 5.1.12. Let $f \in A^*$ and let U be the largest open subset of $\Delta(A)$ on which f vanishes (Lemma 5.1.11). The closed set $\Delta(A) \setminus U$ is called the *support of f* and denoted $\operatorname{supp} f$.

Now the characterisation of spectral sets in terms of A^*, announced above, is as follows.

Proposition 5.1.13. *Let E be a closed subset of $\Delta(A)$. Then E is a spectral set if and only if whenever $f \in A^*$ is such that $\operatorname{supp} f \subseteq E$, then $f(x) = 0$ for all $x \in k(E)$.*

Proof. Suppose first that E is a set of synthesis and let $f \in A^*$ such that $\operatorname{supp} f \subseteq E$. Then f vanishes on $\Delta(A) \backslash E$ and hence $f(x) = 0$ for all $x \in j(E)$. Thus $f(x) = 0$ for all $x \in \overline{j(E)} = k(E)$.

Conversely, if $\overline{j(E)} \neq k(E)$, then by the Hahn–Banach theorem there exists $f \in A^*$ such that $f(x) = 0$ for all $x \in \overline{j(E)}$, whereas $f(y) \neq 0$ for some $y \in k(E)$. Then f vanishes on $\Delta(A) \setminus E$ and hence $\operatorname{supp} f \subseteq E$. This finishes the proof. □

When proceeding to study the local membership principle, it is convenient to introduce the following notation. For $a \in A$ and $M \subseteq A$, let $\Delta(a, M)$ denote the closed subset of $\Delta(A)$ consisting of all $\varphi \in \Delta(A)$ such that $\hat{a}$ does not belong locally to M at φ.

Lemma 5.1.14. *Let A be semisimple and regular and let I be a closed ideal of A. Let $x \in A$ and let φ be an isolated point of $\Delta(x, I)$. In addition, suppose that $\overline{j(\varphi)}$ possesses an approximate identity. Then $\hat{x}$ does not belong locally to $j(\varphi)$ at φ.*

Proof. Towards a contradiction, assume that $\hat{x}$ belongs locally to $\overline{j(\varphi)}$ at φ, and let U be a neighbourhood of φ and $y \in \overline{j(\varphi)}$ such that $\hat{x} = \hat{y}$ on U. Then, because φ is an isolated point of $\Delta(x, I)$, it is an isolated point of $\Delta(y, I)$ and we can choose an open neighbourhood V of φ such that $V \subseteq U$ and $V \cap \Delta(y, I) = \{\varphi\}$. By Corollary 4.2.9, there exists $z \in A$ such that $\hat{z} = 1$ on some neighbourhood of φ and $\operatorname{supp} \hat{z} \subseteq V$. Finally, since $\overline{j(\varphi)}$ has an approximate identity, there exists a sequence $(u_n)_n$ in $j(\varphi)$ such that $\|u_n y - y\| \to 0$ as $n \to \infty$.

Now consider the elements $z_n = u_n z y$, $n \in \mathbb{N}$, of A. Then $\widehat{z_n}$ belongs locally to I at infinity and at every $\psi \in \Delta(A) \setminus V$ since $\operatorname{supp} \hat{z} \subseteq V$. Moreover, $\widehat{z_n}$ belongs locally to I at φ since $\widehat{u_n}$ vanishes in some neighbourhood of φ, and also at every $\psi \in V \setminus \{\varphi\}$ because $V \cap \Delta(y, I) = \emptyset$. It follows that $z_n \in I$ and therefore $zy = \lim_{n \to \infty} z_n \in I$. After all, this means that $\hat{y}$ belongs locally to I at φ since $\hat{z}$ is identically one in some neighbourhood of φ. This contradicts $\varphi \in \Delta(y, I)$ and finishes the proof. □

We conclude this section with the following proposition which is applied in the next section.

Proposition 5.1.15. *Let A be a regular and semisimple commutative Banach algebra. Let I be a closed ideal of A and let $x \in A$ be such that $h(I) \subseteq h(x)$. Then*

(i) *$\Delta(x, I)$ is contained in $h(I) \cap \partial(h(x)) = \partial(h(I)) \cap \partial(h(x))$.*
(ii) *If singletons in $\Delta(A)$ are Ditkin sets, then $\Delta(x, I)$ has no isolated points.*

Proof. (i) By Lemma 5.1.3, x belongs locally to I at each point of $h(x)^0$ and at each point of $\Delta(A) \setminus h(I)$. Thus

$$\Delta(x, I) \subseteq h(I) \cap (\Delta(A) \setminus h(x)^0).$$

However, since $h(I) \subseteq h(x)$,

$$\partial(h(I)) \cap \partial(h(x)) = \left(h(I) \cap \overline{\Delta(A) \setminus h(I)}\right) \cap \left(h(x) \cap \overline{\Delta(A) \setminus h(x)}\right)$$
$$= h(I) \cap h(x) \cap \overline{\Delta(A) \setminus h(x)} = h(I) \cap \partial(h(x))$$
$$= h(I) \cap \left(h(x) \cap (\Delta(A) \setminus h(x)^0)\right)$$
$$= h(I) \cap (\Delta(A) \setminus h(x)^0).$$

So $\Delta(x, I) \subseteq \partial(h(I)) \cap \partial(h(x))$.

(ii) Assume that $\Delta(x, I)$ has an isolated point φ. Because $\{\varphi\}$ is a Ditkin set, it follows from Lemma 5.1.14 that x does not belong locally to $\overline{j(\varphi)} = k(\varphi)$ at φ. But $\varphi \in h(I) \subseteq h(x)$, so that $x \in k(\varphi)$. This contradiction shows (ii). $\square$

5.2 Spectral sets and Ditkin sets

Let A be a commutative Banach algebra. Our objective in this section is the naturally arising problem of which closed subsets of $\Delta(A)$ are sets of synthesis or Ditkin sets and whether these classes of subsets of $\Delta(A)$ are preserved under certain operations, such as forming finite unions. We begin with the latter question which allows a satisfactory answer for Ditkin sets, as the next two results show.

Lemma 5.2.1. *The union of two Ditkin sets is a Ditkin set.*

Proof. Let E_1 and E_2 be Ditkin sets in $\Delta(A)$ and let $E = E_1 \cup E_2$. We have to show that given $x \in k(E)$ and $\varepsilon > 0$, there exists $y \in j(E)$ such that $\|x - xy\| \leq \epsilon$. Now, since $x \in k(E_1)$ and E_1 is a Ditkin set, there exists $y_1 \in j(E_1)$ such that $\|x - xy_1\| \leq \varepsilon/2$. Similarly, since $xy_1 \in k(E_2)$ and E_2 is a Ditkin set, there exists $y_2 \in j(E_2)$ such that $\|xy_1 - xy_1y_2\| \leq \varepsilon/2$. It follows that $y = y_1y_2 \in j(E_1) \cap j(E_2) = j(E)$ and $\|x - xy\| \leq \varepsilon$. $\square$

Theorem 5.2.2. *Let A be a semisimple and regular commutative Banach algebra and suppose that $\emptyset$ is a Ditkin set. Then every closed subset of $\Delta(A)$ which is a countable union of Ditkin sets is again a Ditkin set.*

Proof. Let $(E_i)_i$ be a sequence of Ditkin sets in $\Delta(A)$ such that $E = \bigcup_{i=1}^{\infty} E_i$ is closed. Let $x \in k(E)$. Then, since $\emptyset$ is a Ditkin set, there is a sequence $(u_n)_n$ in A such that $xu_n \to x$ and $\widehat{u_n}$ has compact support for each n. It suffices to show that $xu_n \in \overline{xu_nj(E)}$ for every n. We can therefore assume that $\widehat{x}$ has compact support. Then it will follow from Theorem 5.1.2 that $x \in \overline{xj(E)}$ once we have verified that $\widehat{x}$ belongs locally to $\overline{xj(E)}$ at every $\varphi \in \Delta(A)$.

Fix $\varphi \in \Delta(A)$ and choose a compact neighbourhood U of φ. Since A is regular, there exists $u \in A$ such that $\widehat{u} = 1$ in a neighbourhood of φ and $\operatorname{supp} \widehat{u} \subseteq U$. We show that $xu \in \overline{xj(E)}$. To that end, let $\epsilon > 0$ be given. Then, since all the E_i are Ditkin sets and $x \in k(E_i)$, we can construct by induction a sequence $(y_i)_i$ such that $y_i \in j(E_i)$ for each i, $\|xu - xuy_1\| \leq \epsilon/2$ and

$$\|xu(y_1 \cdot \ldots \cdot y_{i-1}) - xu(y_1 \cdot \ldots \cdot y_{i-1})y_i\| \leq 2^{-i}\epsilon$$

for $i \geq 2$. For each i, let V_i be an open set containing E_i such that $\widehat{y}_i$ vanishes on V_i. Now,

$$E = \bigcup_{i=1}^{\infty} E_i \subseteq \bigcup_{i=1}^{\infty} V_i.$$

Since $E \cap U$ is compact, $E \cap U \subseteq \bigcup_{i=1}^{n} V_i$ for some $n \in \mathbb{N}$. Let $y = y_1 \cdot \ldots \cdot y_n \in j(E \cap U)$. Because $\widehat{u}$ vanishes on $\Delta(A) \setminus U$, it follows that $uy \in j(E)$. Finally, we have

$$\|xu - xuy\| \leq \|xu - xuy_1\| + \|xuy_1 - xuy_1y_2\| + \ldots$$
$$+ \|xu(y_1 \cdot \ldots \cdot y_{n-1}) - xu(y_1 \cdot \ldots \cdot y_{n-1})y_n\|$$
$$\leq \sum_{i=1}^{\infty} 2^{-i}\epsilon = \epsilon.$$

It follows that $xu \in \overline{xj(E)}$ since $xuy \in xj(E)$ and $\epsilon > 0$ was arbitrary. Now, $\widehat{u} = 1$ in a neighbourhood of φ and hence x belongs locally to $\overline{xj(E)}$ at φ. $\quad\square$

In contrast to the behaviour of Ditkin sets, the union of two sets of synthesis need not be a set of synthesis, even when the structure space $\Delta(A)$ is discrete. An example is provided by the Mirkil algebra which we explore in Section 5.4. Theorem 5.2.5 below is close to the strongest available result concerning unions of sets of synthesis. For that, two preparatory lemmas, which appear to be of interest in their own, are required. We continue to assume that A is a regular and semisimple commutative Banach algebra.

Lemma 5.2.3. *Let E_1 and E_2 be closed subsets of $\Delta(A)$. Let $E = E_1 \cup E_2$ and $F = E_1 \cap E_2$, and suppose that F is a Ditkin set. Let I be any closed ideal of A with $h(I) = E$ and let*

$$I_k = \overline{I + j(E_k)}, \quad k = 1, 2.$$

Then $I = I_1 \cap I_2$.

Proof. We only have to show that $I_1 \cap I_2 \subseteq I$. Because I is closed in A it suffices to prove that given $a \in I_1 \cap I_2$ and $\epsilon > 0$, there exists $u \in A$ such that $ua \in I$ and $\|ua - a\| \leq \epsilon$. Note that

$$h(I_k) = h(I) \cap h(j(E_k)) = E_k,$$

$k = 1, 2$, whence $a \in k(E_1) \cap k(E_2) \subseteq k(F)$. As F is a Ditkin set, there exists $u \in j(F)$ such that $\|ua - a\| \leq \epsilon$. Clearly, ua belongs locally to I at infinity.

We show that ua belongs locally to I at every point $\varphi \in \Delta(A)$. Since A is semisimple it then follows that $ua \in I$.

Firstly, ua belongs locally to I at every $\varphi \in \Delta(A) \setminus h(I)$ (Lemma 5.1.3) and at each $\varphi \in \Delta(A) \setminus \mathrm{supp}\,\widehat{ua}$. Thus we are left with points of

$$(E_1 \cap \mathrm{supp}\,\widehat{ua}) \cup (E_2 \cap \mathrm{supp}\,\widehat{ua}).$$

Put $C = E_1 \cap \mathrm{supp}\,\widehat{ua}$, which is a compact set. Then $C \cap E_2 = \emptyset$ because $\widehat{u}$ vanishes in a neighbourhood of $F = E_1 \cap E_2$. Choose a compact neighbourhood V of C such that

$$\emptyset = V \cap E_2 = V \cap h(j(E_2)).$$

We can now apply Theorem 4.2.8, taking the ideal $j(E_2)$ and the compact set V, to deduce the existence of some $x \in j(E_2)$ such that $\widehat{x}(\psi) = 1$ for all $\psi \in V$. Now, since $ua \in I_1 = \overline{I + j(E_1)}$, for every $\delta > 0$ there exist $y_\delta \in I$ and $z_\delta \in j(E_1)$ such that

$$\|ua - (y_\delta + z_\delta)\| \leq \delta \|x\|^{-1}.$$

It follows that $\|xua - x(y_\delta + z_\delta)\| \leq \delta$ as well as

$$xy_\delta \in I \quad \text{and} \quad xz_\delta \in j(E_1) \cap j(E_2) = j(E) \subseteq I.$$

Since I is closed and $\delta > 0$ was arbitrary, we conclude that $xua \in I$. Finally, since $\widehat{x}$ is identically one on V, it follows that ua belongs locally to I at every point of C.

In exactly the same way it is shown that ua belongs locally to I at every point of $E_2 \cap \mathrm{supp}\,\widehat{ua}$.   □

The following lemma shows that, as closed ideals, I_1 and I_2 in the preceding lemma are uniquely determined by the conditions that $I_1 \cap I_2 = I$ and $h(I_k) = E_k$ for $k = 1, 2$.

Lemma 5.2.4. *Let E_1, E_2, and I be as in Lemma 5.2.3. If J_1 and J_2 are closed ideals of A with $h(J_k) = E_k, k = 1, 2$, and $J_1 \cap J_2 = I$, then*

$$J_k = \overline{I + j(E_k)}, \quad k = 1, 2.$$

Proof. As in Lemma 5.2.3, let $I_i = \overline{I + j(E_i)}$ for $i = 1, 2$. We prove that $J_1 \subseteq I_1$, the converse inclusion, $I_1 \subseteq J_1$, being obvious since $I \subseteq J_1$ and $j(E_1)$ is the smallest ideal with hull equal to E_1. Thus, let $a \in J_1$. It is enough to show that given $\epsilon > 0$, there exists $u \in A$ such that $ua \in I_1$ and $\|ua - a\| \leq \epsilon$.

Since $F = E_1 \cap E_2$ is a Ditkin set and $a \in k(F)$, there exists $u \in j(F)$ such that $\|ua - a\| \leq \epsilon$. Let $C = E_1 \cap \mathrm{supp}\,\widehat{ua}$ and choose $v \in j(E_2)$ such that $\widehat{v} = 1$ on some neighbourhood of C (compare the proof of Lemma 5.2.3). Then $ua - vua \in I_1$. In fact, $\widehat{ua - vua}$ belongs locally to I_1 at infinity since $\widehat{u}$ has compact support, and clearly at every $\varphi \in \Delta(A) \setminus h(I)$ as well as at each $\varphi \in \Delta(A) \setminus \text{supp } \widehat{ua}$, and finally also at $\varphi \in C$ since $\hat{v} = 1$ in some neighbourhood of C. By semisimplicity, we get $ua - vua \in I_1$, and this together with

$$vua \in J_1 \cap j(E_2) \subseteq J_1 \cap J_2 = I \subseteq I_1$$

yields that $ua_1 \in I_1$, as required. So $J_1 = I_1$, and similarly $J_2 = I_2$.   □

Using the preceding two lemmas, we can now quickly prove the result on unions of spectral sets alluded to above.

Theorem 5.2.5. *Let A be a semisimple and regular commutative Banach algebra. Suppose that E_1 and E_2 are closed subsets of $\Delta(A)$ such that $E_1 \cap E_2$ is a Ditkin set. Then $E_1 \cup E_2$ is a spectral set if and only if both E_1 and E_2 are spectral sets. In particular, if A satisfies Ditkin's condition at infinity and E_1 and E_2 are disjoint, then $E_1 \cup E_2$ is a spectral set if and only if E_1 and E_2 are spectral sets.*

Proof. First, let E_1 and E_2 be spectral sets. Let $E = E_1 \cup E_2$ and apply Lemma 5.2.3 with $I = \overline{j(E)}$. It follows that

$$\begin{aligned}
\overline{j(E)} &= \overline{I + j(E_1)} \cap \overline{I + j(E_2)} \\
&= \overline{I + k(E_1)} \cap \overline{I + k(E_2)} = k(E_1) \cap k(E_2) \\
&= k(E).
\end{aligned}$$

Thus E is a spectral set.

Conversely, suppose that E is a spectral set and again let $I = \overline{j(E)}$. Then $I = k(E) = k(E_1) \cap k(E_2)$ and $I \subseteq \overline{j(E_i)}$ for $i = 1, 2$. Taking $J_i = k(E_i)$ in Lemma 5.2.4, we see that

$$k(E_i) = \overline{I + j(E_i)} = \overline{j(E_i)},$$

so that E_i is a spectral set for $i = 1, 2$.   □

The notions of set of synthesis and of Ditkin set are local in the sense of the following theorem. Part (i) of this theorem in particular also shows that, for A as in the theorem, the union of two disjoint sets of synthesis is a set of synthesis (compare this with Theorem 5.2.5).

Theorem 5.2.6. *Let A be a regular and semisimple commutative Banach algebra satisfying Ditkin's condition at infinity and let E be a closed subset of $\Delta(A)$.*

(i) *Suppose that each point of E has a closed relative neighbourhood in E which is a set of synthesis for A. Then E is a set of synthesis for A.*

(ii) *Suppose that each point of E has a closed relative neighbourhood in E which is a Ditkin set for A. Then E is a Ditkin set for A.*

Proof. (i) We have to show that every $x \in k(E)$ belongs to $\overline{j(E)}$. Because $\emptyset$ is a Ditkin set, we can assume that $\hat{x}$ has compact support. By Theorem 5.1.2 and Lemma 5.1.3 it then suffices to show that $\hat{x}$ belongs locally to $\overline{j(E)}$ at every $\varphi \in E$.

By hypothesis, there exist a closed subset E_φ of E and an open neighbourhood U_φ of φ in $\Delta(A)$ such that E_φ is a set of synthesis and $U_\varphi \cap E \subseteq E_\varphi$. A being regular, there exists $u \in A$ such that $\operatorname{supp} \hat{u} \subseteq U_\varphi$ and $\hat{u} = 1$ in a neighbourhood of φ in $\Delta(A)$. Since E_φ is a set of synthesis, given any $\epsilon > 0$, there exists $y \in j(E_\varphi)$ with $\|y - x\| \leq \epsilon/\|u\|$. Then $\|yu - xu\| \leq \epsilon$ and $\widehat{yu}$ vanishes in a neighbourhood of E since $\hat{y} = 0$ in a neighbourhood on E_φ, $\hat{u} = 0$ is a neighbourhood of $\Delta(A) \setminus U_\varphi$, and $E \subseteq E_\varphi \cup (\Delta(A) \setminus U_\varphi)$. So $yu \in j(E)$ and hence $xu \in \overline{j(E)}$ since $\epsilon > 0$ was arbitrary. Finally, $\widehat{xu} = \hat{x}$ in a neighbourhood of φ and hence $\hat{x}$ belongs locally to $\overline{j(E)}$ at φ.

(ii) Let $x \in k(E)$ and $\epsilon > 0$ be given. Because $\emptyset$ is a Ditkin set, there exists $u_0 \in A$ such that $\widehat{u_0}$ has compact support and $\|u_0 x - x\| \leq \epsilon/2$. Since $E \cap \operatorname{supp} \widehat{u_0}$ is compact, there exist closed subsets $E_1, \ldots, E_n$ of E such that $E \cap \operatorname{supp} \widehat{u_0} \subseteq \bigcup_{j=1}^n E_j$ and each E_j is a Ditkin set. We now define inductively a sequence $(u_k)_k$ in A such that $u_k \in j(E_k)$ and

$$\|u_k(u_{k-1} \cdot \ldots \cdot u_0)x - (u_{k-1} \cdot \ldots \cdot u_0)x\| \leq \epsilon/2n.$$

Let $u = u_n \cdot \ldots \cdot u_0 \in A$. Then

$$\|ux - x\| \leq \sum_{k=1}^n \|u_k(u_{k-1} \cdot \ldots \cdot u_0)x - (u_{k-1} \cdot \ldots \cdot u_0)x\| + \|u_0 x - x\| \leq \epsilon.$$

Moreover, $\hat{u}$ has compact support disjoint from E. In fact, $u_k \in j(E_k)$ for $1 \leq k \leq n$ and $E \cap \operatorname{supp} \widehat{u_0} \subseteq \bigcup_{j=1}^n E_j$. This shows that E is a Ditkin set. $\square$

Let I be a closed ideal of A and let $q : A \to A/I$ denote the quotient homomorphism. Then the mapping $i : \Delta(A/I) \to \Delta(A)$, defined by $i(\varphi)(x) = \varphi(q(x))$ for $\varphi \in \Delta(A/I)$ and $x \in A$, is a homeomorphism between $\Delta(A/I)$ and the subset $h(I)$ of $\Delta(A)$ (Lemma 4.1.5). It is a challenging issue whether i maintains synthesis properties. As we show in Section 5.5, there is a complete solution to this when $A = L^1(G)$ and I is the kernel of the canonical homomorphism $L^1(G) \to L^1(G/H)$, where G is a locally compact Abelian group and H is a closed subgroup of G. The following theorem, which might well be termed the *injection theorem for spectral sets and Ditkin sets*, comprises what is known in this respect for general commutative Banach algebras.

Theorem 5.2.7. *Let A be a regular and semisimple commutative Banach algebra, I a closed ideal of A, and E a closed subset of $\Delta(A/I)$.*

(i) *If $i(E)$ is a spectral set (respectively, Ditkin set) for A, then E is a spectral set (respectively, Ditkin set) for A/I.*

(ii) *Suppose that $h(I)$ is a spectral set for A. If E is a spectral set for A/I, then $i(E)$ is a spectral set for A.*

(iii) *Suppose that A satisfies Ditkin's condition at infinity and that there exists a constant $c > 0$ such that for any $a \in A$ and $\epsilon > 0$, there exists $b \in A$ such that $\|a - ba\| \leq c\|q(a)\| + \epsilon$ and $\hat{b}$ vanishes in a neighbourhood of $h(I)$. If E is a Ditkin set for A/I, then $i(E)$ is a Ditkin set for A.*

Proof. (i) Suppose that $i(E)$ is a spectral set (respectively, Ditkin set) for A and let $x \in A$ be such that $q(x) \in k(E)$. Then, given $\epsilon > 0$, there exists $y \in j(i(E))$ such that $\|x - y\| < \epsilon$ (respectively, $\|x - xy\| < \epsilon$). Thus

$$\|q(x) - q(y)\| < \epsilon \quad \text{(respectively, } \|q(x) - q(x)q(y)\| < \epsilon \text{).}$$

Also supp $\widehat{q(y)} \subseteq i^{-1}(\text{supp } \hat{y})$, and if $\hat{y}$ vanishes on the open neighbourhood V of $i(E)$ in $\Delta(A)$, then $\widehat{q(y)}$ vanishes on the open neighbourhood $i^{-1}(V)$ of E in $\Delta(A/I)$. This shows that E is a spectral set (respectively, Ditkin set) for A/I.

(ii) We claim that there is a bijective correspondence between the closed ideals J of A with $h(J) = i(E)$ and the closed ideals K of A/I with $h(K) = E$. Given such K, simply take $J = q^{-1}(K)$. Clearly then $h(J) = i(E)$. Conversely, let J be given. We show that $I \subseteq J$. To that end, let $x \in I$ and $\epsilon > 0$ be given. Because $h(I)$ is a set of synthesis, there exists $y \in j(h(I))$ with $\|x - y\| < \epsilon$. Thus $\hat{y}$ has compact support and vanishes in a neighbourhood of $h(J)$ since $h(J) = i(E) \subseteq h(I)$. Corollary 5.1.4 implies that $y \in J$. As $\epsilon > 0$ was arbitrary, we infer that $x \in J$. So $J = q^{-1}(J/I)$. This establishes the above claim. It follows that $k(i(E)) = \overline{j(i(E))}$ since $k(E) = \overline{j(E)}$.

(iii) Let $x \in k(i(E))$, $x \neq 0$, and $\epsilon > 0$. Then $q(x) \in k(E)$ and hence, since E is a Ditkin set for A/I, there exists $u \in A$ such that $\|q(x)q(u) - q(x)\| \leq \epsilon$ and $\widehat{q(u)}$ vanishes on a neighbourhood of E in $\Delta(A/I)$. By (ii), $i(E)$ is a set of synthesis for A. Hence, since $u \in k(i(E))$, there exists $v \in A$ with the properties that $\|v - u\| \leq \epsilon/\|x\|$ and $\hat{v}$ has compact support and vanishes on a neighbourhood of $i(E)$ in $\Delta(A)$. Then

$$\|q(x) - q(x)q(v)\| \leq \|q(x) - q(x)q(u)\| + \|x\| \cdot \|q(u) - q(v)\| \leq 2\epsilon.$$

By hypothesis, there exists $w \in A$ such that

$$\|(x - vx) - w(x - vx)\| \leq c\|q(x - vx)\| + \epsilon$$

and $\hat{w}$ vanishes in a neighbourhood of $i(\Delta(A/I)) = h(I)$. Since A satisfies Ditkin's condition at infinity, we can assume that $\hat{w}$ has compact support. Now, let $y = v + w - vw$. Then

$$\begin{aligned}
\|x - yx\| &= \|(x - vx) - w(x - vx)\| \\
&\leq c\|q(x - vx)\| + \epsilon \\
&\leq \epsilon(2c + 1).
\end{aligned}$$

Finally, $\widehat{y}$ vanishes in a neighbourhood of $i(E)$ in $\Delta(A)$ since $\widehat{w}$ vanishes in a neighbourhood of $i(\Delta(A/I))$ in $\Delta(A)$ and $\widehat{v}$ vanishes in a neighbourhood of $i(E)$ in $\Delta(A)$. Thus $y \in j(i(E))$, and since $\epsilon > 0$ was arbitrary it follows that $x \in \overline{j(i(E))}$. $\square$

It turns out that the hypotheses in part (iii) of Theorem 5.2.7 are fulfilled in the group algebra situation outlined above.

We continue with a simple lemma which shows that for commutative Banach algebras, which satisfy a condition similar to being boundedly regular, a set of synthesis necessarily is a Ditkin set.

Lemma 5.2.8. *Let A be a semisimple and regular commutative Banach algebra and let $E \subseteq \Delta(A)$ be a set of synthesis. Suppose that there exists a constant $c > 0$ such that for every compact subset K of $\Delta(A)$ which is disjoint from E, there exists $y \in j(E)$ such that $\|y\| \leq c$ and $\widehat{y} = 1$ on K. Then E is a Ditkin set.*

Proof. Let $x \in k(E)$ and $\epsilon > 0$ be given. Since E is a set of synthesis, there exists $u \in j(E)$ such that $\|u - x\| \leq \epsilon$. By hypothesis, there exists $y \in j(E)$ satisfying $\|y\| \leq c$ and $\widehat{y} = 1$ on $\operatorname{supp} \widehat{u}$. Then $u = uy$ since $\widehat{u} = \widehat{uy}$ and A is semisimple. It follows that

$$\|x - xy\| \leq \|x - u\| + \|uy - xy\| \leq (1+c)\|x - u\| \leq (1+c)\epsilon.$$

Thus $x \in \overline{xj(E)}$, as was to be shown. $\square$

Definition 5.2.9. A compact subset E of $\Delta(A)$ is said to satisfy *condition (D)* (D referring to Ditkin) if there exists a constant $C > 0$ such that for every neighbourhood U of E, there is $y \in A$ so that $\|y\| \leq C$, $\operatorname{supp} \widehat{y} \subseteq U$, and $\widehat{y} = 1$ in a neighbourhood of E.

The relevance of condition (D) is due to the fact that for unital A it turns out to be equivalent to the condition in Lemma 5.2.8, which ensures that a set of synthesis is already a Ditkin set.

Lemma 5.2.10. *Suppose that A has an identity e and let E be a closed subset of $\Delta(A)$. Then the following are equivalent.*

(i) *E satisfies condition (D).*
(ii) *There exists a constant $c > 0$ such that for every compact subset K of $\Delta(A)$ which is disjoint from E, there exists $a \in j(E)$ with $\|a\| \leq c$ and $\widehat{a} = 1$ on K.*

Proof. (ii) $\Rightarrow$ (i) Let U be an open set containing E. Choose an open set V so that $E \subseteq V$ and $\overline{V} \subseteq U$, and let $K = \Delta(A) \setminus V$. By (ii), there exists $a \in j(E)$ such that $\widehat{a} = 1$ on K and $\|a\| \leq c$. Then the element $x = e - a$ satisfies $\|x\| \leq 1 + c$, $\operatorname{supp} \widehat{x} \subseteq U$, and $\widehat{x} = 1$ is a neighbourhood of E. So (i) holds.
(i) $\Rightarrow$ (ii) is even simpler. $\square$

We now concentrate on finding conditions on an individual closed subset E of $\Delta(A)$ which ensure that E is a set of synthesis. Employing the local membership principle it turns out that a certain condition, which is close to countability of the boundary $\partial(E)$, is sufficient.

Definition 5.2.11. Let X be a topological space. Then X is called *scattered* if every nonempty closed subset F of X has an isolated point; that is, there exist $x \in F$ and an open subset U of X such that $U \cap F = \{x\}$.

Remark 5.2.12. Suppose that X is a countable locally compact Hausdorff space. Then X is scattered. Indeed, if F is any nonempty closed subset of X, then F is a countable union of closed singletons and locally compact, so that by Baire's theorem at least one of these singletons has to be open in F.

Conversely, if X is a Hausdorff space with a countable basis $\mathfrak{U}$ for its topology and if X is scattered, then X has to be countable. This can be seen as follows. Let

$$V = \bigcup \{U \in \mathfrak{U} : U \text{ is countable}\} \quad \text{and} \quad C = X \setminus V.$$

Then V is countable and open. Towards a contradiction, assume that C is nonempty. Then, since X is scattered and C is closed in X, C has an isolated point x. Hence there exists $U \in \mathfrak{U}$ such that $U \cap C = \{x\}$. From

$$U = (U \cap C) \cup (U \cap V) \subseteq \{x\} \cup V$$

we get that U is countable and hence contained in V. So $x \in V$, and this contradiction shows that $X = V$, which is countable.

A classical and powerful theorem, which is often referred to as the *Wiener–Ditkin theorem* (or *Ditkin–Shilov theorem*), asserts that if A is a semisimple and regular commutative Banach algebra which satisfies Ditkin's condition at infinity, then every closed subset of $\Delta(A)$ with scattered boundary is a spectral set. We proceed with the following more general result, which is needed in Section 5.7 and the proof of which is technically somewhat involved.

Theorem 5.2.13. *Let A be a regular and semisimple commutative Banach algebra. Let T be a locally compact Hausdorff space and $\phi : \Delta(A) \to T$ a continuous, surjective, and proper mapping. Suppose that for each $t \in T$, every closed subset of $\phi^{-1}(t)$ is a Ditkin set for A. Let E be a closed subset of $\Delta(A)$ such that $\phi(\partial(E))$ is scattered. Then E is a set of synthesis for A.*

Proof. Let I be a closed ideal of A with $h(I) = E$ and let $x \in k(E)$. We have to show that $x \in I$. Since $\emptyset$ is a Ditkin set, for every $\epsilon > 0$ there exists $y \in A$ such that $\hat{y}$ has compact support and $\|x - xy\| < \epsilon$. It then suffices to show that $xy \in I$ for any such y. Therefore we can assume that $\hat{x}$ has compact support.

Let S denote the set of all $t \in T$ with the property that x does not belong locally to I at at least one point of $\phi^{-1}(t)$. We aim at showing that $S = \emptyset$.

To that end, we first observe that S is closed in T. To see this, let $(s_\alpha)_\alpha$ be a net in S converging to some $t \in T$ and, towards a contradiction, assume that $t \notin S$. For each α, choose $\varphi_\alpha \in \phi^{-1}(s_\alpha)$ such that x does not belong locally to I at φ_α. Fix a compact neighbourhood U of t. Then $\phi^{-1}(U)$ is compact since ϕ is proper. Therefore, after passing to a subnet if necessary, we can assume that $\varphi_\alpha \in \phi^{-1}(U)$ for every α and that $\varphi_\alpha \to \varphi$ for some $\varphi \in \Delta(A)$. Then $\varphi \in \phi^{-1}(t)$ since $\phi(\varphi_\alpha) = s_\alpha \to t$. Hence $\widehat{x}$ belongs locally to I at φ and since $\varphi_\alpha \to \varphi$, the same is true at φ_α for large α. This contradiction shows that S is closed in T.

Because $\widehat{x}$ belongs locally to I at all points of E° and of $\Delta(A) \setminus E$ (Lemma 5.1.3), it follows that

$$\operatorname{supp} \widehat{x} \cap \phi^{-1}(t) \cap \partial(E) \neq \emptyset$$

for every $t \in S$. Thus $S \subseteq \phi(\operatorname{supp} \widehat{x} \cap \partial(E))$ and since $\phi(\partial(E))$ is scattered, so is its compact subset $\phi(\operatorname{supp} \widehat{x} \cap \partial(E))$. Hence S, being closed, is scattered as well. Suppose that $S \neq \emptyset$. Then S has an isolated point t.

Fix an open subset U of T such that $U \cap S = \{t\}$ and let $V = \phi^{-1}(U)$ and $K = \operatorname{supp} \widehat{x} \cap \phi^{-1}(t) \cap \partial(E)$. Then K is compact and V is an open neighbourhood of K. Moreover, choose an open neighbourhood W of K such that $\overline{W} \subseteq V$.

Since A is regular, there exists $a \in A$ such that $\widehat{a} = 0$ on $\Delta(A) \setminus \overline{W}$ and $\widehat{a} = 1$ on a neighbourhood of K (Theorem 4.2.8). By hypothesis, the compact subset K of $\phi^{-1}(t)$ is a Ditkin set. Since $x \in k(K)$, there exists a sequence $(a_n)_n$ in $j(K)$ so that $\|a_n x - x\| \to 0$ and hence $\|a_n a x - a x\| \to 0$.

We claim that $a_n a x \in I$. This follows from the three facts that

(1) $a_n \in j(\partial(E) \cap \operatorname{supp} \widehat{x} \cap \phi^{-1}(t))$,

(2) x belongs locally to I at every point of $\Delta(A) \setminus (\partial(E) \cup \operatorname{supp} \widehat{x})$, at every point of $V \setminus \phi^{-1}(t)$ and at infinity,

(3) a belongs locally to I at all points of $\Delta(A) \setminus \overline{W}$, so at points of $\Delta(A) \setminus V$.

As A is semisimple, we infer that $a_n a x \in I$ for all n and hence $a x \in I$. Now, $\widehat{a} = 1$ in a neighbourhood of K and consequently $\widehat{x}$ belongs locally to I at all points of $\phi^{-1}(t)$. This contradiction shows that $S = \emptyset$ and completes the proof. $\square$

If $T = \Delta(A)$ and ϕ is the identity map, Theorem 5.2.13 reduces to the Wiener–Ditkin theorem. Theorem 5.2.13 in particular applies when E is open and closed in $\Delta(A)$. In this case, however, the hypotheses on A can be weakened. Nevertheless, the proof of the following proposition is very similar to the one of Theorem 5.2.13.

Proposition 5.2.14. *Let A be a regular and semisimple commutative Banach algebra and let E be an open and closed subset of $\Delta(A)$.*

(i) *If A is Tauberian and $x \in \overline{Ax}$ for every $x \in k(E)$, then E is a set of synthesis.*

(ii) *If A satisfies Ditkin's condition at infinity, then E is a Ditkin set.*

Proof. (i) Let $x \in k(E)$ and $I = \overline{j(E)}$. By Lemma 5.1.3, $\widehat{x}$ belongs locally to I at every point of $\Delta(A) \setminus E$ and, since E is open, at every point of $E \subseteq h(x)^\circ$. Because $x \in \overline{Ax}$ and A is Tauberian, there exist $y_k \in A, k \in \mathbb{N}$, such that each $\widehat{y}_k$ has compact support and $y_k x \to x$ as $k \to \infty$. Then $y_k x$ belongs locally to I at every point of $\Delta(A)$ and at infinity. Since A is semisimple, $y_k x \in I$ and hence $x = \lim_{k \to \infty} y_k x \in I$.

(ii) Since E is open and closed in $\Delta(A)$,

$$h(j(E) + j(\Delta(A) \setminus E)) = E \cap (\Delta(A) \setminus E) = \emptyset.$$

Thus, since $\emptyset$ is a Ditkin set, for every $x \in A$ there exist sequences $(u_n)_n \subseteq j(E)$ and $(v_n)_n \subseteq j(\Delta(A) \setminus E)$ such that $x(u_n + v_n) \to x$. Now, let $x \in k(E)$. Then $xv_n = 0$ since A is semisimple and $\widehat{xv_n}$ vanishes on E and on $\Delta(A) \setminus E$. So $x = \lim_{n \to \infty} xu_n \in \overline{x\,j(E)}$, as required. $\qquad\square$

Suppose that spectral synthesis holds for A. Then $x \in \overline{Ax}$ for every $x \in A$. Indeed, let $E = h(x) = h(Ax)$. Since E is a spectral set, $k(E) = \overline{Ax}$ and hence $x \in \overline{Ax}$. The following corollary is now an immediate consequence of Proposition 5.2.14.

Corollary 5.2.15. *Let A be a semisimple and regular commutative Banach algebra. Suppose that A is Tauberian and that $\Delta(A)$ is discrete. Then spectral synthesis holds for A if and only if $x \in \overline{Ax}$ for each $x \in A$.*

As we show in Section 5.5, it follows from Corollary 5.2.15 that spectral synthesis holds for $L^1(G)$, where G is a compact Abelian group.

5.3 Ideals in $C^n[0,1]$

In this section we present what is known about the ideal structure of the regular commutative Banach algebra $C^n[0,1], n \geq 1$. We have observed in Example 2.2.9 that the map $a \to \varphi_a$, where $\varphi_a(f) = f(a)$ for $f \in C^n[0,1]$, provides a homeomorphism from $[0,1]$ onto $\Delta(C^n[0,1])$. In what follows we therefore identify $\Delta(C^n[0,1])$ with $[0,1]$. The algebras $C^n[0,1]$ are the simplest examples of regular semisimple commutative Banach algebras A for which singletons in $\Delta(A)$ fail to be sets of synthesis. More precisely, we show that given any $a \in [0,1]$, there are exactly $n+1$ different closed ideals in $C^n[0,1]$ with hull equal to $\{a\}$. Such ideals with a one-point hull are usually called *primary ideals*. Even though singletons in $\Delta(C^n[0,1])$ fail to be sets of synthesis, it turns out that every proper closed ideal in $C^n[0,1]$ is the intersection of all the closed primary ideals containing it.

Recall that the smallest ideal with hull $\{a\}, j(a)$, consists of all functions in $C^n[0,1]$ which vanish in a neighbourhood of a. We begin with a description of $\overline{j(a)}$.

Theorem 5.3.1. $\overline{j(a)} = \{f \in C^n[0,1] : f^{(i)}(a) = 0 \text{ for } 0 \leq i \leq n\}.$

Proof. Let M denote the set of all such functions on the right-hand side. Then M is closed in $C^n[0,1]$ because the maps $f \to f^{(i)}(a), C^n[0,1] \to \mathbb{C}, 0 \leq i \leq n$, are continuous. Clearly, $j(a) \subseteq M$, and hence $\overline{j(a)} \subseteq M$.

Conversely, let $f \in M$ be given and define a sequence of functions $f_m :$ $[0,1] \to \mathbb{C}, m \in \mathbb{N}$, by

$$f_m(t) = \begin{cases} f(t - \frac{1}{m}) & \text{for } t \in [a + \frac{1}{m}, 1], \\ 0 & \text{for } t \in [a - \frac{1}{m}, a + \frac{1}{m}], \\ f(t + \frac{1}{m}) & \text{for } t \in [0, a - \frac{1}{m}]. \end{cases}$$

We claim that $f_m \in C^n[0,1]$ (and hence $f_m \in j(a)$), and that the derivatives of f_m are given by the formula

$$f_m^{(i)}(t) = \begin{cases} f^{(i)}(t - \frac{1}{m}) & \text{for } t \in [a + \frac{1}{m}, 1], \\ 0 & \text{for } t \in [a - \frac{1}{m}, a + \frac{1}{m}], \\ f^{(i)}(t + \frac{1}{m}) & \text{for } t \in [0, a - \frac{1}{m}]. \end{cases}$$

Obviously, the function f_m is n-times continuously differentiable on the set $[0,1] \setminus \{a - 1/m, a + 1/m\}$ and there its derivatives satisfy the stated formula. Our claim for the whole interval $[0,1]$ now results from the following well-known fact which is a consequence of the mean value theorem. If a continuous function $g : [0,1] \to \mathbb{C}$ is continuously differentiable on $[0,1] \setminus F$, where F is a finite set, and g' admits a continuous extension $h : [0,1] \to \mathbb{C}$, then g is differentiable on $[0,1]$ and $g' = h$.

It remains to show $\|f_m^{(i)} - f^{(i)}\|_\infty \to 0$ for $0 \leq i \leq n$. To this end, fix i and set $g = f^{(i)}$ and $g_m = f_m^{(i)}$, and let $\epsilon > 0$ be given. Because g is uniformly continuous there exists $\delta > 0$ such that $|g(t) - g(s)| \leq \epsilon$ for all $t, s \in [0,1]$ with $|t - s| \leq \delta$. The above formula for $f_m^{(i)}$ then yields that $\|g_m - g\|_\infty \leq \epsilon$ for all m with $m > 1/\delta$. Indeed, if $t \in [a - 1/m, a + 1/m]$, then

$$|g_m(t) - g(t)| = |g(t)| = |g(t) - g(a)| \leq \epsilon,$$

whereas, if $t \in [a + 1/m, 1]$, then

$$|g_m(t) - g(t)| = \left| g\left(t - \frac{1}{m}\right) - g(t) \right| \leq \epsilon,$$

and if $t \in [0, a - 1/m]$, then

$$|g_m(t) - g(t)| = \left| g\left(t + \frac{1}{m}\right) - g(t) \right| \leq \epsilon,$$

as required. $\qquad\qquad\qquad\qquad\qquad\qquad\qquad\qquad\qquad\qquad\qquad\qquad\square$

Our next goal is to determine the closed primary ideals in $C^n[0,1]$ with hull $\{a\}$. Recall that these are just the closed ideals P in $C^n[0,1]$ with

$$\overline{j(a)} \subseteq P \subseteq k(a).$$

For that, we show that $C^n[0,1]/\overline{j(a)}$ is algebraically isomorphic to a certain quotient of $\mathbb{C}[X]$, the ring of complex polynomials in one variable X.

Lemma 5.3.2. *For each $a \in [0,1]$, the quotient algebra $C^n[0,1]/\overline{j(a)}$ is isomorphic to the $(n+1)$-dimensional algebra $\mathbb{C}[X]/J$, where J denotes the ideal in $\mathbb{C}[X]$ generated by X^{n+1}.*

Proof. Define $\phi : C^n[0,1] \to \mathbb{C}[X]/J$ by

$$\phi(f) = \sum_{i=0}^{n} \frac{1}{i!} f^{(i)}(a) X^i + J.$$

ϕ is a homomorphism because, for $f, g \in C^n[0,1]$,

$$
\begin{aligned}
\phi(f)\phi(g) &= \sum_{i,k=0}^{n} \frac{1}{i!k!} f^{(i)}(a) g^{(k)}(a) X^{i+k} + J \\
&= \sum_{k=0}^{n} \left(\sum_{i=0}^{k} \frac{1}{i!(k-i)!} f^{(i)}(a) g^{(k-i)}(a) \right) X^k + J \\
&= \sum_{k=0}^{n} \frac{1}{k!} \left(\sum_{i=0}^{k} \binom{k}{i} f^{(i)}(a) g^{(k-i)}(a) \right) X^k + J \\
&= \sum_{k=0}^{n} \frac{1}{k!} (fg)^{(k)}(a) X^k + J \\
&= \phi(fg).
\end{aligned}
$$

It follows from Theorem 5.3.1 that the kernel of ϕ coincides with $\overline{j(a)}$. Finally, ϕ is surjective since, for any $c_1, \ldots, c_n \in \mathbb{C}$,

$$\sum_{i=0}^{n} c_i X^i + J = \phi\left(\sum_{i=0}^{n} c_i (X-a)^i \right).$$

Thus $C^n[0,1]/\overline{j(a)}$ is algebraically isomorphic to $\mathbb{C}[X]/J$.   □

The ideals of $\mathbb{C}[X]/J$ are easy to determine. For sake of completeness we include the arguments.

Lemma 5.3.3. *Let J be as in Lemma 5.3.2. There are exactly $n+1$ proper ideals in $\mathbb{C}[X]/J$.*

Proof. For $0 \leq k \leq n+1$, let I_k denote the ideal in $\mathbb{C}[X]$ generated by X^k. Then

$$J = I_{n+1} \subsetneqq I_n \subsetneqq \ldots \subsetneqq I_0 = \mathbb{C}[X]$$

since $X^k \in I_k \setminus I_{k+1}, 0 \leq k \leq n$. It therefore suffices to show that every proper ideal I in $\mathbb{C}[X]$ strictly containing J coincides with some I_k. Because $\mathbb{C}[X]$ is a principal ideal domain, $I = \mathbb{C}[X]p$ for some polynomial p of degree ≥ 1. Write

$$p = \sum_{i=m}^{n} c_i X^i + qX^{n+1},$$

where $q \in \mathbb{C}[X]$ and $c_m \neq 0$. Clearly $I \subseteq I_m$, and we claim that conversely $I_m \subseteq I$. We have

$$X^{n-m}p = c_m X^n + X^{n+1}g$$

for some $g \in \mathbb{C}[X]$. Since $J \subseteq I, p \in I$, and $c_m \neq 0$, we obtain that $X^n \in I$ and hence

$$\sum_{i=m}^{n-1} c_i X^i = p - c_n X^n - qX^{n+1} \in I.$$

If $m < n$, by the same argument as before, we conclude from this that $X^{n-1} \in I$. Continuing in this manner, we finally end up with $X^m \in I$. This shows $I_m \subseteq I$. So $I = I_m$, and this establishes the lemma. $\square$

Lemmas 5.3.2 and 5.3.3 now quickly lead to a description of all the closed primary ideals in $C^n[0,1]$.

Theorem 5.3.4. *Given* $a \in [0,1]$, *there are exactly* $n+1$ *closed primary ideals in* $C^n[0,1]$ *with hull equal to* $\{a\}$, *namely the ideals*

$$P_{a,m} = \{f \in C^n[0,1] : f^{(i)}(a) = 0 \text{ for } 0 \leq i \leq m\}, \quad 0 \leq m \leq n.$$

Proof. It is clear that $h(P_{a,m}) = \{a\}$ and that $P_{a,m}$ is properly contained in $P_{a,k}$ for $m > k$. The statement now follows from Lemmas 5.3.2 and 5.3.3. $\square$

We show next that $\overline{j(a)}$ contains an (unbounded) approximate identity. This is a first step towards the main result stated at the outset of this section. Some of the proofs that follow are fairly intricate and employ the local membership principle in a substantial manner. Actually, the ideals $P_{a,m}$ don't possess bounded approximate identities (see Exercise 5.8.12).

Lemma 5.3.5. *There exists a sequence* $(u_m)_m$ *in* $j(a)$ *such that*

$$\lim_{m \to \infty} \|f - fu_m\| = 0$$

for every $f \in \overline{j(a)}$.

Proof. We choose functions g_0 and g_1 in $C^n[0,1]$ with the following properties:

$$g_k^{(i)}(0) = 0 = g_k^{(i)}(1) \text{ for } k = 0, 1 \text{ and all } 1 \le i \le n$$

and

$$g_0(0) = 0 = g_1(1), \quad g_0(1) = 1 = g_1(0).$$

For each $m \in \mathbb{N}$, define $u_m : [0,1] \to \mathbb{C}$ by

$$u_m(t) = \begin{cases} 1 & \text{for } t \in [a + \frac{2}{m}, 1], \\ g_0(m(t-a)+1) & \text{for } t \in [a + \frac{1}{m}, a + \frac{2}{m}], \\ 0 & \text{for } t \in [a - \frac{1}{m}, a + \frac{1}{m}], \\ g_1(m(t-a)+2) & \text{for } t \in [a - \frac{2}{m}, a - \frac{1}{m}], \\ 1 & \text{for } t \in [0, a - \frac{2}{m}]. \end{cases}$$

As in the proof of Theorem 5.3.1, it is straightforward to verify that $u_m \in C^n[0,1]$ and that

$$u_m^{(i)}(t) = \begin{cases} 0 & \text{for } t \in [a + \frac{2}{m}, 1], \\ m^i g_0^{(i)}(m(t-a)+1) & \text{for } t \in [a + \frac{1}{m}, a + \frac{2}{m}], \\ 0 & \text{for } t \in [a - \frac{1}{m}, a + \frac{1}{m}], \\ m^i g_0^{(i)}(m(t-a)+2) & \text{for } t \in [a - \frac{2}{m}, a - \frac{1}{m}], \\ 0 & \text{for } t \in [0, a - \frac{2}{m}], \end{cases}$$

$1 \le i \le n$. Clearly $u_m \in j(a)$, and we show that $\|f - fu_m\| \to 0$ as $m \to \infty$ for every $f \in \overline{j(a)}$. Set

$$c_i = \max(\|g_0^{(i)}\|_\infty, \|g_1^{(i)}\|_\infty),$$

so that $\|u_m^{(i)}\| \le m^i c_i$ for $0 \le i \le n$. Fix f, and for $t \ge 0$ and $0 \le i \le n$, let

$$\epsilon_i(t) = \max\{|f^{(i)}(s)| : s \in [a-t, a+t] \cap [0,1]\}.$$

Because $f^{(i)}(a) = 0$ for $0 \le i \le n$ (Theorem 5.3.1), we infer from the mean value theorem that $\epsilon_{i-1}(t) \le t\,\epsilon_i(t)$ and therefore

$$\epsilon_{i-k}(t) \le t^k\,\epsilon_i(t), \quad 0 \le k \le i \le n.$$

Finally, with $T_m = [a - 2/m, a + 2/m] \cap [0,1]$ and using the above formula for the derivatives $u_m^{(i)}$, we can now estimate the norms $\|(f - fu_m)^{(i)}\|_\infty$ as follows:

$$\|(f - fu_m)^{(i)}\|_\infty = \max_{t \in [0,1]} \left| f^{(i)}(t) - \sum_{k=0}^{i} \binom{i}{k} f^{(k)}(t) u_m^{(i-k)}(t) \right|$$

$$\leq \max_{t\in[0,1]} |f^{(i)}(t) - f^{(i)}(t)u_m(t)|$$

$$+ \sum_{k=0}^{i-1} \binom{i}{k} \max_{t\in[0,1]} |f^{(k)}(t)u_m^{(i-k)}(t)|$$

$$\leq (1 + \|u_m\|_\infty) \max_{t\in T_m} |f^{(i)}(t)|$$

$$+ \sum_{k=0}^{i-1} \binom{i}{k} \|u_m^{(i-k)}\|_\infty \max_{t\in T_m} |f^{(k)}(t)|$$

$$\leq (1 + c_0)\epsilon_i\left(\frac{1}{m}\right) + \sum_{k=0}^{i-1} \binom{i}{k} m^{i-k} c_{i-k}\epsilon_k\left(\frac{2}{m}\right)$$

$$\leq (1 + c_0)\left(\frac{2}{m}\right)^{n-i} \epsilon_n\left(\frac{2}{m}\right)$$

$$+ \sum_{k=0}^{i-1} \binom{i}{k} m^{i-k} c_{i-k} \left(\frac{2}{m}\right)^{n-k} \epsilon_n\left(\frac{2}{m}\right)$$

$$\leq \epsilon_n\left(\frac{2}{m}\right) \cdot 2^n \left(1 + \sum_{k=0}^{i} \binom{i}{k} c_{i-k}\right).$$

Notice that $\epsilon_n(2/m) \to 0$ as $m \to \infty$ because $f^{(n)}(a) = 0$. Consequently $\|(f - fu_m)^{(i)}\|_\infty \to 0$ for all $0 \leq i \leq n$ and hence $\|f - fu_m\| \to 0$.   □

As an immediate consequence of Lemma 5.3.5 and Lemma 5.1.14 we note the following

Corollary 5.3.6. *Let I be a closed ideal in $C^n[0,1]$ and $f \in C^n[0,1]$, and let $\Delta(f, I)$ denote the set of all $a \in [0,1]$ such that f does not belong locally to I at a. If a is an isolated point of $\Delta(f, I)$, then f does not belong locally to $\overline{j(a)}$ at a.*

Lemma 5.3.7. *Let $f \in C^n[0,1]$ and suppose that a is an accumulation point of $h(f)$, the zero set of f. Then $f \in \overline{j(a)}$.*

Proof. According to Theorem 5.3.1, it suffices to show that $f^{(m)}(a) = 0$ for $0 \leq m \leq n$. This is done by induction on m, the case $m = 0$ being clear. Thus, let $m > 1$ and assume that $f^{(i)}(a) = 0$ for $i < m$. By hypothesis, there is a sequence $(a_k)_k$ in $h(f)$ such that $a_k \neq a$ for all k and $a_k \to a$. By Taylor's formula,

$$0 = f^{(m)}(a_k) = \frac{(a_k - a)^m}{m!} f^{(m)}(a + \theta_k(a_k - a)),$$

where $|\theta_k| \leq 1$. It follows that $f^{(m)}(a + \theta_k(a_k - a)) = 0$ for all k and hence $f^{(m)}(a) = 0$ by continuity of $f^{(m)}$.   □

Lemma 5.3.8. *Let I be a closed ideal in $C^n[0,1]$ and a an isolated point of $h(I)$. If $I \subseteq \overline{j(a)}$, then every $f \in \overline{j(a)}$ belongs locally to I at a.*

Proof. Assume that some $f \in \overline{j(a)}$ does not belong locally to I at a. By hypothesis, there is an open set in U in $[0,1]$ such that $U \cap h(I) = \{a\}$. Because $C^n[0,1]$ is regular, f belongs locally to I at every point outside $h(I)$ (Lemma 5.1.3), so in particular at every point of $U \setminus \{a\}$. Hence a is an isolated point of $\Delta(f, I)$. Corollary 5.3.6 now shows that $a \in \Delta(f, \overline{j(a)})$, which is a contradiction.   □

The next lemma generalises Lemma 5.3.8 and is another substantial tool in proving Theorem 5.3.10 below.

Lemma 5.3.9. *Let I and a be as in Lemma 5.3.8, and let $m \le n$ be maximal with the property that*

$$I \subseteq P_{a,m} = \{g \in C^n[0,1] : g^{(i)}(a) = 0 \text{ for } 0 \le i \le m\}.$$

Then every function in $P_{a,m}$ belongs locally to I at a.

Proof. We can assume $m < n$, the case $m = n$ having been dealt with in Lemma 5.3.8. Let $f \in P_{a,m}$ and define $h \in C^n[0,1]$ by

$$h(t) = f(t) - \sum_{i=m+1}^{n} \frac{f^{(i)}(a)}{i!}(t-a)^i, \ t \in [0,1].$$

It is clear that $h^{(k)}(a) = 0$ for $0 \le k \le m$ since $f \in P_{a,m}$. However, for $k \ge m+1$ we also have

$$h^{(k)}(a) = f^{(k)}(a) - \frac{f^{(k)}(a)}{k!}[(t-a)^k]^{(k)}(a) = 0.$$

Theorem 5.3.1 now yields $h \in \overline{j(a)}$, and hence h belongs locally to I at a by Lemma 5.3.8.

Now, fix $f \in P_{a,m}$ and let h be as above. To prove that f belongs locally to I at a, it suffices to show that each of the functions $t \to (t-a)^i$, $m+1 \le i \le n$, belongs locally to I at a. Because $I \not\subseteq P_{a,m+1}$, there exists $g \in I$ such that $g^{(m+1)}(a) \ne 0$. Since $g \in P_{a,m}$, the first paragraph of the proof applies to g and shows that the function h defined by

$$h(t) = g(t) - \sum_{i=m+1}^{n} \frac{g^{(i)}(a)}{i!}(t-a)^i$$

belongs locally to I at a. Now, for $0 \le k \le d = n - (m+1)$, define g_k by

$$g_k(t) = (t-a)^{d-k}g(t) = \frac{g^{m+1}(a)}{(m+1)!}(t-a)^{n-k} + r_k(t) + h_k(t),$$

where $h_k(t) = (t-a)^{d-k}h(t)$ and

$$r_k(t) = (t-a)^{n-k+1} \sum_{i=m+2}^{n} \frac{g^{(i)}(a)}{i!}(t-a)^{i-(m+2)}.$$

Then, for all $0 \le k \le d$, $g_k \in I$ and h_k belongs locally to I at a since h does so. All the derivatives up to order n of r_0 vanish at a, so that $r_0 \in \overline{j(a)}$ by Theorem 5.3.1. Hence r_0 belongs locally to I at a by Lemma 5.3.8. Because the same is true of g_0 and h_0 and since $g^{(m+1)}(a) \ne 0$, we conclude that $(t-a)^n$ belongs locally to I at a. It is now easy to proceed by induction on k. Indeed, if $(t-a)^{n-k}$ belongs locally to I at a, then so does

$$r_{k+1}(t) = (t-a)^{n-k} \sum_{i=m+2}^{n} \frac{g^{(i)}(a)}{i!}(t-a)^{i-(m+2)}$$

and hence also $(t-a)^{n-(k+1)}$ since

$$\frac{g^{(m+1)}(a)}{(m+1)!}(t-a)^{n-(k+1)} = g_{k+1}(t) - r_{k+1}(t) - h_{k+1}(t)$$

and $g^{(m+1)}(a) \ne 0$. Thus f belongs locally to I at a. □

Now we are prepared for the main result of this section.

Theorem 5.3.10. *Every proper closed ideal I of $C^n[0,1]$ is the intersection of all the closed primary ideals containing I.*

Proof. Denote by $\mathcal{P}$ the set of all ideals $P_{a,m}$, $a \in [0,1]$, $0 \le m \le n$, which contain I, and let

$$F = \{a \in [0,1] : \overline{j(a)} = P_{a,n} \in \mathcal{P}\}.$$

Evidently, F is closed in $[0,1]$. Moreover, let $\mathcal{D}$ be the set of all relatively open intervals D in $[0,1]$ such that $D \cap F = \emptyset$ and the boundary $\partial(D)$ of D is contained in F. Then $[0,1] \setminus F = \bigcup\{D : D \in \mathcal{D}\}$ because the connected component of every point in $[0,1] \setminus F$ is an interval which is open and closed in $[0,1] \setminus F$ and hence cannot have a boundary point in $[0,1] \setminus F$. Notice next that due to the condition $\partial(D) \subseteq F$, any two distinct intervals in $\mathcal{D}$ are disjoint.

Now, let $f \in \bigcap\{P : P \in \mathcal{P}\}$ and for $D \in \mathcal{D}$ set

$$\|f\|_D = \max_{0 \le k \le n} \|f^{(k)}|_D\|_\infty.$$

We claim that for any $\delta > 0$, there are only finitely many $D \in \mathcal{D}$ such that $\|f\|_D \ge \delta$. Assuming that this is false, for some $0 \le m \le n$ and $\eta > 0$, there exist infinitely many $D_i \in \mathcal{D}$ and $a_i \in D_i$, $i \in \mathbb{N}$, with $D_i \cap D_k = \emptyset$ for $i \ne k$ and $|f^{(m)}(a_i)| \ge \eta$. Of course, after passing to a subsequence if necessary, we can assume that $a_i \to a$ for some $a \in [0,1]$ and that $a_i \le a$ for all i. The sets D_i being disjoint, we have $a \notin D_i$ for all i. Thus, for each i, there is a boundary point b_i of D_i with $a_i \leq b_i \leq a$. Because $\partial(D_i) \subseteq F$, $f^{(m)}(b_i) = 0$ and hence

$$f^{(m)}(a) = \lim_{i \to \infty} f^{(m)}(b_i) = 0.$$

This contradicts the fact that $|f^{(m)}(a_i)| \geq \eta$ for all i and hence establishes the above claim.

To finish the proof of the theorem, we show that given $\epsilon > 0$, there exists $f_\epsilon \in I$ with $\|f - f_\epsilon\| \leq \epsilon$. By what we have seen in the last paragraph, there exists $r \in \mathbb{N}$ such that $\|f\|_{D_i} < \epsilon/3$ for all $i > r$. Define $f_\epsilon : [0,1] \to \mathbb{C}$ by

$$f_\epsilon(t) = \begin{cases} f(t) & \text{if } t \in \bigcup_{i=1}^r D_i, \\ 0 & \text{if } t \notin \bigcup_{i=1}^r D_i. \end{cases}$$

Then f_ϵ is n-times continuously differentiable on $V = \bigcup_{i=1}^r D_i$ and on $W = [0,1] \setminus \overline{V}$, and $f_\epsilon^{(k)}(t) = f^{(k)}(t)$ for $t \in V$ and $f_\epsilon^{(k)}(t) = 0$ for $t \in W$. The finite set $\partial(V)$ is contained in F, so that $f^{(k)}|_{\partial(V)} = 0$ and $f_\epsilon^{(k)}$ extends continuously from $V \cup W$ to $[0,1]$. It follows that $f_\epsilon \in C^n[0,1]$ and $f_\epsilon^{(k)}(t) = f^{(k)}(t)$ for $t \in \bigcup_{i=1}^r D_i$ and $f_\epsilon^{(k)}(t) = 0$ for $t \notin \bigcup_{i=1}^r D_i$. From this we conclude

$$\|f - f_\epsilon\| = \sum_{k=0}^n \frac{1}{k!} \|(f^{(k)} - f_\epsilon^{(k)})\|_{[0,1]\setminus \bigcup_{i=1}^r D_i}\|_\infty$$

$$= \sum_{k=0}^n \frac{1}{k!} \|f^{(k)}|_{[0,1]\setminus \bigcup_{i=1}^r D_i}\|_\infty$$

$$= \sum_{k=0}^n \frac{1}{k!} \sup\{\|f^{(k)}|_D\|_\infty : D \in \mathcal{D}, D \neq D_i, 1 \leq i \leq r\}$$

$$\leq \frac{\epsilon}{3} \sum_{k=0}^n \frac{1}{k!} \leq \epsilon.$$

It remains to show that $f_\epsilon \in I$. To that end, by Theorem 5.1.2 it suffices to prove $\Delta(f_\epsilon, I) = \emptyset$. Notice first that

$$\Delta(f_\epsilon, I) \subseteq \bigcup_{i=1}^r \overline{D}_i$$

since f_ϵ vanishes on the open set $[0,1] \setminus \bigcup_{i=1}^r \overline{D}_i$. Moreover, f_ϵ belongs locally to I at every $a \notin h(I)$ (Lemma 5.1.3). If a is an accumulation point of $h(I)$ and hence of $h(g)$ for each $g \in I$, then $g^{(k)}(a) = 0$ for all $g \in I$ and $0 \leq k \leq n$ by Lemma 5.3.7, so that $a \in F$. Thus, because $D \cap F = \emptyset$ for $D \in \mathcal{D}$, every $a \in h(I) \cap (\bigcup_{i=1}^r D_i)$ is an isolated point of $h(I)$. An application of Lemma 5.3.9 shows that f belongs locally to I at every $a \in h(I) \cap (\bigcup_{i=1}^r D_i)$, and then so does f_ϵ since f_ϵ and f coincide on $\bigcup_{i=1}^r D_i$. Thus we have seen so far that

$$\Delta(f_\epsilon, I) \subseteq \partial\left(\bigcup_{i=1}^{r} D_i\right) = \bigcup_{i=1}^{r} \partial(D_i),$$

which is a finite subset of F. Therefore, if $a \in \Delta(f_\epsilon, I)$, then $\overline{j(a)} \in \mathcal{P}$ and hence

$$f_\epsilon^{(k)}(a) = f^{(k)}(a) = 0 \text{ for } 0 \leq k \leq n.$$

Thus $f_\epsilon \in \overline{j(a)}$ by Theorem 5.3.1. However, this contradicts Corollary 5.3.6 since $\Delta(f_\epsilon, I)$ is finite. $\qquad\square$

5.4 Spectral synthesis in the Mirkil algebra

Let A be a regular and semisimple commutative Banach algebra. Recall that we have shown in Theorem 5.2.5 that if E and F are closed subsets of $\Delta(A)$ such that $E \cap F$ is a Ditkin set, then $E \cup F$ is a set of synthesis if and only if both E and F are sets of synthesis. This does not remain true if the hypothesis that $E \cap F$ be a Ditkin set is dropped. In fact, the so-called *Mirkil algebra* M, which we study in detail in this section, even has a discrete structure space and nevertheless the union of two disjoint sets of synthesis need not be a set of synthesis and subsets of sets of synthesis need not be sets of synthesis. Moreover, M serves as a counterexample to several other conjectures in spectral synthesis. We start by introducing M and determining its structure space.

In what follows we identity the torus $\mathbb{T}$ with the interval $[-\pi, \pi)$. Multiplication in $\mathbb{T}$ then turns into addition modulo 2π. Let

$$M = \{f \in L^2(\mathbb{T}) : f|_{[-\pi/2, \pi/2]} \text{ is continuous}\}.$$

Clearly, M is a linear space and

$$\|f\| = \sqrt{2\pi}\|f\|_2 + \|f|_{[-\pi/2, \pi/2]}\|_\infty$$

defines a norm on M. With this norm, M is complete. Indeed, let $(f_n)_n$ be a Cauchy sequence in M. Then $f_n \to f$ in $L^2(\mathbb{T})$ and $g_n = f_n|_{[-\pi/2, \pi/2]} \to g$ uniformly on $[-\pi/2, \pi/2]$ for some continuous function g, and the function h, defined by $h(t) = f(t)$ for $t \in [-\pi, \pi) \setminus [-\pi/2, \pi/2]$ and $h(t) = g(t)$ for $t \in [-\pi/2, \pi/2]$, belongs to M and satisfies $\|f_n - h\| \to 0$.

Lemma 5.4.1. *For $f, g \in M$ define $f * g$ on $\mathbb{T}$ by*

$$(f * g)(x) = \frac{1}{2\pi} \int_\mathbb{T} f(x - t)g(t)dt.$$

Then $f * g \in C(\mathbb{T})$, and with this convolution product M becomes a commutative Banach algebra.

Proof. Observe first that, for $x, y \in G$,

$$|(f * g)(x^{-1}) - (f * g)(y^{-1})| \leq \frac{1}{2\pi} \|g\|_2 \|L_x f - L_y f\|_2.$$

The map $x \to L_x f$ from $\mathbb{R}$ into $L^2(\mathbb{T})$ is continuous. Thus we conclude that $f * g$ is continuous. So M is an algebra, and it only remains to show that the norm $\| \cdot \|$ is submultiplicative. For that, note that

$$\|f * g\|_2^2 = \frac{1}{2\pi} \int_{-\pi}^{\pi} \left| \frac{1}{2\pi} \int_{-\pi}^{\pi} f(x - t)g(t)dt \right|^2 dx$$

$$\leq \frac{1}{2\pi} \int_{-\pi}^{\pi} \frac{1}{4\pi^2} \left(\int_{-\pi}^{\pi} |f(x-t)|^2 dt \cdot \int_{-\pi}^{\pi} |g(t)|^2 dt \right) dx$$

$$= \|f\|_2^2 \cdot \|g\|_2^2$$

by Hölder's inequality, and

$$|(f * g)(x)| \leq \frac{1}{2\pi} \|f\|_2 \|g\|_2.$$

Combining these two inequalities yields

$$\|f * g\| = \sqrt{2\pi} \|f * g\|_2 + \|(f * g)|_{[-\pi/2, \pi/2]}\|_\infty$$

$$\leq \left(\sqrt{2\pi} + \frac{1}{2\pi} \right) \|f\|_2 \|g\|_2 \leq 2\pi \|f\|_2 \|g\|_2$$

$$\leq \left(\sqrt{2\pi} \|f\|_2 + \|f\|_{[-\pi/2, \pi/2]}\|_\infty \right) \cdot \left(\sqrt{2\pi} \|g\|_2 + \|g\|_{[-\pi/2, \pi/2]}\|_\infty \right)$$

$$= \|f\| \cdot \|g\|,$$

as required. □

For $n \in \mathbb{Z}$, let e_n denote the function $e_n(t) = e^{int}, t \in [-\pi, \pi)$.

Lemma 5.4.2. *The linear span of the functions* $e_n, n \in \mathbb{Z}$, *is dense in* M.

Proof. By the Stone–Weierstrass theorem, the trigonometric functions are uniformly dense in $C(\mathbb{T})$, and hence also dense in the $\| \cdot \|_2$-norm. Thus the trigonometric functions are dense in $(C(\mathbb{T}), \| \cdot \|)$. It therefore suffices to show that $C(\mathbb{T})$ is dense in M.

To that end, let $f \in M$ and $\varepsilon > 0$ be given. As $C(\mathbb{T})$ is dense in $L^2(\mathbb{T})$, we find $h \in C(\mathbb{T})$ with $\|f - h\|_2 \leq \varepsilon$. Choose $0 < \delta < \pi/2$ such that

$$\delta \left(\|h\|_\infty + \|f|_{[-\pi/2, \pi/2]}\|_\infty \right)^2 \leq \pi \epsilon^2.$$

Define functions $k : [\pi/2, \pi/2 + \delta] \to \mathbb{C}$ by

$$k(t) = f(\pi/2) + \frac{1}{\delta}(t - \pi/2)(h(t) - f(\pi/2))$$

and $l : [-\pi/2 - \delta, -\pi/2] \to \mathbb{C}$ by

$$l(t) = f(-\pi/2) - \frac{1}{\delta}(\pi/2 + t)(h(t) - f(-\pi/2)).$$

Finally, we define g on $[-\pi, \pi)$ by

$$g(t) = \begin{cases} h(t) & \text{for } |t| \geq \pi/2 + \delta, \\ f(t) & \text{for } |t| \leq \pi/2, \\ k(t) & \text{for } \pi/2 \leq t \leq \pi/2 + \delta, \\ l(t) & \text{for } -\pi/2 - \delta \leq t \leq -\pi/2. \end{cases}$$

Then, because $g(t) = h(t)$ for $|t| \geq \pi/2 + \delta$ and $g(t) = f(t)$ for $|t| \leq \pi/2$, it follows that, with $I = [-\pi/2 - \delta, -\pi/2] \cup [\pi/2, \pi/2 + \delta]$,

$$\|f - g\| = \sqrt{2\pi}\|f - g\|_2 \leq \sqrt{2\pi}(\epsilon + \|(g - h)1_I\|_2)$$

$$\leq \sqrt{2\pi}\left(\epsilon + \sqrt{\frac{\delta}{\pi}}\left(\|h\|_\infty + \|f|_{[-\pi/2,\pi/2]}\|_\infty\right)\right)$$

$$\leq 2\epsilon\sqrt{2\pi}.$$

This shows that $C(\mathbb{T})$ is dense in M. $\qquad\square$

We now identify the structure space of M.

Theorem 5.4.3. *For $n \in \mathbb{Z}$, define $\varphi_n : M \to \mathbb{C}$ by*

$$\varphi_n(f) = \frac{1}{2\pi}\int_\mathbb{T} f(t)e_{-n}(t)dt, \ f \in M.$$

Then $\varphi_n \in \Delta(M)$ and the map $n \to \varphi_n$ is a homeomorphism from $\mathbb{Z}$ onto $\Delta(M)$.

Proof. It is easy to check that $\varphi_n \in \Delta(M)$. Moreover, since $\varphi_n(e_m) \neq 0$ if and only if $m = n$, the map $n \to \varphi_n$ is injective. Observe next that given $\varphi \in \Delta(M)$, there exists $n \in \mathbb{Z}$ so that $\varphi = \varphi_n$. Indeed, $e_n * e_m = \delta_{nm}e_n$ and therefore

$$\varphi(e_n)\varphi(e_m) = \delta_{nm}\varphi(e_n)$$

for all $n, m \in \mathbb{Z}$, and $\varphi(e_n) \neq 0$ for some n since the trigonometric polynomials are dense in M (Lemma 5.4.2). Thus, for such $n \in \mathbb{N}$,

$$\varphi(e_m) = \delta_{nm} = \varphi_n(e_m)$$

for all $m \in \mathbb{Z}$, and this implies that $\varphi = \varphi_n$ because φ and φ_n agree on the trigonometric polynomials.

It remains to verify that $\Delta(M)$ is discrete. However, this is obvious since, for each $n \in \mathbb{N}$,

$$U(\varphi_n, e_n, 1) = \{\varphi \in \Delta(M) : |\varphi(e_n) - \varphi_n(e_n)| < 1\}$$
$$= \{\varphi_m \in \Delta(M) : |\varphi_m(e_n) - 1| < 1\}$$
$$= \{\varphi_n\}$$

since $\varphi_m(e_n) = \delta_{nm}$ for all $m \in \mathbb{Z}$. $\square$

Corollary 5.4.4. *M is semisimple and regular.*

Proof. If $f \in M$ is such that $\widehat{f} = 0$, then $\langle f, e_n \rangle = 0$ for all $n \in \mathbb{Z}$. Because the functions $e_n, n \in \mathbb{Z}$, form an orthonormal basis of $L^2(\mathbb{T})$, $f = 0$ in $L^2(\mathbb{T})$ and then also $f(t) = 0$ for all $t \in [-\pi/2, \pi/2]$ since f is continuous on $[-\pi/2, \pi/2]$. So M is semisimple.

Let E be any subset of $\mathbb{Z}$ and $n \in \mathbb{Z} \backslash E$. Then the function $f = e_n$ satisfies $\widehat{f}(\varphi_n) = 1$ and $\widehat{f}(\varphi_k) = 0$ for all $k \neq n$. Thus M is regular. $\square$

In order to find spectral sets E in $\mathbb{Z} = \Delta(M)$, an explicit description of $j(E)$ is required.

Lemma 5.4.5. *Let E be any subset of $\mathbb{Z}$. Then $j(E)$ equals the linear span of all functions e_n, where $n \in \mathbb{Z} \backslash E$.*

Proof. Clearly, if $n \notin E$, then $e_n \in j(E)$ since $\varphi_m(e_n) = 0$ for all $m \neq n$, so for all $m \in E$.

Conversely, let $f \in j(E)$ and let $F = \{n \in \mathbb{Z} : \varphi_n(f) \neq 0\}$. Then F is a finite subset of $\mathbb{Z} \backslash E$. Let $p = \sum_{n \in F} \varphi_n(f) e_n$. Then, for each $m \in \mathbb{Z}$,

$$\varphi_m(p) = \sum_{n \in F} \varphi_n(f) \varphi_m(e_n) = \sum_{n \in F} \delta_{nm} \varphi_n(f) = \varphi_m(f).$$

Since M is semisimple, $f = p$ which is of the desired form. $\square$

Corollary 5.4.6. *A subset E of $\mathbb{Z}$ is a spectral set for M if and only if $k(E)$ is contained in the closed linear span of all functions e_n, where $n \in \mathbb{Z} \backslash E$.*

Proof. Because M is semisimple and regular, E is a spectral set if and only if $k(E) = \overline{j(E)}$. The statement now follows immediately from the preceding lemma. $\square$

Corollary 5.4.7. *Let $E \subseteq \mathbb{Z}$ and $m \in \mathbb{Z}$. Then E is a spectral set for M if and only if $E + m = \{n + m : n \in E\}$ is a spectral set for M.*

Proof. Of course, it suffices to show that if E is a spectral set, then so is $E + m$. Let $f \in k(E + m)$ and let $g \in M$ be defined by $g(t) = f(t)e^{-imt}$. Then, for each $n \in E$,

$$\varphi_n(g) = \frac{1}{2\pi} \int_{\mathbb{T}} f(t)e^{-i(n+m)t} dt = \varphi_{n+m}(f) = 0,$$

whence $g \in k(E)$. Since E is spectral set, given $\epsilon > 0$, by Corollary 5.4.6 there exist $n_1, \ldots, n_r \in \mathbb{Z} \setminus E$ and $c_1, \ldots, c_r \in \mathbb{C}$ such that

$$\left\| g - \sum_{j=1}^{r} c_j e_{n_j} \right\| \leq \epsilon.$$

Thus, with $h = \sum_{j=1}^{r} c_j e_{n_j}$,

$$\| f - h e_m \| = \| (g - h) e_m \| = \| g - h \| \leq \epsilon.$$

Since $n_j + m \in \mathbb{Z} \setminus (E + m)$, $h e_m = \sum_{j=1}^{r} c_j e_{n_j + m} \in j(E + m)$. Since $\epsilon > 0$ was arbitrary, it follows that $f \in \overline{j(E + m)}$. So $E + m$ is a spectral set.   □

We continue to identify $\Delta(M)$ with $\mathbb{Z}$.

Theorem 5.4.8. *Let E and F be subsets of $\mathbb{Z}$ such that $E \subseteq F$ and $F \setminus E$ is finite. Then E is a set of synthesis for M if and only if F is a set of synthesis for M.*

Proof. Suppose first that E is a spectral set. To show that F is a spectral set, proceeding inductively, it suffices to treat the case where $F = E \cup \{m\}$ for some $m \in \mathbb{Z} \setminus E$. Let $f \in k(F)$ and $\epsilon > 0$ be given. There exists $g \in j(E)$ such that $\| f - g \| \leq \epsilon$. Let $h = g - \widehat{g}(m) e_m \in M$. Then $\widehat{h}$ has finite support and vanishes on F. Indeed, $\widehat{h}(m) = 0$ and $\widehat{h}$ vanishes on E since both $\widehat{g}$ and $\widehat{e_m}$ vanish on E as $m \notin E$. Moreover

$$\| f - h \| \leq \| f - g \| + |\widehat{g}(m)| \cdot \| e_m \|$$
$$= \| f - g \| + |\widehat{f}(m) - \widehat{g}(m)|(1 + \sqrt{2\pi})$$
$$\leq \| f - g \|(2 + \sqrt{2\pi}).$$

Because $h \in j(F)$ and $\| f - g \| \leq \epsilon$ and $\epsilon > 0$ is arbitrary, it follows that F is a spectral set.

To prove the converse statement of the theorem, we can again assume that $F = E \cup \{m\}$ for some $m \in \mathbb{Z}$. Thus, suppose that F is a spectral set and let $f \in k(E)$ and $\epsilon > 0$ be given. Then $f - f * e_m \in k(F)$, and hence there exists $g \in j(F)$ such that $\| (f - e_m * f) - g \| \leq \epsilon$. Since $\widehat{e_m * f}(n) = \widehat{e_m}(n) \widehat{f}(n) = 0$ for all $n \in E$ and $\widehat{e_m}$ has support $\{m\}$, it follows that $g + e_m * f \in j(E)$. This finishes the proof.   □

Of course, Theorem 5.4.8 can be reformulated as follows. Let E and F be subsets of $\mathbb{Z}$ such that the symmetric difference of E and F is finite. Then E is a set of synthesis if and only F is a set of synthesis. We now prove an analogous result for Ditkin sets.

Theorem 5.4.9. *Let E and F be subsets of $\mathbb{Z}$ such that $F \subseteq E$.*

(i) *If F is a Ditkin set, so is E.*

(ii) *If E is a Ditkin set and $E \setminus F$ is finite, then F is a Ditkin set.*

Proof. (i) Because, by Theorem 5.2.2, a closed countable union of Ditkin sets is a Ditkin set, it suffices to show that if F is a Ditkin set and $m \in \mathbb{Z} \setminus F$, then $E = F \cup \{m\}$ is a Ditkin set.

Let $f \in k(E)$ and $\epsilon > 0$ be given. Since F is a Ditkin set, there exists $u \in j(F)$ such that $\|f - u * f\| \leq \epsilon$. Then $u - u * e_m \in j(E)$ since $u \in j(F)$ and $\widehat{e_m}(m) = 1$. Also $\widehat{f * e_m} = 0$ since $\hat{f}(m) = 0$ and $\widehat{e_m}(n) = 0$ for all $n \neq m$. So $f * e_m = 0$ and therefore

$$\|f - (u - u * e_m) * f\| = \|f - u * f\| \leq \epsilon.$$

This shows that E is a Ditkin set.

As in the proof of Theorem 5.4.8, to show (ii) we can assume that $E = F \cup \{m\}$, where $m \in \mathbb{Z} \setminus F$. Let $f \in k(F)$ and $\epsilon > 0$. Then $f - e_m * f \in k(E)$ and hence there exists $g \in j(E)$ such that

$$\|(f - e_m * f) - g * (f - e_m * f)\| \leq \epsilon.$$

Let $h = g * e_m - g - e_m \in M$. Then $\|f - h * f\| \leq \epsilon$ and $\hat{h}$ has finite support and vanishes on E, as desired. $\qquad\square$

Next we present useful characterisations of spectral sets in terms of the dual space M^* of M. To that end, we need some insight into how M^* looks. Let

$$E = L^2(\mathbb{T}) \oplus C([-\pi/2, \pi/2])$$

and equip E with the norm $\|(g, h)\| = \|g\|_2 + \|h\|_\infty$, $g \in L^2(\mathbb{T}), h \in C([-\pi/2, \pi/2])$. The mapping $f \to (f, f|_{[-\pi/2, \pi/2]})$ is an isometric isomorphism of M onto a closed linear subspace of E. Thus, by the Hahn–Banach theorem, every element of M^* is the restriction to M of some element of E^*. Recall that

$$E^* = L^2(\mathbb{T})^* \oplus C([-\pi/2, \pi/2])^* = L^2(\mathbb{T}) \oplus M([-\pi/2, \pi/2])$$

with norm given by

$$\|(g, \mu)\| = \max(\|g\|_2, |\mu|(\mathbb{T}))$$

for $g \in L^2(\mathbb{T})$ and $\mu \in M([-\pi/2, \pi/2])$. Here $|\mu|(\mathbb{T})$ denotes the total variation norm of the measure μ. Thus every element of M^* can be represented as a measure ν on $\mathbb{T}$ which admits a representation $\nu = g dx + \mu$, where $g \in L^2(\mathbb{T})$ and μ is a Borel measure on $\mathbb{T}$ supported on $[-\pi/2, \pi/2]$. The evaluation of ν at $f \in M$ is given by

$$\langle \nu, f \rangle = \frac{1}{2\pi} \int_{\mathbb{T}} f(z) \overline{g(z)} dz + \int_{-\pi/2}^{\pi/2} f(z) d\bar{\mu}(z).$$

Clearly $\|\nu\|_{M^*} \leq \max(\|g\|_2, |\mu|(\mathbb{T}))$.

For $\nu \in M^*$, set $\widehat{\nu}(n) = \langle \nu, e_n \rangle, n \in \mathbb{Z}$, and define the spectrum of ν to be the set $\sigma(\nu) = \{n \in \mathbb{Z} : \widehat{\nu}(n) \neq 0\}$. Every $g \in L^2(\mathbb{T})$ defines an element ν_g of M^* by

$$\langle \nu_g, f \rangle = \frac{1}{2\pi} \int_{\mathbb{T}} f(z)\overline{g(z)}dz.$$

The following lemma and Proposition 5.4.11 provide characterisations of spectral sets by means of spectra of elements in M^*.

Lemma 5.4.10. *For a subset E of $\mathbb{Z}$, the following conditions are equivalent.*

(i) *E is a spectral set for M.*
(ii) *$\langle \mu, f \rangle = 0$ for every $f \in k(E)$ and every $\mu \in M^*$ with $\sigma(\mu) \subseteq E$.*

Proof. (i) $\Rightarrow$ (ii) Let $f \in k(E)$ and $\mu \in M^*$ with $\sigma(\mu) \subseteq E$. Because E is a spectral set, by Lemma 5.4.5 $k(E)$ is the closed linear span of functions $e_n, n \in \mathbb{Z} \setminus E$. Thus there exists a sequence $(p_j)_j$ of trigonometric polynomials such that $p_j \to f$ in M. Then, for each j,

$$p_j(t) = \sum_{n \in \mathbb{Z} \setminus E} \widehat{p}_j(n)e_n(t) \qquad .$$

and, since $\sigma(\mu) \subseteq E$,

$$\langle \mu, p_j \rangle = \sum_{n \in \mathbb{Z} \setminus E} \widehat{p}_j(n)\langle \mu, e_n \rangle = \sum_{n \in \mathbb{Z} \setminus E} \widehat{p}_j(n)\widehat{\mu}(n) = 0.$$

So $\langle \mu, f \rangle = \lim_{j \to \infty} \langle \mu, p_j \rangle = 0$.

(ii) $\Rightarrow$ (i) Towards a contradiction, assume that $\overline{j(E)} \neq k(E)$. Then there exists $f \in k(E)$ such that $\delta = \inf\{\|g - f\| : g \in j(E)\} > 0$. By the Hahn–Banach theorem there exists $\mu \in M^*$ such that $\langle \mu, f \rangle = \delta$ and $\mu|_{j(E)} = 0$. By Lemma 5.4.5, $j(E)$ is the linear span of functions $e_n, n \in \mathbb{Z} \setminus E$. Thus $0 = \langle \mu, e_n \rangle = \widehat{\mu}(n)$ for each $n \in \mathbb{Z} \setminus E$, whence $\sigma(\mu) \subseteq E$. By hypothesis (ii), $\langle \mu, f \rangle = 0$, which is a contradiction. □

Proposition 5.4.11. *For a subset E of $\mathbb{Z}$ the following conditions are equivalent.*

(i) *E is a spectral for M.*
(ii) *If $\mu \in M^*$ is such that $\sigma(\mu) \subseteq E$, then μ is the w^*-limit of functionals in the linear span of all $\mu_{e_n}, n \in E$.*

Proof. (i) $\Rightarrow$ (ii) Let $\mu \in M^*$ such that $\sigma(\mu) \subseteq E$. Because E is a spectral set, Lemma 5.4.10 implies that $\langle \mu, f \rangle = 0$ for every $f \in k(E)$. Towards a contradiction, assume that μ does not belong to the w^*-closed subspace F generated by the $\mu_{e_n}, n \in E$. By the Hahn–Banach theorem, there exists $f \in M$ satisfying $\langle \mu, f \rangle \neq 0$ and $0 = \langle \mu_{e_n}, f \rangle = \widehat{f}(n)$ for all $n \in E$. Thus $f \in k(E)$, whereas $\langle \mu, f \rangle \neq 0$. This contradiction proves (ii).

(ii) $\Rightarrow$ (i) By Lemma 5.4.10 we have to show that $\langle \mu, f \rangle = 0$ for every $f \in k(E)$ and $\mu \in M^*$ with $\sigma(\mu) \subseteq E$. By hypothesis (ii), it suffices to verify that $\langle \nu, f \rangle = 0$ whenever ν is of the form

$$\nu = \sum_{n \in E} \alpha_n \mu_{e_n},$$

where $\alpha_n \neq 0$ for only finitely many n. However, since $f \in k(E)$,

$$\langle \nu, f \rangle = \sum_{n \in E} \alpha_n \langle \mu_{e_n}, f \rangle = \frac{1}{2\pi} \sum_{n \in E} \alpha_n \int_{-\pi}^{\pi} e^{int} f(t) dt = \sum_{n \in E} \alpha_n \widehat{f}(n) = 0.$$

This proves (ii) $\Rightarrow$ (i). $\square$

Theorem 5.4.12. *Every subset of $4\mathbb{Z}$ is a spectral set for M.*

Proof. Let E be a nonempty subset of $4\mathbb{Z}$ and $\nu = g dx + \mu \in M^*$ with $\sigma(\nu) \subseteq E$. Let p be any trigonometric polynomial and let $q(t) = p(t + \pi/2)$. Then

$$\widehat{q}(n) = e^{in\pi/2} \int_{-\pi}^{\pi} p(t) e^{-int} dt = e^{in\pi/2} \widehat{p}(n)$$

for all $n \in \mathbb{Z}$, and hence $\widehat{q}(n) = \widehat{p}(n)$ for all $n \in 4\mathbb{Z}$. Because $\widehat{\nu}(n) = 0$ for all $n \notin E$ and $E \subseteq 4\mathbb{Z}$, it follows that

$$\int_{-\pi}^{\pi} q(t) d\nu(t) = \sum_{n \in 4\mathbb{Z}} \widehat{q}(n) \widehat{\nu}(-n) = \sum_{n \in 4\mathbb{Z}} \widehat{p}(n) \widehat{\nu}(-n) = \int_{-\pi}^{\pi} p(t) d\nu(t).$$

The trigonometric polynomials are uniformly dense in $C(\mathbb{T})$. Thus

$$\int_{-\pi}^{\pi} f(t) d\nu(t) = \int_{-\pi}^{\pi} f(t + \pi/2) d\nu(t)$$

for all $f \in C(\mathbb{T})$, and this implies $\nu(S) = \nu(S + \pi/2)$ for every Borel subset S of $\mathbb{T}$. On the other hand, since μ is supported on $[-\pi/2, \pi/2]$, it follows that $\nu(S) = \int_S g(x) dx$ whenever $S \subseteq [0, \pi/2)$. Combining these two facts we see that $\nu \in L^2(\mathbb{T})$. Setting

$$\nu_N = \sum_{n=-N}^{N} \widehat{\nu}(n) e_n \in L^2(\mathbb{T})$$

for $N \in \mathbb{N}$, we deduce that

$$\|\nu - \nu_N\|_{M^*} \leq \|\nu - \nu_N\|_2 \to 0$$

as $N \to \infty$. Thus ν is contained in the norm closure of the linear subspace of M^* spanned by the functions e_n, $n \in E$. Proposition 5.4.11 now shows that E is a set of synthesis. $\square$

The next two lemmas are straightforward and most likely well-known. However, we include the proofs for completeness.

Lemma 5.4.13. *Let $f \in M$ be defined by*

$$f(t) = \begin{cases} 1 & \text{if } |t| \leq \pi/2, \\ -1 & \text{if } \pi/2 < |t| \leq \pi. \end{cases}$$

Then $\widehat{f}(0) = 0$ and, for $n \in \mathbb{Z} \setminus \{0\}$,

$$\widehat{f}(n) = \frac{2}{n\pi} \sin\left(\frac{n\pi}{2}\right).$$

In particular, $f \in k(2\mathbb{Z})$.

Proof. For $n \in \mathbb{Z}, n \neq 0$, we have

$$\widehat{f}(n) = \frac{1}{2\pi} \left(-\int_{-\pi}^{-\pi/2} e^{-int} dt + \int_{-\pi/2}^{\pi/2} e^{-int} dt - \int_{\pi/2}^{\pi} e^{-int} dt \right)$$

$$= \frac{1}{2\pi i n} \left(e^{in\pi/2} - e^{in\pi} - e^{-in\pi/2} + e^{in\pi/2} + e^{-in\pi} - e^{-in\pi/2} \right)$$

$$= \frac{1}{\pi i n} \left(e^{in\pi/2} - e^{-in\pi/2} \right)$$

$$= \frac{2}{n\pi} \sin\left(\frac{n\pi}{2}\right).$$

It is clear that $\widehat{f}(0) = 0$.   □

Lemma 5.4.14. *Let $\mu = \delta_{-\pi/2} + \delta_{\pi/2} \in M^*$. Then*

$$\widehat{\mu}(n) = 2\cos\left(\frac{n\pi}{2}\right)$$

for every $n \in \mathbb{Z}$. In particular, $\sigma(\mu) = 2\mathbb{Z}$.

Proof. For each $n \in \mathbb{Z}$,

$$\widehat{\mu}(n) = \int_{\mathbb{T}} e^{-int} d\mu(t) = e^{in\pi/2} + e^{-in\pi/2} = 2\cos\left(\frac{n\pi}{2}\right).$$

Since $\cos(n\pi/2) \neq 0$ if and only if $n \in 2\mathbb{Z}$, we get that $\sigma(\mu) = 2\mathbb{Z}$.   □

Corollary 5.4.15. *Let f and μ be as in Lemmas 5.4.13 and 5.4.14, respectively. Then $\langle \mu, f - p * f \rangle = 2$ for every trigonometric polynomial p.*

Proof. Notice first that $\langle \mu, f \rangle = f(-\pi/2) + f(\pi/2) = 2$. Then

$$\langle \mu, f - p * f \rangle = 2 - \langle \mu, p * f \rangle = 2 - \sum_{n \in \mathbb{Z}} \widehat{p}(n) \widehat{f}(n) \widehat{\mu}(n)$$

$$= 2 - \sum_{n \in 2\mathbb{Z}} \widehat{p}(n) \widehat{f}(n) \widehat{\mu}(n) = 2$$

since $\sigma(\mu) = 2\mathbb{Z}$ and $f \in k(2\mathbb{Z})$.   □

Corollary 5.4.16. *The union of two disjoint sets of synthesis for M need not be a set of synthesis. For example, $2\mathbb{Z} = 4\mathbb{Z} \cup (4\mathbb{Z} + 2)$ fails to be of synthesis even though both $4\mathbb{Z}$ and $4\mathbb{Z} + 2$ are sets of synthesis.*

Proof. By Theorem 5.4.12, $4\mathbb{Z}$ is a spectral set and hence so is $4\mathbb{Z} + 2$ by Corollary 5.4.7. However, $2\mathbb{Z}$ is not of synthesis. Indeed, let f and μ be as in Lemmas 5.4.13 and 5.4.14. Then $f \in k(2\mathbb{Z})$ and $\sigma(\mu) = 2\mathbb{Z}$ and $\langle \mu, f \rangle = 2$ (Corollary 5.4.15). Lemma 5.4.10 now implies that $2\mathbb{Z}$ is not of synthesis.   □

Theorem 5.4.17. *Let E be a finite subset of $\mathbb{Z} = \Delta(M)$. Then E is a set of synthesis, but E fails to be a Ditkin set.*

Proof. We first show that E is a set of synthesis. Since the trigonometric polynomials p are dense in M (Lemma 5.4.2) and $\widehat{p}$ has finite support for any such p, $\emptyset$ is a spectral set. Thus let E be nonempty and let $f \in k(E)$ and $\mu \in M^*$ with $\sigma(\mu) \subseteq E$. To show that $\langle \mu, f \rangle = 0$, let $\epsilon > 0$ be given. There exists a trigonometric polynomial $p = \sum_{n=-N}^{N} c_n e_n$ such that $\|f - p\| \leq \epsilon$. Of course, we can assume that $E \subseteq \{-N, \ldots, N\}$. Because $\sigma(\mu) \subseteq E$, $\langle \mu, p \rangle = \sum_{n \in E} c_n \langle \mu, e_n \rangle$. Moreover, for each $n \in E$,

$$\varphi_n(p) = \sum_{j=-N}^{N} c_j \varphi_n(e_j) = c_n$$

and hence, since $f \in k(E)$,

$$|c_n| \leq |\varphi_n(f)| + |\varphi_n(p - f)| \leq \|p - f\| \leq \epsilon.$$

Denoting by $|E|$ the cardinality of E, it follows that

$$|\langle \mu, f \rangle| \leq |\langle \mu, f - p \rangle| + \sum_{n \in E} |c_n| \cdot |\langle \mu, e_n \rangle|$$

$$\leq \epsilon \|\mu\| (1 + |E| + |E| \sqrt{2\pi}).$$

Thus $\langle \mu, f \rangle = 0$ since $\epsilon > 0$ was arbitrary.

Turning to the second statement of the theorem, by Theorem 5.4.9(ii) it suffices to show that $\emptyset$ is not a Ditkin set. To that end, let f and μ be as in Lemmas 5.4.13 and 5.4.14. Then, by Corollary 5.4.15, $\langle \mu, f - p * f \rangle = 2$ and hence $\|f - p * f\| \geq 2/\|\mu\|$ for every trigonometric polynomial p. Using again that the trigonometric polynomials are dense in M, it follows that $f \notin \overline{f * M}$. Thus $\emptyset$ fails to be a Ditkin set.   □

For group algebras $L^1(G)$, where G is a locally compact Abelian group, it is an open question whether every set of synthesis is a Ditkin set. This question is open even for the group of integers. In the context of general regular and semisimple commutative Banach algebras A, Theorem 5.4.17 shows in a dramatic manner that singletons in $\Delta(A)$ which are spectral sets need not be Ditkin sets.

5.5 Spectral sets and Ditkin sets for $L^1(G)$

A famous theorem due to Malliavin [84] states that spectral synthesis fails for any noncompact locally compact Abelian group G; that is, there exists a closed subset E of $\widehat{G} = \Delta(L^1(G))$ with the property that $k(E)$ is not the only closed ideal with hull equal to E. Therefore, there is profound interest in producing sets of synthesis as well as Ditkin sets for $L^1(G)$. This is the purpose of this section. Our main tools are results obtained in Sections 5.2 and 4.4.

Theorem 5.5.1. *Let G be a locally compact Abelian group.*

(i) *$L^1(G)$ satisfies Ditkin's condition at infinity.*
(ii) *Let K be a compact subset of $\widehat{G}$ and $\epsilon > 0$. Then there exists $f \in L^1(G)$ such that $\widehat{f} = 1$ on K, $\widehat{f}$ has compact support and $\|f\|_1 < 1 + \epsilon$.*

Proof. (i) Because $L^1(G)$ has an approximate identity, it suffices to show that $L^1(G)$ is Tauberian. Let F denote the space of all functions $f \in L^2(G)$ such that $\widehat{f}$ equals almost everywhere a continuous function with compact support. By the Plancherel theorem, F is dense in $L^2(G)$. Observe that if $g, h \in F$, then $\widehat{g} * \widehat{h} \in C_c(\widehat{G})$, $gh \in L^1(G)$, and the Fourier transform $\widehat{gh}$ of gh equals $\widehat{g} * \widehat{h}$. Indeed, by the Parseval identity (Corollary 4.4.12) and Corollary 4.4.13,

$$\widehat{gh}(\overline{\alpha}) = \langle \alpha g, \overline{h} \rangle = \langle \widehat{\alpha g}, \overline{\widehat{h}} \rangle$$
$$= \langle L_\alpha \widehat{g}, \overline{\widehat{h}} \rangle = \int_{\widehat{G}} \widehat{g}(\overline{\alpha}\beta) \widehat{h}(\overline{\beta}) d\beta$$
$$= (\widehat{g} * \widehat{h})(\overline{\alpha})$$

for every $\alpha \in \widehat{G}$ (the same argument was used in the proof of Theorem 4.4.14). Now, every $f \in L^1(G)$ can be written as a pointwise product $f = gh$ where $g, h \in L^2(G)$. Since F is dense in $L^2(G)$, it follows that the set $\{gh : g, h \in F\}$ is dense in $L^1(G)$. On the other hand,

$$\{gh : g, h \in F\} \subseteq \{f \in L^1(G) : \widehat{f} \in C_c(\widehat{G})\}.$$

Thus the set on the right-hand side is dense in $L^1(G)$, as required.

(ii) Since $L^1(G)$ is regular (Theorem 4.4.14), by Corollary 4.2.10 there exists $g \in L^1(G)$ such that $\widehat{g}|_K = 1$ and $\widehat{g}$ has compact support. Now $L^1(G)$ has an approximate identity of norm bound 1. So there exists $u \in L^1(G)$ with $\|g - g * u\|_1 < \epsilon/3$ and $\|u\|_1 = 1$. As $L^1(G)$ is Tauberian, we find $v \in L^1(G)$ so that $\widehat{v}$ has compact support and $\|v - u\|_1 < \epsilon/3 \min\{1, 1/\|g\|_1\}$. Let $f = g + v - g * v$. Then

$$\|f\|_1 \leq \|v\|_1 + \|g - g * v\|_1$$
$$\leq 1 + \|v - u\|_1 + \|g - g * u\|_1 + \|g\|_1 \|v - u\|_1$$
$$< 1 + \epsilon.$$

Moreover, $\widehat{f} = 1$ on K and $\widehat{f}$ has compact support.   □

In the following, if not otherwise stated, G is an arbitrary locally compact Abelian group. Theorem 5.5.1, combined with earlier results, yields some interesting consequences. Because $L^1(G)$ has an approximate identity, $f \in \overline{f * L^1(G)}$ for every $f \in L^1(G)$. From Theorem 5.5.1 and Lemma 5.1.9 we conclude the following

Corollary 5.5.2. *Every proper closed ideal of $L^1(G)$ is contained in some maximal modular ideal.*

Because $\Delta(L^1(G)) = \widehat{G}$ is discrete when G is compact, Theorem 5.5.1 and Corollary 5.2.15 imply the following

Corollary 5.5.3. *Let G be a compact Abelian group. Then spectral synthesis holds for $L^1(G)$.*

Corollary 5.5.4. *Let $f \in L^1(G)$ and let I denote the closed linear subspace of $L^1(G)$ generated by all the translates $L_x f, x \in G$. Then the following conditions are equivalent.*

(i) $I = L^1(G)$.
(ii $\widehat{f}(\alpha) \neq 0$ for all $\alpha \in \widehat{G}$.

Proof. Suppose that $I \neq L^1(G)$. Then, by Proposition 1.4.7, I is a proper closed ideal of $L^1(G)$ and hence, by Corollary 5.5.2, there exists a maximal modular ideal M of $L^1(G)$ such that $M \supseteq I$. By Theorem 2.7.2, M is of the form
$$M = \{g \in L^1(G) : \widehat{g}(\alpha) = 0\}$$
for some $\alpha \in \widehat{G}$. Thus $\widehat{f}(\alpha) = 0$, and this shows that (ii) implies (i).

Conversely, let $\widehat{f}(\alpha) = 0$ for some $\alpha \in \widehat{G}$ and recall that $\widehat{L_x f}(\alpha) = \overline{\alpha(x)}\,\widehat{f}(\alpha)$ for every $x \in G$ (Lemma 2.7.3(i)). Since the function $g \to \widehat{g}(\alpha)$ is continuous on $L^1(G)$, we conclude that $I \subseteq \ker \varphi_\alpha$. Thus (i) $\Rightarrow$ (ii). $\square$

Our next goal is to show that cosets of closed subgroups of $\widehat{G}$ are Ditkin sets. For that, we need a sequence of preparatory lemmas. For $C > 1$, let $\mathcal{F}_C(G)$ denote the set of all functions $f \in L^1(G)$ such that $\|f\|_1 < C$ and $\widehat{f}(\chi) = 1$ for all χ in some neighbourhood (depending on f) of the trivial character 1_G in $\widehat{G}$.

Lemma 5.5.5. *For every compact subset K of G and $\epsilon > 0$, there exists $f \in \mathcal{F}_C(G)$ such that $\|L_y f - f\|_1 < \epsilon$ for all $y \in K$.*

Proof. For a Borel set S in G, let $|S|$ denote the Haar measure of S. Define an open neighbourhood U of 1_G in $\widehat{G}$ by
$$U = \{\alpha \in \widehat{G} : |\alpha(x) - 1| < \epsilon/2C \text{ for all } x \in K\}.$$

Choose a compact symmetric neighbourhood V of 1_G in $\widehat{G}$ such that $V \subseteq U$, and then choose another such neighbourhood W of 1_G such that $VW \subseteq U$ and

$|VW| < C|V|$. Note that such a W exists because V is compact, $C > 1$, and the Haar measure is regular. Denoting by $g \to \widehat{g}$ the Plancherel isomorphism between $L^2(G)$ and $L^2(\widehat{G})$ (Theorem 4.4.10), let $u, v \in L^2(G)$ be such that $\widehat{u} = 1_V$ and $\widehat{v} = 1_{VW}$. We claim that the function

$$f = \frac{1}{|V|} uv \in L^1(G)$$

has all the desired properties. To that end, note first that

$$\|f\|_1 \leq \frac{1}{|V|} \|u\|_2 \|v\|_2 = \frac{1}{|V|} \|\widehat{u}\|_2 \|\widehat{v}\|_2$$

$$= \frac{1}{|V|} \|1_V\|_2 \|1_{VW}\|_2 = \left(\frac{|VW|}{|V|} \right)^{1/2} < C^{1/2}$$

$$< C.$$

Next, observe that $\widehat{f}(\alpha) = 1$ for all $\alpha \in W$. Indeed, if $\alpha \in W$, then

$$|V| \widehat{f}(\alpha) = \int_G u(x) v(x) \overline{\alpha(x)} dx = \langle u \cdot \overline{\alpha}, \overline{v} \rangle$$

$$= \langle \widehat{u \cdot \overline{\alpha}}, \widehat{v} \rangle = \int_{\widehat{G}} \widehat{u}(\alpha\gamma) \overline{\widehat{v}(\gamma^{-1})} d\gamma$$

$$= \int_{\widehat{G}} 1_V(\alpha\gamma) 1_{VW}(\gamma^{-1}) d\gamma = \int_V 1_{VW}(\gamma^{-1}\alpha) d\gamma$$

$$= |V|.$$

For each $y \in G$, we have

$$|V| \cdot \|L_y f - f\|_1 \leq \int_G |u(y^{-1}x) - u(x)| \cdot |v(y^{-1}x)| dx$$

$$+ \int_G |u(x)| \cdot |v(y^{-1}x) - v(x)| dx$$

$$\leq \|L_y u - u\|_2 \|v\|_2 + \|u\|_2 \|L_y v - v\|_2.$$

Now, let $y \in K$. Then, since $VW \subseteq U$,

$$\|L_y v - v\|_2^2 = \int_{\widehat{G}} |\alpha(y) - 1|^2 |\widehat{v}(\alpha)|^2 d\alpha$$

$$= \int_{VW} |\alpha(y) - 1|^2 d\alpha$$

$$< \left(\frac{\epsilon}{2C} \right)^2 |VW|,$$

and similarly

$$\|L_y u - u\|_2^2 < \left(\frac{\epsilon}{2C} \right)^2 |V|.$$

Combining these inequalities we obtain, for all $y \in K$,

$$\|L_y f - f\|_1 \leq \frac{1}{|V|} \cdot \frac{\epsilon}{2C} \left(|V|^{1/2} \|v\|_2 + \|u\|_2 |VW|^{1/2} \right)$$
$$< \frac{\epsilon}{C} \left(\frac{|VW|}{|V|} \right)^{1/2} < \epsilon C^{-1/2}$$
$$< \epsilon,$$

because $|VW| < C|V|$ and $C > 1$. This finishes the proof. $\hspace{1em}\square$

Lemma 5.5.6. *Given $f \in L^1(G)$ and $\epsilon > 0$, there exists $g \in \mathcal{F}_C(G)$ such that*

$$\|f * g - \widehat{f}(1_G)g\|_1 < \epsilon,$$

and hence also

$$\|f * g\|_1 < C|\widehat{f}(1_G)| + \epsilon.$$

Proof. For any $g \in L^1(G)$, we have

$$\|f * g - \widehat{f}(1_G)g\|_1 \leq \int_G |f(y)| \cdot \|L_y g - g\|_1 dy.$$

Choose a compact subset K of G such that

$$\int_{G \setminus K} |f(y)| dy < \frac{\epsilon}{3C}.$$

For any $g \in \mathcal{F}_C(G)$, since $\|L_y g - g\|_1 < 2C$, it follows that

$$\|f * g - \widehat{f}(1_G)g\|_1 < \max_{y \in K} \|L_y g - g\|_1 \int_K |f(y)| dy + \frac{2\epsilon}{3}.$$

Now, by Lemma 5.5.5 there exists $g \in \mathcal{F}_C(G)$ such that

$$\|f\|_1 \cdot \max_{y \in K} \|L_y g - g\|_1 < \frac{\epsilon}{3}.$$

For such g, we obtain $\|f * g - \widehat{f}(1_G)g\|_1 < \epsilon$. This in turn implies

$$\|f * g\|_1 < C|\widehat{f}(1_G)| + \epsilon,$$

since $\|g\|_1 < C$. $\hspace{1em}\square$

Let H be a closed subgroup of G. In the remainder of this section, we always identify $\widehat{G/H}$ with its annihilator in $\widehat{G}$, that is, the closed subgroup of $\widehat{G}$ consisting of all $\alpha \in \widehat{G}$ such that $\alpha(H) = \{1\}$. Remember that T_H denotes the homomorphism from $L^1(G)$ onto $L^1(G/H)$ defined by $T_H(f)(xH) = \int_H f(xh)dh$ for $f \in L^1(G)$ and almost all $x \in G$.

Lemma 5.5.7. *Let H be a closed subgroup of G and let $f \in L^1(G)$ and $\epsilon > 0$. Then there exists a measure $\mu \in M(G)$ with the following properties.*

(i) $\|\mu\| < 1 + \epsilon$ and $\widehat{\mu} = 1$ on a neighbourhood of $\widehat{G/H}$ in $\widehat{G}$.
(ii) $\|\mu * f\|_1 < \|T_H f\|_1 + \epsilon$.

Proof. We first establish the lemma for $f \in C_c(G)$. Choose $C > 1$ such that $C < 1 + \epsilon$ and $(C-1)\|T_H f\|_1 < \epsilon/2$. By Lemma 5.5.6 there exists $h \in L^1(H)$ such that $\|h\|_1 < C$, $\widehat{h} = 1$ in a neighbourhood V of the trivial character in $\widehat{H}$ and

$$\int_G \left| \int_H f(xt)h(t^{-1})dt \right| dx < C\|T_H f\|_1 + \varepsilon/2 < \|T_H f\|_1 + \epsilon.$$

Let $\mu = \mu_h \in M(G)$. Then $\|\mu\| = \int_H |h(t)|dt < 1 + \epsilon$ and

$$\|\mu * f\|_1 = \int_G \left| \int_G f(xy^{-1})d\mu(y) \right| dx = \int_G \left| \int_H f(xt^{-1})h(t)dt \right| dx$$

$$= \int_G \left| \int_H f(xt)h(t^{-1})dt \right| dx$$

$$< \|T_H f\|_1 + \epsilon.$$

Moreover, for each $\gamma \in \widehat{G}$,

$$\widehat{\mu}(\gamma) = \int_H \overline{\gamma(t)}h(t)dt = \widehat{h}(\gamma|_H).$$

Thus $\widehat{\mu} = 1$ on the open neighbourhood $\{\alpha \in \widehat{G} : \alpha|_H \in V\}$ of $\widehat{G/H}$ in $\widehat{G}$.

Let now f be an arbitrary element of $L^1(G)$ and select $g \in C_c(G)$ such that $\|g - f\|_1 < \epsilon/(4+2\epsilon)$. By the first part of the proof there exists a measure μ satisfying (i) and $\|\mu * g\|_1 < \|T_H g\|_1 + \epsilon/2$. Then

$$\|\mu * f\|_1 \leq \|\mu * g\|_1 + \|\mu * (f - g)\|_1$$

$$< \|T_H g\|_1 + \|\mu\| \cdot \|f - g\|_1 + \frac{\epsilon}{2}$$

$$\leq \|T_H f\|_1 + (1 + \|\mu\|)\|f - g\|_1 + \frac{\epsilon}{2}$$

$$\leq \|T_H f\|_1 + (2 + \epsilon)\|f - g\|_1 + \frac{\epsilon}{2}$$

$$< \|T_H f\|_1 + \epsilon.$$

This finishes the proof of the lemma. $\square$

We are now in a position to prove that cosets of closed subgroups in $\widehat{G}$ are Ditkin sets. Actually, a stronger result is shown.

In passing, we make the simple observation that if E is a closed subset of $\widehat{G}$ and $\alpha \in \widehat{G}$, then E is a spectral set (respectively, Ditkin set) if and only if αE is a spectral set (respectively, Ditkin set). This follows readily from the two facts that for $f, g \in L^1(G)$ and $\gamma \in \widehat{G}$, $\widehat{f}(\alpha\gamma) = \overline{\alpha}\widehat{f}(\gamma)$ and $\gamma(f * g) = (\gamma f) * (\gamma g)$.

Theorem 5.5.8. *Let G be a locally compact Abelian group, Γ a closed subgroup of $\widehat{G}$, and $\alpha \in \widehat{G}$. For every $\delta > 0$, $k(\alpha\Gamma)$ possesses an approximate identity of norm bound $2 + \delta$ which is contained in $j(\alpha\Gamma)$. In particular, $\alpha\Gamma$ is a Ditkin set for $L^1(G)$.*

Proof. By the remark preceding the theorem, we can assume that $\alpha = 1_G$. The closed subgroup Γ of $\widehat{G}$ is of the form $\Gamma = \widehat{G/H}$, where H is the annihilator of Γ in $\widehat{G}$; that is, $H = \{x \in G : \gamma(x) = 1 \text{ for all } \gamma \in \Gamma\}$ (Corollary A.5.3). Let $f \in L^1(G)$ be such that $\widehat{f}$ vanishes on Γ. Then

$$\widehat{T_H f}(\gamma) = \int_{G/H} \overline{\gamma(xH)} \left(\int_H f(xh)dh \right) d(xH) = \int_G f(x)\gamma(x)dx = 0$$

for all $\gamma \in \Gamma$, whence $T_H f = 0$.

Given $\epsilon > 0$, we find $u \in L^1(G)$ such that $\|u\| \leq 1$, $\widehat{u}$ has compact support and $\|f - u * f\|_1 < \epsilon$. Because $T_H(u * f) = T_H u * T_H f = 0$, by Lemma 5.5.7 there exists $\mu \in M(G)$ such that $\widehat{\mu} = 1$ in a neighbourhood of Γ, $\|\mu\| \leq 1 + \delta$ and $\|\mu * u * f\|_1 < \epsilon$. Then

$$\|f - (u - \mu * u) * f\|_1 < 2\epsilon,$$

and the Fourier transform of $\mu * u - u$ vanishes on a neighbourhood of Γ. Let $g = \mu * u - u$. Then $g \in j(\Gamma)$ and

$$\|g\|_1 \leq \|u\|_1(\|\mu\| + 1) < 2 + \delta.$$

This proves the theorem.   □

From Theorem 5.5.8, Lemma 5.5.7, and Theorem 5.2.6 we can now derive the injection theorem for spectral sets and Ditkin sets for L^1-algebras.

Theorem 5.5.9. *Let G be a locally compact Abelian group, H a closed subgroup of G, and embed $\widehat{G/H}$ into $\widehat{G}$. Let E be a closed subset of $\widehat{G/H}$.*

(i) *E is a set of synthesis for $L^1(G/H)$ if and only if E is a set of synthesis for $L^1(G)$.*
(ii) *E is a Ditkin set for $L^1(G/H)$ if and only if E is a Ditkin set for $L^1(G)$.*

Proof. It follows from Theorem 5.2.6(i) that if E is a set of synthesis (respectively, Ditkin set) for $L^1(G)$, then E is a set of synthesis (respectively, Ditkin set) for $L^1(G/H) = L^1(G)/\ker T_H$. Moreover, the converse conclusion of (i) follows from part (ii) of Theorem 5.2.6 because $h(\ker T_H) = \widehat{G/H}$ is a Ditkin set by Theorem 5.5.8.

Finally, suppose that E is a Ditkin set for $L^1(G/H)$. Since $\widehat{G/H}$ is a Ditkin set for $L^1(G)$ (Theorem 5.5.8) and $L^1(G)$ satisfies Ditkin's condition at infinity (Theorem 5.5.1), it follows from Theorem 5.2.6(iii) that E is a Ditkin set for $L^1(G)$ provided that there exists a constant $C > 1$ with the following property. For every $f \in L^1(G)$ and $\epsilon > 0$, there exists $g \in L^1(G)$ such that $\hat{g}$ vanishes in a neighbourhood of $\widehat{G/H}$ in $\hat{G}$ and $\|f - f * g\| \leq C\|T_H(f)\|_1 + \epsilon$.

In the situation at hand, we can simply take any $C > 1$. In fact, this can be seen as follows. By Lemma 5.5.7 there exists $\mu \in M(G)$ such that $\hat{\mu} = 1$ in a neighbourhhod of $\widehat{G/H}$ in $\hat{G}$ and

$$\|\mu * f\|_1 < C\,\|T_H(f)\|_1 + \epsilon.$$

Choose $0 < \delta < C\|T_H(f)\|_1 + \epsilon - \|\mu * f\|_1$. Then there exists $u \in L^1(G)$ with $\|u\|_1 \leq 1$ and $\|f - f * u\|_1 < \delta$. Let $g = u - \mu * u \in L^1(G)$. Then

$$\begin{aligned}
\|f - f * g\|_1 &\leq \|f - f * u\|_1 + \|\mu * f * u\|_1 \\
&< \delta + \|\mu * f\|_1 \\
&< C\,\|T_H(f)\|_1 + \epsilon,
\end{aligned}$$

as required.   □

Corollary 5.5.10. *Let* $\Gamma_1, \ldots, \Gamma_m$ *be closed subgroups of* $\hat{G}$ *and let* $\gamma_1, \ldots, \gamma_m$ *be characters of* G. *For each* $j = 1, \ldots, m$, *let* $m_j \in \mathbb{N}_0$ *and for* $1 \leq k \leq m_j$, *let* Δ_{jk} *be an open subgroup of* Γ_j *and* $\delta_{jk} \in \hat{G}$. *Then the subset*

$$E = \bigcup_{j=1}^{m} \gamma_j \left(\Gamma_j \setminus \delta_{jk} \Delta_{jk} \right)$$

of $\hat{G}$ *is a Ditkin set for* $L^1(G)$.

Proof. Finite unions of Ditkin sets are Ditkin sets (Lemma 5.2.1). Therefore we can assume that $m = 1$. Moreover, if F is a Ditkin set then so is γF for every $\gamma \in \hat{G}$. Thus we can further assume that E is of the form $E = \Gamma \setminus \bigcup_{k=1}^{n} \delta_k \Delta_k$, where Γ is a closed subgroup of $\hat{G}$ and for each $k = 1, \ldots, n$, Δ_k is an open subgroup of Γ and $\delta_k \in \hat{G}$. Then E is open and closed in Γ. Let H be the annihilator of Γ in G, so that $\Gamma = \widehat{G/H}$. Since $L^1(G/H)$ satisfies Ditkin's condition at infinity, the open and closed subset E of $\widehat{G/H}$ is a Ditkin set for $L^1(G/H)$ (Proposition 5.2.14). Then assertion (ii) of Theorem 5.5.9 shows that E is a Ditkin set for $L^1(G)$.

Subsets E of $\hat{G}$ of the form in Corollary 5.5.10 play a vital role in Section 5.6 in that the ideals $k(E)$, for such E, will turn out to be precisely the closed ideals of $L^1(G)$ with bounded approximate identities.

5.6 Ideals with bounded approximate identities in $L^1(G)$

Let G be a locally compact Abelian group. In this section we take up the study of closed ideals in $L^1(G)$ with bounded approximate identities. We start with the simple observation that if μ is an idempotent measure in $M(G)$ (that is, $\mu * \mu = \mu$), then $\mu * L^1(G)$ is a closed ideal with bounded approximate identity. Indeed, $\mu * L^1(G)$ is closed in $L^1(G)$ since $\mu * f_n \to g$ in $L^1(G)$ implies that $\mu * g = g$. Moreover, if $(u_\alpha)_\alpha$ is a bounded approximate identity in $L^1(G)$, then $(\mu * u_\alpha)_\alpha$ is one in $\mu * L^1(G)$. We show that, when G is compact, conversely every closed ideal with bounded approximate identity is of this form. In the general case such ideals turn out to give rise to idempotent measures on the Bohr compactification $b(G)$ of G which we have introduced and investigated in Section 2.10. We remind the reader that $b(G)$ is a compact Abelian group and, by Corollary 2.10.14, the discrete group $\widehat{b(G)}$ is isomorphic to $\widehat{G}_d$, the dual group $\widehat{G}$ of G with the discrete topology. This isomorphism is given by $\gamma \to \widehat{\gamma}$, where $\widehat{\gamma}(\varphi_x) = \gamma(x)$ for $\gamma \in \widehat{G}$ and $x \in G$.

Proposition 5.6.1. *Let G be a locally compact Abelian group and $b(G)$ the Bohr compactification of G. Let I be a closed ideal of $L^1(G)$ with bounded approximate identity. Then there exists an idempotent measure $\mu \in M^1(b(G))$ such that $\widehat{\mu}$ coincides with the characteristic function of $\widehat{G}_d \setminus h(I)$.*

Proof. By assumption, I has an approximate identity of norm bound $c > 0$, say. Thus, given any $f \in I$ and $\epsilon > 0$, there exists $u \in I$ such that $\|u\|_1 \leq c$ and $\|u * f - f\|_1 < \epsilon$.

If F is any finite subset of $\widehat{G} \setminus h(I)$, then there exists $f \in I$ such that $\widehat{f}(\gamma) = 1$ for all $\gamma \in F$. In fact, this is a consequence of Theorem 4.2.8 and regularity of $L^1(G)$ (Theorem 4.4.14). For u as above and all $\gamma \in F$, it follows that

$$|\widehat{u}(\gamma) - 1| = |\widehat{u * f}(\gamma) - \widehat{f}(\gamma)| \leq \|u * f - f\|_1 < \epsilon.$$

Now, let

$$A(F, \epsilon) = \{u \in I : \|u\|_1 \leq c, |\widehat{u}(\gamma) - 1| < \epsilon \text{ for all } \gamma \in F\}.$$

By the preceding, $A(F, \epsilon)$ is nonempty for every finite subset F of $\widehat{G} \setminus h(I)$ and $\epsilon > 0$. We embed $A(F, \epsilon)$ into $M^1(b(G)) \subseteq C(b(G))^* = AP(G)^*$ (compare Section 2.10) by making correspond the measure $u(x)dx$ to the function u. Thus, for any $g \in AP(G)$,

$$\langle u, g \rangle = \int_G g(x)u(x)dx.$$

This embedding is isometric; that is, for all $u \in L^1(G)$ we have

$$\|u\|_1 = \sup\{|\langle u, g \rangle| : g \in AP(G), \|g\|_\infty \leq 1\}.$$

To verify this, it is sufficient to consider $u \in C_c(G)$ and to show that given any $\delta > 0$, there exists $g \in AP(G)$ such that $\|g\|_\infty = 1$ and $\|u\|_1 \leq |\langle u, g \rangle| + 2\delta$. For that, let

$$C_\delta = \{x \in G : |u(x)| \geq \delta/m_G(\operatorname{supp} u)\},$$

where m_G denotes the Haar measure of G, and let $u_\delta = u \cdot 1_{C_\delta}$. Then $\|u - u_\delta\|_1 \leq \delta$. Since C_δ is compact, the injective continuous homomorphism ϕ from G into $b(G)$ maps C_δ homeomorphically onto the closed subset $\phi(C_\delta)$ of $b(G)$. Thus the assignment $\phi(x) \rightarrow |u(x)|/u(x)$, $x \in C_\delta$, defines a continuous function of absolute value one on $\phi(C_\delta)$. Extending this function to a norm one function in $C(b(G))$, we see that there exists $g \in AP(G)$ such that $\|g\|_\infty = 1$ and $g(x) = |u(x)|/u(x)$ for all $x \in C_\delta$. It follows that

$$\|u\|_1 - \delta \leq \int_{C_\delta} |u(x)| dx = \int_{C_\delta} u(x) g(x) dx$$
$$\leq |\langle u_\delta, g \rangle - \langle u, g \rangle| + |\langle u, g \rangle|$$
$$\leq \|u_\delta - u\|_1 \cdot \|g\|_\infty + |\langle u, g \rangle|$$

and hence $\|u\|_1 \leq |\langle u, g \rangle| + 2\delta$.

Let $B(F, \epsilon)$ denote the closure of $A(F, \epsilon)$ in $M(b(G))$ in the w^*-topology $\sigma(M(b(G)), C(b(G)))$. By continuity (in this topology), every $\mu \in B(F, \epsilon)$ satisfies $|\hat{\mu}(\gamma) - 1| \leq \epsilon$ for all $\gamma \in F$. Moreover, since $A(F, \epsilon) \subseteq I$, by continuity again, $\hat{\mu}(\gamma) = 0$ for all $\gamma \in h(I)$. The set $B(F, \epsilon)$ is compact in the topology $\sigma(M(b(G)), C(b(G)))$. In fact, $B(F, \epsilon)$ is a closed subset of the compact ball of radius c around the origin in $M(b(G))$. Now, the family

$$\{A(F, \epsilon) : F \subseteq \widehat{G} \setminus h(I) \text{ finite}, \epsilon > 0\}$$

obviously has the finite intersection property. Thus the family of all sets $B(F, \epsilon)$ also has the finite intersection property. It follows that the set

$$\bigcap \{B(F, \epsilon) : F \subseteq \widehat{G} \setminus h(I) \text{ finite}, \epsilon > 0\}$$

is nonempty. Now any measure μ on $b(G)$ which belongs to all of the sets $B(F, \epsilon)$ has the property that $\hat{\mu}(\gamma) = 0$ for all $\gamma \in h(I)$ and $\hat{\mu}(\gamma) = 1$ for all $\gamma \in \widehat{G} \setminus h(I)$. In particular, μ is an idempotent.

Corollary 5.6.2. *Let G be a compact Abelian group and I a closed ideal of $L^1(G)$. Then I has a bounded approximate identity if and only if $I = \mu * L^1(G)$ for some idempotent measure $\mu \in M(G)$.*

Proof. Let I have a bounded approximate identity. Then, by Proposition 5.6.1, there exists an idempotent measure μ on $b(G) = G$ such that $\hat{\mu}$ equals the characteristic function of $\widehat{G} \setminus h(I)$. Now,

$$h(\mu * L^1(G)) = \{\alpha \in \widehat{G} : \hat{\mu}(\alpha) = 0\} = h(I).$$

Because G is compact, every subset of $\widehat{G}$ is a set of synthesis for $L^1(G)$ (Corollary 5.5.2). It follows that $I = \mu * L^1(G)$.

As pointed out at the outset of this section, the converse is true for arbitrary locally compact Abelian groups.

According to Proposition 5.6.1 the description of the hull of ideals with bounded approximate identities in $L^1(G)$ is related to the description of the sets $S(\mu) = \{\gamma \in \widehat{G}_d : \widehat{\mu}(\gamma) \neq 0\}$ associated to idempotent measures μ on the compact group $b(G)$. We therefore proceed to study idempotent measures on groups.

Let G be a locally compact Abelian group. For $\mu \in M(G)$ and $f \in C^b(G)$, the measure $f \cdot \mu$ is defined by $\langle f \cdot \mu, g \rangle = \langle \mu, fg \rangle$, $g \in C_0(G)$. Let H be a compact subgroup of G and denote by m_H the normalized Haar measure of H, considered as a measure on G. If γ is a character of G, then $\gamma \cdot m_H$ is an idempotent measure and $\widehat{\gamma \cdot m_H}$ equals the characteristic function of the subset $\gamma \cdot \widehat{G/H}$ of $\widehat{G}$. Indeed, for each $\alpha \in \widehat{G}$,

$$\widehat{\gamma \cdot m_H}(\alpha) = \int_G \overline{\alpha(x)}\gamma(x)dm_H(x) = \int_H \overline{\alpha(h)}\gamma(h)dm_H(h),$$

which, by the orthogonality relations, equals one if $\alpha|_H = \gamma|_H$ and is zero whenever $\alpha|_H \neq \gamma|_H$.

The most decisive result in this context is *Cohen's idempotent theorem* (Theorem 5.6.6) which states that any idempotent in $M(G)$ is a finite linear combination with integer coefficients of such measures $\gamma \cdot m_H$. To approach Cohen's idempotent theorem, it turns out to be useful to consider, more generally, the set

$$F(G) = \{\mu \in M(G) : \widehat{\mu} \text{ is integer valued}\}.$$

Then $F(G)$ contains the idempotents and is closed under addition and convolution and also under multiplication by characters. For $\mu \in F(G)$, the range of $\widehat{\mu}$ is finite. Moreover, note that if $\mu_1, \mu_2 \in F(G)$ and $\mu_1 \neq \mu_2$, then $\|\mu_1 - \mu_2\| \geq \|\widehat{\mu}_1 - \widehat{\mu}_2\|_\infty \geq 1$.

The following proposition reduces the study of measures in $F(G)$ to compact groups G.

Proposition 5.6.3. *If $\mu \in F(G)$, then the support group of μ, the smallest closed subgroup of G whose complement is a μ-null set, is compact.*

Proof. Assume that $\mu \neq 0$ and let H be the support group of μ. Then μ can be regarded as an element of $F(H)$. We show that $\widehat{H}$ is discrete and hence H is compact.

Let γ be any nontrivial character of H. We claim that $\gamma \cdot \mu \neq \mu$. Assume that $\gamma \cdot \mu = \mu$ and let $L = \{h \in H : \gamma(h) = 1\}$. Then L is a proper subgroup of H since $\gamma \neq 1_H$. The uniqueness theorem for the Fourier–Stieltjes transform implies that $\gamma(x) = 1$ $|\mu|$-almost everywhere. It follows that $\operatorname{supp}\mu \subseteq L$, which contradicts $L \neq H$. Thus $\gamma \cdot \mu \neq \mu$ and hence $\|\gamma \cdot \mu - \mu\| \geq 1$. Choose a compact subset C of H such that $|\mu|(H \setminus C) < 1/4$, and define an open neighbourhood of 1_H by

$$V = \left\{\gamma \in \widehat{H} : |\gamma(x) - 1| < \frac{1}{3\|\mu\|} \text{ for all } x \in C\right\}.$$

Then, for $\gamma \in V$,

$$\|\gamma \cdot \mu - \mu\| \leq \int_H |\gamma(x) - 1| d|\mu|(x)$$
$$= \int_C |\gamma(x) - 1| d|\mu|(x) + \int_{H \setminus C} |\gamma(x) - 1| d|\mu|(x)$$
$$\leq \frac{1}{3} + 2|\mu(H \setminus C)| < 1.$$

It follows that V consists only of the trivial character of H. So $\widehat{H}$ is discrete, and hence H is compact (Remark 4.4.15). □

Lemma 5.6.4. *Let G be a compact Abelian group and let $\mu \in F(G), \mu \neq 0$. Let M be a subset of $M(G)$ such that every measure in M is of the form $\gamma \cdot \mu$ for some $\gamma \in \widehat{G}$. If ν is an accumulation point of M in the w^*-topology on $M(G)$, then*

$$\|\nu\| \leq \|\mu\| - \frac{1}{16\|\mu\|}.$$

Proof. Assuming that $\nu \neq 0$, we have $0 < \|\nu\| \leq \|\mu\|$ since $\|\gamma \cdot \mu\| = \|\mu\|$ and the ball of radius $\|\mu\|$ in $M(G)$ is w^*-closed. Fix any $c > 0$ such that $c < \|\nu\|/\|\mu\|$. Then there exist $f \in C(G)$ with $\|f\|_\infty \leq 1$ and $\int_G f(x)d\nu(x) > c\|\mu\|$, and hence the set

$$V = \left\{\lambda \in M(G) : \operatorname{Re}\left(\int_G f(x)d\lambda(x)\right) > c\|\mu\|\right\}$$

is a w^*-open neighbourhood of ν in $M(G)$. Since ν is an accumulation point of M, there are $\gamma_1, \gamma_2 \in \widehat{G}$ such that $\gamma_1 \cdot \mu \neq \gamma_2 \cdot \mu$ and $\gamma_1 \cdot \mu, \gamma_2 \cdot \mu \in M \cap V$. Let

$$m_j = \operatorname{Re}\left(\int_G f(x)\gamma_j(x)d\mu(x)\right), \quad j = 1, 2,$$

and $d\mu = \theta d|\mu|$, where θ is a measurable function of absolute value one. Write $(f\gamma_j\theta)(x) = g_j(x) + ih_j(x), j = 1, 2$, where g_j and h_j are real-valued. Then, since $\|f\gamma_j\theta\|_\infty \leq 1$,

$$\|\mu\|^2 \geq \left|\int_G (g_j + i|h_j|)(x)d|\mu|(x)\right|^2$$

$$= \left(\int_G g_j(x) d|\mu|(x) \right)^2 + \left(\int_G |h_j(x)| d|\mu|(x) \right)^2$$

$$= m_j^2 + \left(\int_G |h_j(x)| d|\mu|(x) \right)^2$$

and hence

$$\int_G |h_j(x)| d|\mu|(x) \le \left(\|\mu\|^2 - m_j^2 \right)^{1/2}$$

for $j = 1, 2$. Now, for $j = 1, 2$, $m_j > c\|\mu\|$ since $\gamma_j \cdot \mu \in V$, and hence

$$\int_G |1 - f\gamma_j \theta(x)| d|\mu|(x) \le \int_G (1 - g_j(x)) d|\mu|(x) + \int_G |h_j(x)| d|\mu|(x)$$

$$\le \|\mu\| - m_j + \left(\|\mu\|^2 - m_j^2 \right)^{1/2}$$

$$\le \|\mu\| \left(1 - c + (1 - c^2)^{1/2} \right).$$

Since $\gamma_1 \cdot \mu \ne \gamma_2 \cdot \mu$ and $\hat{\mu}$ is integer-valued, we get

$$1 \le \|\gamma_1 \cdot \mu - \gamma_2 \cdot \mu\| \le \int_G |\gamma_1(x) - \gamma_2(x)| d|\mu|(x)$$

$$\le \int_G \left(|\gamma_1(x) - (\gamma_1 \gamma_2 f\theta)(x)| + |(\gamma_1 \gamma_2 f\theta)(x) - \gamma_2(x)| \right) d|\mu|(x)$$

$$\le 2\|\mu\| \left(1 - c + (1 - c^2)^{1/2} \right).$$

This inequality implies

$$\frac{1}{4\|\mu\|^2} \le \left((1 - c) + (1 - c^2)^{1/2} \right)^2$$

$$= 2(1 - c) \left(1 + (1 - c^2)^{1/2} \right)$$

$$< 4(1 - c),$$

and hence

$$c < 1 - \frac{1}{16\|\mu\|^2}.$$

This holds for all $0 < c < \|\nu\|/\|\mu\|$. Thus we finally obtain that

$$\|\nu\| \le \|\mu\| - \frac{1}{16\|\mu\|},$$

as was to be shown.   □

Definition 5.6.5. A measure $\mu \in M(G)$ is said to be *canonical* if μ is of the form

$$\mu = \sum_{j=1}^{m} n_j(\gamma_j \cdot m_H),$$

where H is a compact subgroup of G, $n_1, \ldots, n_m$ are integers, and $\gamma_1, \ldots, \gamma_m$ are characters of G.

Two measures μ and ν in $M(G)$ are called *mutually singular* if they are concentrated on disjoint sets (we then write $\mu \perp \nu$). In that case, $\|\mu + \nu\| = \|\mu\| + \|\nu\|$.

With Lemma 5.6.4 at our disposal, we can now prove Cohen's theorem. We remind the reader that assuming G to be compact is no restriction since the support group of any $\mu \in F(G)$ is compact by Proposition 5.6.3.

Theorem 5.6.6. *Let G be a compact Abelian group. Then each $\mu \in F(G)$ is a finite sum of mutually singular canonical measures.*

Proof. Let $0 \neq \mu \in F(G)$ and set

$$M = \left\{ \gamma \cdot \mu : \gamma \in \widehat{G}, \int_G \gamma(x) d\mu(x) \neq 0 \right\}.$$

Let $\overline{M}$ denote the w^*-closure of M in $M(G)$. Then $0 \notin \overline{M}$, because

$$|\langle \gamma \cdot \mu, 1_G \rangle| = \left| \int_G \gamma(x) d\mu(x) \right| \geq 1$$

for all $\gamma \cdot \mu \in M$ since $\widehat{\mu}$ is integer-valued. Let $\delta = \inf\{\|\lambda\| : \lambda \in \overline{M}\}$. Then the sets

$$\{\lambda \in \overline{M} : \|\lambda\| \leq \delta + \frac{1}{n}\},$$

$n \in \mathbb{N}$, are all nonempty and w^*-compact. It follows that there exists $\nu \in \overline{M}$ such that $\|\nu\| = \delta$. In particular, $\delta > 0$ since $0 \notin \overline{M}$. Let

$$N = \left\{ \gamma \cdot \nu : \gamma \in \widehat{G}, \int_G \gamma(x) d\nu(x) \neq 0 \right\}.$$

Then N is contained in $\overline{M}$. Indeed, if $\gamma \cdot \nu \in N$ and $(\gamma_\alpha)_\alpha$ is a net in $\widehat{G}$ such that $\gamma_\alpha \cdot \mu \to \nu$, then $(\gamma \gamma_\alpha) \cdot \mu \to \gamma \cdot \nu$ in the w^*-topology, and since $\int_G \gamma(x) d\nu(x) \neq 0$, we have $\langle (\gamma \gamma_\alpha) \cdot \mu, 1_G \rangle \neq 0$ and hence $\gamma_\alpha \cdot \mu \in M$ eventually. Consequently, $\nu \in \overline{M}$. But now N must be finite, because otherwise N has a w^*-accumulation point σ, and then $\|\sigma\| < \|\nu\|$ by Lemma 5.6.4 and $\sigma \in \overline{M}$, contradicting the fact that ν is an element of $\overline{M}$ with minimal norm.

We now construct the support group of ν. Let

$$\Gamma = \left\{ \gamma \in \widehat{G} : \int_G \gamma(x) d\nu(x) \neq 0 \right\},$$

and define an equivalence relation on Γ by $\gamma_1 \sim \gamma_2$ if and only if $\gamma_1 \cdot \nu = \gamma_2 \cdot \nu$. Because N is finite, Γ consists of only finitely many equivalence classes, $\Gamma_1, \ldots, \Gamma_m$ say. For $1 \leq i \leq m$, let

$$H_i = \{x \in G : \gamma(x) = \gamma'(x) \text{ for all } \gamma, \gamma' \in \Gamma_i\}.$$

Then H_i is a closed subgroup of G. We show next that $|\nu|(G \setminus H_i) = 0$. Since

$$H_i = \{x \in G : (\delta^{-1}\gamma)(x) = 1 \text{ for all } \gamma, \delta \in \Gamma_i\},$$

$\widehat{G/H_i}$ is the subgroup of $\widehat{G}$ generated by all these elements $\delta^{-1}\gamma, \gamma, \delta \in \Gamma_i$. Thus, if $\lambda \in \widehat{G/H_i}$ and hence $\lambda = \prod_{l=1}^{r} \delta_l^{-1}\gamma_l$ with $\gamma_l, \delta_l \in \Gamma_i$ and α is an arbitrary element of $\widehat{G}$, then by definition of Γ_i,

$$\widehat{\nu}(\alpha\lambda) = \int_G \alpha(x) \prod_{l=1}^{r} (\delta_l^{-1}\gamma_l)(x) d\nu(x)$$

$$= \int_G \alpha(x) \prod_{l=1}^{r-1} (\delta_l^{-1}\gamma_l)(x) d((\delta_r^{-1}\gamma_r) \cdot \nu)(x)$$

$$= \int_G \alpha(x) \prod_{l=1}^{r-1} (\delta_l^{-1}\gamma_l)(x) d\nu(x)$$

$$= \ldots = \int_G \alpha(x) d\nu(x)$$

$$= \widehat{\nu}(\alpha).$$

Thus, for every $\lambda \in \widehat{G/H_i}$, the Fourier–Stieltjes transform of $(1_G - \lambda) \cdot \nu$ vanishes on all of $\widehat{G}$. This implies that $|\nu|(G \setminus H_i) = 0$, $1 \leq i \leq m$. Now, let $H = \bigcap_{i=1}^{m} H_i$. It follows that $|\nu|(G \setminus H) = 0$.

We claim that ν has only a finite number of nonzero Fourier coefficients on H. For that, observe that $\Gamma|_H$ is finite since $\Gamma = \bigcup_{i=1}^{m} \Gamma_i$ and $\Gamma_i \subseteq \gamma_i \cdot \widehat{G/H_i} \subseteq \gamma_i \cdot \widehat{G/H}$ for every $\gamma_i \in \Gamma_i$. Now, if $\chi \in \widehat{H}$ is such that $\int_H \chi(x) d\nu(x) \neq 0$, then χ is the restriction to H of some $\gamma \in \Gamma$ since $|\nu|(G \setminus H) = 0$. This proves the claim. It follows that ν is a finite sum

$$\nu = \sum_{i=1}^{m} n_i(\overline{\chi}_i \cdot m_H) = \sum_{i=1}^{m} n_i(\overline{\gamma}_i \cdot m_H),$$

where $n_i \in \mathbb{Z}$ and $\gamma_i \in \Gamma_i$, $i = 1, \ldots, m$.

To prove that μ is a finite sum of mutually singular canonical measures, we distinguish two cases. First, if ν is not an accumulation point of M, then $\nu = \gamma \cdot \mu$ for some $\gamma \in \widehat{G}$, so $\mu = \overline{\gamma} \cdot \nu$ is canonical. Thus we can assume that ν is an accumulation point of M. Because ν is absolutely continuous with respect to m_H, it follows that $\nu = \gamma \cdot \mu|_H$ for some $\gamma \in \widehat{G}$. Let $\mu_1 = \overline{\gamma} \cdot \nu$. Then $\mu = \mu_1 + (\mu - \mu_1)$ and $\mu_1 \perp (\mu - \mu_1)$ since $\mu_1 = \mu|_H$. Thus

$$\|\mu\| = \|\mu_1\| + \|\mu - \mu_1\| \geq 1 + \|\mu - \mu_1\|,$$

so $\|\mu - \mu_1\| \leq \|\mu\| - 1$. Since $\mu - \mu_1 \in F(G)$, the same argument can now be applied to $\mu - \mu_1$, and so on. Because the norm is decreased by at least one at each step, this process must stop after a finite number of steps. $\square$

In order to describe the support set $S(\mu)$ of $\mu \in F(G)$ in the most convenient manner, we have to introduce the coset ring of an Abelian group.

Definition 5.6.7. The *coset ring* of an Abelian group G, denoted $\mathcal{R}(G)$, is the smallest Boolean algebra of subsets of G containing the cosets of all subgroups of G. That is, $\mathcal{R}(G)$ is the smallest family of subsets of G which contains all the cosets of subgroups of G and which is closed under forming finite unions, finite intersections, and complements.

Let G be a topological Abelian group. The *closed coset ring* of G, $\mathcal{R}_c(G)$, is defined to be

$$\mathcal{R}_c(G) = \{E \in \mathcal{R}(G) : E \text{ is closed in } G\}.$$

The following proposition is a first indication of the importance of the coset ring in our context.

Proposition 5.6.8. *Let G be a compact Abelian group and $\mu \in F(G)$. Then the set*

$$S(\mu) = \{\alpha \in \widehat{G} : \widehat{\mu}(\alpha) \neq 0\}$$

belongs to $\mathcal{R}(\widehat{G})$.

Proof. For $k \in \mathbb{Z}^* = \mathbb{Z} \setminus \{0\}$, let $S(\mu)_k = \{\alpha \in \widehat{G} : \widehat{\mu}(\alpha) = k\}$. Since the range of $\widehat{\mu}$ is finite, it suffices to show that $S(\mu)_k \in \mathcal{R}(\widehat{G})$ for each k.

Assume first that μ is a canonical measure. Thus μ is of the form $\mu = \sum_{j=1}^{n} n_j(\gamma_j \cdot m_H)$, where H is a closed subgroup of G, the n_j are nonzero integers, and the γ_j are characters of G such that $\gamma_j|_H \neq \gamma_i|_H$ for $j \neq i$. Then

$$\widehat{\mu}(\alpha) = \sum_{j=1}^{n} n_j \widehat{\gamma_j \cdot m_H}(\alpha) = \sum_{j=1}^{n} n_j \int_H \gamma_j(h)\overline{\alpha(h)}dh.$$

The orthogonality relations for characters of H now imply that $S(\mu)_{n_j} = \gamma_j \cdot \widehat{G/H} \in \mathcal{R}(\widehat{G})$ and $S(\mu)_k = \emptyset$ if $k \neq n_j$ for all $j = 1, \ldots, n$.

Now, by Theorem 5.6.6, an arbitrary $\mu \in F(G)$ can be represented as a finite sum of (mutually singular) canonical measures. To establish the statement of the proposition for μ, we can therefore proceed by induction on the number of summands in such a sum. For that, it is enough to show that if $\mu_1, \mu_2 \in F(G)$ are such that $S(\mu_j)_k \in \mathcal{R}(\widehat{G})$ for all $k \in \mathbb{Z}^*$ and $j = 1, 2$, then $S(\mu_1 + \mu_2)_k \in \mathcal{R}(\widehat{G})$ for each $k \in \mathbb{Z}^*$. However, $S(\mu_1 + \mu_2)_k$ is the union of the three sets

$$\left(S(\mu_1)_k \setminus \bigcup_{l \in \mathbb{Z}^*, l \neq k} S(\mu_2)_l\right), \left(S(\mu_2)_k \setminus \bigcup_{l \in \mathbb{Z}^*, l \neq k} S(\mu_1)_l\right)$$

$$\text{and} \bigcup_{l \in \mathbb{Z}^*, l \neq k} \left(S(\mu_1)_l \bigcap S(\mu_2)_{k-l}\right),$$

and this shows $S(\mu_1 + \mu_2)_k \in \mathcal{R}(\widehat{G})$. □

We proceed with a characterisation of the sets in the coset ring $\mathcal{R}(G)$. The proof of the following proposition is postponed to the Appendix (Proposition A.6.1).

Proposition 5.6.9. *Let G be an Abelian group. A subset E of G belongs to $\mathcal{R}(G)$ if and only if E is of the form*

$$E = \bigcup_{i=1}^{n} \left(C_i \setminus \bigcup_{j=1}^{n_i} C_{ij} \right), \quad n, n_i \in \mathbb{N},$$

where C_i and C_{ij} are (possibly void) cosets of subgroups of G.

Remark 5.6.10. Let H and K be subgroups of G and $a, b \in G$ such that $aH \cap bK \neq \emptyset$. Then there exists $h \in H$ such that

$$aH \setminus bK = ah(H \setminus (H \cap K)),$$

and $H \cap K$ has infinite index in H whenever $aH \setminus bK$ is infinite. Thus Proposition 5.6.9 can be reformulated as follows. A subset E of G belongs to $\mathcal{R}(G)$ if and only if E can be written as

$$E = F \cup \bigcup_{i=1}^{m} \left(a_i H_i \setminus \bigcup_{j=1}^{m_i} b_{ij} K_{ij} \right),$$

where F is finite, H_i is a subgroup of G, and K_{ij} is a subgroup of infinite index in H_i, $1 \leq i \leq m, 1 \leq j \leq m_i$.

If G is a compact Abelian group, Propositions 5.6.9 and 5.6.1 combined with results obtained in Sections 5.5 and 1.4 already allow a description in terms of the coset ring $\widehat{G}$ of those closed ideals in $L^1(G)$ which possess bounded approximate identities. When G is noncompact, however, one needs to identify the closed subsets of $\mathcal{R}(\widehat{G})$. This is fairly difficult. The result is the following theorem the proof of which is also given in the Appendix (Theorem A.6.8).

Theorem 5.6.11. *Let G be an Abelian topological group and $E \in \mathcal{R}(G)$. Then $\overline{E} \in \mathcal{R}(G)$, and E is closed if and only if E can be written as*

$$E = \bigcup_{j=1}^{m} \left(C_j \setminus \bigcup_{l=1}^{m_j} C_{jl} \right),$$

where C_j and C_{jl} are (possibly void) closed cosets in G and C_{jl} is contained in C_j and open in C_j.

Theorem 5.6.11 is the main tool to establish the following description of closed ideals in $L^1(G)$ with bounded approximate identities. Using Theorem 5.6.12, an alternative description can be given in terms of idempotent measures on quotient groups G/H and the pullbacks to $L^1(G)$ of the associated ideals in $L^1(G/H)$ (Exercise 5.8.32).

Theorem 5.6.12. *Let G be a locally compact Abelian group and I a closed ideal of $L^1(G)$. Then I has a bounded approximate identity if and only if $I = k(E)$ for some $E \in \mathcal{R}_c(\widehat{G})$; that is, a set E of the form*

$$E = \bigcup_{j=1}^{m} \gamma_j \left(\Gamma_j \setminus \bigcup_{k=1}^{m_j} \delta_{jk} \Delta_{jk} \right),$$

where Γ_j is a closed subgroup of $\widehat{G}$, Δ_{jk} is an open subgroup of Γ_j, and γ_j and δ_{jk} are characters of G $(1 \leq j \leq m, 1 \leq k \leq m_j)$.

Proof. Suppose first that I has a bounded approximate identity. By Proposition 5.6.1, $h(I) \in \mathcal{R}(\widehat{G})$. Since $h(I)$ is closed in $\widehat{G}$, by Theorem 5.6.11, $h(I) = E$ where E is of the stated form. Moreover, by Corollary 5.5.10, E is a Ditkin set and hence a spectral set. Thus $I = k(E)$.

Conversely, assume that E is as in the theorem and let

$$E_j = \Gamma_j \setminus \bigcup_{k=1}^{m_j} \delta_{jk} \Delta_{jk}, \ 1 \leq j \leq m.$$

Then $k(E) = \bigcap_{j=1}^{m} k(\gamma_j E_j)$ and by Lemma 1.4.9 it suffices to show that each $k(\gamma_j E_j)$ has a bounded approximate identity. Thus we can assume that $E = \gamma F$ with $F = \Gamma \setminus \bigcup_{l=1}^{n} \delta_l \Delta_l$, where Γ is a closed subgroup of $\widehat{G}$, Δ_l is an open subgroup of Γ, and $\delta_l \in \widehat{G}, 1 \leq l \leq n$. Since $\alpha(g * f) = (\alpha g) * (\alpha f)$ for $f, g \in L^1(G)$ and $\alpha \in \widehat{G}$ and $k(\gamma F) = \{\overline{\gamma} f : f \in k(F)\}$, the ideal $k(\gamma F)$ has a bounded approximate identity precisely when $k(F)$ does so. Since F is a spectral set and $h(\sum_{l=1}^{n} k(\Gamma \setminus \delta_l \Delta_l)) = F$, it follows that

$$k(F) = k \left(\bigcap_{l=1}^{n} (\Gamma \setminus \delta_l \Delta_l) \right) = \overline{\sum_{l=1}^{n} k(\Gamma \setminus \delta_l \Delta_l)}.$$

Thus, by Lemma 1.4.9, $k(F)$ has a bounded approximate identity whenever each ideal of the form $k(\Gamma \setminus \delta \Delta)$, where Δ is an open subgroup of Γ and $\delta \in \widehat{G}$, has a bounded approximate identity. Now, either $\Gamma \setminus \delta\Delta = \Gamma$ or $\delta \in \Gamma$ and then $\Gamma \setminus \delta\Delta = \delta(\Gamma \setminus \Delta)$. Therefore it remains to observe that both $k(\Gamma)$ and $k(\Gamma \setminus \Delta)$ have bounded approximate identities. For $k(\Gamma)$ this follows from Theorem 5.5.8. In the second case, $\Gamma = \widehat{G/H}$ and $\Delta = \widehat{G/K}$ where H and K are closed subgroups of G such that $H \subseteq K$ and K/H is compact since Δ is open in Γ. With $m_{K/H}$ denoting the normalized Haar measure of K/H, we have

$$k(\Gamma \setminus \Delta)/k(\Gamma) = (\delta_{eH} - m_{K/H}) * L^1(G/H),$$

and the ideal on the right has a bounded approximate identity. Since $k(\Gamma)$ has a bounded approximate identity, it follows that $k(\Gamma \setminus \Delta)$ also has a bounded approximate identity (Lemma 1.4.8). □

5.7 On spectral synthesis in projective tensor products

The question of how spectral synthesis of the projective tensor product $A \widehat{\otimes}_\pi B$ of two commutative Banach algebras A and B is related to spectral synthesis of A and B, appears to be quite delicate. Although it is not difficult to see that if spectral synthesis holds for $A \widehat{\otimes}_\pi B$, then it holds for A and B (Theorem 5.7.2), the converse problem, however, is far from being settled. In this section we present those results we are aware of in this context.

Recall that given $\psi \in \Delta(B)$, there exists a unique continuous homomorphism $\phi_\psi : A \widehat{\otimes}_\pi B \to A$ such that $\phi_\psi(a \otimes b) = \psi(b)a$ for all $a \in A$ and $b \in B$ (Lemma 2.11.5).

Lemma 5.7.1. *Let E and F be closed subsets of $\Delta(A)$ and $\Delta(B)$, respectively, and let $\psi \in F$. Then*

$$\phi_\psi(j(E \times F)) \subseteq j(E).$$

Proof. Let $x \in j(E \times F)$. There exists a compact subset C of $\Delta(A \widehat{\otimes} B) = \Delta(A) \times \Delta(B)$ such that $\widehat{x} = 0$ outside of C and $C \cap (E \times F) = \emptyset$. Now,

$$\operatorname{supp} \widehat{\phi_\psi(x)} \subseteq \{\varphi \in \Delta(A) : \varphi \otimes \psi \in C\}.$$

This implies that $E \cap \operatorname{supp} \widehat{\phi_\psi(x)} = \emptyset$ since $C \cap (E \times F) = \emptyset$. So $\widehat{\phi_\psi(x)}$ has compact support disjoint from E. $\qquad\square$

Theorem 5.7.2. *Let A and B be regular commutative Banach algebras and suppose that $A \widehat{\otimes}_\pi B$ is semisimple. Let E and F be closed subsets of $\Delta(A)$ and $\Delta(B)$, respectively. If $E \times F$ is a set of synthesis (Ditkin set) for $A \widehat{\otimes}_\pi B$, then E is a set of synthesis (Ditkin set) for A, and likewise for F. In particular, if spectral synthesis holds for $A \widehat{\otimes}_\pi B$, then it holds for both A and B.*

Proof. Let $a \in A$ and $\epsilon > 0$ be given. Choose $\psi \in F$ and $b \in B$ such that $\psi(b) = 1$. If $E \times F$ is a set of synthesis, there exists $x \in j(E \times F)$ such that $\|x - a \otimes b\| \leq \epsilon$. Then $\phi_\psi(x) \in j(E)$ by Lemma 5.7.1 and

$$\|a - \phi_\psi(x)\| = \|\phi_\psi(a \otimes b) - \phi_\psi(x)\| \leq \|a \otimes b - x\| \leq \epsilon.$$

If $E \times F$ is a Ditkin set, there exists $y \in j(E \times F)$ such that $\|a \otimes b - (a \otimes b)y\| \leq \epsilon$. Then

$$\begin{aligned}
\|a - a\phi_\psi(x)\| &= \|\phi_\psi(a \otimes b) - \phi_\psi(a \otimes b)\phi_\psi(x)\| \\
&= \|\phi_\psi(a \otimes b - (a \otimes b)x)\| \\
&\leq \|a \otimes b - (a \otimes b)x\| \\
&\leq \epsilon.
\end{aligned}$$

Since $\phi_\psi(x) \in j(E)$ and $\epsilon > 0$ was arbitrary, we obtain that $a \in \overline{j(E)}$ in the first case and $a \in \overline{aj(E)}$ in the second case. This proves the theorem. $\qquad\square$

One of the special situations in which a converse to Theorem 5.7.2 can be established is when $\Delta(B)$, say, is discrete.

Theorem 5.7.3. *Let A and B be regular commutative Banach algebras and suppose that $A \widehat{\otimes}_\pi B$ is semisimple and satisfies Ditkin's condition at infinity. In addition, let $\Delta(B)$ be discrete. Then spectral synthesis holds for $A \widehat{\otimes}_\pi B$ if and only if it holds for A.*

Proof. According to Theorem 5.7.2, we only have to show that if spectral synthesis holds for A, then it holds for $A \widehat{\otimes}_\pi B$. Thus let T be a closed subset of $\Delta(A \widehat{\otimes}_\pi B)$ and let $x \in k(T)$. Write T as

$$T = \bigcup_{\psi \in \Delta(B)} (E_\psi \times \{\psi\}),$$

where each E_ψ is a closed subset of $\Delta(A)$.

Because $A \widehat{\otimes}_\pi B$ satisfies Ditkin's condition at infinity, given $\epsilon > 0$, there exists $y \in A \widehat{\otimes}_\pi B$ such that $\|x - xy\| \leq \epsilon$ and $\widehat{y}$ has compact support. Therefore it is enough to show that $x \in \overline{j(T)}$ for every $x \in k(T)$ such that $\widehat{x}$ has compact support. Then, since $\Delta(B)$ is discrete,

$$\operatorname{supp} \widehat{x} = \bigcup_{i=1}^n (S_i \times \{\psi_i\}),$$

where $\psi_i \in \Delta(B)$ and S_i is a compact subset of $\Delta(A)$, $1 \leq i \leq n$. Now, since $A \widehat{\otimes}_\pi B$ is semisimple and regular (Lemma 4.2.20), there exists a partition of unity associated to these data; that is, there are $u_1, \ldots, u_n \in A \widehat{\otimes}_\pi B$ such that $\operatorname{supp} \widehat{u}_i \subseteq \Delta(A) \times \{\psi_i\}$ and $x = \sum_{i=1}^n x u_i$ (Corollary 4.2.12). It suffices to show that $x u_i \in \overline{j(T)}$ for each $1 \leq i \leq n$. Thus we can henceforth assume that

$$x \in k((E_\psi \times \{\psi\}) \cup (\Delta(A) \times (\Delta(B) \setminus \{\psi\})))$$

for some $\psi \in \Delta(B)$. Define a closed ideal I of $A \widehat{\otimes}_\pi B$ by

$$I = k(\Delta(A \widehat{\otimes}_\pi B) \setminus (\Delta(A) \times \{\psi\})) = k(\Delta(A) \times (\Delta(B) \setminus \{\psi\})).$$

The ideal $A \widehat{\otimes}_\pi k(\Delta(B) \setminus \{\psi\})$ of $A \widehat{\otimes} B$ has the same hull as I. Because $A \widehat{\otimes}_\pi B$ is semisimple and regular) and $\emptyset$ is a Ditkin set for $A \widehat{\otimes}_\pi B$, every open and closed subset of $\Delta(A \widehat{\otimes}_\pi B)$ is a set of synthesis (Proposition 5.2.12). It follows that

$$I = A \widehat{\otimes}_\pi k(\Delta(B) \setminus \{\psi\}).$$

So the element $x \in I$ can be expressed as $x = \sum_{i=1}^\infty a_i \otimes b_i$, where $a_i \in A$ and $b_i \in k(\Delta(B) \setminus \{\psi\})$. Since B is semisimple, the ideal $k(\Delta(B) \setminus \{\psi\})$ is one-dimensional and therefore there exists $b \in B$ such that $\psi(b) \neq 0$ and $b_i = \alpha_i b$, $\alpha_i \in \mathbb{C}$, for each $i \in \mathbb{N}$. Thus $x = a \otimes b$ with $a = \sum_{i=1}^\infty \alpha_i a_i$. Now, $a \in k(E_\psi)$ since $x \in k(E_\psi \times \{\psi\})$ and $\psi(b) \neq 0$. Since E_ψ is a spectral set, $a \in \overline{j(E_\psi)}$ and so $x = a \otimes b \in \overline{j(E_\psi \times \{\psi\})}$. Finally, since $\widehat{x} = 0$ on $\Delta(A) \times (\Delta(B) \setminus \{\psi\})$, it follows that $x \in \overline{j(E)}$, as required.

Theorem 5.7.3 applies to L^1-algebras of vector-valued functions on compact groups as follows.

Corollary 5.7.4. *Let G be a compact Abelian group and A a semisimple and regular commutative Banach algebra. Suppose that A is Tauberian and has a bounded approximate identity. Then spectral synthesis holds for $L^1(G, A)$ if and only if it holds for A.*

Proof. Recall first that $L^1(G, A) = L^1(G) \widehat{\otimes}_\pi A$ is semisimple by Theorem 2.11.6 because L^1-spaces share the approximation property. Thus, if spectral synthesis holds for $L^1(G, A)$ then it holds for A (Theorem 5.7.2).

For the converse, note that G is compact and hence $\Delta(L^1(G)) = \widehat{G}$ is discrete. Therefore, by Theorem 5.7.3, it suffices to observe that $L^1(G, A)$ satisfies Ditkin's condition at infinity. Because both $L^1(G)$ and A are Tauberian, $L^1(G) \widehat{\otimes}_\pi A$ is Tauberian as well. Moreover, $L^1(G) \widehat{\otimes}_\pi A$ has a bounded approximate identity since this is true of $L^1(G)$ and A (Lemma 1.5.3). It follows that $L^1(G) \widehat{\otimes}_\pi A$ satisfies Ditkin's condition at infinity since this is true of any commutative Banach algebra which is Tauberian and has an approximate identity. $\square$

Using Malliavin's deep theorem mentioned at the beginning of this section, Corollary 5.7.4 admits the following generalisation. Let G be a locally compact Abelian group and let A be as in Corollary 5.7.4. Then spectral synthesis holds for $L^1(G, A)$ if and only if G is compact and spectral synthesis holds for A. Indeed, if spectral synthesis holds for $L^1(G) \widehat{\otimes}_\pi A$, then the same is true of $L^1(G)$ (Theorem 5.7.2) and Malliavin's theorem forces G to be compact.

Lemma 5.7.5. *Suppose that A and B are Tauberian commutative Banach algebras, and let E and F be closed subsets of $\Delta(A)$ and $\Delta(B)$, respectively. Then*

$$j(E) \otimes B + A \otimes j(F) \subseteq \overline{j(E \times F)}.$$

Proof. It suffices to show that if

$$x = \sum_{k=1}^{n} a_k \otimes b_k + \sum_{l=1}^{m} c_l \otimes d_l,$$

where $a_k \in j(E), b_k \in B, c_l \in A$, and $d_l \in j(F)$, then $x \in \overline{j(E \times F)}$. Of course, we can assume that $a_k \neq 0$ for all k and $d_l \neq 0$ for all l. Because both A and B are Tauberian, given $\epsilon > 0$, there exist $v_k \in B$ and $u_l \in A$ such that $\widehat{v}_k$ and $\widehat{u}_l$ have compact support in $\Delta(B)$ and $\Delta(A)$, respectively, and

$$\|v_k - b_k\| \leq \frac{\epsilon}{2n\|a_k\|} \quad \text{and} \quad \|u_l - c_l\| \leq \frac{\epsilon}{2m\|d_l\|}$$

$(1 \leq k \leq n, 1 \leq l \leq m)$. Let

$$y = \sum_{k=1}^{n} a_k \otimes v_k + \sum_{l=1}^{m} u_l \otimes d_l.$$

Then $\|y - x\| \leq \epsilon$ and $\hat{y}$ has compact support and vanishes on a neighbourhood of $E \times F$. Thus $y \in j(E \times F)$, and since $\varepsilon > 0$ was arbitrary, we conclude that $x \in \overline{j(E \times F)}$. □

Lemma 5.7.6. *Let F be a closed subset of $\Delta(A)$ and $\varphi \in \Delta(A)$.*

(i) *Every $x \in k(\{\varphi\} \times F)$ has a representation of the form*

$$x = \sum_{j=1}^{\infty} (a_j \otimes b_j) + e \otimes b,$$

where $a_j \in k(\{\varphi\})$, $b_j \in B$, $e \in A$, and $b \in k(F)$.
(ii) $k(\{\varphi\} \times F) = k(\varphi) \widehat{\otimes}_\pi B + A \widehat{\otimes}_\pi k(F)$.

Proof. (i) Let $x \in k(\{\varphi\} \times F)$. Then $\sum_{j=1}^{\infty} y_j \otimes b_j$, where $y_j \in A$, $b_j \in B$, and $\sum_{j=1}^{\infty} \|y_j\| \cdot \|b_j\| < \infty$. Choose $e \in A$ such that $\varphi(e) = 1$ and write $y_j = a_j + \lambda_j e$, where $a_j \in k(\varphi)$ and $\lambda_j \in \mathbb{C}$. Then, since $|\lambda_j| = |\varphi(y_j)| \leq \|y_j\|$ and $\|a_j\| = \|y_j - \lambda_j e\| \leq \|y_j\|(1 + \|e\|)$,

$$x = \sum_{j=1}^{\infty} a_j \otimes b_j + e \otimes \sum_{j=1}^{\infty} \lambda_j b_j.$$

Let $b = \sum_{j=1}^{\infty} \lambda_j b_j$. Then, for each $\psi \in F$,

$$\psi(b) = \psi\left(\sum_{j=1}^{\infty} \varphi(x_j) b_j \right) = (\varphi \otimes \psi)(x) = 0.$$

Thus $b \in k(F)$, and this establishes the desired representation of x.
   (ii) Because $k(\varphi) \widehat{\otimes}_\pi B$ and $A \widehat{\otimes}_\pi k(F)$ are both contained in $k(\{\varphi\} \times F)$, the statement follows from (i). □

Corollary 5.7.7. *Let A and B be regular and Tauberian commutative Banach algebras and suppose that $A \widehat{\otimes}_\pi B$ is semisimple. Let $\{\varphi\}$ be a spectral set for A and $F \subseteq \Delta(B)$ a spectral set for B. Then $\{\varphi\} \times F$ is a spectral set for $A \widehat{\otimes}_\pi B$.*

Proof. It follows from part (ii) of Lemma 5.7.6 and the hypotheses that

$$\begin{aligned}
k(\{\varphi\} \times F) &= k(\varphi) \widehat{\otimes}_\pi B + A \otimes k(F) \\
&= \overline{j(\varphi)} \widehat{\otimes}_\pi B + A \otimes \overline{j(F)} \\
&\subseteq \overline{j(\varphi) \otimes B} + \overline{A \otimes j(F)}.
\end{aligned}$$

Since A and B are Tauberian, by Lemma 5.7.5 $j(\varphi) \otimes B$ and $A \otimes j(F)$ are both contained in $\overline{j(\{\varphi\} \times F)}$. Thus $k(\{\varphi\} \times F) \subseteq \overline{j(\{\varphi\} \times F)}$, and so $\{\varphi\} \times F$ is a spectral set because $A \widehat{\otimes}_\pi B$ is regular and semisimple. □

Theorem 5.7.8. *Let A and B be unital regular commutative Banach algebras. Suppose that $A \widehat{\otimes}_\pi B$ is semisimple, and let E be a closed subset of $\Delta(A \widehat{\otimes}_\pi B)$ such that the set*

$$\{\varphi \in \Delta(A) : (\{\varphi\} \times \Delta(B)) \cap \partial(E) \neq \emptyset\}$$

is scattered in $\Delta(A)$. Then E is a set of synthesis for $A \widehat{\otimes}_\pi B$ provided that one of the following conditions (i) *or* (ii) *is satisfied.*

(i) *For each closed subset F of $\Delta(B)$ and each $\varphi \in \Delta(A)$, the set $\{\varphi\} \times F$ is a Ditkin set for $A \widehat{\otimes}_\pi B$.*

(ii) *Each closed subset F of $\Delta(B)$ is a set of synthesis and satisfies condition* (D) *of Definition 5.2.9, and likewise for each singleton $\{\varphi\}$, $\varphi \in \Delta(A)$.*

Proof. We apply Theorem 5.2.13 to $A \widehat{\otimes}_\pi B$, $T = \Delta(A)$ and the projection

$$\phi : \Delta(A \widehat{\otimes}_\pi B) = \Delta(A) \times \Delta(B) \to \Delta(A), \quad (\varphi, \psi) \to \varphi.$$

Then the assertion of the theorem follows at once if condition (i) is satisfied.

So assume that (ii) holds. We claim that then (i) holds. Let $\varphi \in \Delta(A)$ and let F be a closed subset of $\Delta(B)$. By Corollary 5.7.7, $\{\varphi\} \times F$ is a set of synthesis. To conclude that $\{\varphi\} \times F$ is actually a Ditkin set, in virtue of Lemmas 5.2.8 and 5.2.10 it suffices to show that $\{\varphi\} \times F$ satisfies condition (D).

Thus let U be an open neighbourhood of φ in $\Delta(A)$ and V an open neighbourhood of F in $\Delta(B)$. By hypothesis (D) holds for both F and $\{\varphi\}$. Therefore there exist constants $c, d > 0$ and elements $a \in A$ and $b \in B$ with the following properties: $\|a\| \leq c$, $\operatorname{supp} \widehat{a} \subseteq U$, $\widehat{a} = 1$ near φ and $\|b\| \leq d$, $\operatorname{supp} \widehat{b} \subseteq V$, $\widehat{b} = 1$ near F. Then the element $x = a \otimes b$ of $A \widehat{\otimes}_\pi B$ satisfies $\|x\| \leq d$, $\operatorname{supp} \widehat{x} \subseteq U \times V$ and $\widehat{x} = 1$ in a neighbourhood of $\{\varphi\} \times F$. So (ii) $\Rightarrow$ (i), and this finishes the proof of the theorem. $\square$

Let X be a compact Hausdorff space, E a closed subset of X, and U an open set containing E. Choose an open set V such that $E \subseteq V$ and $\overline{V} \subseteq U$. By Urysohn's lemma, there exists $f \in C(X)$ such that $f = 1$ on $\overline{V}$, $f = 0$ on $X \setminus U$, and $\|f\|_\infty = 1$. Thus condition (D) is trivially satisfied, and because every closed subset of X is a spectral set, we have the following straightforward consequence of Theorem 5.7.8.

Corollary 5.7.9. *Let X be a compact Hausdorff space and A a regular and semisimple commutative Banach algebra with identity. Suppose that $\Delta(A)$ is scattered and that every singleton $\{\varphi\}$, $\varphi \in \Delta(A)$, is a spectral set and satisfies condition* (D). *Then spectral synthesis holds for $A \widehat{\otimes}_\pi C(X)$.*

Proof. We only have to note that $A \widehat{\otimes}_\pi C(X)$ is semisimple because A and $C(X)$ are semisimple and $C(X)$ has the approximation property. $\square$

It is worth pointing out in this context that spectral synthesis need not hold for the projective tensor product $C(X) \widehat{\otimes}_\pi C(Y)$, where X and Y are compact Hausdorff spaces. In fact, it is an essential step in Varopoulos's proof [129] of Malliavin's theorem that spectral synthesis fails for $C(G) \widehat{\otimes}_\pi C(G)$ when G is any infinite compact group.

5.8 Exercises

Exercise 5.8.1. Let A be a unital regular semisimple commutative Banach algebra and let $E, F \subseteq \Delta(A)$ be sets of synthesis for A. Show that $E \cup F$ is a set of synthesis if and only if

$$\overline{j(E) \cap j(F)} = \overline{j(E)} \cap \overline{j(F)}.$$

It is an interesting question of when, for arbitrary closed subsets E and F of $\Delta(A)$, this equality holds.

Exercise 5.8.2. Let A be a commutative Banach algebra and $\varphi \in \Delta(A)$. A linear functional D on A is called a *point derivation at* φ if $D(xy) = \varphi(x)D(y) + \varphi(y)D(x)$ for all $x, y \in A$. Suppose that the singleton $\{\varphi\}$ is a set of synthesis. Prove that A does not admit a nonzero point derivation at φ. (Hint: Let D be a point derivation at φ and let J be the ideal consisting of all $a \in A$ such that $\varphi(a) = D(a) = 0$. Show that the hull of J equals $\{\varphi\}$.)

Exercise 5.8.3. Let X be a locally compact Hausdorff space. Prove that the following three conditions are equivalent.
  (i) X is scattered.
  (ii) The one-point compactification of X is scattered.
  (iii) Every $f \in C_0(X)$ has countable range.

Let A be a regular and semisimple commutative Banach algebra satisfying Ditkin's condition at infinity. For a closed subset E of $\Delta(A)$, let Δ_E denote the set of all $\varphi \in \Delta(A)$ with the property that there exists $x \in k(E)$ such that $\hat{x}$ does not belong locally to $\overline{j(E)}$ at φ. The set Δ_E is called the *difference spectrum of* E. Retain this situation and notation in the next exercises.

Exercise 5.8.4. For $\varphi \in \Delta_E$, show that
  (i) φ belongs to the boundary $\partial(E)$ of E in $\Delta(A)$.
  (ii) φ has no closed relative neighbourhood in E which is a set of synthesis for A.

Exercise 5.8.5. For closed subsets E_1 and E_2 of $\Delta(A)$, prove
  (i) $\Delta_{E_1 \cap E_2} \subseteq \Delta_{E_1} \cup \Delta_{E_2} \cup (\partial(E_1) \cap \partial(E_2))$.
  (ii) $\Delta_{E_1} \cup \Delta_{E_2} \subseteq \Delta_{E_1 \cup E_2} \cup (E_1 \cap E_2)$.
(Hint: For (ii), let $\varphi \in \Delta_{E_1 \cap E_2}$ and assume that $\varphi \notin \partial(E_1) \cap \partial(E_2)$. Show that $\varphi \in \partial(E_j)$ for exactly one $j \in \{1, 2\}$ and that then $\varphi \in \Delta_{E_j}$.)

Exercise 5.8.6. Use the fact that for $n \geq 3$, $S^{n-1} \subseteq \mathbb{R}^n = \Delta(L^1(\mathbb{R}^n))$ is not a set of synthesis for $L^1(\mathbb{R}^n)$ to show that $\Delta_{S^{n-1}} = S^{n-1}$.

Exercise 5.8.7. Let I be a closed ideal of A and let $x \in A$ be such that $\hat{x}$ belongs locally to I at every $\varphi \in \Delta(A) \setminus E$. Show that $\overline{aj(E)} \subseteq I$.

Exercise 5.8.8. Suppose that there exists a Ditkin set F such that $\Delta_E \subseteq F \subseteq E$. Prove that E is a set of synthesis.
(Hint: Conclude from Exercise 5.8.7 that $\overline{xj(\Delta_E)} \subseteq \overline{j(E)}$ for every $x \in k(E)$.)

For a closed subset E of $\Delta(A)$, let Γ_E denote the set of all $\varphi \in \Delta(A)$ with the property that there exists $x \in \overline{j(E)}$ such that $\hat{x}$ does not belong locally to $\overline{xj(E)}$.

Exercise 5.8.9. Suppose that there exists a Ditkin set F such that $\Gamma_E \subseteq F \subseteq E$. Use Exercise 5.8.8 to prove that E is a Ditkin set.

Exercise 5.8.10. Let $\mathrm{Lip}_\alpha[0,1]$ be the algebra of Lipschitz functions of order α on $[0,1]$ and identify $\Delta(\mathrm{Lip}_\alpha[0,1])$ with the interval $[0,1]$. Let E be a closed subset of $[0,1]$ such that $E \neq \emptyset$ and $E \neq [0,1]$ and let $d(t,E) = \inf\{|t-s| : s \in E\}$, $t \in [0,1]$. Let $0 < \alpha \leq 1$ and define $f \in \mathrm{Lip}_\alpha[0,1]$ by $f(t) = d(t,E)^\alpha$. Let $\|\cdot\|_\alpha$ denote the norm on $\mathrm{Lip}_\alpha[0,1]$ (Exercise 1.6.11). Show that $\|f - g\|_\alpha \geq 1$ for every $g \in j(E)$. Thus E fails to be a set of synthesis for $\mathrm{Lip}_\alpha[0,1]$.

Exercise 5.8.11. Identify functions on $\mathbb{T}$ with 2π-periodic functions on $\mathbb{R}$. Let $C^1(\mathbb{T})$ denote the subalgebra of $C(\mathbb{T})$ consisting of all continuously differentiable functions, equipped with the norm $\|f\| = \|f\|_\infty + \|f'\|_\infty$. Determine $\Delta(C^1(\mathbb{T}))$ and all the closed primary ideals of $C^1(\mathbb{T})$ (compare the results for $C^1[0,1]$ in Section 5.3).

Exercise 5.8.12. Let $C^1[0,1]$ be the Banach algebra of all continuously differentiable functions $f : [0,1] \to \mathbb{C}$ with the norm $\|f\| = \|f\|_\infty + \|f'\|_\infty$. For $a \in [0,1]$, let

$$I(a) = \{f \in C^1[0,1] : f(0) = 0\}.$$

Show that $I(a)$ cannot possess a bounded approximate identity.

Exercise 5.8.13. Let $n \in \mathbb{N}$ and E a closed subset of $[0,1] = \Delta(C^n[0,1])$. Use the results of Section 5.3 and Leibniz' rule to show that $k(E)^{n+1} \subseteq \overline{j(E)}$. Moreover, show that if E is a singleton, then $n+1$ is the smallest integer k such that $f^k \in \overline{j(E)}$.

Exercise 5.8.14. Let M be the Mirkil algebra as investigated in Section 5.4 and let M_e be its unitisation. Prove that a subset E of $\mathbb{Z} = \Delta(M)$ is a set of synthesis for M if and only if $E \cup \{\infty\} \subseteq \Delta(M_e)$ is a set of synthesis for M_e.

Exercise 5.8.15. Let M be the Mirkil algebra and let E be a subset of $\mathbb{Z} = \Delta(M)$. Let $f \in k(E)$ and suppose that f is continuous. Show that $f \in \overline{f * j(E)}$.
(Hint: Let $(k_n)_{n \in \mathbb{N}}$ be the Fejér kernel; that is,

$$k_n(t) = \sum_{j=-n}^{n} \left(1 - \frac{|j|}{n+1}\right) e^{ijt}, \quad t \in [-\pi, \pi].$$

Use that $\widehat{k_n}$ has finite support and that, by Fejér's theorem, $k_n * f$ converges uniformly to f as $n \to \infty$.)

Exercise 5.8.16. Let M be the Mirkil algebra. Show that for every subset E of $\underline{\Delta(M)}$ and $f \in k(E)$, $f * f \in \overline{(f * f) * j(E)}$. Conclude that $k(E) * k(E) \subseteq \overline{j(E)}$.

Exercise 5.8.17. For $n \in \mathbb{N}_0$, let I_n be the closed ideal of $A(\mathbb{D})$ defined by

$$I_n = \{f \in A(\mathbb{D}) : f^{(j)}(0) = 0 \text{ for } 0 \leq j \leq n\}.$$

Show that $h(I_n) = \{0\}$ and I_{n+1} is a proper subset of I_n for all $n \in \mathbb{N}_0$. Since $\bigcap_{n=0}^{\infty} I_n = \{0\}$, this yields that there does not exist a minimal primary ideal with hull equal to $\{0\}$.

Exercise 5.8.18. Extend Corollary 5.5.4 as follows. Let G be a locally compact Abelian group and $F \subseteq L^1(G)$, and let I denote the translation invariant closed linear subspace of $L^1(G)$ generated by F. Then the following two conditions are equivalent.
  (i) $I = L^1(G)$.
  (ii) For each $\alpha \in \widehat{G}$, there exists $f \in F$ such that $\widehat{f}(\alpha) \neq 0$.
(Hint: For (ii) $\Rightarrow$ (i), let C be any compact subset of $\widehat{G}$ and show the existence of some $f \in I$ such that $\widehat{f}(\alpha) \neq 0$ for all $\alpha \in C$. Then use that $L^1(G)$ is Tauberian.)

Exercise 5.8.19. Let G be a locally compact Abelian group. For a bounded continuous function f on G and a constant c the statement '$f(x) \to c$ as $x \to \infty$' means that for every $\epsilon > 0$ there exists a compact subset C of G such that $|f(x) - c| \leq \epsilon$ for all $x \in G \setminus C$.
  Now, let $f \in L^\infty(G)$ and $g \in L^1(G)$ such that $\widehat{g}(\alpha) \neq 0$ for all $\alpha \in \widehat{G}$. Let $c \in \mathbb{C}$ and suppose that

$$(g * f)(x) \to c\,\widehat{g}(1_G) \text{ as } x \to \infty.$$

Prove that then, for all $h \in L^1(G)$,

$$(h * f)(x) \to c\,\widehat{h}(1_G) \text{ as } x \to \infty.$$

(Hint: First reduce to the case $c = 0$, and then employ Corollary 5.5.5.)

Exercise 5.8.20. Let G be a compact Abelian group, $f \in L^1(G)$, and $\epsilon > 0$. Apply Corollary 5.5.4 to show that there exist $\alpha_1, \ldots, \alpha_n \in \widehat{G}$ and $c_1, \ldots, c_n \in \mathbb{C}$ such that

$$\left\| f - f * \sum_{j=1}^{n} c_j \alpha_j \right\|_1 \leq \epsilon.$$

Exercise 5.8.21. Let G be a compact Abelian group and identify $\Delta(L^2(G))$ with $\widehat{G}$ (Exercise 2.12.31). Every $f \in L^2(G)$ can be written as an L^2-convergent series $f = \sum_{\chi \in \widehat{G}} \widehat{f}(\chi)\overline{\chi}$. Let E be any subset of $\widehat{G}$.

  (i) Show that $f \in k(E)$ if and only if $\widehat{f}(\chi) = 0$ for all $\chi \in \widehat{G} \setminus E$ and $f \in j(E)$ if and only if $f = \sum_{j=1}^{n} \widehat{f}(\chi_j)\overline{\chi_j}$, where $\chi_1, \ldots, \chi_n \in \widehat{G} \setminus E$.
  (ii) Use (i) to show that E is a Ditkin set.

Exercise 5.8.22. Let G be a locally compact Abelian group and ω a weight function on G such that $\widehat{G}(\omega) = \widehat{G}$. Let Γ be a compact open subgroup of $\widehat{G}$. Prove that Γ is a Ditkin set for $L^1(G, \omega)$.
(Hint: The subgroup $H = \{x \in G : \gamma(x) = 1 \text{ for all } \gamma \in \Gamma\}$ of G is compact and open. Verify that $f * 1_H = 0$ and that $\widehat{1_H} = |H| \cdot 1_\Gamma$.)

Exercise 5.8.23. Let $f, g \in l^1(\mathbb{Z})$ and suppose that there exists $\delta > 0$ such that $|\widehat{f}(z)| \geq \delta$ for all $z \in \mathbb{T} = \Delta(l^1(\mathbb{Z}))$ such that $\widehat{g}(z) \neq 0$. Show that $g = h * f$ for some $h \in l^1(\mathbb{Z})$. Note that this generalizes Wiener's theorem (Corollary 2.2.11).
(Hint: Let $I = l^1(\mathbb{Z}) * f$, the ideal generated by f. Then $h(I) \cap \operatorname{supp} \widehat{g} = \emptyset$ and hence $g \in I$.)

Exercise 5.8.24. Let C be a convex compact subset of $\mathbb{R}^n$, $n \in \mathbb{N}$. Prove that C is a set of synthesis for $L^1(\mathbb{R}^n)$. For that, one can assume without loss of generality that $0 \in C$. Then proceed as follows. For $f \in L^1(\mathbb{R}^n)$ such that $\widehat{f}|_C = 0$ and each $0 < \alpha < 1$, define $f_\alpha \in L^1(\mathbb{R}^n)$ by $f_\alpha(x) = f((1/\alpha)x), x \in \mathbb{R}^n$, and show that

  (i) $\widehat{f_\alpha}$ vanishes in a neighbourhood of C.
  (ii) $\|f_\alpha - f\|_1 \to 0$ as $\alpha \to 1$.

Exercise 5.8.25. Use the same method as in Exercise 5.8.24 to show that the set $\{y \in \mathbb{R}^n : \|y\| \geq 1\}$ is a set of synthesis for $L^1(\mathbb{R}^n)$. This example, Exercise 5.8.24, and Schwartz' result that the sphere $S^{n-1} \subseteq \mathbb{R}^n$ fails to be a set of synthesis for $L^1(\mathbb{R}^n)$ if $n \geq 3$, show that the intersection of two sets of synthesis need not be a set of synthesis.

Exercise 5.8.26. A closed subset E of $\mathbb{R}^2$ is called a *polyhedral set* if its boundary is a union of countably many translates of closed subsets F_j, where each F_j is a closed subset with countable boundary of some one-dimensional subgroup of $\mathbb{R}$. Use the injection theorem for Ditkin sets for $L^1(\mathbb{R}^2)$ and Theorem 5.2.2 to prove that every polyhedral set in $\mathbb{R}^2$ is a Ditkin set for

$L^1(\mathbb{R}^2)$. In particular, sets of the form $[a, b] \times [c, d]$, $a < b, c < d$, are Ditkin sets for $L^1(\mathbb{R}^2)$.

Exercise 5.8.27. Let E be an infinite and proper subset of $\mathbb{Z}$. Prove that E belongs to $\mathcal{R}(\mathbb{Z})$ (if and) only if there are finitely many arithmetic progressions $S_1, \ldots, S_n$; that is, $S_j = \{p_j + nq_j : n \in \mathbb{Z}\}, 0 \le p_j < q_j, p_j, q_j \in \mathbb{Z}, 1 \le j \le n$, such that the symmetric difference $E \triangle (S_1 \cup \ldots \cup S_n)$ is finite.
(Hint: Given two arithmetic progressions $c_1 + d_1\mathbb{Z}$ and $c_2 + d_2\mathbb{Z}$, let d be the least common multiple of d_1 and d_2 and express $c_1 + d_1\mathbb{Z} \setminus c_2 + d_2\mathbb{Z}$ in terms of cosets of $d\mathbb{Z}$.)

Exercise 5.8.28. Use Exercise 5.8.27 and the description of the closed subgroups of $\mathbb{R}$ and of $\mathbb{T}$ to determine the closed coset rings $\mathcal{R}_c(\mathbb{R})$ and $\mathcal{R}_c(\mathbb{T})$.

Exercise 5.8.29. Let G be a compact Abelian group and let I be a closed ideal of $L^1(G)$ with bounded approximate identity. Prove that I has a bounded approximate identity consisting of functions of the form $u = \sum_{j=1}^n \lambda_j \chi_j$, where $\chi_1, \ldots, \chi_n \in \widehat{G}$ and $\lambda_1, \ldots, \lambda_n \in \mathbb{C}$.

Exercise 5.8.30. Let G be a locally compact Abelian group and I a closed ideal of $L^1(G)$. Suppose that P is a bounded projection from $L^1(G)$ onto I such that $P(L_x f) = L_x(P(f))$ for all $f \in L^1(G)$ and $x \in G$. Use Exercise 2.12.57 and results of Section 5.6 to show that $h(I) \in \mathcal{R}_c(G)$.

Let G be a locally compact Abelian group and H a closed subgroup of G. Recall from Section 1.3 that $T_H : L^1(G) \to L^1(G/H)$ denotes the homomorphism defined by

$$T_H f(xH) = \int_H f(xh)dh, \quad f \in L^1(G),$$

and that the ideal $\ker T_H$ has a bounded approximate identity.

Exercise 5.8.31. Let G be a locally compact Abelian group, Γ a closed subgroup of $\widehat{G}$, and H the closed subgroup of G such that $\Gamma = \widehat{G/H}$. Let Δ be an open subgroup of Γ and $\gamma \in \Gamma$, and let I be a closed ideal of $L^1(G)$ with $h(I) = \Gamma \setminus \gamma\Delta$. Show that there exists a (unique) finite idempotent measure μ on G/H such that $I = T_H^{-1}(\mu * L^1(G/H))$.

Exercise 5.8.32. Let G be a locally compact Abelian group and I a closed ideal in $L^1(G)$. Use Exercise 5.8.31 and Theorem 2.6.12 to prove that I has a bounded approximate identity if and only if I is of the form

$$I = \bigcap_{j=1}^n \chi_j \cdot T_H^{-1}(\mu_j * L^1(G/H_j)),$$

where, for $1 \le j \le n$, χ_j is a character of G, H_j is a closed subgroup of G, and μ_j is a finite idempotent measure on G/H_j.

Exercise 5.8.33. Let G be a locally compact group and suppose that the Fourier algebra $A(G)$ possesses an approximate identity. Identify $\Delta(A(G))$ with G (Theorem 2.9.4).

(i) Let $(u_\alpha)_\alpha$ be the net constructed in Exercise 2.12.54 and consider functions of the form $v - u_\alpha v$, $v \in A(G)$, to show that $\{e\}$ is a Ditkin set for $A(G)$.

(ii) Deduce from (i) that singletons in G are Ditkin sets for $A(G)$.

Let A be a semisimple and regular commutative Banach algebra. A closed subset E of $\Delta(A)$ is called a *weak spectral set* if every element of the quotient algebra $k(E)/\overline{j(E)}$ is nilpotent; that is, for every $x \in k(E)$ there exists $m \in \mathbb{N}$, depending on x, such that $x^m \in \overline{j(E)}$. We say that *weak spectral synthesis holds* for A if every closed subset of $\Delta(A)$ is a weak spectral set. We have seen in Exercises 5.8.13 and 5.8.16 that weak spectral synthesis holds for $C^n[0,1]$ and for the Mirkil algebra even though for both of them spectral synthesis fails to hold. Moreover, given $E \subseteq \Delta(A)$, in both cases there exists $m \in \mathbb{N}$ such that $a^m \in \overline{j(E)}$ for all $a \in k(E)$. This latter fact is a general phenomenon as the following exercise shows.

Exercise 5.8.34. Let A be as above and let $E \subset \Delta(A)$ be a weak spectral set. For $n \in \mathbb{N}$, let $S_n = \{x \in k(E) : x^n \in \overline{j(E)}\}$.

(i) Show that there exist $m \in \mathbb{N}$, $a \in A$ and $\epsilon > 0$ such that

$$\{y \in A : \|y - a\| < \epsilon\} \subseteq S_m.$$

(ii) Let m, a and ϵ be as in (i) and let $x \in A$ be arbitrary. Then $a + (1/k)x \in S_m$ for sufficiently large k. Conclude that $x \in S_m + S_m$.

(iii) Show that $S_m + S_m \subseteq S_{2m}$ and hence $A = S_{2m}$.

As mentioned earlier, it is an open question whether the union of two sets of synthesis is again a set of synthesis. However, the union of two weak spectral sets turns out to be a weak spectral set. In particular, a finite union of spectral sets is at least a weak spectral set. For a weak spectral set E, by Exercise 5.8.34, there exists a smallest $k \in \mathbb{N}$ such that $a^k \in \overline{j(E)}$ for all $a \in k(E)$. Denote this number by $\xi(E)$.

Exercise 5.8.35. Let E and F be weak spectral sets for A. Prove that $E \cup F$ is a weak spectral set and $\xi(E \cup F) \leq \xi(E) + \xi(F)$.

Exercise 5.8.36. Let E be a closed subset of $\Delta(A)$ such that $k(E)$, as a closed ideal in A, is generated by finitely many elements $x_1, \ldots, x_n$. Suppose that there exists $m \in \mathbb{N}$ such that $x_j^m \in \overline{j(E)}$ for $1 \leq j \leq n$. Show that $x^{nm} \in \overline{j(E)}$ for every $x \in k(E)$. So E is a weak spectral set and $\xi(E) \leq nm$.

Let A be a commutative Banach algebra. A sequence $(e_n)_{n \in \mathbb{N}}$ in A is called an *orthogonal basis* for A if it satisfies the following conditions.

(1) $e_n e_m = \delta_{nm} e_n$ for all $n, m \in \mathbb{N}$.

(2) Every $x \in A$ has a unique representation of the form $x = \sum_{n=1}^{\infty} c_n e_n$, where $c_n \in \mathbb{C}$ and convergence means norm convergence of the corresponding sequence of partial sums.

Exercise 5.8.37. Let A be a commutative Banach algebra with orthogonal basis $(e_n)_n$. For each $n \in \mathbb{N}$, define $\varphi_n : A \to \mathbb{C}$ by

$$\varphi_n \left(\sum_{k=1}^{\infty} c_n e_n \right) = c_n.$$

Prove that $\varphi_n \in \Delta(A)$ and that the map $n \to \varphi_n$ is a homeomorphism between $\mathbb{N}$ and $\Delta(A)$. Deduce that A is regular and semisimple.
(Hint: For surjectivity of the map, show that given $\varphi \in \Delta(A)$, there exists precisely one $n \in \mathbb{N}$ such that $\varphi(e_n) \neq 0$.)

Exercise 5.8.38. Let A be as in the preceding exercise. Show that spectral synthesis holds for A; that is,

$$I = k(h(I)) = \bigcap \{\ker \varphi_n : \varphi_n(I) = \{0\}\}$$

for every closed ideal I of A.

Exercise 5.8.39. Let A be a commutative Banach algebra with orthogonal basis and let I be a nonzero closed ideal of A. Use Exercise 5.8.38 to show that I has an (unbounded) approximate identity.

Exercise 5.8.40. Show that the following examples provide Banach algebras with orthogonal bases.
   (i) The convolution algebras $L^p(G)$, $1 \leq p < \infty$, where G is a first countable compact Abelian group.
   (ii) The sequence algebras $c_0(\mathbb{N})$ and $l^p(\mathbb{N})$, $1 \leq p < \infty$, with componentwise operations.
   (iii) The Hardy spaces $H^p(\mathbb{D}^\circ)$, $1 < p < \infty$, with the Hadamard product (see Exercise 1.6.13). To see this, note that if f and g in $H^p(\mathbb{D}^\circ)$ are represented by the power series $\sum_{n=0}^{\infty} a_n z^n$ and $\sum_{n=0}^{\infty} b_n z^n$, respectively, then $f \bullet g$ is represented by the power series $\sum_{n=0}^{\infty} a_n b_n z^n$.

5.9 Notes and references

Since usually spectral synthesis does not hold for a semisimple and regular commutative Banach algebra, it is a major issue to study the classes of spectral sets and of Ditkin sets and to establish permanence properties. This is the main concern of Section 5.2. Theorem 5.2.5 on unions of sets of synthesis appears, in the case of the Fourier algebra of a locally compact group G, in Warner [131], and was shown by Reiter [103] for disjoint sets. In this context, see also [94]. Injection theorems for sets of synthesis and Ditkin sets can be found in [105]. Theorem 5.2.13 extends Theorem 1.2 of [8], which in turn is a generalisation of the Wiener–Ditkin theorem.

The ideal theory of $C^n[0,1]$, as presented in Section 5.3, is classical and goes back to Gelfand, Raikov, and Shilov [41]. The algebra M, discussed in Section 5.4, was invented by Mirkil [90] to provide an example of a regular and semisimple commutative Banach algebra with discrete structure space, for which nevertheless spectral synthesis fails. The Mirkil algebra was studied further by Atzmon [9] and Warner [133]. Somewhat surprisingly, it also turned out to serve as a counterexample to the union conjecture in that $\Delta(M)$ contains two disjoint sets of synthesis the union of which is not a set of synthesis.

The literature on spectral synthesis problems for the group algebra $L^1(G)$ of a locally compact Abelian group is enormous. We therefore confine ourselves to just a few highlights and historical remarks and otherwise refer the reader to the monographs [11], [46], [54], [55], [105], and [113]. Theorem 5.5.1(ii) was shown by Rudin [113, Theorem 2.6.8] using structure theory of locally compact Abelian groups. The proof presented here was given in [10]. Kaplansky [70] showed that singletons in $\widehat{G} = \Delta(L^1(G))$ are sets of synthesis. This result was improved by Helson [50] to the effect that closed subsets of $\widehat{G}$ with scattered boundary are spectral sets. The first and best accessible example of a non-spectral set is due to Schwartz [118]. For $n \geq 3$, the unit sphere S^{n-1} in $\mathbb{R}^n$ fails to be a spectral set for $L^1(\mathbb{R}^n)$. On the other hand, $S^1 \subseteq \mathbb{R}^2$ is a set of synthesis for $L^1(\mathbb{R}^2)$ [51]. Spectral synthesis holds for $L^1(G)$ when G is compact. It is Malliavin's celebrated achievement that conversely spectral synthesis fails for $L^1(G)$ whenever G is noncompact [84]. More or less complete accounts of Malliavin's proof are given in all of the books mentioned above. An alternative approach to Malliavin's theorem was found by Varopoulos [129]. A main step in his proof is the failure of spectral synthesis for the projective tensor product $C(G) \widehat{\otimes}_\pi C(G)$ for any infinite compact Abelian group.

There are two major open questions in the spectral synthesis of $L^1(G)$. The first one is whether any set of synthesis in $\widehat{G}$ actually is a Ditkin set. The second one is whether the union of two sets of synthesis is again a set of synthesis. Note that because finite unions of Ditkin sets are Ditkin sets (Lemma 5.2.1), an affirmative answer to the first question would imply an affirmative answer to the second.

In Section 5.5 we have applied the general results of Section 5.2 and results of Section 4.4 on $L^1(G)$ to produce sets of synthesis and Ditkin sets in $\widehat{G}$. The identification of the closed ideals in $L^1(G)$ with bounded approximate identities, which we have exposed in Section 5.6, is mainly due to Liu, van Rooij, and Wang [80], building on a description of the closed sets in the coset ring of $\widehat{G}$ established by Gilbert [43] and, independently and with a much simpler proof, by Schreiber [117]. A major ingredient is Cohen's idempotent theorem [21], the simple proof of which was found by Ito and Amemiya [59].

Very little is known regarding spectral synthesis in projective tensor products $A \widehat{\otimes}_\pi B$ in terms of A and B, although $\Delta(A \widehat{\otimes}_\pi B)$ is canonically homeomorphic to $\Delta(A) \times \Delta(B)$. We have presented some partial results in Section 5.7. Theorem 5.7.8 generalizes results of [8].

At this point, it seems in order to mention a few, mostly recent, developments extending results on $L^1(G)$. Let G be an arbitrary locally compact group. Then closed subgroups of $G = \Delta(A(G))$ are sets of synthesis for $A(G)$ [127]. Using Malliavin's theorem and a deep theorem of Zelmanov [144] on the existence of infinite Abelian subgroups in infinite compact groups, it was shown in [68], under a mild additional hypothesis on G (the existence of an approximate identity in the weakest possible sense) that spectral synthesis holds for the Fourier algebra $A(G)$ (if and) only if G is discrete. In [67] an injection theorem for Ditkin sets of Fourier algebras was proved. Moreover, for an amenable locally compact group G, in [33] a complete description of the closed ideals with bounded approximate identities in $A(G)$ was established.

Weak spectral synthesis in commutative Banach algebras, as first studied by Warner [132] (compare Exercises 5.8.34 to 5.8.36), has since gained some attention (see [95], [100], [65], and [66]) since there are several examples of weak spectral sets which fail to be spectral sets, such as the sphere $S^{n-1} \subset \mathbb{R}^n = \Delta(L^1(\mathbb{R}^n))$ for $n \geq 3$, and also algebras, such as Lipschitz algebras, for which weak spectral synthesis holds, whereas spectral synthesis fails. Concerning the concept of difference spectrum, which we have briefly addressed in Exercises 5.8.4 and 5.8.5, we refer to [115], [125], [105], and [95].

A

Appendix

The purpose of this short appendix is twofold. In the first three sections, which concern point set topology, functional analysis, and measure theory, we simply repeat some standard notation and concepts and list a number of results that are used throughout the book. No proofs are given and we tend to be very brief because the reader will have a firm background in these areas. However, because we cannot expect the reader to be familiar with basic harmonic analysis, a different viewpoint is taken in the remaining three sections where we treat Abelian topological groups. Although we just cite the existence and uniqneness of Haar measure on a locally compact group, we prove a number of facts about convolution of functions since the group algebra $L^1(G)$ serves as a prominent example in the book and also the Hilbert space $L^2(G)$ is used substantially. In Section 5 we deduce the Pontryagin duality theorem from the Plancherel formula and point out the bijection between closed subgroups of G and closed subgroups of the dual group $\widehat{G}$. Finally, in Section 6 we describe the coset ring of an Abelian group and the closed sets in the coset ring of a topological Abelian group.

A.1 Topology

Let X be a topological space. Then $C(X)$ denotes the set of all continuous complex-valued functions on X and $C^b(X)$ the subspace of all bounded functions in $C(X)$. For $f \in C(X)$, the support of f, supp f, is the closure of the set of all $x \in X$ at which $f(x) \neq 0$. The set of all functions in $C(X)$ with compact support is denoted $C_c(X)$. A function f on X is said to *vanish at infinity* if for each $\epsilon > 0$, there exists a compact subset K_ϵ of X such that $|f(x)| < \epsilon$ for all $x \in X \setminus K_\epsilon$. Then $C_0(X)$ stands for the set of all $f \in C(X)$ which vanish at infinity. Clearly, $C_c(X) \subseteq C_0(X) \subseteq C^b(X)$ and all these spaces coincide with $C(X)$ when X is compact. Also, all these spaces are algebras under pointwise operations. On $C^b(X)$ we can introduce the supremum norm defined by $\|f\|_\infty = \sup\{|f(x)| : x \in X\}$. This norm turns $C^b(X)$ and $C_0(X)$

into Banach spaces. If X is a locally compact Hausdorff space, then $C_c(X)$ is dense in $C_0(X)$. This is a consequence of the Stone–Weierstrass theorem (Theorem A.1.3 below).

A topological space X is called *normal* if it is Hausdorff and for each pair $\{A, B\}$ of disjoint closed subsets of X there exist open subsets U and V of X such that $A \subseteq U$, $B \subseteq V$ and $U \cap V = \emptyset$.

Theorem A.1.1. *(Urysohn's lemma)*

(i) *Let X be a normal topological space, and let A and B be disjoint closed subsets of X. Then there exists a continuous function $f : X \to [0, 1]$ such that $f|_A = 1$ and $f|_B = 0$.*
(ii) *Let X be a locally compact Hausdorff space, and let C be a compact subset of X and U an open set containing C. Then there exists $f \in C_c(X)$ with $f|_C = 1$, $0 \le f(x) \le 1$ for all $x \in X$ and $\operatorname{supp} f \subseteq U$.*

Theorem A.1.2. *(Tietze's extension theorem) A Hausdorff space X is normal if and only if every real valued function, which is defined and continuous on a closed subset of X, admits a continuous extension to all of X.*

A family F of complex-valued functions on a topological space X is said to *strongly separate the points* of X if for each $x \in X$, there exists $f \in F$ with $f(x) \neq 0$, and for each $x, y \in X$ with $x \neq y$, there exists $g \in F$ such that $g(x) \neq g(y)$. The family F is said to be self-adjoint if it contains with a function f the conjugate complex function $\overline{f}$.

Theorem A.1.3. *(Stone–Weierstrass theorem) Let X be a locally compact Hausdorff space, and let A be a self-adjoint subalgebra of $C_0(X)$. Suppose that A strongly separates the points of X. Then A is uniformly dense in $C_0(X)$.*

Theorem A.1.4. *(Arzela–Ascoli) Let X be a locally compact Hausdorff space and $F \subseteq C_0(X)$. Suppose that F satisfies the following two conditions.*

(i) *The set $F(x) = \{f(x) : f \in F\}$ is bounded for every $x \in X$.*
(ii) *F is equicontinuous; that is, for each $x \in X$ and $\epsilon > 0$, there exists a neighbourhood U of x such that $|f(y) - f(x)| < \epsilon$ for all $f \in F$ and $y \in U$.*

Then F is relatively compact in $(C_0(X), \|\cdot\|_\infty)$.

Theorem A.1.5. *(Baire's category theorem) Let X be either a locally compact Hausdorff space or a complete metric space.*

(i) *If X is the union of countably many closed subsets, then one of them contains a nonempty open set.*
(ii) *The intersection of a countable collection of dense open subsets of X is dense in X.*

Let $\{X_\lambda : \lambda \in \Lambda\}$ be a family of topological spaces. Let X be a nonempty set and for each $\lambda \in \Lambda$, let $f_\lambda : X \to X_\lambda$ be a mapping. Then there exists a weakest (or coarsest) topology on X with respect to which all the mappings f_λ are continuous. This topology can be characterized by the universal property that for any topological space Y and any mapping $f : Y \to X$, f is continuous if and only if $f_\lambda \circ f : Y \to X_\lambda$ is continuous for all $\lambda \in \Lambda$. In the special case where X is the Cartesian product of the sets X_λ and, for each λ, p_λ is the projection from X onto X_λ, this topology is called the *product topology* on X.

Theorem A.1.6. *(Tychonoff's theorem) Let X be the product of topological spaces X_λ, $\lambda \in \Lambda$. Then X is compact in the product topology if and only if all X_λ are compact.*

A compact space C is a compactification of a topological space X if there exists a continuous injective mapping from X onto a dense subset of C. Let X be a locally compact Hausdorff space. Then there exists a compact Hausdorff space $\widetilde{X}$ together with an embedding $j : X \to \widetilde{X}$ such that $\widetilde{X} \setminus j(X)$ is a singleton. $\widetilde{X}$ is uniquely determined up to homeomorphisms and is called the *one-point compactification* of X. The space $\widetilde{X}$ can be constructed as follows. Let $\widetilde{X} = X \cup \{\infty\}$ as a set and take the open sets in $\widetilde{X}$ to be the open sets in X together with the complements in $\widetilde{X}$ of the compact subsets of X.

Note that each $f \in C_0(X)$ extends to a continuous function on $\widetilde{X}$, also denoted f, by setting $f(\infty) = 0$.

Let X be a compact space and Y a Hausdorff space. If f is a continuous and injective mapping from X into Y, then f is a homeomorphism from X onto its range $f(X)$.

Proposition A.1.7. *If f is a continuous open map of a locally compact Hausdorff space X onto a Hausdorff space Y and if K is a compact subset of Y, then there exists a compact subset C of X such that $f(C) = K$.*

Proposition A.1.8. *Let X be a locally compact Hausdorff space. A subset Y of X is locally compact (in the induced topology) if and only if there exist a closed subset A of X and an open subset V of X such that $Y = A \cap V$. In particular, a dense subset of X is locally compact if and only if it is open in X.*

Occasionally, a topology is introduced by designating the closed subsets rather than the open subsets. The procedure is as follows. A *closure operation* on a set X is an assignment $A \to \overline{A}$ from $\mathcal{P}(X)$, the collection of all subsets of X, to itself such that $\overline{\emptyset} = \emptyset$, $A \subseteq \overline{A} = \overline{\overline{A}}$, and $\overline{A \cup B} = \overline{A} \cup \overline{B}$ for all $A, B \subseteq X$. If such a closure operation is given, there exists a unique topology on X such that for each $A \subseteq X$, $\overline{A}$ equals the closure of A in X with respect to this topology.

A.2 Functional analysis

Let E and F be normed linear spaces (over the complex number field $\mathbb{C}$). Note that a linear transformation T from E into F is continuous if and only if it is bounded. The set $\mathcal{B}(E, F)$ of bounded linear transformations $T : E \to F$ is itself a normed linear space with the norm given by

$$\|T\| = \sup\{\|Tx\| : x \in E, \|x\| \leq 1\},$$

and $\mathcal{B}(E, F)$ is complete if F is a Banach space. It is common to write $\mathcal{B}(E)$ instead of $\mathcal{B}(E, E)$. Composition of bounded linear operators turns $\mathcal{B}(E)$ into a Banach algebra. For $T \in \mathcal{B}(E, F)$, the *adjoint* T^* of T is the linear map from F^* into E^* defined by $(T^*g)(x) = g(Tx)$ for all $g \in F^*$ and $x \in E$. Clearly, $T^* \in \mathcal{B}(F^*, E^*)$.

For a normed space E, let E^* denote the *dual space* of E; that is, $E^* = \mathcal{B}(E, \mathbb{C})$, the vector space of all continuous linear functionals on E. Thus E^* is a Banach space when equipped with the norm

$$\|f\| = \sup\{|f(x)| : x \in E, \|x\| \leq 1\},$$

$f \in E^*$. The space E embeds isometrically into the second dual space E^{**} as follows. For each $x \in E$, define $\widehat{x} : E^* \to \mathbb{C}$ by $\widehat{x}(f) = f(x)$ for $f \in E^*$. Then $\widehat{x} \in E^{**}$, and it is a consequence of the Hahn–Banach theorem (Theorem A.2.1 below) that $\|\widehat{x}\| = \|x\|$.

The *weak topology* $\sigma(E, E^*)$ on E is the coarsest topology with respect to which all the functionals $f \in E^*$ are continuous on E. Similarly, the *weak*-topology* (or *w*-topology*) $\sigma(E^*, E)$ is the coarsest topology on E^* with respect to which all the linear functionals $\widehat{x}$ on E^*, $x \in E$, are continuous. Thus a neighbourhood basis of $f_0 \in E^*$ in the w^*-topology is formed by the sets

$$U(f_0, F, \epsilon) = \{f \in E^* : |f(x) - f_0(x)| < \epsilon \text{ for all } x \in F\},$$

where $\epsilon > 0$ and F is any finite subset of E.

We now collect some fundamental results about dual spaces and bounded linear operators.

Theorem A.2.1. *(Hahn–Banach) Let E be a normed space and F a (not necessarily closed) linear subspace of E. If f is a bounded linear functional on F, then there exists $g \in E^*$ such that $g(x) = f(x)$ for all $x \in F$ and $\|g\| = \|f\|$.*

Corollary A.2.2. *If F is a linear subspace of E and x is an element of E which is not contained in the closure of F, then there exists $g \in E^*$ such that $g|_F = \{0\}$ and $g(x) \neq 0$.*

Theorem A.2.3. *(Banach–Alaoglu) Let E be a normed space. Then the unit ball $E_1^* = \{f \in E^* : \|f\| \leq 1\}$ of E^* is w^*-compact.*

However, E_1^* is compact in the norm topology only if E is finite-dimensional.

Corollary A.2.4. *If M is a w^*-closed linear subspace of E^* and $f \in E^* \setminus M$, then there exists $x \in E$ such that $f(x) \neq 0$ but $g(x) = 0$ for all $g \in M$.*

Theorem A.2.5. *(Closed graph theorem) Let E and F be Banach spaces, and let $T : E \to F$ be a linear map. Then the following conditions on T are equivalent.*

(i) T *is continuous.*
(ii) *The graph $G_T = \{(x, Tx) : x \in E\}$ of T is closed in $E \times F$.*
(iii) *If $x_n \to 0$ in E and $Tx_n \to y$ in F, then $y = 0$.*

Theorem A.2.6. *(Open mapping theorem) Let E and F be Banach spaces, and let $T : E \to F$ be a continuous linear mapping. If T is surjective, then T is open. In particular, if $T \in \mathcal{B}(E, F)$ is bijective, then $T^{-1} \in \mathcal{B}(F, E)$.*

Corollary A.2.7. *If a vector space E is a Banach space with respect to two norms, say $\|\cdot\|_1$ and $\|\cdot\|_2$, and if there is a constant c such that $\|x\|_2 \leq c\|x\|_1$ for all $x \in E$, then the two norms are equivalent, that is, there is a constant d such that $\|x\|_1 \leq d\|x\|_2$ for all $x \in E$.*

Theorem A.2.8. *(Uniform boundedness principle) Let E be a Banach space, F a normed space, and $\{T_\lambda : \lambda \in \Lambda\}$ a family of continuous linear maps from E into F. Suppose that $\{T_\lambda x : \lambda \in \Lambda\}$ is bounded in F for each $x \in E$. Then there exists a constant $C \geq 0$ such that $\|T_\lambda\| \leq C$ for all $\lambda \in \Lambda$.*

Theorem A.2.9. *(Krein–Milman) Let E be a locally convex topological vector space and let C be a nonempty convex subset of E. If C is compact, then C is the closed convex hull of the set of its extreme points.*

Let E and F be complex vector spaces. The algebraic tensor product $E \otimes F$ of E and F can be introduced in different ways. However, it is uniquely determined up to isomorphism by the following universal property. Given any complex vector space G and a bilinear map $T : E \times F \to \mathbb{C}$, there exists a unique linear map $S : E \otimes F \to G$ such that $S(x \otimes y) = T(x, y)$ for all $x \in E$ and $y \in F$.

Suppose that E and F are Banach spaces. A basic natural requirement for a norm γ on $E \otimes F$ is to satisfy $\gamma(x \otimes y) = \|x\| \cdot \|y\|$ for all $x \in E$ and $y \in F$. Such a norm is called a *cross-norm*. We now introduce the two cross-norms which play a role in this book.

Let $\mathcal{B}^2(E^* \times F^*, \mathbb{C})$ be the space of all bounded bilinear maps from $E^* \times F^*$ into $\mathbb{C}$, equipped with the norm given by

$$\|T\| = \sup \{|T(f, g)| : f \in E_1^*, g \in F_1^*\}.$$

Then $\mathcal{B}^2(E^* \times F^*, \mathbb{C})$ is complete. Given $x \in E$ and $y \in F$, let $B_{x,y}$ denote the element of $\mathcal{B}^2(E^* \times F^*, \mathbb{C})$ defined by

$$B_{x,y}(f, g) = f(x)g(y), \quad f \in E^*, \quad g \in F^*.$$

Then there is an injective linear map from $E \otimes F$ into $\mathcal{B}^2(E^* \times F^*, \mathbb{C})$ sending $x \otimes y$ to $B_{x,y}$. The norm ϵ on $E \otimes F$, inherited from $\mathcal{B}^2(E^* \times F^*, \mathbb{C})$, is called the *injective tensor norm*. So

$$\epsilon(u) = \sup\left\{\sum_{j=1}^{n} f(x_j)g(y_j) : f \in E_1^*, g \in F_1^*\right\},$$

where the supremun is taken over all representations $u = \sum_{j=1}^{n} x_j \otimes y_j$ of u. Obviously, ϵ is a cross-norm. The completion of $E \otimes F$ in $\mathcal{B}^2(E^* \times F^*, \mathbb{C})$ is called the *injective tensor product* of E and F and is denoted $E \widehat{\otimes}_\epsilon F$.

The *projective tensor norm* π on $E \otimes F$ is defined by

$$\pi(u) = \inf\left\{\sum_{j=1}^{n} \|x_j\| \cdot \|y_j\| : u = \sum_{j=1}^{n} x_j \otimes y_j\right\},$$

where the infimum is taken over all such representations of u. Then $\pi(x \otimes y) = \|x\| \cdot \|y\|$ for all $x \in E$ and $y \in F$. Actually, π is the largest cross-norm on $E \otimes F$. The completion of $E \otimes F$ with respect to π is called the *projective tensor product* of E and F and denoted $E \widehat{\otimes}_\pi F$.

Proposition A.2.10. *Let E and F be Banach spaces and let $u \in E \widehat{\otimes}_\pi F$ and $\epsilon > 0$. Then there exist bounded sequences $(x_n)_n$ in E and $(y_n)_n$ in F such that the series $\sum_{n=1}^{\infty} x_n \otimes y_n$ converges to u and $\sum_{n=1}^{\infty} \|x_n\| \cdot \|y_n\| < \pi(u) + \epsilon$. In particular, for any $u \in E \widehat{\otimes}_\pi F$,*

$$\pi(u) = \inf\left\{\sum_{n=1}^{\infty} \|x_n\| \cdot \|y_n\| : u = \sum_{n=1}^{\infty} x_n \otimes y_n, \sum_{n=1}^{\infty} \|x_n\| \cdot \|y_n\| < \infty\right\},$$

where the infimum is taken over all such representations of u.

It follows from Proposition A.2.10 that every element u of $E \widehat{\otimes}_\pi F$ has a representation of the form $u = \sum_{j=1}^{\infty} x_j \otimes y_j$, where $\sum_{j=1}^{\infty} \|x_j\| \cdot \|y_j\| < \infty$ and $u(f,g) = \sum_{j=1}^{\infty} f(x_j)g(y_j)$ for $f \in E^*$ and $g \in F^*$.

The following proposition provides a useful description of the projective tensor product.

Proposition A.2.11. *Let E and F be Banach spaces. There exists an isometric isomorphism from $\mathcal{B}(E, F^*)$ onto $(E \widehat{\otimes}_\pi F)^*$ with the property that*

$$\langle T, \sum_{j=1}^{n} x_j \otimes y_j \rangle = \sum_{j=1}^{n} \langle Tx_j, y_j \rangle$$

for any $T \in \mathcal{B}(E, F^*)$, $x_1, \ldots, x_n \in E$ and $y_1, \ldots, y_n \in F$.

Let $u = \sum_{n=1}^{\infty} x_n \otimes y_n \in E \widehat{\otimes}_\pi F$, $\sum_{n=1}^{\infty} \|x_n\| \cdot \|y_n\| < \infty$. Then the series $\sum_{n=1}^{\infty} B_{x_n, y_n}$ converges in $\mathcal{B}^2(E^* \times F^*, \mathbb{C})$. This element of $\mathcal{B}^2(E^* \times F^*, \mathbb{C})$ does not depend on the representation of u. Indeed, if $\sum_{n=1}^{\infty} x_n \otimes y_n = 0$ in $E \widehat{\otimes}_\pi F$, then $\sum_{n=1}^{\infty} \langle Sx_n, y_n \rangle = 0$ for all $S \in \mathcal{B}(E, F^*)$ by Proposition A.2.11, and hence, taking for S the operator defined by $Sx = f(x)g$ for $x \in E$,

$$\sum_{n=1}^{\infty} B_{x_n,y_n}(f,g) = g\left(\sum_{n=1}^{\infty} f(x_n)y_n\right) = \sum_{n=1}^{\infty} f(x_n)g(y_n) = 0$$

for all $f \in E^*$ and $g \in F^*$. Thus we have a natural continuous linear mapping $u \to B_u = \sum_{n=1}^{\infty} B_{x_n,y_n}$ from $E \widehat{\otimes}_\pi F$ into the injective tensor product $E \widehat{\otimes}_\epsilon F$.

This map is not injective in general, and the problem of when it is injective is closely related to the approximation property of a Banach space. A Banach space E has the *approximation property* if for each compact subset C of E and each $\epsilon > 0$, there exists a finite rank operator S on E with $\|Sx - x\| \leq \epsilon$ for all $x \in C$. The class of Banach spaces with the approximation property includes all spaces $C_0(X)$ for X a locally compact Hausdorff space, all Banach spaces with a Schauder basis, $c_0(I)$ and $l^p(I)$, $1 \leq p < \infty$, for any index set I, spaces $L^p(\mu)$, $1 \leq p < \infty$, for any measure μ and the disc algebra. Also, if E^* has the approximation property, then so does E. The first example of a Banach space which was shown to not have the approximation property, is $\mathcal{B}(l^2, l^2)$. A very good reference to the approximation property and tensor products of Banach spaces in general is [114].

Theorem A.2.12. *For a Banach space E, the following two conditions are equivalent.*

(i) *E has the approximation property.*
(ii) *The natural mapping from $E \widehat{\otimes}_\pi F$ into $E \widehat{\otimes}_\epsilon F$ is injective for every Banach space F.*

A.3 Measure and integration

In the following, if μ is a positive measure on a set X, $L^p(X,\mu)$ or $L^p(\mu)$, for short, denotes the set of (equivalence classes of) p-integrable complex-valued μ-measurable functions on X.

Theorem A.3.1. *(Hölder's inequality) Let μ be a positive measure and let $1 \leq p \leq \infty$, and $(1/p) + (1/q) = 1$. If $f \in L^p(\mu)$ and $g \in L^q(\mu)$, then $fg \in L^1(\mu)$ and*

$$\|fg\|_1 \leq \|f\|_p \|g\|_q.$$

Theorem A.3.2. *(Minkowski's inequality) Let μ be a positive measure, $1 \leq p \leq \infty$, and $f,g \in L^p(\mu)$. Then $f + g \in L^p(\mu)$ and*

$$\|f + g\|_p \leq \|f\|_p + \|g\|_p.$$

It follows that $L^p(\mu)$ with the norm $\|\cdot\|_p$ is a Banach space, and $L^2(\mu)$ is a Hilbert space for the scalar product $\langle f, g \rangle = \int_X f(x)\overline{g(x)}d\mu(x)$.

A consequence of Hölder's inequality is that every $g \in L^q(\mu)$ defines a bounded linear functional F_g of $L^p(\mu)$ by

$$\langle F_g, f \rangle = \int_X f(x)g(x)d\mu(x)$$

for all $f \in L^p(\mu)$.

Theorem A.3.3. *Let μ be a positive measure and suppose that either $1 < p < \infty$ or that $p = 1$ and μ is σ-finite. Let q be the conjugate index to p. Then, for each $F \in L^p(\mu)^*$, there exists a unique $g \in L^q(\mu)$ such that $F = F_g$. The map $g \rightarrow F_g$ is an isometric isomorphism from $L^q(\mu)$ onto $L^p(\mu)^*$.*

Theorem A.3.4. *Let μ be a positive regular Borel measure on a locally compact Hausdorff space X. Then, for $1 \leq p < \infty$, $C_c(X)$ is dense in $L^p(X, \mu)$.*

Theorem A.3.5. *(Riesz representation theorem) Let X be a locally compact Hausdorff space. For each $\mu \in M(X)$, define $F_\mu \in C_0(X)^*$ by*

$$\langle F_\mu, f \rangle = \int_X f(x)d\mu(x), \quad f \in C_0(X).$$

Then the map $\mu \rightarrow F_\mu$ is an isometric isomorphism from $M(X)$ onto $C_0(X)^*$.

Let X and Y be locally compact Hausdorff spaces and let μ and ν be positive regular Borel measures on X and Y, respectively. Then the assignment

$$f \rightarrow \int_X \left(\int_Y f(x, y)d\nu(y) \right) d\mu(x) = \int_Y \left(\int_X f(x, y)d\mu(x) \right) d\nu(y)$$

defines a positive linear functional F on $C_c(X \times Y)$. Hence, by the Riesz representation theorem, there exists a unique positive regular Borel measure, denoted $\mu \times \nu$, on $X \times Y$ such that

$$\langle F, f \rangle = \int_{X \times Y} f(x, y)d(\mu \times \nu)(x, y)$$

for all $f \in C_c(X \times Y)$.

Theorem A.3.6. *(Fubini's theorem) Let X and Y be locally compact Hausdorff spaces and let μ and ν be positive regular Borel measures on X and Y, respectively. Let $f \in L^1(\mu \times \nu)$ and suppose that there exist σ-finite Borel sets A and B of X and Y, respectively, such that f vanishes on $(X \times Y) \setminus (A \times B)$. Then*

$$\int_X \left(\int_Y f(x, y)d\nu(y) \right) d\mu(x) = \int_{X \times Y} f(x, y)d(\mu \times \nu)(x, y)$$

$$= \int_Y \left(\int_X f(x, y)d\mu(x) \right) d\nu(y).$$

We now briefly discuss vector-valued integration. Let (X, μ) be a measure space and E a Banach space. A function $f : X \to E$ is said to be a μ-measurable simple function if it is of the form $f(x) = \sum_{j=1}^{n} 1_{M_j}(x) a_j$, where the a_j are elements of E, the M_j are disjoint μ-measurable subsets of X with $\mu(M_j) < \infty$ and 1_{M_j} denotes the characteristic function of M_j, $j = 1, \ldots, n$. The integral $\int_X f(x) d\mu(x)$ of such a μ-measurable simple function is defined to be the element $\sum_{j=1}^{n} \mu(M_j) a_j$ of E. An arbitrary function $f : X \to E$ is called μ-measurable if there exists a sequence of μ-measurable simple functions converging to f almost everywhere. Such an f can then be defined to be Bochner integrable if the scalar-valued function $x \to \|f(x)\|$ is integrable. In this case, the Bochner integral of f is

$$\int_X f(x) d\mu(x) = \lim_{n \to \infty} \int_X f_n(x) d\mu(x),$$

where $(f_n)_n$ is any sequence of μ-measurable simple functions such that

$$\int_X \|f(x) - f_n(x)\| d\mu(x) \to 0.$$

The space of E-valued Bochner integrable functions is denoted $L^1(\mu, E)$ or $L^1(X, E)$, if the measure μ is understood.

A.4 Haar measure and convolution on locally compact groups

A locally compact group G is always understood to be a group which is also a locally compact Hausdorff space and for which the map $(x, y) \to xy^{-1}$ of the product space $G \times G$ to G is continuous.

For a function f on G and $x \in G$, the left and right translation $L_x f$ and $R_x f$ of f are defined by $L_x f(y) = f(x^{-1} y)$ and $R_x f(y) = f(yx)$ for $y \in G$, respectively. A nonzero positive regular Borel measure μ on G is called a *left Haar measure* if it satisfies $\mu(xE) = \mu(E)$ for all Borel sets E and all $x \in G$. This left invariance condition is equivalent to $\int_G L_x f(y) d\mu(y) = \int_G f(y) d\mu(y)$ for all $f \in L^1(G, \mu)$. Likewise, a right Haar measure is defined.

Theorem A.4.1. *On any locally compact group G there exists a left invariant (right invariant) Haar measure. If μ and λ are two left Haar measures on G, then there exists a constant $c > 0$ such that $\mu = c\lambda$.*

If the Haar measure is fixed, we most times denote the Haar measure of a set M by $|M|$.

Remark A.4.2. Let μ be a Haar measure on G.

(1) Then $\mu(U) > 0$ for every nonempty open set and $\int_G f(x) dx > 0$ for each $f \in C_c^+(G)$ which is not identically zero.

(2) The support of μ equals G, and $\mu(G) < \infty$ if and only if G is compact.

Example A.4.3. (1) On $\mathbb{R}$ (more generally $\mathbb{R}^n$) Lebesgue measure is a Haar measure.

(2) On any discrete group counting measure, that is, $\mu(f) = \sum_{x \in G} f(x)$ is a left and right invariant Haar measure.

(3) On the circle group $\mathbb{T}$ a Haar measure is given by

$$\mu(f) = \frac{1}{2\pi} \int_0^{2\pi} f(e^{it}) dt,$$

$f \in C(\mathbb{T})$, where dt is Lebesgue measure on $[0, 2\pi]$.

(4) If G_1 and G_2 are two locally compact groups with left Haar measures μ_1 and μ_2, respectively, then the product measure $\mu_1 \times \mu_2$ is a left Haar measure on the product group $G_1 \times G_2$.

Let dx be a left Haar measure on G. For every $x \in G$ the measure

$$f \to \int_G f(yx^{-1}) dy, \quad f \in C_c(G),$$

is left invariant. So there is a unique positive number $\Delta(A)$ such that

$$\int_G f(yx^{-1}) dy = \Delta(x) \int_G f(y) dy$$

for all $f \in C_c(G)$. The function $\Delta : x \to \Delta(x)$ is a continuous homomorphism from G into $\mathbb{R}_+^\times$, the multiplicative group of positive real numbers.

The function Δ is called the *modular function* of G and G is called *unimodular* if $\Delta(x) = 1$ for all $x \in G$.

Abelian groups are unimodular, and so are compact groups because $\{1\}$ is the only compact subgroup of $\mathbb{R}_+^\times$.

Proposition A.4.4. *Let $1 \le p < \infty$ and let $f \in L^p(G)$. Given $\epsilon > 0$, there exists a neighbourhood U of e in G such that $\|L_x f - L_y f\|_p < \epsilon$ for all $x, y \in G$ such that $x^{-1} y \in U$.*

Proof. Because $C_c(G)$ is dense in $L^p(G)$, we find $g \in C_c(G)$ with $\|f - g\|_p < \epsilon/3$. Choose a compact neighbourhood V of e in G. Then $0 < |V \cdot \operatorname{supp} g| < \infty$ since $V \cdot \operatorname{supp} g$ is compact and has nonempty interior. Since g is uniformly continuous, there exists a symmetric neighbourhood U of e in G such that $U \subseteq V$ and

$$|g(x) - g(y)| < \frac{\epsilon}{3} \cdot |V \cdot \operatorname{supp} g|^{-1/p}$$

for all $x, y \in G$ with $y^{-1} x \in U$. For such x and y, we have

$$\|L_x g - L_y g\|_p^p = \int_G |g(x^{-1}t) - g(y^{-1}t)|^p dt$$

$$= \int_{V \cdot \operatorname{supp} g} |g(x^{-1}yt) - g(t)|^p dt$$

$$\leq \left(\frac{\epsilon}{3}\right)^p |V \cdot \operatorname{supp} g|^{-1} |V \cdot \operatorname{supp} g| = \left(\frac{\epsilon}{3}\right)^p,$$

and this implies

$$\|L_x f - L_y f\|_p \leq \|L_x(f - g)\|_p + \|L_x g - L_y g\|_p + \|L_y(f - g)\|_p$$

$$= 2\|f - g\|_p + \|L_x g - L_y g\|_p,$$

which, by the above and the choice of g, is $< \epsilon$. □

Proposition A.4.5. *Let $1 < p < \infty$ and let q be the conjugate index to p; that is, $(1/p) + (1/q) = 1$. Suppose that $f \in L^p(G)$ and $g \in L^q(G)$. Then $f * \breve{g} \in C_0(G)$ and $\|f * \breve{g}\|_\infty \leq \|f\|_p \|g\|_q$.*

Proof. For each $x \in G$, we have

$$\int_G |f(xy)\breve{g}(y^{-1})| dy = \int_G |L_{x^{-1}}f(y)| \cdot |g(y)| dy \leq \|L_{x^{-1}}f\|_p \|g\|_q$$

by Hölder's inequality. So $f * \breve{g}$ is defined everywhere and bounded on G by $\|f\|_p \|g\|_q$. For x and y in G, Hölder's inequality gives

$$|f * \breve{g}(x) - f * \breve{g}(y)| \leq \|L_{x^{-1}}f - L_{y^{-1}}f\|_p \|g\|_q.$$

The map $t \to L_t f$ from G into $L^p(G)$ is continuous (Proposition A.4.4), and therefore we obtain that $f * \breve{g}$ is continuous.

To prove that $f * \breve{g}$ vanishes at infinity, note first that $f * \breve{g} \in C_c(G)$ whenever $f, g \in C_c(G)$. If $f \in L^p(G)$ and $g \in L^q(G)$ then, since $C_c(G)$ is dense in $L^r(G)$ for each $1 \leq r < \infty$, there exist sequences $(f_n)_n$ and $(g_n)_n$ in $C_c(G)$ such that $\|f - f_n\|_p \to 0$ and $\|g - g_n\|_q \to 0$. Then, for all $x \in G$,

$$|f * \breve{g}(x) - f_n * \breve{g}_n(x)| \leq |(f - f_n) * \breve{g}(x)| + |f_n * (\breve{g} - \breve{g}_n)(x)|$$

$$\leq \|f - f_n\|_p \|g\|_q + \|f_n\|_p \|g - g_n\|_q,$$

which tends to 0 as $n \to \infty$. It follows that $f * \breve{g} \in C_0(G)$. □

Proposition A.4.6. *Let G be a locally compact group. For every relatively compact symmetric open neighbourhood V of e in G, let $u_V \in L^1(G)$ be such that $u_V \geq 0$, $\|u_V\|_1 = 1$ and $u_V = 0$ almost everywhere on $G \backslash V$. Then, given $f \in L^p(G)$, $1 \leq p < \infty$, and $\epsilon > 0$,*

$$\|u_V * f - f\|_p < \epsilon$$

for all sufficiently small V.

Proof. Since $C_c(G)$ is dense in $L^p(G)$, we can choose $g \in C_c(G)$ such that $\|f - g\|_p < \epsilon/3$. For g it follows that

$$\|u_V * g - g\|_p^p = \int_G \left| \int_G u_V(xy)g(y^{-1})dy - g(x) \right|^p dx$$

$$= \int_G \left| \int_G u_V(y)L_y g(x)dy - g(x) \right|^p dx$$

$$= \int_G \left| \int_G u_V(y)[L_y g(x) - g(x)]dy \right|^p dx$$

$$\leq \int_G \left(\int_G u_V(y)|L_y g(x) - g(x)|dy \right)^p dx$$

$$\leq |V \cdot \operatorname{supp} g| \cdot \sup\{\|L_y g - g\|_\infty^p : y \in V\}.$$

Now, since the map $y \to L_y g$ from G into $L^p(G)$ is continuous, we find a neighbourhood W of e in G such that, for all $y \in W$,

$$\|L_y g - g\|_\infty \leq \frac{\epsilon}{3|V \cdot \operatorname{supp} g|^{1/p}}.$$

Together with the above estimate we get for all $V \subseteq W$,

$$\|u_V * f - f\|_p \leq \|u_V * (f - g)\|_p + \|u_V * g - g\|_p + \|g - f\|_p$$
$$\leq (\|u_V\|_1 + 1)\|f - g\|_p + \frac{\epsilon}{3},$$

which is $< \epsilon$ since $\|u\|_1 = 1$.   □

In Proposition A.4.6, u_V can, for instance, be taken to be $|V|^{-1}1_V$.

Proposition A.4.7. *Suppose that $1 \leq p \leq \infty$, $g \in L^p(G)$, and $f \in L^1(G)$. Then $f * g(x)$ is defined for almost all $x \in G$, and we have $f * g \in L^p(G)$ and $\|f * g\|_p \leq \|f\|_1 \|g\|_p$. If $p = \infty$, then $f * g(x)$ is defined for all $x \in G$ and $f * g$ is continuous.*

Proof. Assume first that $p = \infty$. By Hölder's inequality, the integral $f * g(x) = \int_G f(xy)g(y^{-1})dy$ converges for every $x \in G$ and satisfies $|f * g(x)| \leq \|f\|_1 \|g\|_\infty$. Moreover, for all $x, y \in G$,

$$|f * g(x^{-1}) - f * g(y^{-1})| \leq \int_G |L_x f(t) - L_y f(t)| \cdot |g(t^{-1})| dt$$
$$\leq \|g\|_\infty \|L_x f - L_y f\|_1.$$

Proposition A.4.4 shows that $f * g$ is continuous.

Now let $1 \leq p < \infty$ and let q be the conjugate exponent to p. Since the map $y \to L_y g$ from G into $L_p(G)$ is continuous (Proposition A.4.4) and bounded by $\|g\|_p$, the L^p-valued function $y \to f(y)L_y g$ is Bochner integrable and satisfies, for each $h \in L^q(G)$,

$$\int_G \left(\int_G f(y)L_y g\, dy \right)(x)h(x)dx = \left\langle \int_G f(y)L_y g\, dy, h \right\rangle$$

$$= \int_G f(y)\langle L_y g, h \rangle dy$$

$$= \int_G \int_G f(y)g(y^{-1}x)h(x)dx\,dy.$$

It follows from Fubini's theorem and Hölder's inequality that the order of integration can be reversed. Since $h \in L^q(G)$ is arbitrary, we conclude that

$$\left(\int_G f(y)L_y g \right)(x) = \int_G f(y)g(y^{-1}x)dy = f * g(x)$$

for almost all $x \in G$. Finally, using this, we get

$$\|f * g\|_p = \left(\int_G \left| \left(\int_G f(y)L_y g \right)(x) \right|^p dx \right)^{1/p}$$

$$= \left\| \int_G f(y)L_y g \right\|_p \leq \int_G \|f(y)L_y g\|_p\, dy$$

$$= \int_G |f(y)| \cdot \|L_y g\|_p\, dy$$

$$= \|g\|_p \|f\|_1.$$

This finishes the proof of the proposition.   □

Let H be a closed normal subgroup of a locally compact group G. For $f \in C_c(G)$, define the function $T_H f$ on G/H by $T_H f(xH) = \int_H f(xh)dh$, $x \in G$. Then $T_H f \in C_c(G/H)$ and T_H maps $C_c(G)$ onto $C_c(G/H)$. Given a left Haar integral on G/H, the assignment $f \to \int_{G/H} T_H f(xH)d(xH)$ defines a left Haar integral on G. Hence there exists a unique left Haar measure dx on G such that

$$\int_{G/H} \left(\int_H f(xh)dh \right) d(xH) = \int_G f(x)dx.$$

This formula is called *Weil's formula*. If two of the left Haar integrals on G, H and G/H are given, the third can always be normalized so that Weil's formula holds. It follows from Weil's formula that $T_H(f * g) = T_H(f) * T_H(g)$ and $\|T_H f\|_1 \leq \|f\|_1$ for all $f, g \in C_c(G)$. Hence T_H extends to a continuous homomorphism from $L^1(G)$ into $L^1(G/H)$, also denoted T_H. Since $\|T_H f\|_1$ equals the quotient norm, T_H is actually surjective.

Theorem A.4.8. *Let H be a closed normal subgroup of G and let $f \in L^1(G)$.*

(i) There exists a set S of measure zero in G/H such that the function $h \to f(xh)$ is in $L^1(H)$ for each $x \in G$ with $xH \notin S$.

(ii) The function $xH \to \int_H f(xh)dh$, which is defined on $G/H \setminus S$, is integrable.

(iii) If left Haar measures on H, G/H, and G are chosen so that Weil's formula holds, then

$$\int_{G/H} \left(\int_H f(xh)dh \right) d(xH) = \int_G f(x)dx.$$

The formula in (iii) is often called the *extended Weil formula*. A function $f \in L^1(G)$ is in the kernel of the homomorphism $T_H : L^1(G) \to L^1(G/H)$ if and only if $\int_H f(xh)dh = 0$ for almost all $x \in G$.

A.5 The Pontryagin duality theorem

The famous Pontryagin duality theorem asserts that there is a canonical topological isomorphism between a locally compact Abelian group and its second dual group. In this section we deduce this duality theorem from the results in Sections 4.4 and 2.7. In the sequel, G will always denote a locally compact Abelian group and $\widehat{G}$ the dual group of G.

Proposition A.5.1. *For $x \in G$, $\epsilon > 0$, and a compact subset Γ of $\widehat{G}$, let*

$$V(x, \Gamma, \epsilon) = \{y \in G : |\gamma(y) - \gamma(x)| < \epsilon \text{ for all } \gamma \in \Gamma\}.$$

Then $V(x, \Gamma, \epsilon)$ is open in G, and the sets $V(x, \Gamma, \epsilon)$ form a neighbourhood basis of x in G.

Proof. Let $(y_\alpha)_\alpha$ be a net in $G \setminus V(x, \Gamma, \epsilon)$ converging to some $y \in G$. For each α, there exists $\gamma_\alpha \in \Gamma$ with $|\gamma_\alpha(y_\alpha) - \gamma_\alpha(x)| \geq \epsilon$. Since Γ is compact, passing to a subnet if necessary, we can assume that $\gamma_\alpha \to \gamma$ for some $\gamma \in \Gamma$. Since the function $(t, \lambda) \to \lambda(t)$ on $G \times \widehat{G}$ is continuous (Lemma 2.7.4), it follows that $|\gamma(y) - \gamma(x)| \geq \epsilon$. This shows that $y \notin V(x, \Gamma, \epsilon)$. So $V(x, \Gamma, \epsilon)$ is open in G.

Since $V(x, \Gamma, \epsilon) = xV(e, \Gamma, \epsilon)$, it remains to show that if U is an open neighbourhood of e in G, then there exist a compact subset Γ of $\widehat{G}$ and $\epsilon > 0$ such that $V(e, \Gamma, \epsilon) \subseteq U$. To that end, choose symmetric open neighbourhoods V and W of e in G such that $W \subseteq V$, $V^2 \subseteq U$, $\overline{W^2} \subseteq V$ and V is relatively compact. Let

$$f = \|1_W * 1_W\|_2^{-1}(1_W * 1_W).$$

Then $f \in C_c(G)$, $f(x) \geq 0$ for all $x \in G$ and $\operatorname{supp} f \subseteq V$. If $x \in G \setminus U$, then $\operatorname{supp} f$ and $\operatorname{supp}(L_x f)$ are disjoint and therefore

$$\|f - L_x f\|_2^2 = \|f\|_2^2 + \|L_x f\|_2^2.$$

In particular, if $x \in G$ is such that $\|f - L_x f\|_2 \leq 1$, then $x \in U$. Now, choose $u \in L^1(G)$ such that $u \geq 0$, $\|u\|_1 = 1$ and $\|u * f - f\|_2 < 1/3$ (Proposition A.4.6). Since the $\| \cdot \|_2$-norm is translation invariant, it follows that, for all $x \in G$,

$$\|u * L_x f - L_x f\|_2 = \|L_x(u * f - f)\|_2 = \|u * f - f\|_2 < frac13.$$

Thus, if $x \in G$ satisfies $\|u * f - f\|_2 \leq 1/3$, then

$$\|f - L_x f\|_2 \leq \|f - u * f\|_2 + \|u * f - u * L_x f\|_2 + \|u * L_x f - f\|_2 < 1,$$

and hence $x \in U$. Applying the regular representation (Section 4.4), we have

$$\begin{aligned}
\|u * f - u * L_x f\|_2 &= \|\lambda_u f - \lambda_{L_x u} f\|_2 \\
&\leq \|\lambda_u - \lambda_{L_x u}\| \cdot \|f\|_2 \\
&= \|\widehat{u} - \widehat{L_x u}\|_\infty.
\end{aligned}$$

So, if $x \in G$ is such that $\|\widehat{u} - \widehat{L_x u}\|_\infty \leq 1/3$, then $x \in U$ by the above.

Since $\widehat{u} \in C_0(\widehat{G})$, there exists a compact subset Γ of $\widehat{G}$ such that $|\widehat{u}(\gamma)| < 1/6$ for all $\gamma \in \widehat{G} \setminus \Gamma$. Let $x \in V(e, \Gamma, 1/3$. Then, for $\gamma \in \Gamma$,

$$\left| \widehat{u}(\gamma) - \widehat{L_x u}(\gamma) \right| = |\widehat{u}(\gamma)| \cdot |1 - \overline{\gamma(x)}| \leq \|u\|_1 \cdot |1 - \overline{\gamma(x)}| < \frac{1}{3}$$

since $\|u\|_1 = 1$, and if $\gamma \in \widehat{G} \setminus \Gamma$, then

$$\left| \widehat{u}(\gamma) - \widehat{L_x u}(\gamma) \right| = |\widehat{u}(\gamma)| \cdot |1 - \overline{\gamma(x)}| < \frac{1}{3}$$

as $|\widehat{u}(\gamma)| < \frac{1}{6}$. Therefore, if $x \in V(e, \Gamma, \frac{1}{3})$, then

$$\left| \widehat{u}(\gamma) - \widehat{L_x u}(\gamma) \right| < \frac{1}{3}$$

for all $\gamma \in \Gamma$, and this implies $x \in U$ by what we have seen above.   □

Let $\widehat{\widehat{G}}$ denote the second dual of G, that is, the dual group of $\widehat{G}$. Each $x \in G$ defines an element $\widehat{x}$ of $\widehat{\widehat{G}}$ by setting $\widehat{x}(\alpha) = \alpha(x)$ for $\alpha \in \widehat{G}$.

Theorem A.5.2. *(Pontryagin duality theorem) Let G be a locally compact Abelian group. The map $\wedge : x \to \widehat{x}$ is a topological isomorphism from G onto the second dual group $\widehat{\widehat{G}}$.*

Proof. Clearly, $\wedge$ is a homomorphism. If x and y are elements of G such that $\widehat{x} = \widehat{y}$, then for all $f \in L^1(G)$ and $\alpha \in \widehat{G}$,

$$\widehat{L_x f}(\alpha) = \overline{\alpha(x)} \widehat{f}(\alpha) = \overline{\alpha(y)} \widehat{f}(\alpha) = \widehat{L_y f}(\alpha).$$

Since the Gelfand homomorphism of $L^1(G)$ is injective, it follows that $L_x f = L_y f$ for all $f \in L^1(G)$ and this implies $x = y$. Proposition A.5.1 shows that $\wedge$ is a homeomorphism of G onto its range $\wedge(G) \subseteq \widehat{\widehat{G}}$. In particular, $\wedge(G)$ is a locally compact group. Since a locally compact subset of a Hausdorff space is the intersection of an open set and a closed set, it follows that $\wedge(G)$ is open in its closure $\overline{\wedge(G)}$. Now, $\overline{\wedge(G)}$ is a topological group and an open subgroup of a topological group is automatically closed. So $\wedge(G)$ is closed in $\widehat{\widehat{G}}$.

Towards a contradiction, assume that $\wedge(G) \neq \widehat{\widehat{G}}$. Since $L^1(\widehat{G})$ is a regular commutative Banach algebra, there exists $g \in L^1(\widehat{G})$, $g \neq 0$, such that $\widehat{g}(\widehat{x}) = 0$ for all $x \in G$. As $g \neq 0$, we find $h \in C_0(\widehat{G})$ with $\int_{\widehat{G}} g(\alpha)h(\alpha)d\alpha \neq 0$. Now, the Gelfand homomorphism maps $C^*(G)$ onto $C_0(\widehat{G})$ (Section 4.4). Thus there exists $T \in C^*(G)$ with $\widehat{T} = h$. Since $L^1(G)$ is dense in $C^*(G)$ and $C_c(G)$ is dense in $L^1(G)$ in the $\|\cdot\|_1$-norm and hence in the C^*-norm, there exists $f \in C_c(G)$ with

$$\int_{\widehat{G}} g(\alpha)\widehat{f}(\alpha)d\alpha = \int_{\widehat{G}} g(\alpha)\widehat{\lambda_f}(\alpha)d\alpha \neq 0.$$

Fubini's theorem yields

$$\int_{\widehat{G}} g(\alpha)\widehat{f}(\alpha)d\alpha = \int_{\widehat{G}} g(\alpha)\left(\int_G f(x)\overline{\alpha(x)}dx\right)d\alpha$$

$$= \int_G f(x)\left(\int_{\widehat{G}} g(\alpha)\overline{\alpha(x)}d\alpha\right)dx$$

$$= \int_G f(x)\left(\int_{\widehat{G}} g(\alpha)\overline{\widehat{x}(\alpha)}d\alpha\right)dx$$

$$= \int_G f(x)\widehat{g}(\widehat{x})dx,$$

which is zero since $\widehat{g}$ vanishes on $\wedge(G)$. This contradiction shows that $\wedge$ is surjective and finishes the proof of the duality theorem. $\square$

For any subset Γ of $\widehat{G}$, let $A(G, \Gamma)$ denote the *annihilator* of Γ in G, that is,

$$A(G, \Gamma) = \{x \in G : \gamma(x) = 1 \text{ for all } \gamma \in \Gamma\}.$$

Similarly, the annihilator of a subset M of G is defined to be

$$A(\widehat{G}, M) = \{\alpha \in \widehat{G} : \alpha(x) = 1 \text{ for all } x \in M\}.$$

Clearly, $A(G, \Gamma)$ is a closed subgroup of G and $A(\widehat{G}, M)$ is a closed subgroup of $\widehat{G}$.

Corollary A.5.3. *Suppose that Γ is a closed subgroup of $\widehat{G}$. Then*

$$\Gamma = \{\alpha \in \widehat{G} : \alpha(A(G, \Gamma)) = \{1\}\}.$$

Proof. The subgroup Γ of $\widehat{G}$ identifies canonically with a closed subgroup of $(G/A(G,\Gamma))^\wedge$ which separates the points of $G/A(G,\Gamma)$. Therefore, we can assume that $A(G,\Gamma) = \{e\}$.

Since $\Delta(L^1(\widehat{G}/\Gamma)) = \widehat{G}/\Gamma$, it suffices to show that if α is a character of $\widehat{G}$ with $\alpha|_\Gamma = 1$, then $\alpha = 1_{\widehat{G}}$. By the duality theorem, there exists $x \in G$ such that $\alpha = \widehat{x}$. Then $\gamma(x) = \alpha(x) = 1$ for all $\gamma \in \Gamma$ and hence $x = e$ by hypothesis. So $\alpha = 1$ on all of G. $\qquad\qquad\square$

It follows from the preceding corollary that the map $\Gamma \to A(G,\Gamma)$ is a bijection between the closed subgroups of G and the closed subgroups of $\widehat{G}$.

Corollary A.5.4. *Let G be a locally compact Abelian group. If $\mu \in M(G)$ and*

$$\widehat{\mu}(\alpha) = \int_G \overline{\alpha(x)} d\mu(x) = 0$$

for all $\alpha \in \widehat{G}$, then $\mu = 0$.

Proof. If $f \in L^1(\widehat{G})$, then $\widehat{f} \in C_0(\widehat{\widehat{G}})$ and hence, since $x \to \widehat{x}$ is a homeomorphism from G to $\widehat{\widehat{G}}$, the function $x \to \widehat{f}(\widehat{x})$ belongs to $C_0(G)$. By Fubini's theorem,

$$\int_G \widehat{f}(\widehat{x}) d\mu(x) = \int_G \left(\int_{\widehat{G}} f(\alpha) \overline{\widehat{x}(\alpha)} d\alpha \right) d\mu(x)$$

$$= \int_{\widehat{G}} f(\alpha) \left(\int_G \overline{\alpha(x)} d\mu(x) \right) d\alpha$$

$$= \int_{\widehat{G}} f(\alpha) \widehat{\mu}(\alpha) d\alpha,$$

whence $\int_G \widehat{f}(\widehat{x}) d\mu(x) = 0$ for all $f \in L^1(\widehat{G})$. Thus, denoting by ν the image of μ under the homeomorphism $x \to \widehat{x}$,

$$\int_{\widehat{\widehat{G}}} \widehat{f}(\chi) d\nu(\chi) = 0$$

for all $f \in L^1(\widehat{G})$.

However, the image of $L^1(\widehat{G})$ under the Gelfand homomorphism is norm-dense in $C_0(\widehat{\widehat{G}})$ (Lemma 2.7.3(iii)). It follows that $\int_{\widehat{\widehat{G}}} g(\chi) d\nu(\chi) = 0$ for all $g \in C_0(\widehat{\widehat{G}})$ and this implies $\nu = 0$ and hence $\mu = 0$. $\qquad\square$

In passing we recall the notion of a compactly generated topological group. For a subset M of G and $n \in \mathbb{N}$, let M^n denote the set of all n-fold products $x_1 x_2 \cdot \ldots \cdot x_n$ of elements x_j of M. Suppose that G is a topological group. Then G is said to be *compactly generated* if there exists a compact subset C of G such that $G = \bigcup_{n=1}^{\infty} C^n$. If C is any compact symmetric neighbourhood of the identity of G, then $\bigcup_{n=1}^{\infty} C^n$ is an open compactly generated subgroup of G. Conversely, every compactly generated open subgroup of G arises in this manner.

Clearly, $\mathbb{R}$ and $\mathbb{Z}$ are compactly generated, and so is the direct product of two compactly generated topological groups. In particular, groups of the form $\mathbb{R}^n \times \mathbb{Z}^m \times K$, where $n, m \in \mathbb{N}_0$ and K is a compact group, are compactly generated. The following structure theorem, for the proof of which we refer to the literature, says that within the class of locally compact Abelian groups these groups are the only compactly generated ones.

Theorem A.5.5. *Let G be a compactly generated locally compact Abelian group. Then G is topologically isomorphic to a direct product $\mathbb{R}^n \times \mathbb{Z}^m \times K$, where $n, m \in \mathbb{N}_0$ and K is a compact Abelian group.*

A.6 The coset ring of an Abelian group

Let G be a locally compact Abelian group. In Section 5.6 we have described explicitly the closed ideals in $L^1(G)$ with bounded approximate identities. As an essential tool we have used a characterisation of the closed sets in the coset ring of an Abelian topological group. This characterisation, Theorem A.6.9 below, was established by Gilbert [43] and, independently and with a much simpler proof, by Schreiber [117]. Accordingly, our presentation follows very closely the one of [117]. Schreiber's approach, in turn, is based on a result due to Cohen [22] (Proposition A.6.5). We start with the relevant definitions.

The *coset ring* of an Abelian group G, denoted $\mathcal{R}(G)$, is the smallest Boolean algebra of subsets of G containing the cosets of all subgroups of G. That is, $\mathcal{R}(G)$ is the smallest family of subsets of G which contains all the cosets of subgroups of G and which is closed under the processes of forming finite unions, finite intersections and complements.

Suppose now that G is a topological Abelian group. Then the *closed coset ring* of G, $\mathcal{R}_c(G)$, is defined to be

$$\mathcal{R}_c(G) = \{E \in \mathcal{R}(G) : E \text{ is closed in } G\}.$$

We start with a description of the sets in $\mathcal{R}(G)$.

Proposition A.6.1. *Let G be an Abelian group. A subset E of G belongs to $\mathcal{R}(G)$ if and only if E is of the form*

$$E = \bigcup_{i=1}^{n} \left(C_i \setminus \bigcup_{j=1}^{n_i} C_{ij} \right), \quad n, n_i \in \mathbb{N},$$

where C_i and C_{ij} are (possibly void) cosets of subgroups of G.

Proof. Let $\mathcal{E}$ denote the collection of all such sets E. By definition of $\mathcal{R}(G)$ and since $\mathcal{E}$ is closed under forming finite unions, it suffices to show that if $E, F \in \mathcal{E}$, then $E \cap F \in \mathcal{E}$ and $E \setminus F \in \mathcal{E}$. Let E be as above and let

$$F = \bigcup_{k=1}^{m} \left(D_k \setminus \bigcup_{l=1}^{m_k} D_{kl} \right), \quad m, m_k \in \mathbb{N},$$

where D_k and D_{kl} are cosets of subgroups of G (or empty). Since

$$E \cap F = \bigcup_{i=1}^{n} \bigcup_{k=1}^{m} \bigcap_{j=1}^{n_i} \bigcap_{l=1}^{m_k} ((C_i \setminus C_{ij}) \cap (D_k \setminus D_{kl})),$$

it will follow that $E \cap F \in \mathcal{E}$ once we have shown that if C, C', D, D' are cosets in G, then $(C \setminus C') \cap (D \setminus D') \in \mathcal{E}$. However, for that we only have to observe that

$$(C \setminus C') \cap (D \setminus D') = (C \cap D) \setminus (C' \cup D')$$

and that $C \cap D$ is either empty or a coset. Turning to complements, note that

$$(C \setminus C') \setminus (D \setminus D') = (C \setminus (C' \cup D)) \cup ((C \cap D') \setminus C')$$

belongs to $\mathcal{E}$. Finally, with the above notation, we have

$$E \setminus F = \bigcup_{i=1}^{n} \bigcap_{k=1}^{m} \left(\bigcap_{j=1}^{n_i} \bigcup_{l=1}^{m_k} ((C_i \setminus C_{ij}) \setminus (D_k \setminus D_{kl})) \right).$$

Because $\mathcal{E}$ is closed under forming finite unions and intersections, we conclude that $E \setminus F \in \mathcal{E}$. $\square$

Remark A.6.2. Let H and K be subgroups of G and $a, b \in G$ such that $aH \cap bK \neq \emptyset$. Then there exists $h \in H$ such that

$$aH \setminus bK = ah(H \setminus (H \cap K)),$$

and $H \cap K$ has infinite index in H whenever $aH \setminus bK$ is infinite. Thus Proposition A.6.1 can be reformulated as follows. A subset E of G belongs to $\mathcal{R}(G)$ if and only if E can be written as

$$E = F \cup \bigcup_{i=1}^{m} \left(a_i H_i \setminus \bigcup_{j=1}^{m_i} b_{ij} K_{ij} \right),$$

where F is finite, H_i is a subgroup of G and K_{ij} is a subgroup of infinite index in H_i, $1 \leq i \leq m, 1 \leq j \leq m_i$.

The following lemma is used to show that homomorphisms map coset rings into coset rings (Theorem A.6.6).

Lemma A.6.3. *Let G be an Abelian group, H a subgroup of G, and $K_1, \ldots,$ K_n cosets in G. Then the set*

$$E = \{x \in G : xH \subseteq K_1 \cup \ldots \cup K_n\}$$

belongs to $\mathcal{R}(G)$.

Proof. We prove this by induction on $n \in \mathbb{N}$. If $n = 1$, then either $E = \emptyset$ or some coset of H is contained in K_1. In the latter case, since K_1 is a coset, E is a coset and hence $E \in \mathcal{R}(G)$.

Assume the statement is true for n and let

$$E = \{x \in G : xH \subseteq K_1 \cup \ldots \cup K_{n+1}\}.$$

If $E \neq \emptyset$ then, replacing E by a suitable translate, we can assume that $H \subseteq K_1 \cup \ldots \cup K_{n+1}$. Set $H_i = H \cap K_i$ and let K_i be a coset of the subgroup G_i of G ($i = 1, \ldots, n+1$). Since $xH = \bigcup_{i=1}^{n+1} xH_i$ for all $x \in G$, we have

$$E = \bigcap_{i=1}^{n+1} \{x \in G : xH_i \subseteq K_1 \cup \ldots \cup K_{n+1}\}$$

$$= \bigcap_{i=1}^{n+1} \left(\{x \in G : xH_i \subseteq K_i\} \bigcup \left\{ x \in G : xH_i \subseteq \bigcup_{j \neq i} K_j \right\} \right)$$

$$= \bigcap_{i=1}^{n+1} \left(G_i \cup \left\{ x \in G : xH_i \subseteq \bigcup_{j \neq i} K_j \right\} \right),$$

which belongs to $\mathcal{R}(G)$ by the induction hypothesis. $\square$

Let $\mathcal{F}(G)$ be the space of all finite linear combinations of characteristic functions 1_A, where A is a coset of a subgroup of G. Observe the following simple facts.

(1) The intersection of two cosets is a coset.
(2) $1_{A \cap B} = 1_A 1_B$.
(3) $1_{A \cup B} = 1_A + 1_B - 1_A 1_B$.
(4) $1_{G \setminus A} = 1_G - 1_A$.

It follows from (1), (2) and (3) that $\mathcal{F}(G)$ is an algebra of functions on G.

The next proposition is shown in [22]. For sake of brevity, we refrain from giving the proof. As pointed out by Schreiber, the following corollary is actually equivalent to Proposition A.6.4.

Proposition A.6.4. *Let $f \in \mathcal{F}(G)$, and let $B_1, \ldots, B_r$ be the finite family of sets in G on which f takes on its different values. Then the Boolean algebra generated by $B_1, \ldots, B_r$ and all of their translates contains a finite collection $\{K_1, \ldots, K_s\}$ of cosets in G such that the Boolean algebra generated by $\{K_1, \ldots, K_s\}$ contains every B_k.*

Corollary A.6.5. *Let $E \in \mathcal{R}(G)$. Then the Boolean algebra generated by E and all of its translates contains a finite collection $\{K_1, \ldots, K_r\}$ of cosets in G such that the Boolean algebra generated by $\{K_1, \ldots, K_r\}$ contains E.*

Proof. Recall that E can be written as a finite union of finite intersections of sets of the form $K \setminus L$, where K and L are cosets and L may be empty (Proposition A.6.1). The characteristic function of $K \setminus L$ is equal to $1_K - 1_{K \cap L}$. Since these functions are in $\mathcal{F}(G)$ and $\mathcal{F}(G)$ is an algebra, we have $1_E \in \mathcal{F}(G)$. Now, the sets of constancy of 1_E are just E and $G \setminus E$, so that the statement follows.   □

Even though we need the following theorem only in the special case where G^* is a quotient group of G and ϕ the quotient homomorphism, we present it in slightly more generality.

Theorem A.6.6. *Let G and G^* be Abelian groups and let $\phi : G \to G^*$ be a homomorphism. If $E \in \mathcal{R}(G)$, then $\phi(E) \in \mathcal{R}(G^*)$.*

Proof. Because ϕ preserves unions and translations we only need to consider sets of the form $E = H \setminus \bigcup_{i=1}^n K_i$, where H is a subgroup of G and $K_1, \ldots, K_n$ are cosets in H of subgroups of H.

Let $N = \ker \phi$ and let $q : H \to H/(H \cap N)$ be the quotient homomorphism. Then there is an injective homomorphism $j : H/(H \cap N) \to G^*$ such that $\phi|_H = j \circ q$. We show that $q(E) \in \mathcal{R}(H/(H \cap N))$. Since j is an injective homomorphism, it then follows that $\phi(E) = j(q(E)) \in \mathcal{R}(G^*)$.

Therefore it suffices to show that if G is an Abelian group, H is a subgroup of G and $K_1, \ldots, K_n$ are cosets in G then, with $q : G \to G/H$ the quotient homomorphism,

$$q\left(G \setminus \bigcup_{i=1}^n K_i\right) \in \mathcal{R}(G/H).$$

This is equivalent to showing that the complement

$$F = \{\xi \in G/H : q^{-1}(\xi) \subseteq K_1 \cup \ldots \cup K_n\}$$

belongs to $\mathcal{R}(G/H)$. Let

$$E = \{x \in G : q(x) \in K_1 \cup \ldots \cup K_n\}.$$

Then $q(E) = F$ and $E \in \mathcal{R}(G)$ by Lemma A.6.3. We show that this implies that $q(E) \in \mathcal{R}(G/H)$.

If $E \neq \emptyset$, then E is a union of cosets of H, and the same is true of every member of the Boolean algebra $\mathcal{A}$ generated by E and all its translates. By Corollary A.6.5, $\mathcal{A}$ contains a finite collection $\mathcal{C}$ of cosets such that the Boolean algebra $\mathcal{B}$ generated by $\mathcal{C}$ contains E. The quotient homomorphism q induces a Boolean algebra homomorphism on $\mathcal{A}$, and hence on $\mathcal{B}$. It follows that $q(E) \in \mathcal{R}(G/H)$. This finishes the proof.   □

Lemma A.6.7. *Let G be an Abelian topological group and G_0 a dense subgroup of G. Suppose that $K_1, \ldots, K_n$ are cosets in G_0 and let $E = G_0 \setminus \bigcup_{i=1}^{n} K_i$. Then there exists an open subgroup H of G such that $\overline{E}$ is a union of cosets of H.*

Proof. Let K_i be a coset of the subgroup $G_i, i = 1, \ldots, n$. Let $\mathcal{S}$ denote the smallest (and necessarily finite) collection of subgroups of G which contains all G_i, $i = 0, 1, \ldots, n$, and is closed under forming intersections. Since G_0 is dense in G and $G_0 \in \mathcal{S}$, there exists $K \in \mathcal{S}$ which is minimal with respect to the property that $\overline{K}$ is open G. Then there is a (possibly void) subset I of $\{1, \ldots, n\}$ such that

(1) $K = G_0 \cap \left(\bigcap_{i \in I} G_i \right)$.
(2) $i \in F$ and $G_i = G_j$ implies $j \in I$.

Set $H = \overline{K}$, and let C be any coset of H. We have to show that either $C \cap \overline{E} = \emptyset$ or $C \subseteq \overline{E}$. To that end, suppose that $x \in C \cap E$ (equivalently, $C \cap \overline{E} \neq \emptyset$ since C is open in G). Then xK is a dense subset of $C = xH$. Let $L_i = K_i \cap xK, i = 1, \ldots, n$. We claim that even

$$xK \setminus \bigcup_{i=1}^{n} K_i = xK \setminus \bigcup_{i=1}^{n} L_i$$

is dense in C. Since xK is dense in C, it suffices to verify that $\bigcup_{i=1}^{n} L_i$ is nowhere dense in G. If $i \in I$ then K is a subgroup of G_i. Thus $L_i = \emptyset$ since $xK \not\subseteq K_i$. If $i \notin I$ then L_i is either void or a coset of $K \cap G_i$. By the choice of K and I, $\overline{K \cap G_i}$ is not open, so $K \cap G_i$ is nowhere dense and hence so is L_i. Since $x \in G_0$ and $K \subseteq G_0$, it follows that

$$C = \overline{\left(xK \setminus \bigcup_{i=1}^{n} K_i \right)} \subseteq \overline{\left(G_0 \setminus \bigcup_{i=1}^{n} K_i \right)} = \overline{E},$$

as was to be shown. $\square$

The preceding lemmas now lead to the characterisation of closed sets in the coset ring at which we were aiming.

Theorem A.6.8. *Let G be an Abelian topological group and $E \in \mathcal{R}(G)$. Then $\overline{E} \in \mathcal{R}(G)$ and E is closed if and only if E can be written*

$$E = \bigcup_{j=1}^{m} \left(C_j \setminus \bigcup_{l=1}^{m_j} C_{jl} \right),$$

where C_j and C_{jl} are (possibly void) closed cosets in G and C_{jl} is contained in C_j and open in C_j.

Proof. Clearly, a set E of the above form is closed and an element of $\mathcal{R}(G)$. Conversely, let $E \in \mathcal{R}(G)$ and suppose first that E is of the form $E = G_0 \setminus \bigcup_{l=1}^{n} K_l$, where G_0 is a subgroup of G and the K_l are cosets contained in G_0. By Lemma 5.6.13 there exists an open subgroup H of $\overline{G}_0$ such that $\overline{E}$ is a union of cosets of H. If $q : \overline{G}_0 \to \overline{G}_0/H$ is the quotient homomorphism then, by Lemma 5.6.12, $q(E) \in \mathcal{R}(\overline{G}_0/H)$, say

$$q(E) = \bigcup_{j=1}^{m} \left(D_j \setminus \bigcup_{l=1}^{m_j} D_{jl} \right),$$

where the D_j and D_{jl} are cosets in $\overline{G}_0/H$ (Lemma 5.6.9). Moreover, $q(E) = q(\overline{E})$ since q is continuous and $\overline{G}_0/H$ is discrete. Thus

$$\overline{E} = q^{-1}(q(\overline{E})) = q^{-1}(q(E)) = \bigcup_{j=1}^{m} \left(q^{-1}(D_j) \setminus \bigcup_{l=1}^{m_j} q^{-1}(D_{jl}) \right),$$

and each $q^{-1}(D_{jl})$ and $q^{-1}(D_j)$ is open in $\overline{G}_0$. This proves that $\overline{E} \in \mathcal{R}(G)$.

Now let E be an arbitrary set in $\mathcal{R}(G)$. Then $E = E_1 \cup \ldots \cup E_m$, where each E_i is a translate of a set of the type considered above. It follows that $\overline{E} = \overline{E}_1 \cup \ldots \cup \overline{E}_m \in \mathcal{R}(G)$ and $\overline{E}$ has the desired form. □

References

1. E. Albrecht, *Decomposable systems of operators in harmonic analysis*, Toeplitz Centennial, Birkhäuser, Basel, 1982, 19-35.
2. M. Altman, *Contracteurs dans les algèbres de Banach*, C.R. Acad. Sci. Paris Sér. A - B **274** (1972) 399-400.
3. R. Arens, *Inverse-producing extensions of normed algebras*, Trans. Amer. Math. Soc. **88** (1958), 536-548.
4. R. Arens, *The maximal ideals of certain function algebras*, Pacific J. Math. **8** (1958), 641-648.
5. R. Arens, *Extensions of Banach algebras*, Pacific J. Math. **10** (1960), 1-16.
6. R. Arens, *The group of invertible elements of a commutative Banach algebra*, Studia Math. (ser. Specjalna) Zeszyt **1** (1963), 21-23.
7. R. Arens and A.P. Calderon, *Analytic functions of several Banach algebra elements*, Ann. Math.(2) **62** (1955), 204-216.
8. A. Atzmon, *Spectral synthesis in regular Banach algebras*, Israel J. Math. **8** (1970), 197-212.
9. A. Atzmon, *On the union of sets of synthesis and Ditkin's condition in regular Banach algebras*, Bull. Amer. Math. Soc. **2** (1980), 317-320.
10. G.F. Bachelis, W.A. Parker and K.A. Ross, *Local units in $L^1(G)$*, Proc. Amer. Math. Soc. **31** (1972), 312-313.
11. J. Benedetto, *Spectral synthesis*, Teubner, Stuttgart, 1975.
12. A. Beurling, *Sur les intégrales de Fourier absolument convergentes et leurs applications à une transformation fonctionelle*, Congrès des Math. Scand., Helsingfors, 1938.
13. S.J. Bhatt and H.V. Dedania, *Uniqueness of the uniform norm and adjoining an identity in Banach algebras*, Proc. Indian Acad. Sci. Math. Sci. **105** (1995), 405-409.
14. S.J. Bhatt and H.V. Dedania, *Banach algebras with unique uniform norm*, Proc. Amer. Math. Soc. **124** (1996), 579-584.
15. S.J. Bhatt and H.V. Dedania, *Banach algebras with unique uniform norm. II*, Studia Math. **147** (2001), 211-235.
16. S.J. Bhatt and H.V. Dedania, *A Beurling algebra is semisimple: an elementary proof*, Bull. Austral. Math. Soc. **66** (2002), 91-93.
17. S.J. Bhatt and H.V. Dedania, *Beurling algebras and uniform norms*, Studia Math. **160** (2004), 179-183.

18. H. Bohr, *Almost periodic functions*, Chelsea, New York, 1947.

19. F. Bonsall and J. Duncan, *Complete normed algebras*, Springer, Berlin, 1973.

20. A. Browder, *Introduction to function algebras*, Benjamin, New York and Amsterdam, 1969.

21. P.J. Cohen, *On a conjecture of Littlewood and idempotent measures*, Amer. J. Math. **82** (1960), 191-212.

22. P.J. Cohen, *Homomorphisms of group algebras*, Amer. J. Math. **82** (1960), 213-226.

23. J.B. Conway, *Functions of one complex variable*, Graduate Texts in Mathematics **11**, Springer-Verlag, New York, 1973.

24. C. Corduneanu, *Almost periodic functions*, Interscience, New York, 1968.

25. H.G. Dales, *Banach algebras and automatic continuity*, Clarendon Press, Oxford, 2000.

26. Y. Domar, *Harmonic analysis based on certain commutative Banach algebras*, Acta Math. **96** (1956), 1-66.

27. R.S. Doran and J. Wichmann, *Approximate identities and factorization in Banach modules*, LNM **768**, Springer, New York, 1979.

28. C.F. Dunkl and D.E. Ramirez, *Topics in harmonic analysis*, Appleton-Century-Crofts, New York, 1971.

29. H.A.M. Dzinotyiweyi, *Weighted function algebras on groups and semigroups*, Bull. Austral. Math. Soc. **33** (1986), 307-318.

30. R.E. Edwards, *Integration and harmonic analysis on compact groups*, London Math. Soc. Lecture Notes Series **8**, Cambridge University Press, London-New York, 1972.

31. P. Enflo, *A counterexample to the approximation problem in Banach spaces*, Acta Math. **130** (1973), 309-317.

32. P. Eymard, *L'algèbre de Fourier d'un groupe localement compact*, Bull. Soc. Math. France **92** (1964), 181-235.

33. B. Forrest, E. Kaniuth, A.T. Lau and N. Spronk, *Ideals with bounded approximate identities in Fourier algebras*, J. Funct. Anal. **203** (2003), 288-306.

34. S. Friedberg, *The Fourier transform is onto only when the group is finite*, Proc. Amer. Math. Soc. **27** (1971), 421-422.

35. S. Frunza, *A characterization of regular Banach algebras*, Rev. Roum. Math. Pure Appl. **18** (1973), 1057-1059.

36. T.W. Gamelin, *Uniform algebras*, Prentice Hall, Englewood Cliffs, NJ, 1969.

37. B.R. Gelbaum, *Tensor products and related questions*, Trans. Amer. Math. Soc. **103** (1962), 525-548.

38. I.M. Gelfand, *Normierte Ringe*, Mat. Sbornik **9** (1941), 3-24.

39. I.M. Gelfand and M.A. Naimark, *On the embedding of normed rings into a ring of operators in Hilbert space*, Mat. Sb. **12** (1943), 197-213.

40. I.M. Gelfand, D.A. Raikov and G.E. Shilov, *Commutative normed rings*, Uspekhi Mat. Nauk **1** (1946), 48-146. AMS Transl. **5** (1957), 115-220.

41. I.M. Gelfand, D.A. Raikov and G.E. Shilov, *Commutative normed rings*, Chelsea, New York, 1964.

42. I.M. Gelfand and G.E. Shilov, *Über verschiedene Methoden der Einführung der Topologie in die Menge der maximalen Ideale eines normierten Ringes*, Mt. Sb. **9** (1941), 25-39.

43. J.E. Gilbert, *On projections of $L^\infty(G)$ onto translation-invariant subspaces*, Proc. London Math. Soc. **19** (1969), 69-88.

44. A.M. Gleason, *A characterization of maximal ideals*, J. Analyse Math. **18** (1967), 171-172.

45. C.C. Graham, *The Fourier transform is onto only when the group is finite*, Proc. Amer. Math. Soc. **38** (1973), 365-366.

46. C.C. Graham and O.C. McGehee, *Essays in commutative harmonic analysis*, Springer, Berlin-Heidelberg-New York, 1979.

47. R.C. Gunning and H. Rossi, *Analytic functions of several complex variables*, Prentice Hall, Englewood Cliffs, NJ, 1965.

48. A. Hausner, *Ideals in a certain Banach algebra*, Proc. Amer. Math. Soc. **8** (1957), 246-249.

49. A. Hausner, *The Tauberian theorem for groups algebras of vector-valued functions*, Pacific J. Math. **7** (1957), 1603-1610.

50. H. Helson, *Spectral synthesis of bounded functions*, Ark. Mat. **1** (1952), 497-502.

51. C.S. Herz, *Spectral synthesis for the circle*, Ann. Math. **68** (1958), 709-712.

52. C.S. Herz, *The spectral theory of bounded functions*, Trans. Amer. Math. Soc. **94** (1960), 181-232.

53. C.S. Herz, *Harmonic synthesis for subgroups*, Ann. Inst. Fourier (Grenoble) **23** (1973), 91-123.

54. E. Hewitt and K.A. Ross, *Abstract harmonic analysis. I*, Springer, Berlin-Heidelberg-New York, 1963.

55. E. Hewitt and K.A. Ross, *Abstract harmonic analysis. II*, Springer, Berlin-Heidelberg-New York, 1970.

56. L. Hörmander, *An introduction to complex analysis in several-variables*, van Nostrand, Princeton, NJ, 1966.

57. T. Husain, *Orthogonal Schauder bases*, Marcel Dekker, New York, 1991.

58. J. Inoue and S. E. Takahasi, *A remark on the largest regular subalgebra of a Banach algebra*, Proc. Amer. Math. Soc. **116** (1992), 961-962.

59. T. Itô and I. Ameniya, *A simple proof of the theorem of P.J. Cohen*, Bull. Amer. Math. Soc. **70** (1964), 774-776.

60. N. Jacobson, *A topology for the set of primitive ideals in an arbitrary ring*, Proc. Nat. Acad. Sci. U.S.A. **31** (1945), 333-338.

61. B.E. Johnson, *The uniqueness of the (complete) norm topology*, Bull. Amer. Math. Soc. **73** (1967), 537-539.

62. G.P. Johnson, *Spaces of functions with values in a Banach algebra*, Trans. Amer. Math. Soc. **92** (1959), 411-429.

63. J.P. Kahane, *Séries de Fourier absolument convergentes*, Springer, New York, 1970.

64. J.P. Kahane and W. Zelazko, *A characterization of maximal ideals in commutative Banach algebras*, Studia Math. **29** (1968), 339-343.

65. E. Kaniuth, *Weak spectral synthesis for the projective tensor product of commutative Banach algebras*, Proc. Amer. Math. Soc. **132** (2004), 2959-2967.

66. E. Kaniuth, *Weak spectral synthesis in commutative Banach algebras*, J. Funct. Anal. **254** (2008), 987-1002.

67. E. Kaniuth and A.T. Lau, *A separation property of positive definite functions on locally compact groups and applications to Fourier algebras*, J. Funct. Anal. **175** (2000), 89-110.

68. E. Kaniuth and A.T. Lau, *Spectral synthesis for $A(G)$ and subspaces of $VN(G)$*, Proc. Amer. Math. Soc. **129** (2001), 3253-3263.

69. R. Kantrowitz and M.M. Neumann, *The greatest regular subalgebra of certain Banach algebras of vector valued functions*, Rend. Circ. Mat. Palermo(2) **43** (1994), 435-446.

70. I. Kaplansky, *Primary ideals in group algebras*, Proc. Nat. Acad. Sciences USA **35** (1949), 133-136.

71. I. Kaplansky, *Normed algebras*, Duke Math. J. **16** (1949), 399-418.

72. Y. Katznelson, *An introduction to harmonic analysis*, Dover, New York, 1976.

73. T.W. Körner, *A cheaper Swiss cheese*, Studia Math. **83** (1986), 33-36.

74. R. Larsen, *An introduction to the theory of multipliers*, Springer, Berlin, 1971.

75. R. Larsen, *Banach algebras*, Marcel Dekker, New York, 1973.

76. K.B. Laursen and M.M. Neumann, *An introduction to local spectral theory*, Clarendon Press, Oxford, 2000.

77. A. Lebow, *Maximal ideals in tensor products of Banach algebras*, Bull. Amer. Math. Soc. **74** (1968), 1020-1022.

78. G.M. Leibowitz, *Lectures on complex function algebras*, Scott, Foresman, Glenview, IL, 1970.

79. H. Leptin, *Sur l'algèbre de Fourier d'un groupe localement compact*, C.R. Acad. Sci. Paris Sér. A **266** (1968), 1180-1182.

80. T.S. Liu, A. van Rooij and J.K. Wang, *Projections and approximate identities for ideals in group algebras*, Trans. Amer. Math. Soc. **175** (1973), 469-482.

81. L.H. Loomis, *An introduction to abstract harmonic analysis*, van Nostrand, New York, 1953.

82. E.R. Lorch, *The theory of analytic functions in normed Abelian vector rings*, Trans. Amer. Math. Soc. **54** (1943), 414-425.

83. Yu. I. Lyubich, *Introduction to the theory of Banach representations of groups*, Birkhäuser, Basel, 1988.

84. P. Malliavin, *Impossibilitè de la synthèse spectral sur les groupes abéliens non compacts*, Inst. Hautes Études Sci. Publ. Math. **2** (1959), 61-68.

85. R. McKissick, *A non-trivial normal sup norm algebra*, Bull. Amer. Math. Soc. **69** (1963), 391-395.

86. S. Mazur, *Sur les anneaux linéaires*, C.R. Acad. Sci. Paris **207** (1938), 1025-1027.

87. S.N. Mergelyan, *Uniform approximation to functions of a complex variable*, Uspekhi Mat. Nauk (N.S.) **7** (1952), 31-122 (in Russian); English translation, Amer. Math. Soc. Transl. **101**.

88. M.J. Meyer, *The sprectral extension property and extension of multiplicative linear functionals*, Proc. Amer. Math. Soc. **112** (1991), 855-861.

89. H. Milne, *Banach space properties of uniform algebras*, Bull. London Math. Soc. **4** (1972), 323-326.

90. H. Mirkil, *A counter example to discrete spectral synthesis*, Compos. Math. **14** (1960), 269-273.

91. C. Morschel, *Reguläre kommutative Banachalgebren*, Diplomarbeit, Universität Paderborn, 1997.

92. R.D. Mosak, *Banach algebras*, The University of Chicago Press, Chicago, 1975.

93. T.K. Muraleedharan and K. Parthasarathy, *On Ditkin sets*, Colloq. Math. **69** (1995), 271-274.

94. T.K. Muraleedharan and K. Parthasarathy, *On unions and intersections of sets of synthesis*, Proc. Amer. Math. Soc. **123** (1995), 1213-1316.

95. T.K. Muraleedharan and K. Parthasarathy, *Difference spectrum and spectral synthesis*, Tohoku Math. J. **51** (1999), 65-73.

96. M.A. Naimark, *Normed rings*, Noordhoff, Groningen, 1964.

97. M.M. Neumann, *Commutative Banach algebras and decomposable operators*, Monatsh. Math. **113** (1992), 227-243.

98. M.M. Neumann,*Banach algebras, decomposable convolution operators, and a spectral mapping property*, in: Function spaces, K. Jarosz (ed.), Marcel Dekker, New York, 1995, 307-323.

99. T.W. Palmer, *Banach algebras and the general theory of *-algebras, Vol. I*, Cambridge University Press, Cambridge, UK, 1994.

100. K. Parthararathy and S. Varma, *On weak spectral synthesis*, Bull. Austral. Math. Soc. **43** (1991), 279-282.

101. T.J. Ransford, *A short proof of Johnson's uniqueness-of-norm theorem*, Bull. London Math. Soc. **21** (1989), 487-488.

102. H. Reiter, *Investigations in harmonic analysis*, Trans. Amer. Math. Soc. **73** (1952), 401-427.

103. H. Reiter, *Contributions to harmonic analysis. VI*, Ann. Math. **77** (1963), 552-562.

104. H. Reiter, *L^1-algebras and Segal algebras*, LNM **231**, Springer, New York, 1971.

105. H. Reiter and J.D. Stegeman, *Classical harmonic analysis and locally compact groups*, Oxford University Press, Oxford, 2000.

106. C.E. Rickart, *The uniqueness of norm problem in Banach algebras*, Ann. Math.(2) **51** (1950), 615-628.

107. C.E. Rickart, *On spectral permanence for certain Banach algebras*, Proc. Amer. Math. Soc. **4** (1953), 191-196.

108. C.E. Rickart, *General theory of Banach algebras*, van Nostrand, Princeton, NJ, 1960.

109. M. Roitman and Y. Sternfeld, *When is a linear functional multiplicative ?*, Trans. Amer. Math. Soc. **267** (1981), 111-124.

110. H.P. Rosenthal, *On the existence of approximate identities in ideals of group algebras*, Ark. Mat. **7** (1967), 185-191.

111. H.L. Royden, *Function algebras*, Bull. Amer. Math. Soc. **69** (1963), 281-298.

112. W. Rudin, *Analyticity and the maximum modulus principle*, Duke Math. J. **29** (1953), 449-457.

113. W. Rudin, *Fourier analysis on groups*, Interscience, New York, 1962.

114. R.A. Ryan, *Introduction to tensor products of Banach spaces*, Springer Monographs in Mathematics, Springer, New York, 2002.

115. D.L. Salinger and J.D. Stegeman, *Difference spectra of ideals for non-metrizable groups*, J. London Math. Soc.(2) **26** (1982), 531-540.

116. R. Schatten, *A theory of cross-spaces*, Ann. Math. Studies **26**, Princeton University Press, Princeton, NJ, 1950.

117. B.M. Schreiber, *On the coset ring and strong Ditkin sets*, Pacific J. Math. **32** (1970), 805-812.

118. L. Schwartz, *Sur une propriété de synthèse spectrale dans les groupes non-compacts*, C.R. Acad. Sci. Paris Ser. A **227** (1948), 424-426.

119. G.E. Shilov, *Sur la théorie des idéaux dans les anneaux normés de fonctions*, C.R. (Doklady) Acad. Sci. URSS (N.S.) **27** (1940), 900-903.

120. G.E. Shilov, *On the extension of maximal ideals*, C.R. (Doklady) Acad. Sci. URSS (N.S.) **29** (1940), 83-84.

121. G.E. Shilov, *On normed rings possessing one generator*, Mat. Sb. **21** (1947), 25-47.

122. G.E. Shilov, *On regular normed rings*, Trudy Mat. Inst. Steklov **21** (1947), 118 pp. (Russian. English summary).

123. G.E. Shilov, *On the decomposition of a commutative normed ring into a direct sum of ideals*, Mat. Sb. **32** (1954), 353-364 (in Russian). (English translation: Amer. Math. Soc. Translations (II), **1** (1955)).

124. A. Soltysiak, *The strong spectral extension property does not imply the multiplicative Hahn–Banach property*, Studia Math. **153** (2002), 297-301.

125. J.D. Stegeman, *Some problems in spectral synthesis*, in: *Harmonic Analysis Iraklion 1978 Proceedings*, 194-203, LNM **781**, Springer, New York, 1980.

126. E.L. Stout, *The theory of uniform algebras*, Bogden & Quigley, New York, 1971.

127. M. Takesaki and N. Tatsuuma, *Duality and subgroups. II*, J. Funct. Anal. **11** (1972), 184-190.

128. J. Tomiyama, *Tensor products of commutative Banach algebras*, Tohoku Math. J. (2) **12** (1960), 147-154.

129. N.Th. Varopoulos, *Tensor algebras and harmonic analysis*, Acta Math. **119** (1967), 51-111.

130. C.R. Warner, *A generalization of the Shilov-Wiener Tauberian theorem*, J. Funct. Anal. **4** (1969), 329-331.

131. C.R. Warner, *A class of spectral sets*, Proc. Amer. Math. Soc. **57** (1976), 99-102.

132. C.R. Warner, *Weak spectral synthesis*, Proc. Amer. Math. Soc. **99** (1987), 244-248.

133. C.R. Warner, *Spectral synthesis in the Mirkil algebra*, J. Math. Anal. Appl. **167** (1992), 172-181.

134. A. Weil, *L'intégration dans les groupes topologiques et ses applications*, Hermann, Paris, 1953.

135. J.G. Wendel, *Left centralizes and isomorphisms of group algebras*, Pacific J. Math. **2** (1952), 251-261.

136. J. Wermer, *On algebras of continuous functions*, Proc. Amer. Math. Soc. **4** (1953), 866-869.

137. J. Wermer, *Banach algebras and several complex variables*, Springer, Berlin-Heidelberg-New York, 1976.

138. J. Wichmann, *Bounded approximate units and bounded approximate identities*, Proc. Amer. Math. Soc. **41** (1973), 547-550.

139. J.H. Williamson, *Remarks on the Plancherel and Pontryagin theorems*, Topology **1** (1967), 73-80.

140. W. Zelazko, *A characterization of multiplicative linear functionals in complex Banach algebras*, Studia Math. **30** (1968), 83-85.

141. W. Zelazko, *Concerning extensions of multiplicative linear functionals in Banach algebras*, Studia Math. **31** (1968), 495-499.

142. W. Zelazko, *On a certain class of non-removable ideals in Banach algebras*, Studia Math. **44** (1972), 87-92.

143. W. Zelazko, *Banach algebras*, Elsevier, Amsterdam, 1973.

144. E.I. Zelmanov, *On periodic compact groups*, Israel J. Math. **77** (1992), 83-95.

Index

abstract spectral theorem, 71
adjoining an identity, 6
algebra
  C^*-, 66
  L^1-group, 18
  $*$-, 5
  Banach, 2
  Banach $*$-, 5
  Beurling, 20
  commutative, 2
  Fourier, 107
  group C^*-, 96
  Lipschitz, 37
  Mirkil, 278
  multiplier, 27
  normed, 1
  normed $*$-, 5
  unital, 2
  Volterra, 37
  with involution, 5
almost periodic compactification, 119
almost periodic function, 115
annihilator, 334
approximate identity, 6
  bounded, 6
  left, 6
  right, 6
approximate unit
  bounded, 6
  left, 6
  right, 6
approximation property, 122, 325
Arzela–Ascoli theorem, 320

automatic continuity
  of homomorphisms, 50
  of involution, 51

Baire's category theorem, 320
Banach algebra
  boundedly regular, 250
  extension, 222
  faithful, 27
  finitely generated, 60
  greatest regular subalgebra, 208
  radical, 50
  regular, 198
  semisimple, 50
  Tauberian, 257
  weakly regular, 231
Banach–Alaoglu theorem, 322
Beurling algebra, 20
Bishop boundary, 188
Bohr compactification, 119
boundary, 158
  Bishop, 188
  Shilov, 160, 162
  topological, 14

character, 89
  ω-bounded generalised, 99
  orthogonality relation, 129
closed coset ring, 336
  of $\mathbb{R}$, 314
  of $\mathbb{T}$, 314
closed graph theorem, 323
closure operation, 321
Cohen's factorization theorem, 248